Progress in Botany/Fortschritte der Botanik 46

46

Progress in Botany

Morphology · Physiology · Genetics
Taxonomy · Geobotany

Fortschritte der Botanik

Morphologie · Physiologie · Genetik
Systematik · Geobotanik

Editors/Herausgeber

Karl Esser, Bochum
Klaus Kubitzki, Hamburg
Michael Runge, Göttingen
Eberhard Schnepf, Heidelberg
Hubert Ziegler, München

Springer-Verlag
Berlin Heidelberg New York Tokyo 1984

With 33 Figures

ISBN 3-540-13731-9 Springer-Verlag Berlin Heidelberg New York Tokyo
ISBN 0-387-13731-9 Springer-Verlag New York Heidelberg Berlin Tokyo

Printed in Germany

Offsetprinting and bookbinding: Brühlsche Universitätsdruckerei, Giessen
2131/3130-543210

Contents

List of Editors

Section A: Professor Dr. E. SCHNEPF, Zellenlehre, Universität Heidelberg, Im Neuenheimer Feld 230, D 6900 Heidelberg

Section B: Professor Dr. H. ZIEGLER, Institut für Botanik und Mikrobiologie, Technische Universität München, Arcisstr. 21, D 8000 München 2

Section C: Professor Dr. Dr. h.c. K. ESSER, Lehrstuhl für Allgemeine Botanik, Ruhr-Universität, Postfach 10 21 48, D 4630 Bochum 1

Section D: Professor Dr. K. KUBITZKI, Institut für Allgemeine Botanik und Botanischer Garten, Universität Hamburg, Ohnhorststr. 18, D 2000 Hamburg 52

Section E: Professor Dr. M. RUNGE, Lehrstuhl für Geobotanik, Systematisch-Geobotanisches Institut der Universität, Untere Karspüle 2, D 3400 Göttingen

Section F: Professor Dr. M. RUNGE, Lehrstuhl für Geobotanik, Systematisch-Geobotanisches Institut der Universität, Untere Karspüle 2, D 3400 Göttingen

A. Morphology

I. Cytology

a) Special Cytology: Cytology and Morphogenesis of the Prokaryotic Cell

By FRANK MAYER

1. Cell Shape and Variation of Cell Structure

a) Progress in Preparation Techniques

α) Structural Studies on Frozen-Hydrated Bacteria. Conventional specimen preparation methods for electron microscopy involve steps which originate artifacts. Cryoelectron microscopy, i.e., high-resolution observation of sectioned, frozen-hydrated biological material, overcomes these artifacts. This approach was pioneered by GLAESER and his group (TAYLOR 1978). DUBOCHET et al. (1983) studied how the structures of some typical Gram-positive and Gram-negative bacteria appear in frozen-hydrated sections. They found that these unstained sections provided a true high-resolution image. They confirmed an earlier finding (ACHTERRATH and LICKFELD 1972, BURDETT and ROGERS 1970, EBERSOLD et al. 1981, NANNINGA 1973) that so called mesosomes do not exist in exponentially growing *Staphylococcus aureus* cells and that these structures are artifacts generated immediately after the onset of osmium fixation. The bacterial nucleoid appeared in exponentially growing bacteria as an ill-defined central region and became more clearly visible when bacteria were in the stationary phase.

β) Low Denaturation Embedding for Thin Sections. KELLENBERGER et al. (1980) presented newly designed cross-linked embedding resins ("Lowicryl" resins). Application of these resins involves infiltration and curing at temperatures down to -50° C and can be expected to be of great importance not only for good structural preservation of bacterial fine structure but also for maintenance of antigenic properties of macromolecules in immunoelectron microscopy.

b) New Isolates

α) Box-shaped (Square) Bacteria. Box-shaped (or square) bacteria show an ubiquitous distribution in hypersaline environments (STOECKENIUS 1981, WALSBY 1980). Recent investigations (JAVOR et al. 1982, KESSEL and COHEN 1982) suggested the existence of more than one type of square bacteria in the populations studied. NICHOLSON and FOX (1983) provided molecular evidence for a close phylogenetic relationship among the box-shaped bacteria *Halobacterium vallismortis*, *Halobacterium marismortue* and two other extremely halophilic isolates. 5 S rRNA sequence data indicated that *Halobacterium cutirubrum* and *Halococcus morrhuae* are quite distinct from the four isolates mentioned above.

β) Another Flat Archaebacterium: a Plate-Shaped Methanogen. WILDGRUBER et al. (1982) described another flat archaebacterium. Compared to typical box-shaped bacteria, this organism grows at much lower ionic strength. It belongs to the methanogens; growth occurs on H_2/CO_2 or on formate; acetate is required; the doubling time was found to be 7 h at 40°C.

Progress in Botany, Vol. 46

On the basis of DNA-RNA hybridization it was found to represent a new family, the *Methanoplanaceae* within the order *Methonomicrobiales*.

γ) A new Genus of Filamentous Methane Bacteria. HUSER et al. (1982) isolated a filamentous bacterium from a mesophilic sewage digester. They assumed that this organism is, in fact, the "acetate organism" described by ZEHNDER et al. (1980). The filament forms characteristic bundles. As the bacterium was found to decarboxylate acetate, yielding methane and carbon dioxide, the name *Methanothrix* was proposed for this new genus (type species: *M. soehngenii*).

δ) True Branching in Filamentous Cyanobacteria. The morphology of the branching filamentous cyanobacteria (at the light microscopic level) has been described thoroughly (LEE 1927, MARCENKO 1961, MARTIN and WYATT 1974, RIPPKA et al. 1979, SCHWABE 1960, SPEARING 1937). THURSTON and INGRAM (1971) performed an ultrastructural study on *Fischerella ambigua;* they showed that this organism possesses a mechanism of cell division quite unlike that seen in unbranched, filamentous bacteria. Recently, NIERZWICKI et al. (1982) provided further evidence that, in *Mastigocladus laminosus,* septum formation in older cells may occur parallel to the long axis of the filament, thereby confirming that true branching took place.

c) Assembly of Rickettsia

Rickettsiae are obligatory intracellular parasites. They are generally believed to maintain their morphological integrity throughout their life cycle. However, HASE (1983) obtained other findings when he investigated the assembly process of *Rickettsia tsutsugamushi* in irradiated L cells. In his studies, nascent forms, 3 days after infection, lacked limiting membranes. During "maturation," they assumed a round shape.

Faint membranous components became visible in fuzzy zones covering the surface of these round structures. The rickettsial double membranes were assembled along the fuzzy zones. This assembly was paralleled by formation of a filamentous network in one zone (f area) of the parasite-host cell system and manufacture of rickettsial ribosomes in another zone (g area).

2. Bacterial Cell Aggregates

a) A Cell Conglomerate: *Conglomeromonas largomobilis*

SKERMAN et al. (1983) described a Gram-negative rod-shaped organism which exhibits unicellular and multicellular phases of growth. Unicellular-phase cells were motile, with two different types of flagella per cell. Multicellular conglomerates arose from single cells which lost motility, became optically refractile, and reproduced by multiplanar centripetal septation. This process was enhanced by sodium ions. Under suitable conditions, conglomerates dissociated into single cells which produced clear colonies with, initially, only a few unicellular-phase cells.

b) A Thermophilic Methanogenic Consortium

BOCHEM et al. (1982) began to study a consortium which consists of methanogenic bacterial "tissue" colonized by other bacteria, and which grows at 60° C. The system converts acetic acid to methane, a process which appears to be very important for methanogenesis both in nature and in digesters. Thin sections of the consortium showed three morphologically distinct layers; the outer zone exhibited resemblance to a pseudoparenchyma and consisted of "macrocysts" and coccoid cells, similar to *Methanosarcina* cells. The inner zone was built up of loosely packed ovoid cells; the third zone showed internal cavities containing thermophilic rod-shaped bacteria. The cavities were the exclusive site for gas formation. However, the nonmethanogenic nature of the rods was substantiated by pure culture isolations.

3. The outermost Bacterial Surface Structures

a) Chroococcacean Cyanobacteria

VAARA (1982), in the first comparative study on the nature and occurrence of the outermost surface structures in chroococcacean cyanobacteria, showed that all the structures are similar to those of other bacteria. He found glycocalyx layers (see DREWS 1982), sheaths, and loosely packed material between sheath and the cells, mostly consisting of acidic polysaccharides. Some outermost layers showed hexagonally arranged subunits, some strains possessed fimbriae. None of the strains produced spinae.

b) An External Layer in *Brucella*

In view of the biological importance of the capsule, OBERTI et al. (1982) undertook a study to demonstrate the existence of an external layer in *Brucella* by action of antibodies. Labeling with conjugated specific antibodies showed that the capsule incorporates and transports antigens belonging probably to the outer membrane. Staining with ruthenium red showed that it contains acidic mucopolysaccharides.

c) Extracellular Slime and Swarming

STAHL et al. (1983) found copious quantities of slime surrounding swarming *Proteus mirabilis*. The slime appeared to provide a matrix through which the cells could migrate.

4. Cell Wall Type, Composition, Assembly, and Degradation

a) Identification of the Gram Type

Although Gram staining is a routine technique (BARTHOLOMEW and MITTWER 1952, BENIANS 1920, BEVERIDGE 1981, SALTON 1963, SCHERRER 1963, WEBB 1948), the actual effect of the stain on cells is poorly understood. Instead of $KJ\text{-}J_2$, BEVERIDGE and DAVIES (1983) and DAVIES et al. (1983) used potassium trichlor (η^2-ethylene) platinum (II) (TPt) and stained *Bacillus subtilis* and *Escherichia coli*. They found that TPt produced an anion in aqueous solution which was compatible with crystal violet (CV) of the Gram stain. Thus, the insoluble complex formed by the interaction of TPt with CV contained platinum as an electron-opaque marker for electron microscopy and allowed the Gram staining mechanism to be followed by electron microscopy. It could be shown that a CV-TPt complex was formed within the cells and at the cell surfaces. In *B. subtilis*, the decolorization step dissolved the precipitate from the cell surface, but the internal complex remained within the cell. This was not the case for *E. coli*; the decolorization step removed both surface-bound and cellular CB-TPt. During its removal, the outer membrane was sloughed off until only the murein sacculus and plasma membrane remained. Probably, the plasma membrane was also disturbed, but it was retained within the cell by the murein sacculus.

b) Regular Arrays

Regular arrays of hexagonally or tetragonally arranged macromolecules are present on the surface of cell walls in a variety of bacteria (GLAUERT and THORNLEY 1969, SLEYTR 1978). Such arrays were recently found also for several strains of *Lactobacillus* (KAWATA et al. 1974, MASUDA and KAWATA 1979, 1980, 1981, 1983), for *Bacillus polymyxa* (BURLEY and MURRAY 1983) and for *Deinococcus radiodurans* (THOMPSON et al. 1982). Chemical characterization of such arrays should lead to a better understanding of the molecular mechanism of their functions.

c) Cell Wall Assembly

EDELSTEIN et al. (1980) and GIBSON et al. (1983), when analyzing the initiation of wall assembly sites in *Streptococcus faecium*, found that the volume of the two completed poles produced by a single site was constant and thus independent of growth rate, and that the initiation of new sites could not be related to the timing of rounds of chromosome synthesis but did occur at a constant cell volume. In their investigation, GIBSON et al. (1983) described that chromosome replication is necessary for the terminal events of growth site development, which result in the division of a site into two separate poles.

d) Cell Wall Degradation

Electron microscopic studies performed by WECKE et al. (1982) demonstrated that, under acid conditions, the cell walls of *Staphylococcus aureus* were degraded by "attack from the inside." They found that lysozyme acted less as a muralytic enzyme but more as an activator of preexisting autolytic wall enzymes.

5. Bacteriophage Adsorption to Gram-positive Bacteria

Teichoic acids form part of the receptor for many of the phages that infect Gram-positive bacteria (ARCHIBALD 1980). GIVAN et al. (1982) found that two different phages (SP 50 and Ø 25) specific for *Bacillus subtilis* required different receptors. Glucosyl substituents on teichoic acid were necessary for phage Ø 25 binding, together with high proportions of teichoic acid in the wall, whereas binding of phage SP 50 was not highly dependent on the presence of glucosyl substituents. Such differences might be of value in the use of the phages as probes for the localization and distribution of teichoic acid in the wall.

6. Cell Adhesion

a) General Remarks

Recent investigations demonstrate that bacterial function in a variety of natural habitats depends strongly on microenvironments (BOCHEM et al. 1982). These microenvironments result from adhesion of bacteria to solid surfaces (CHENG et al. 1981a,b). The investigations center around bacterial colonization of surfaces either inorganic in nature (soil particles etc.) or of animal or plant origin (tissue of the oral cavity, the intestinal tract, of decaying plant material, etc.).

b) Cell Adherence Mediated by Pili or Fimbriae

JONES and ISAACSON (1983), in a review on proteinaceous bacterial adhesins and their receptors, emphasized that the terms "fimbriae" and "pili" denote morphology, but not function. They proposed that the terms fimbriae and pili be retained only as morphological descriptions of the adhesins. In addition, these authors suggest that the terms pili and fimbriae be applied to appendages involved in gene transfer and bacterial attachment to surfaces associated with colonization, respectively.

α) Conjugative Pili. F-Pili are elaborated by bacterial hosts which carry the conjugative plasmid F and are responsible for establishing the contact between donor and recipient which is necessary for plasmid transfer. Many F-gene products appear to be involved in the assembly of F-Pili (MANNING and ACHTMAN 1979, WILLETTS and SKURRAY 1980). Recently, MOORE et al. (1981a,b) reported the detection of a pool of unassembled F-pilin in inner membranes from F_{lac} and F_{lac} tra mutant cells. Now, SOWA et al. (1983) studied the synthesis and turnover of F-pilin under conditions which have been reported to affect the expression of F-piliation. The data are consistent with a model in which pilin subunits are assembled transiently into pili, conserved by retraction, and made available for subsequent reassembly.

WOROBEC et al. (1983) were able to localize the major antigenic determinant of EDP 208 pili at the N-terminus of the pilus protein. Using the amino acid sequence for an N-terminal dodecapeptide (ET 1), FROST et al. (1983) synthesized that peptide chemically, conjugated it to bovine serum albumin, and subjected it to immunological studies. The experiments showed strong reactivity between the synthetic peptide and antiserum raised against whole pili. Antiserum raised against the synthetic peptide was reactive against intact pilin.

β) Adhesion to Eukaryotic Cells Mediated by Fimbriae. Numerous studies have demonstrated that bacterial fimbriae are involved in the process of bacterial adhesion to eukaryotic cells (DUGUID 1968, EISENSTEIN et al. 1980, ESHDAT et al. 1978, JANN et al. 1981, KORHONEN et al. 1981, SALIT and GOTSCHLICH 1977a,b, WEVERS et al. 1980). Different mechanisms of fimbriae-mediated adhesion have been detected by means of hemagglutination (HA): mannose-sensitive HA mediated by common "type I pili" and mannose-resistant HA mediated by fimbriae isolated from several *E.coli* strains (VOSBECK and HUBER 1982), indicating that HA is mediated by different interactions and several target cell specificity patterns. METT et al. (1983) found that mannose-resistant adhesion to tissue culture cells can be classified into different subtypes. These adhesions appear to be carbohydrate-mediated. KORHONEN et al. (1983) investigated type 3 fimbriae of *Klebsiella* sp.; they performed a molecular characterization of these fimbriae and their role in bacterial adhesion to plant roots. They proposed that type 3 fimbriae are involved in the establishment of the plant-bacterium association concerning nitrogen-fixing *Klebsiella* strains living on the roots of nonleguminous plants.

c) Cell Adherence Mediated by Nonfimbrial Adhesins

RAUVALA (1983) and SUTHERLAND (1983) published reviews on cell surface carbohydrates and on microbial exopolysaccharides and their role in microbial adhesion in aqueous systems. A role for the adhesion of bacteria has been postulated in the pathogenesis of plants and animals. Enteropathogenic strains of *E. coli* were investigated in detail by SCOTLAND et al. (1983). They found that the majority of all strains analyzed from different outbreaks of diarrhoea were nonfimbriate. Various workers have suggested the involvement of polysaccharides, slime layers or lectins. ROSENBERG et al. (1983) postulated, as a hypothesis for the explanation of bacterial adherence, that adherence primarily depends on hydrophobic interactions. An understanding of the adhesive mechanisms is likely to be valuable for the development of vaccines. Recently, LAMED et al. (1983) performed a characterization of a cellulose-binding, cellulase-containing complex in *Clostridium thermocellum.* Their results are consistent with the contention that this cellulose-binding factor (CBF) comprises a discrete, multisubunit complex which exhibits separate antigenic and multiple cellulase activities. It appears that CBF is responsible for the adherence of the cells to cellulose, and that it constitutes a major part of the cellulolytic apparatus.

7. Bacterial Membranes

a) Membranes of Photosynthetic Bacteria

VARGA and STAEHELIN (1983) investigated the spatial differentiation in photosynthetic and nonphotosynthetic membranes of *Rhodopseudomonas palustris.* They found that the cytoplasmic membrane and the photosynthetic intracytoplasmic membranes were spatially differentiated into regions of extremely high intramembrane-particle density and areas of lower density of these particles. The high density areas were always seen in association with photosynthetic membrane stacks. This differentiation was also seen in those areas of the cytoplasmic membrane which adhered to the underlying intracytoplasmic membrane, implying that the cytoplasmic membrane too is differentiated for photosynthesis in this region. These results supplement data reviewed by DREWS (1978), KAPLAN (1978, 1981) and KAPLAN and ARNTZEN (1982). VARGA and STAEHELIN (1983) also described an additional interesting finding, i.e., a distinct periodicity in the sizes of intramembrane particles found in the stacked regions. This suggests that, as light levels decrease, subunits of discrete size are being added to a core particle.

KAUFMANN et al. (1982) described the differentiation of the membrane system in cells of *Rhodopseudomonas capsulata* after transition from chemotrophic to phototrophic growth conditions. The intracytoplasmic chromatophore - membrane fraction was found to increase during adaptation 9-fold in surface area per cell, 42-fold in bacteriochlorophyll content, 7-fold in reaction center content, and 6-fold in the size of the photosynthetic unit. An ultrastructural study (KUNKEL 1982) of four cyanobacteria indicated the presence of previously undescribed "thylakoid centers" from which thylakoids radiate. These centers are peripherally located; they are cylinders 30 nm wide by 320 nm long, consisting of globular subunits oriented in nonparallel stacked arrays. Thylakoids are attached to the outer surface of the cylinder along its longitudinal axis. This finding adds new information for a better understanding of one of the least understood phenomena in cyanobacteria, the biogenesis of photosynthetic membranes (ALLEN 1968, GUGLIELMI et al. 1981, JOST 1965, PANKRATZ and BOWEN 1963, STANIER and COHEN-BAZIRE 1977). The fine structure of the photosynthetic membranes of five bacteriochlorophyll b - containing bacteria has recently been investigated (ENGELHARDT et al. 1983) by electron microscopy and digital image analysis. The authors found that all five species had the photosynthetic complexes hexagonally arrayed in the membrane. The basic organization, i.e., a ring, probably containing the light-harvesting (LH) polypeptides, surrounding a core (the "reaction center") appeared to be almost identical for all five species. GEST and FAVINGER (1983) observed in an enrichment dark-green, slimly mats. This "green" organism, designated as *Heliobacterium chlorum*, has a constellation of properties that is unique to the degree that it can be considered to represent a hitherto unknown genus. It contains a "new" form of bacteriochlorophyll ($Bchl_{Gg}$). Its subcellular localization is uncertain; neither chlorosomes nor extensively developed intracytoplasmic membranes of the kind produced by most photosynthetic bacteria were observed.

b) Membranes in Non Photosynthetic Bacteria

Azotobacter vinolandii contains intracytoplasmic membrane invaginations as reported by several authors (HILL et al. 1972, OPPENHEIM and MARCUS 1970, PANGBORN et al. 1962, PATE et al. 1973, WYSS et al. 1961). Up to now, the control of internal membrane development remained controversial. POST et al. (1982) got data which suggest that it is oxygen which controls changes in the ratio of intracytoplasmic to cytoplasmic membrane surface areas under conditions of nitrogen fixation, whereas in ammonium-assimilating cells this ratio stayed almost constant. An additional report (POST et al. 1983) supplied data suggesting that intracytoplasmic and cytoplasmic membranes of this organism do not differ with respect to respiratory activities. Probably, both types of membranes represent differently localized parts of an otherwise identical membrane system.

8. Organization of the Bacterial Nucleoid

a) Folding of Prokaryotic DNA

A great deal of progress has been made toward knowledge of DNA compaction in eukaryotic chromatin (FELSENFELD 1978, GARRETT 1979). However, the situation for bacterial DNA is less well analyzed. SLOOF et al. (1983) isolated and characterized nucleoids from *Bacillus licheniformis*. They found that DNA is organized in independent, negatively superwisted domains. In contrast to the DNA organization in bacterial nucleoids isolated under high-salt conditions and in the presence of detergents (STONINGTON and PETTIJOHN 1971, WORCEL and BURGI 1972), the domains of supertwisted DNA in the low-salt-isolated nucleoids were found to be maintained by protein-DNA interactions. A major role for nascent RNA in restraining supertwisted DNA was not observed.

b) Cell Wall - DNA Association

Investigators in several laboratories confirmed the existence of DNA-membrane complexes (FIRSCHEIN 1972, HEIDRICH and OLSEN 1975, HOROWITZ et al. 1979, OLSEN et al. 1974, WORCEL and BURGI 1972) in various kinds of bacteria. KOCH et al. (1981) provided an explanation for the origin of wall-associated DNA in *Bacillus subtilis*. They proposed an improved replicon model based on the assumption that both the origin and terminus of replication were bound to cell ends, whereas the replication fork was bound to the cell cylinder. DOYLE et al. (1983) reported results supporting this view, which suggest that there is a stable cell pole DNA attachment in *B. subtilis*.

So far, endomembranes have not yet been found in nuclear bodies. However, ELLINGER et al. (1982), when investigating folded chromosomes from *Anacystis nidulans*, observed vesicle-like particles containing DNA; the DNA was shown to be attached to thylakoid membranes.

c) Cosegregation of Cell Wall and DNA

SCHLAEPPI and KARAMATA (1982) observed a cosegregation of cell wall and DNA in *Bacillus subtilis* by continously labeling cell wall or DNA or both cell wall and DNA. In a former study, SCHLAEPPI et al. (1982) had demonstrated zonal insertion, during wall synthesis, by establishing that the cell wall of *B. subtilis* consists of a limited number of large "subunits" which segregate during cell growth.

9. Bacterial Motility

a) Motility Mediated by Flagella

α) Mechanical Problems. The helical flagellar filaments of bacteria are constructed from subunits (flagellin) by self-assembly. They are polymorphic, and change from one helical form to another in a variety of circumstances. CALLADINE (1982), in a detailed review, discussed two of the aspects connected with this fact, i.e., the general problem of how to construct a helical rod from identical subunits, and the problem concerning the hydrodynamic performance of a helical filament which can switch waveform when subjected to mechanical overload.

β) Coordination of Flagella. In a recent investigation, ISHIHARA et al. (1983) analyzed the direction of rotation of flagellar hooks of mutants (and, hence, flagella in wild-type cells) of *E. coli*, using filamentous, nonseptate cells. They found that the directions of rotation on a given cell were not correlated; reversals of the flagellar motors occurred asynchronously. The bias of the motors (the fraction of time spent spinning counterclockwise) changed with time. Variations in bias were correlated, provided that the motors were within a few micrometers of one another. Thus, the signal appeared to have a limited range. These observations are in favor of a former model (BLOCK et al. 1982, KHAN and MACNAB 1980) and do not support a second postulate (KOSHLAND 1977, 1979), i.e., the global nature of the signal and a synchronous reversal of the flagellar motors in the same cell.

γ) Growth of Flagella. As shown previously (EMERSON et al. 1970, IINO 1969, SUZUKI and IINO 1977), bacterial flagella grow in vivo by the addition of flagellin molecules to the distal end of the filaments. Possibly, the flagellin molecules are transported to the tip through the central hole of the filament (BODE et al. 1972, SHIRAKIHARA and WAKABAYASHI 1979, SLEYTR and GLAUERT 1973, WAGENKNECHT et al. 1981). IKEDA et al. (1983) investigated a short-flagella mutant of *Salmonella thyphimurium* and observed excretion of flagellin into the medium. They proposed that such mutants should be useful in clarifying the mechanism of flagellin transportation, especially the energetics and the exact path of flagellin secretion.

b) A Model for Gliding Motility

DWORKIN et al. (1983) and KELLER et al. (1983) analyzed motility of *Myxococcus xanthus*. They presented evidence to support a model that gliding motility of this organism is driven by surface tension at the interfaces that surround the cells, thus providing the driving force for active and regulated movement.

c) Motility of Mycoplasma

Three pathogenic Mycoplasma species which exhibit nonflagellar gliding motility possess specialized terminal regions (BREDT 1973, TULLY et al. 1983). WALL et al. (1983) assumed that these regions might be implicated in motility of these microorganisms which have no cell wall. The nature of the underlying mechanism of motility remains obscure.

References

ACHTERRATH, M., LICKFELD, K.G.: Number, size and location of *Staphylococcus* mesosomes - a function of fixation parameters, 220-221. In: Proceedings of the Fifth European Congress on Electron Microscopy, Manchester. London: Institute of Physics 1972. - ALLEN, M.M.: J. Bacteriol. 96, 842-852 (1968). - ARCHIBALD, A.R.: Phage receptors in Gram-positive bacteria, 7-26. In: Virus Receptors, eds. L.L. RANDALL, L. PHILIPSON, Part 1, Bacterial Viruses (receptors and recognition, Series B, Vol.7). London: Chapman and Hall 1980.

BARTHOLOMEW, J.W., MITTWER, T.: Bacteriol. Rev. 16, 1-29 (1952). - BENIANS, T.H.C.: J. Pathol. Bacteriol. 23, 401-412 (1920). - BEVERIDGE, T.J.: Int. Rev. Cytol. 72, 229-317 (1981). - BEVERIDGE, T.J., DAVIES, J.A.: J. Bacteriol. 156, 846-858 (1983). - BOCHEM, H.P., SCHOBERTH, S. M., SPREY, B., WENGLER, P.: Can. J. Microbiol. 28, 500-510 (1982). - BODE, W., ENGEL, J., WINKLAIR, D.: Eur. J. Biochem. 26, 313-327 (1972). - BREDT, W.: Ann. N.Y. Acad. Sci. 225, 246-250 (1973). - BURDETT, I. D.J., ROGERS, H.J.: J. Ultrastruct. Res. 30, 354-367 (1970). - BURLEY, S.K., MURRAY, R.G.E.: Can. J. Microbiol. 29, 775-780 (1983).

CALLADINE, C.R.: Construction of bacterial flagellar filaments, and aspects of their conversion to different helical forms, 33-51. In: Prokaryotic and Eukaryotic Flagella, eds. W.B. AMOS, J.G. DUCKETT. Society for Experimental Biology Symposium No. XXXV. Cambridge: Cambridge University Press 1982. - CHENG, K.-J., FAY, J.P., COLEMAN, R.N., MILLIGAN, L.P., COSTERTON, J.W.: Appl. Environ. Microbiol. 41, 298-305 (1981a). - CHENG, K.-J., IRVIN, R.T., COSTERTON, J.W.: Can. J. Microbiol. 27, 461-490 (1981b).

DAVIES, J.A., ANDERSON, G.K., BEVERIDGE, T.J., CLARK, H.C.: J. Bacteriol. 156, 837-845 (1983). - DOYLE, R.J., KOCH, A.L., CARSTENS, P.H. B.: J. Bacteriol. 153, 1521-1527 (1983). - DREWS, G.: Curr. Top. Bioenerg. 8, 161-207 (1978); - Prog. Bot. 44, 14-21 (1982). - DUBOCHET,

J., McDOWALL, A.W., MENGE, B., SCHMID, E.N., LICKFELD, K.G.: J. Bacteriol. 155, 381-390 (1983). - DUGUID, J.P.: Arch. Immunol. Ther. Exp. 16, 173-188 (1968). - DWORKIN, M., KELLER, K.H., WEISBERG, D.: J. Bacteriol. 155, 1367-1371 (1983).

EBERSOLD, H.R., CORDIER, J.-L., LÜTHY, P.: Arch. Microbiol. 130, 19-22 (1981). - EDELSTEIN, E.M., ROSENZWEIG, M.S., DANEO-MOORE, L., HIGGINS, M.L.: J. Bacteriol. 143, 499-505 (1980). - EISENSTEIN, B.I., BEACHEY, E.H., OFEK, I.: Infect. Immun. 28, 154-159 (1980). - ELLINGER, A., DWORSKY, P., WEISSHÄUPL, V.: Z. Allg. Mikrobiol. 22, 17-27 (1982). - EMERSON, S.U., TOKUYASU, K., SIMON, M.I.: Science 169, 190-192 (1970). - ENGELHARDT, H., BAUMEISTER, W., SAXTON, W.O.: Arch. Microbiol. 135, 169-175 (1983). - ESHDAT, Y., OFEK, I., YASHOUV-GAN, Y., SHARON, N., MIRELMAN, D.: Biochem. Biophys. Res. Commun. 85, 1551-1559 (1978).

FELSENFELD, G.: Nature (London) 271, 115-122 (1978). - FIRSCHEIN, W.: J. Mol. Biol. 70, 383-397 (1972). - FROST, L.S., ARMSTRONG, G.D., FINLAY, B.B., EDWARDS, B.F.P., Paranchych, W.: J. Bacteriol. 153, 950-954 (1983).

GARRETT, R.A.: Int. Rev. Biochem. 25, 179-203 (1979). - GEST, H., FAVINGER, J.L.: Arch. Microbiol. 136, 11-16 (1983). - GIBSON, C.W., DANEO-MOORE, L., HIGGINS, M.L.: J. Bacteriol. 155, 351-356 (1983). - GIVAN, A.L., GLASSEY, K., GREEN, R.S., LANG, W.K., ANDERSON, A.J., ARCHIBALD, A.R.: Arch. Microbiol. 133, 318-322 (1982). - GLAUERT, A. M., THORNLEY, M.J.: Annu. Rev. Microbiol. 23, 159-198 (1969). - GUGLIELMI, G., COHEN-BAZIRE, G., BRYANT, D.A.: Arch. Microbiol. 129, 181-189 (1981).

HASE, T.: J. Bacteriol. 154, 976-979 (1983). - HEIDRICH, H.G., OLSEN, W.L.: J. Cell Biol. 67, 444-460 (1975). - HILL, S., DROZD, J.W., POSTGATE, J.R.: J. Appl. Chem. Biotechnol. 22, 541-558 (1972). - HOROWITZ, S., DOYLE, R.J., YOUNG, F.E., STREIPS, U.N.: J. Bacteriol. 138, 799-802 (1979). - HUSER, B.A., WUHRMANN, K., ZEHNDER, A.J.B.: Arch. Microbiol. 132, 1-9 (1982).

IINO, T.: J. Gen. Microbiol. 56, 227-239 (1969). - IKEDA, T., KAMIYA, R., Yamaguchi, S.: J. Bacteriol. 153, 506-510 (1983). - ISHIHARA, A., SEGALL, J.E., BLOCK, S.M., BERG, H.C.: J. Bacteriol. 155, 228-237 (1983).

JANN, K., JANN, B., SCHMIDT, G.: FEMS Microbiol. Lett. 11, 21-25 (1981). - JAVOR, B., REQUADT, C., STOECKENIUS, W.: J. Bacteriol. 151, 1532-1542 (1982). - JONES, G.W., ISAACSON, R.E.: Crit. Rev. Microbiol. 10, 229-260 (1983). - JOST, M.: Arch. Mikrobiol. 50, 211-245 (1965).

KAPLAN, S.: Control and kinetics of photosynthetic membrane development, 809-839. In: The Photosynthetic Bacteria, ed. R. CLAYTON, W. SISTROM. New York: Plenum 1978; - Photochem. Photobiol. 34, 769-774 (1981). - KAPLAN, S., ARNTZEN, C.J.: Photosynthetic membrane structure and function. 65-153. In: Photosynthesis: Comparative Aspects of Bacteria and Green Plants, ed. I. GOVINJEE. New York: Academic Press 1982. - KAUFMANN, N., REIDL, H.-H., GOLECKI, J.R., GARCIA, A.F., DREWS, G.: Arch. Microbiol. 131, 313-322 (1982). - KAWATA, T., MASUDA, K., YOSHINO, K., FUJIMOTO, M.: Jpn. J. Microbiol. 18, 469-476 (1974). - KELLENBERGER, E., CARLEMALM, E., VILLINGER, W., ROTH, J., GARAVITO, R.M.: Low Denaturation Embedding for Electron Microscopy of Thin Sections. 59 S. Waldkraiburg: Chemische Werke Lowi 1980. - KELLER, K.H., GRADY, M., DWORKIN, M.: J. Bacteriol. 155, 1372-1381 (1983). - KESSEL, M., COHEN, Y.: J. Bacteriol. 150, 851-860 (1982). - KHAN, S., MACNAB, R.M.: J. Mol Biol. 138, 563-597 (1980). - KOCH, A.L., HIGGINS,

M.L., DOYLE, R.J.: J. Gen. Microbiol. 123, 151-161 (1981). - KORHONEN, T.K., LEFFLER, H., SVANBORG EDEN, C.: Infect. Immun. 32, 796-804 (1981). - KORHONEN, T., TARKKA, E., RANTA, H., HAAHTELA, K.: J. Bacteriol. 155, 860-865 (1983). - KOSHLAND, D.E., Jr.: Science 196, 1055-1063 (1977); - Physiol. Rev. 59, 811-862 (1979). - KUNKEL, D.D.: Arch. Microbiol. 133, 97-99 (1982).

LAMED, R., SETTER, E., BAYER, E.A.: J. Bacteriol. 156, 826-836 (1983). - LEE, S.: Bot. Gaz. 83, 420-424 (1927).

MANNING, P.A., ACHTMAN, M.: Cell to cell interactions in conjugating *E.coli* : the involvement of the cell envelope, 409-447. In: Bacterial Outer Membranes, ed. M. Inouye. New York: Wiley 1979. - MARCENKO, E.: Acta Bot. Croatica 20/21, 47-74 (1961). - MARTIN, T.C., WYATT, J.T.: J. Phycol. 10, 57-65 (1974). - MASUDA, K., KAWATA, T.: Microbiol. Immunol. 23, 941-953 (1979); - ibid. 24, 299-308 (1980); - J. Gen. Microbiol. 124, 81-90 (1981). - FEMS Microbiol. Lett. 20, 145-150 (1983). - METT, H., KlOETZLEN, L., VOSBECK, K.: J. Bacteriol. 153, 1038-1044 (1983). - MOORE, D., SOWA, B.A., IPPEN-IHLER, K.: J. Bacteriol. 146, 251-259 (1981a); - Mol. Gen. Genet. 184, 260-264 (1981b).

NANNINGA, N.: Freeze-fracturing of microorganisms: physical and chemical fixation of *Bacillus subtilis*, 151-180. In: Freeze-Etching, Techniques and Applications, eds. L. BENEDETTI, P. FAVARD. Paris: Société Francaise de Microscopie Electronique 1973. - NICHOLSON, D.E., FOX, G.E.: Can. J. Microbiol. 29, 52-59 (1983). - NIERZWICKI, S.A., MARATEA, D., BALKWILL, D.L., HARDIE, L.P., MEHTA, V.B., STECENS Jr., S.E.: Arch. Microbiol. 133, 11-19 (1982).

OBERTI, J., CARAVANO, R., ROUX, J.: Can J. Microbiol. 28, 1300-1303 (1982). - OLSEN, W.L., HEIDRICH, H.G., HENNING, K., HOFSCHNEIDER, P. H.: J. Bacteriol. 118, 646-653 (1974). - OPPENHEIM, J., MARCUS, L.: J. Bacteriol. 101, 286-291 (1970).

PANGBORN, J., MARR, A.G., ROBRISH, S.A.: J. Bacteriol. 84, 669-678 (1962). - PANKRATZ, H.S., BOWEN C.C.: Am J. Bot. 50, 387-399 (1963) PATE, J.L., SHAH, V.K., BRILL, W.J.: J. Bacteriol. 114, 1346-1350 (1973). - POST, E., GOLECKI, J.R., OELZE, J.: Arch. Microbiol. 133, 75-82 (1982). - POST, E., VAKALOPOULOU, E., OELZE, J.: Arch. Microbiol. 134, 265-269 (1983).

RAUVALA, H.: Trends Biochem. Sci. 8, 323-325 (1983). - RIPPKA, R., DERUELLES, J., WATERBURY, J.B., HERDMAN, M., STANIER, R.Y.: J. Gen. Microbiol. 111, 1-61 (1979). - ROSENBERG, M., ROSENBERG, E., JUDES, H., WEISS, E.: FEMS Microbiol. Lett. 20, 1-5 (1983).

SALIT, I.E., GOTSCHLICH, E.C.: J. Exp. Med. 146, 1169-1181 (1977a); - ibid. 146, 1182-1194 (1977b). - SALTON, M.R.J.: J. Gen. Microbiol. 30, 223-235 (1963). - SCHERRER, R.: J. Gen. Microbiol. 31, 135-145 (1963). - SCHLAEPPI, J.-M., KARAMATA, D.: J. Bacteriol. 152, 1231-1240 (1982). - SCHLAEPPI, J.-M., POOLEY, H.M., KARAMATA, D.: J. Bacteriol. 149, 329-337 (1982). - SCHWABE, G.H.: Hydrology 22, 759-792 (1960). - SCOTLAND, S.M., RICHMOND, J.E., ROWE, B.: FEMS Microbiol. Lett. 20, 191-195 (1983). - SHIRAKIHARA, Y., WAKABAYASHI, T.: J. Mol. Biol. 131, 485-507 (1979). - SKERMAN, V.B.D., SLY, L.I., WILLIAMSON, M.-L.: Int. J. Sys. Bacteriol. 33, 300-308 (1983). - SLEYTR, U.B.: Int. Rev. Cytol. 53, 1 64 (1978). - SLEYTR, U.B., GLAUERT, A.M.: Nature (London) 241, 542-543 (1973). - SLOOF, P., MAAGDELIJN, A., BOSWINKEL, E.: J. Mol. Biol. 163, 277-297 (1983). - SOWA, B.A., MOORE, D., IPPEN-IHLER, K.: J. Bacteriol. 153, 962-968 (1983). - SPEARING, J.K.: Arch. Protistenk. 89, 209-278 (1937). - STAHL, S.J., STEWART, K.R., WILLIAMS, F.D.: J. Bacteriol. 154, 930-937 (1983). - STANIER,

R.Y., COHEN-BAZIRE, G.: Annu. Rev. Microbiol. 31, 225-274 (1977). - STOECKENIUS, W.: J. Bacteriol. 148, 352-360 (1981). - STONINGTON, O. G., PETTIJOHN, D.E.: Proc. Natl. Acad. Sci. USA 68, 6-9 (1971). - SUTHERLAND, I.W.: Critic. Rev. Microbiol. 10, 173-201 (1983). - SUZUKI, T., IINO, T.: J. Bacteriol. 129, 527-529 (1977).

TAYLOR, K.A.: J. Microsc. (Oxf.) 112, 115-125 (1978). - THOMPSON, B. G., MURRAY, R.G.E.: Can. J. Microbiol. 28, 1081-1088 (1982). - THURSTON, E.L., INGRAM, L.O.: J. Phycol. 7, 203-210 (1971). - TULLY, J.G., TAYLOR-ROBINSON, D., ROSE, D.L., COLE, R.M., BOVE, J.M.: Int. J. Syst. Bacteriol. 33, 387-396 (1983).

VAARA, T.: Can. J. Microbiol. 28, 929-941 (1982). - VARGA, A.R., STAEHELIN, L.A.: J. Bacteriol. 154, 1414-1430 (1983). - VOSBECK, K., HUBER, U.: Eur. J. Clin. Microbiol. 1, 22-28 (1982).

WAGENKNECHT, T., DeROSIER, D., SHAPIRO, L., WEISSBORN, A.: J. Mol. Biol. 151, 439-465 (1981). - WALL, F., PFISTER, R.M., SOMERSON, N.L.: J. Bacteriol. 154, 924-929 (1983). - WALSBY, A.E.: Nature (London) 283, 69-71 (1980). - WEBB, M.: J. Gen. Microbiol. 2, 260-274 (1948). - WECKE, J., LAHAV, M., GINSBURG, I., GIESBRECHT, P.: Arch. Microbiol. 131, 116-123 (1982). - WEVERS, P., PICKEN, R., SCHMIDT, G., JANN, B., JANN, K., GOLECKI, J.R., KIST, M.: Infect. Immun. 29, 685-691 (1980). - WILDGRUBER, G., THOMM, M., KÖNIG, H., OBER, K., RICCHIUTO, T., STETTER, K.O.: Arch. Microbiol. 132, 31-36 (1982). - WILLETTS, N., SKURRAY, R.: Annu. Rev. Genet. 14, 41-76 (1980). - WORCEL, A., BURGI, E.: J. Mol. Biol. 71, 127-147 (1972). - WOROBEC, E.A., TANEJA, A.K., HODGES, R.S., PARANCHYCH, W.: J. Bacteriol. 153, 955-961 (1983). - WYSS, O., MARILYN, P., NEUMANN, G., SOCOLOFSKY, M.: J. Biophys. Biochem. Cytol. 10, 555-565 (1961).

ZEHNDER, A.J.B., HUSER, B.A., BROCK, T.D., WUHRMANN, K.: Arch. Microbiol. 124, 1-11 (1980).

Professor Dr. FRANK MAYER
Institut für Mikrobiologie
der Universität Göttingen
Abteilung Mikromorphologie
Grisebachstraße 8
D-3400 Göttingen, FRG

b) Cytology and Morphogenesis of the Fungal Cell

By ECKART JELKE

1. Books, Monographs, etc.

The translation of the second edition of *Introduction to Fungi* into German, entitled *Pilze* (WEBSTER 1983), points out some additional aspects of genetics, reproduction and developmental cycles. The fourth revised and enlarged edition of the compendium of *Mykologie* (MÜLLER and LOEFFLER 1982) has appeared. DEACON (1980) gives an *Introduction to Modern Mycology*. The taxonomy and biology of hyphomycetes describes SUBRAMANIAN (1983). JONES (1983) reports in *Recent Advances in Aquatic Mycology*. Fungi of phytopathological significance are treated: rust fungi by SCOTT and CHAKRAVORTY (1982), zoosporic plant pathogens by BUCZACKI (1983), the biology of *Botrytis* by COLEY-SMITH et al. (1981), fungal wild diseases of plants by MACE et al. (1981). Development and/or special differentiation states are discussed: conidial fungi by COLE and SAMSON (1979a,b), MINTER et al. (1982, 1983), *Physarum* by SAUER (1982), zoospores of uniflagellate Phycomycetes by LANGE and OLSON (1979), basidium and basidiocarp by WELLS and WELLS (1982). ALDRICH and DANIEL (1982) together with 40 contributors treat comprehensively the *Cell Biology of Physarum and Didymium*, and HARLEY and SMITH (1983) present a thorough review in *Mycorrhizal Symbiosis*. ESSER (1981) reports reports on some aspects of the contribution of basic research to the practical application of fungal processes.

2. Morphogenesis of Spores

a) Zoospores (OLSON et al. 1981, LANGE and OLSON 1983)

α) *Ultrastructural Organization of Zoospores*. Zoospore substructure is frequently used as the base for describing new genera and for a reevaluation of the phylogenetic and taxonomic relations among zoosporic fungi (POWELL 1978, LANGE and OLSON 1979, BARR 1981, Progr. Bot. 44, 345, 1982).

Concerning the flagellar rootlet system, the kinetosome and transiiton zone, the zoospore of Plasmodiophoromycetes differs from all the other zoosporic fungi. There is more similarity to certain protozoa (BARR and ALLAN 1982). The spatial arrangements and the structure of the microbody-lipid globule complexes (MLCs) differ in a number of Chytridiomycetedes (POWELL 1978). In Polyphagus the central pair of axonemal microtubules continues into the kinetosome (POWELL 1981). HEATH et al. (1982) discuss examples of some root arrangements in terms of their probable functions including spore shape generation. Associations of microtubules of the rootlet systems with rumposomes are reported (LUCAROTTI 1981, POWELL 1981).

β) Zoospore Formation (Progr. Bot. 35, 28, 1973; 39, 22, 1981). OLSON et al. (1981) report that in all species of uniflagellate fungi and in several biflagellate fungi the sporangial cytoplasm is divided into spores by means of progressive fusion of small vesicles. Astonishing is the speed by which the process can be accomplished (LANGE and OLSON 1983).

In *Allomyces catenoides* - one of several model systems of zoosporogenesis - the nuclei at the begin of spore formation lie around the periphery of the sporangial cytoplasm and an array of microtubules radiates from the two kinetosomes into the cytoplasm. Small vesicles appear in the cytoplasm as the electron-dense material disappears from the inclusions of dense bodies. After development of flagella - a process in which both the flagellum membrane and the plasmalemma of the flagellar invagination are formed by deposition and fusion of the same vesicles - the final cleavage of the cytoplasm takes place by incorporation of small vesicles or flattened cisternae along the cleavage furrow (OLSON et al. 1981). Whether the cytoplasmic microtubules hold the cleavage vesicles roughly equidistant from nuclei and are involved in the determination of the portion of cytoplasm which will be incorporated into spore, is discussed (HEATH 1976, ARMBRUSTER 1982a,b). The origin of the cleavage vesicles is frequently undecided. However, in some cases it could be suggested that the Golgi apparatus produces one or more types of these vesicles. In *Chytridium* it produces in succession four vesicle types. Microtubules near dictyosomes and cleavage furrows suggest that cytoskeletal components might be involved in the precise orientation of Golgi-apparatus-derived vesicles in the cytoplasm (TAYLOR and FULLER 1981). MIYATA et al. 1980 demonstrate that in *Phytophthora* radiolabeled sitosterol moves directly via cleavage vesicles to zoospore membranes.

b) Sporangiospores (BENJAMIN 1979, COLE and SAMSON 1979b)

Methods: Fewer sporangiospores are lost from mature sporangiophores by the freeze-drying method employed for scanning electron microscopy by KITZ et al. (1982). ELLIS and AJELLO (1982) describe a procedure for stimulating sporulation of non-sporulating isolates of *Saksenaea* and recommend this procedure for use with other nonsporulating zygomycetes.

HAMMILL and SECOR (1983) report on the nuclear number in eight strains of *Mucor mucedo* with an average number of 15-20 nuclei per spore. In *Cokeromyces* numerous recurved pedicles arise from the sporangiophore terminal vesicle. Each pedicle produces a multispored sporangiole (JEFFRIES and YOUNG 1983). The habitual conspicuous different sporangiolar arrangements in two species of *Mycotropha* represent two possibilities to realize a maximal utilization of space on the sporangiolar vesicles for spore production (BRAIN and YOUNG 1979). In contrast to an earlier misinterpretation, the monosporous sporangiolum is a simply reduced sporangium and is not a conidium (COLE and SAMSON 1979b).

Spores of *Piptocephalis* are cleaved out of the merosporangial protoplast by simultaneous invagination of the plasmalemma. Between the opposing plasmalemma at each cleavage site an electron-transparent abscissions zone extends across the merosporangium. After detachment of the mature spores circular scars appear on each spore (BAKER 1979). The septum in the sporangiophores of *Tieghemiomyces* has a lenticular central pore closed by a discoid, globose body on each side of the septum (BRAIN et al. 1982).

During development of the zygospore of *Mortierella*, on suspensor acts as a basal supporting element from which investing hyphae develop

(ANSELL and YOUNG 1983). The verrucose ornamented zygosporangium of *Cokeromyces* is the fusion product of two homogametangia. It contains a zygospore with a finely sculptured wall (JEFFRIES and YOUNG 1983). In *Syzygites* the osmotic potential, and not the carbohydrate concentration of the medium, is most important in regulating zygospore formation (KAPLAN and GOOS 1982).

The formation of an azygospore in *Conidiobolus* begins by protoplasmic condensation at a particular point in the hyphal body. The young spherical prespore contains more or less the same cytoplasm as the hypha and its wall structure is the same as that of the mycelium (LATGÉ et al. 1982).

c) Ascospores

Electron microscopy has shown that in all Ascomycotina ascopore development involves the compartmentalization of ascus cytoplasm by a pair of concentric membranes into uninucleate portions (BECKETT 1981). However, the mechanisms of spore delimitation in Euascomycetes and Hemiascomycetes are different.

α) Ascospore Formation in Euascomycetes. In Euascomycetes a double membrane sac, termed the ascus vesicle, is formed around the priphery of the ascus near the ascus plasmalemma. Spore delimitation results from the invaginations of the ascus vesicle around each nucleus after the last nuclear division. In contrast to some proposals on the origin of these ascospore-delimiting membranes, reviewed for instance by ELLIS (1981 a,b) in *Thermoascus* and ROSING in *Chaetomium* (1982), the spore-delimiting membranes arise from the ascus plasmalemma. In *Chaetomium brasiliense* the nucleus-associated organelles (NAOs) are presumed to be involved in spore delimitation. In *Podospora anserina*, however, NAOs are not necessary (ZICKLER and SIMONET 1980).

β) Ascospore Formation in Hemiascomycetes. In Hemiascomycetes the ascospore-delimiting membranes are formed on the cytoplasmic side of a modified NAO during meiotic nuclear division. In *Saccharomyces* during the second meiotic division all four NAOs develop an outer plaque and after this structural modification of the NAOs the initiation of the spore-delimiting membranes takes place adjacent to the outer plaques. In the case of a two-spored ascus, only one NAO is modified thus indicating that the morphogenesis of spores is regulated by the number of NAOs producing an outer plaque (DAVIDOV et al. 1980). BILINSKI and MILLER (1980, 1983) point out that in cells of an apomictic strain during the differentiation from the vegetative to the sporulating state zinc is translocated from the vacuolar compartment to nucleoli and NAOs. In some further Hemiascomycetes the spore morphogenesis starts also with a disc-shaped vesicle in a cytoplasmic area adjacent to the outer plaque of the NAO (ASHTON and MOENS 1979). Similarly, the ascospore-delimiting membranes (prespore membranes) are formed in *Schizosaccharomyces* in the interval between meiosis I and II (TANAKA and HIRATA 1982). It has been shown that in some species the formation of the spore-delimiting membranes starts by fusion of cytoplasmic vesicles. Staining results suggest that these membranes do not come from the ascus plasmalemma, but from other membrane systems, such as the endoplasmic reticulum.

Ascosporogenesis in *Tuber magnatum* is intermediate between the general pattern of Euascomycetes and that of Hemiascomycates (BERTA and FUSCONI 1983). Although a double-membrane system adjacent to the ascus plasmalemma resembles ascus vesicles, it does not invaginate to delimit the ascospore.

γ) Spore-wall Formation. Spore-wall formation in all Ascomycotina so far examined starts by deposition of spore primary wall material in the space between the spore-delimiting membranes (ROSING 1982). The subsequent phases of wall development vary according to the species (BECKETT 1981, MARMIROLI et al. 1981). A considerable expansion of the perispore forming a honeycomb ornamental layer is characteristic of *Tuber magnatum* (BERTA and FUSCONI 1983).

Recently numerous publications report on light, scanning and transmission electron microscopical observations of ascospore appendages of marine ascomycetes particularly because of the taxonomic significance of this morphological feature (JONES and MOSS 1980a,b, JONES et al. 1983, KOCH et al. 1983).

PORTER (1982) has followed the development of ascospores and their appendages in *Trichomaris*. At the earliest developmental stages observed the appendages are distinct from epispore. They appear to be deposited or condensed within the inflated spore sac.

Another characteristic attribute of several ascospores, the germ pore, describes BECKETT (1979a,b).

d) Basidiospores

In a review on ulstrastructure and cytochemistry of basidial and basidiospore development McLAUGHLIN (1982) points out that most of ultrastructural investigation of the basidium has dealt with the holobasidium, but only the Agaricales and Aphyllophorales have been carefully examined. A comprehensive report of the cytoplasmic events in phragmobasidial development is still lacking.

α) Holobasidium Formation. The holobasidium, in the initiation stage an enlarging hyphal tip, contains at first few types of organelles. The absence of apex vesicles in the young, growing basidium of *Coprinus cinereus* may indicate a slow growth rate. In the late prefusion stage glycogen is evident at the basidial base, and apical and basal vacuoles are present. The basidial wall resembles that of subhymenial cells. The septum at the base of basidium has a dolipore and an outer cap of differentiated cytoplasm (McLAUGHLIN 1982). Karyogamy takes place in a central position of basidium and is presumably initiated by the NAOs (THIELKE 1976). In a report on meiotic divisions in basidia of about 20 species of Agaricales, THIELKE (1982) concludes that within the Agaricales the mechanisms of nuclear divisions are essentially similar.

β) Phragmobasidium Formation. The development of the phragmobasidium is more complicated. The recent knowledge of ultrastructure, morphogenesis, wall formation, shape and surface structures of teliospores of Uredinales are summarized by LITTLEFIELD and HEATH (1979), HIRATSUKA and SATO (1982), McLAUGHLIN (1982), those of Ustilaginales by DEML and OBERWINKLER (1981), DEML et al. (1981), GARDNER et al. (1983a,b), and HESS and Gardner (1983).

The formation of metabasidium begins with the teliospore germination by emerging of a germ tube from the germ pore region. During this process the outer portion of teliospore wall disintegrates (MIMS 1981). The innermost layer of the teliospore wall is continuous with that of the growing promycelium or metabasidium (LITTLEFIELD and HEATH 1979, RAMBERG and McLAUGHLIN 1980, MIMS 1981). *Gymnosporangium*, a favorable object because it germinates readily and more or less synchronously in the laboratory, is examined by MIMS (1981). In this organism vesicles are abundant at the tip of the rapidly elongating promycelium. After the promycelium has reached a typical length meiosis occurs.

γ) Meiosis. A thoroughly documented account of the ulstrastructure of meiosis and the meiotic NAO cycle in a rust fungus *(Puccinia malvacearum)* is presented by O'DONNELL and McLAUGHLIN (1981a,b,c). The NAO discs and the middle piece are considerably larger than those reported for holobasidiomycetes. The outer zone of the disc consits of parallely aligned 15-nm filaments. In the early metaphase I an electron-dense collateral disc (co-disc) is inserted eccentrically within each NAO. In division II co-discs and middle pieces are absent.

δ) Septation. Following meiosis the promycelium becomes a phragmobasidium by septation. Septa which form during reduction divisions within phragmobasidia show an absence of septal pore apparatus (McLAUGHLIN 1982). In this paper also septation within diverse phragmobasidia is discussed. Teliospores of *Uromyces koa* produce unusual modified basidia (GARDNER 1981).

ε) Sterigmata Formation. Sterigmata appear to result from renewed tip growth at the margins of the holobasidial apex and at a restricted locus of each phragmobasidial cell. Numerous small cytoplasmic vesicles are evident in the developing sterigmata (SUNDBERG 1978, MIMS 1981, McLAUGHLIN 1982). The vesicles appear to be derived from primitively stacked Golgi cisternae. In contrast to the sterigmal formation in several phargmobasidiomycetes in which the sterigmata arise endogenously, in *Coprinus* and *Boletus* the outer layer of basidial wall is preserved during the sterigmal initiation.

ζ) Basidiospore Formation. Early basidiospore development is described by McLAUGHLIN (1977) for *Coprinus*. Four ontogenetic stages correlate with changes in orientation of growth. During the early growth phase, an electron-opaque structure - the hilar appendix body - appears. It may be involved with the hilar appendix and with the discharge mechanism. The hilar appendix consisted of a projecting, structurally modified, multilayered wall thickening. One layer - an electron-transparent zone - appears to be the site of droplet formation prior to basidiospore discharge. The "ampoule effect" of a basidial vacuole (CORNER 1948) seems to be an unsatisfactory model of the cellular processes in basidiosporogenesis.

η) Spore-wall Formation. The ontogenesis and ultrastructure of spore walls in higher Basidiomycetes are examined by OLAH and REISINGER (1981). The authors see a possibility of generalization: Three definitive layers constitute the spore-wall structure. The first-formed layer, designated with the letter C, is in close contact with the plasmalemma. It becomes more and more electron-translucent as a result of the translocation of the materials utilized in the formation of the fundamental wall layers B and A. In certain cases the sterigmal wall surrounds the nascent and mature spore with a transitional space between the sterigmal wall and the newly formed spore wall. The spore walls of *Agaricus* show a densely staining external layer which corresponds to a chitin or chitosan melanin complex. This association may explain in part the resistance of these spores to chitinase digestion (GARCIA MENDOZA et al. 1979).

RAST et al. (1981) report in a review on the chemical composition of fungal melanins and present studies on the ultrastructure of chemically isolated native and synthetic mushroom melanins. Considering their results and the presence of granular melanin in a large variety of fungi, reported in the literature, they postulate the presence of intracellular compartments for the processing of granular melanin in fungi. They might be analogous to the melanosomes of animals.

e) Conidiospores

α) Conidium Formation. (HAMMILL 1981, COLE and KENDRICK 1981, Progr. Bot. 35, 29, 1973; 39, 23, 1977).

Methods. COLE (1981a,b) informs on techniques for examining developmental and ultrastructural aspects of conidial fungi.

COLE and SAMSON (1979a,b) give a comprehensive analysis of the various modes of conidium formation in the Deuteromycotina. To distinguish the developmental categories, they use three major criteria: conidial wall differentiation; order of production and arrangement of conidia on conidiogenous cell; process of conidiogenous cell proliferation. MADELIN (1979) proposes the hypothesis that "the mode of blastic ontogeny is related to the juvenility or maturity of the wall at the conidiogenous locus" and that distinct enteroblastic and holoblastic developments might represent extremities of a continuum which reflects diverse degrees and rates of wall stratification. The realization of one or other mode appears to be dependent on external conditions (i.e., on field or laboratory conditions). MINTER et al. (1982, 1983) attempt a much more clear explanation of phialidic conidiogenesis on the basis of a definition of conidial ontogeny. It is concluded that the terms phialidic, blastic and thallic are at present ill defined and are unsuitable for fundamental classifications of conidial fungi.

Pseudobasidiospora forms conidia similar in shape to that of many basidiospores in consequence of an unequal development of the conidium base (CAMPBELL 1979). *Spiniger* appears to be the only species in which the mode of formation of denticles is different from that of conidia. Denticles formation is enteroblastic, and that of conidia holoblastic (HANLIN 1982). A profuse rough endoplasmic reticulum system in the phialides, conidial initiales and conidia of *Sphaerostilbe* may account for the origin of the mucus which held together the subsequent pushed-out conidia (GLOVER and HANLIN 1981). *Humicola* develops two distinct types of conidia. One type is formed holoblastically. The other one is a typical chlamydospore (ELLIS 1982). Changes in rodlet orientation during blastic conidiogenesis appear to reflect changing stresses applied to the outer wall layer (COLE and SAMSON 1979a,b). Negative staining of separated rodlet layer fragments shows that the individual rodlets have a hollow central core (DEMPSEY and BEEVER 1979).

β) Microcycle Conidation. (SMITH et al. 1981). A feature common to most forms of microcycle conidation is the increase in size of the spore by spherical growth during the treatment which suppresses the germtube formation (SMITH 1978).

Microcycle conidation does not result simply from growth restriction. In *Paecilomyces* additionally to temperature shifts, glutamate and CO_2 are necessary for optimal microcycle conidiation (ANDERSON et al. 1978). Elemental sulfur decreases the cellular respiration and oxidative phosphorylation in *Phomopsis* and inhibits the growth but enhances the formation of sporulated pycnidia whose conidia develop directly secondary pycnidia (PEZET et al. 1983).

During the heat treatment of *Neurospora crassa* conidia the respiratory pathways are changed and mitochondria are progressively disorganized (MICHEA-HAMZEHPOUR et al. 1980). Cytochrome oxidase activity decreases during heat induction of microcycle conidation. After shift-down to normal temperature, cytochrome oxidase activity gradually recovers during conidia germination. A protection of this enzyme activity by Tween 20 during the elevated temperature is not sufficient to polarize the growth in the conidia. Only the temperature shift-down enables the outgrowth of the conidiogenous germ tubes (TON-THAT et al. 1983). This phenomenon may be related to the establishment of the decreasing apical gradient of pH value associated with the onset of polarity at germination (TURIAN 1980).

In the microcycle conidation of *Penicillium italicum* all conidia produce many well-developed penicilli but no conidia are formed and growth becomes vegetative if the pH value in the medium is not increased, initiating a synchronous conidiogenesis (VAN GESTEL 1983). SEVIOUR (1981) describes the process of microcycle conidation of *Acremonium* in submerged culture.

3. Germination (Progr. Bot. 35, 29 (1973); 39, 23 (1977))

a) Spores

Germination means for most if not all fungal spores the formation of new cell walls. Two well-defined stages are involved: (I) Swelling, which is not only a consequence of passive hydration but above all an active, isotropic growth and development of a spherical wall around the spore protoplast. In the case of the zoosporic fungi this stage is reduced to the de novo formation of a cell wall during the abrupt transition from a naked cell to a cyst. (II) Outgrowth or emergence of a germ tube. Prior to stage I there is an initiation and activation phase. The entire process seems to be a cascade phenomenon, regulated by chemical and/or physical activators, auto-activators, autoinhibitors, and perhaps cyclic nucleotides (COTTER 1981).

α) Germination Stage I: Sporangiospores of *Mucor* begin to grow exponentially (BARTNICKI-GARCIA 1981). The formation of a new wall under the existing sporangiospore wall is demonstrated also for *Rhizopus* (JOHRI 1982). Hydration of uredospores of *Puccinia* causes no gross ultrastructural changes of the cytoplasm (KNIGHTS et al. 1982). The surface rodlet pattern of spores of *Syncephalasstrum* becomes dispersed during germination and areas free from rodlets appear (HOBOT and GULL 1981).

In zoospores, germination stage I (encystment) is a short phase of wall synthesis by preformed enzymes (BARTNICKI-GARCIA 1981). Chitin synthetase of *Blastocladiella* zoospores is mobilized to the cell surface via release from the gamma particles. The vesicles resemble chitosomes and produce in vitro chitin microfibrils (CANTINO and MILLS 1979, MILLS and CANTINO 1981). After fusion of these vesicles with the cytoplasmic membrane over 90% of the chitin synthetase activity is located at the cytoplasmic membrane (DALLEY and SONNEBORN 1982, BARSTOW and POMMERVILLE 1980). Exogenous nutrients are needed or not, depending on species of *Piptocephalis*, for germination (McDANIEL and HINDAL 1982). In calcium-triggered, encysting Phytophthora zoospores calcium-enriched electron-dense deposits reflect secretory processes during encystment (HEMMES and PINTO DA SILVA 1980). In *Blastocladiella* zoospores a "zoospore maintenance factor" (ZMF) was found, which is released into the medium by the organism during the sporulation phase which regulates the encystment by blocking this process (GOTTSCHALK and SONNEBORN 1981, 1982a). This ZMF is a thiolated cyclic nucleotide (GOTTSCHALK and SONNEBORN 1982b). So called ISG (in situ germination) mutants of *Verticillium alboatrum* have conidia which germinate on phialides presumably because they do not produce a conidium germination inhibitor (which prevent conidia from germination on phialides) or are insensitive to such an inhibitor (HASTIE and GADD 1981).

The axonemal tubulins of *Blastocladiella* zoospores do not contribute to the pool of soluble, colchicine-binding cytoplasmic tubulins which increase by 49% after encystment. Preexisting free pools and disassembled cytoplasmic tubulins provide an adequate supply of tubulins for assembly of new microtubules during germination. The axonemal and cytoplasmic tubulins of *Blastocladiella* therefore appear to be different (ABE and LOVETT 1982).

A range of ions and a number of organic compounds were tested to determine their ability to induce encystment or germination of zoospores of *Phytophthora*. Some compounds induce only encystment and others only germination (BYRT et al. 1982a,b). Nuclear divisions in a synchronously germinating population of *Phytophthora* zoospore cysts appear highly synchronous at least to the 8-nucleate stage (HOOLEY et al. 1982).

β) Germination Stage II: "The switch from stage I to stage II probably begins with the introduction of a polarization signal which directs wall-synthesizing vesicles to move preferentially to a small target area on the germ sphere or cyst, where the germ-tube would emerge" (BARTNICKI-GARCIA 1981). The establishment of an electrical potential gradient may be contributing to this process, as shown by TURIAN (1980). Also the cyclic nucleotide system seems to be involved (KÜNKEL and RÖMER 1980, ORLOWSKI 1980). Cyclic AMP, added to sporangiospores of *Mucor* spp. during germination stage I, prevented the outgrowth of germ tubes (ORLOWSKI 1980, BARTNICKI-GARCIA 1981). Breakdown of glycogen particles and an increase in the amount of microvesicles and endoplasmic reticulum accompanies germ tube emergence (SALAKO 1981, KNIGHTS et al. 1982). Wetting the spores of *Fuligo* with a surfactant results in a rapid, synchronous and complete germination. There is evidence for an autoinhibitor of germination (NELSON and ORLOWSKI 1981). A specific adenosine triphosphatase (ATPase) in mitochondria of *Botryodiplodia* is lacking in dormant spores but is elaborated rapidly during spore germination. Only one subunit of this enzyme is synthesized on mitochondrial ribosomes and the 11 remainder subunits are produced in the cytoplasm. Prior to germination mitochondria contain only two or three of the cytoplasmic subunits. The de novo synthesis of the other subunits early in germination appears necessary for assembly and function of this spore mitochondrial ATPase (WENZLER and BRAMBL 1981).

Cytochalasin A inhibits spore germination and causes large swollen spores in several fungi (THOMAS 1978). The *α, β*-unsaturated keto moiety of cytochalasin A seems to be essential for its biological effect (GROVE and SWEIGARD 1980).

b) Zoosporangia

The patterns of spore release correlate in saprolegniaceous fungi with sporangial wall widths. The thick wall of *Achlya* sporangia presumably is needed to resist the turgor pressure generated for spore expulsion. The thinner sporangial wall of *Dictyuchus* opens manifoldly because each spore produces its own exit pore. This mode of sporangial germination is prepared early in formation of primary spore initiales by deposition of wall vesicles on one region of the sporangial wall per nucleus. In *Brevilegnia*, which has the thinnest sporangial wall, the wall expands generally, becomes amorphous and breaks to release groups or individual spores (ARMBRUSTER 1982b; ARMBRUSTER and JOHNSON jr. 1983).

In *Phytophthora cactorum* opening of the sporangium and zoospore discharge is understood as being a passive reaction to change of osmotic gradient and turgor pressure inside the sporangium (GISI et al. 1979, SENN and GISI 1982). In contrast to that, MIYATA (1981) shows that in *Phytophthora capsici* activation of the cell metabolism with new mRNA-dependent protein synthesis and oxidative phosphorylation is necessary and that the mechano-chemical forces of the microtubule systems are concerned with zoospore release.

During the germination of the resisting sporangia of *Synchytrium* an expanding translucent body appears to rupture in irregular patterns the sporangial wall through which the germination vesicle protrudes. The vesicle expands and forms a new wall layer inside the vesicle wall. Spores release the sporangium through an irregular slit in the wall (LANGE and OLSON 1981).

Acknowledgement. I am very grateful to Prof. Dr. M. Girbardt for critical reading of the manuscript. The invaluable help of Mrs. I. Lange, Mrs. I. Bähring, and Mrs. H. Münkel in processing the literature and in careful typing is gratefully acknowledged.

References

ABE, S.S., LOVETT, J.S.: Arch. Microbiol. 131, 323-329 (1982). - ALDRICH, H.C., DANIEL, J.W.: Cell Biology of Physarum and Didymium, Vol. 1, 480 pp, Vol. 2, 400 pp, London: Academic Press 1982. - ANDERSON, J.G., ARYEE, V., SMITH, J.E.: FEMS Microbiol. Lett. 3, 57-60 (1978). - ANSELL, P.J., YOUNG, T.W.K.: Mycologia 75, 64-69 (1983). - ARMBRUSTER, B.L.: Mycologia 74, 433-459 (1982a); - ibid. 74, 975-999 (1982b), - ARMBRUSTER, B.L., JOHNSON, T.W., Jr.: Can. J. Bot. 61, 545-550 (1983). - ASHTON, M.L., MOENS, P.B.: Can. J. Bot. 57, 1259-1284 (1979).

BAKER, K.K.: Can. J. Bot. 57, 2460-2465 (1979). - BARR, D.J.S.: BioSystems 14, 359-370 (1981). - BARR, D.J.S., ALLAN, P.M.E.: Can. J. Bot. 60, 2496-2504 (1982). - BARSTOW, W.E., POMMERVILLE, J.: Arch. Microbiol. 128, 179-189 (1980). - BARTNICKI-GARCIA, S.: Cell wall construction during spore germination in Phycomycetes, 533-556. In: The Fungal Spore: Morphogenetic Controls, eds. G. TURIAN, H.R. HOHL, London, New York, Toronto, Sydney, San Francisco: Academic Press 1981. - BECKETT, A.: Trans. Br. Mycol. Soc. 72, 269-276 (1979a); - ibid. 72, 320-322 (1979b); - Ascospore formation, 107-129. In: The Fungal Spore: Morphogenetic Controls, eds. G. TURIAN, H.R. HOHL, London, New York, Toronto, Syndney, San Francisco: Academic Press 1981. - BENJAMIN, R.K., in: The Whole Fungus, Vol. 2, ed. B. KENDRICK, Ottawa: National Museums of Canada 1979. - BERTA, G., FUSCONI, A.: Trans. Br. Mycol. Soc. 80, 201-207 (1983). - BILINSKI, C.A., MILLER, J.J.: J. Bacteriol. 143, 343-348 (1980); - Can. J. Genet. Cytol. 25, 415-419 (1983). - BRAIN, A.P.R., YOUNG, T.W.K.: Microbios 25, 93-106 (1979). - BRAIN, A.P.R., JEFFRIES, P., YOUNG, T.W.K.: Mycologia 74, 173-181 (1982). - BUCZACKI, S.T.: Zoosporic Plant Pathogens, 324 pp. London: Academic Press 1983. - BYRT, P.N., IRVING, H.R., GRANT, B.R.: J. Gen. Microbiol. 128, 1189-1198 (1982a); - ibid. 128, 2343-2351 (1982b).

CAMPBELL, C.K., SMITH, M.D.: Mycopathologia 78, 145-150 (1982). - CAMPBELL, R.: Trans. Br. Mycol. Soc. 72, 207-212 (1979). - CANTINO, E.C., MILLS, G.L.: The gamma particles in Blastocladiella emersonii: what is it? 441-484. In: Virus and Plasmids in Fungi, ed. P.A. LEMKE. New York: Dekker 1979. - COLE, G.T.: Techniques for examining developmental and ultrastructural aspects of conidial fungi, 557-634. In: Biology of Conidial Fungi, eds. G.T. COLE, B. KENDRICK. New York: Academic Press 1981a; - Scanning Electron Microsc. 3, 304-312 (1981b); - Mycologia 75, 93-116 (1983). - COLE, G.T., KENDRICK, B.: Biology of

Conidial Fungi, Vol. 1, 480 pp., Vol. 2, 672 pp. New York, London, Toronto, Syndney, San Francisco: Academic Press 1981. - COLE, G.T., SAMSON, R.A.: Patterns of Development in Conidial Fungi. London, San Francisco, Melbourne: Pitman 1979a; - Patterns of sporangiospore development in the Zygomycetes, 131-161. In: Patterns of Development in Conidial Fungi, eds. G.T. COLE, R.A. SAMSON, London-San Francisco-Melbourne: Pitman 1979b. - COLEY-SMITH, J.R., VERHOEFF, K., JARVIS, W. R.: The Biology of Botrytis, 336 pp. New York: Academic Press 1981. - CORNER, E.H.: New Phytol. 47, 159-194 (1948). - COTTER, D.A.: Spore activation, 385-411. In: The Fungal Spore: Morphogenetic Controls, eds. G. TURIAN, H.R. HOHL, London, New York, Toronto, Sydney, San Francisco: Academic Press 1981.

DALLEY, N.E., SONNEBORN, D.R.: Biochim. Biophys. Acta 686, 65-76 (1982). - DAVIDOV, L.S., GOETSCH, L., BYERS, B.: Genetics 94, 581-595 (1980). - DEACON, J.W.: Introduction to Modern Mycology, 197 pp., Halstad: Blackwell 1980. - DEML, G., OBERWINKLER, F.: Mycologia 73, 392-398 (1981). - DEML, G., NEBEL, M., OBERWINKLER, F.: Can. J. Bot. 59, 122-128 (1981). - DEMPSEY, G.P., BEEVER, R.E.: J. Bact. 140, 1050-1062 (1979).

ELLIS, D.H.: Trans. Br. Mycol. Soc. 76, 457-466 (1981a); - ibid. 76, 467-478 (1981b); - ibid. 78, 129-139 (1982). - ELLIS, J.J., AJELLO, L.: Mycologia 74, 144-145 (1982). - ESSER, K.: Dev. Ind. Microbiol. 22, 19-37 (1981).

GARCIA-MENDOZA, C., LEAL, J.A., NOVAES-LEDIEU, M.: Can. J. Microbiol. 25, 32-39 (1979). - GARDNER, D.E.: Can. J. Bot. 59, 939-946 (1981). - GARDNER, J.S., ALLEN, J.V., HESS, W.M., TRIPATHI, R.K.: Mycologia 75, 333-336 (1983a). - GARDNER, J.S., HESS, W.M., TRIPATHI, R.K.: J. Bacteriol. 154, 502-504 (1983b). - GISI, U., SCHWINN, F.J., OERTLI, J.J.: Trans. Br. Mycol. Soc. 72, 437-446 (1979). - GLOVER, S.U., HANLIN, R.T.: Am. J. Bot. 68, 685-696 (1981). - GOTTSCHALK, W.K., SONNEBORN, D.R.: Exp. Mycol. 5, 1-14 (1981); - Devel. Biol. 93, 165-180 (1982a); - J. Cell. Biol. 95, 48a (1982b). - GROVE, S.N., SWEIGARD, J.A.: Exp. Mycol. 4, 239-250 (1980).

HAMMILL, T.M.: Conidiogenesis, 151-172. In: Fungal Spore: Morphogenetic Controls, eds. G. TURIAN, H.R. HOHL, London, New York, Toronto, Sydney, San Francisco: Academic Press 1981. - HAMMILL, T.M., SECOR, D.L.: Mycologia 75, 648-655 (1983). - HANLIN, R.T.: Mycologia 74, 236-241 (1982). - HARLEY, J.L. (FRS), SMITH, S.E.: Mycorrhizal Symbiosis, 496 pp. London: Academic Press 1983. - HASTIE, A.C., GADD, G.M.: FEMS Microbiol. Lett. 12, 139-141 (1981). - HEATH, J.B.: Ultrastructure of freshwater Phycomycetes, 603-650. In: Recent Advances in Aquatic Mycology, ed. E.B.G. JONES, London: Elek 1976. - HEATH, J.B., HEATH, M. C., HERR, F.: Motile systems in fungi, 563-587. In: Prokaryotic and Eukaryotic Flagella, eds. W.B. AMOS, J.G. DUCKETT, Cambridge: Cambridge Univ. Press 1982. - HEMMES, D.E., PINTO DA SILVA, P.: Biol. Cell. 37, 235-240 (1980). - HESS, W.M., GARDNER, J.S.: J. Bacteriol. 154, 499-501 (1983). - HIRATSUKA, Y., SATO, S.: Morphology and taxonomy of rust fungi, 1-36. In: The Rust Fungi, eds. K.J. SCOTT, A.K. CHAKRAVORTY, London, New York, Paris, San Diego, San Francisco, Sao Paulo, Sydney, Tokyo, Toronto: Academic Press 1982. - HOBOT, J.A., GULL, K.: Protoplasma 107, 339-343 (1981). - HOOLEY, P., FYFE, A.M., MALTESE, C.E., SHAW, D.S.: Trans. Br. Mycol. Soc. 79, 563-566 (1982).

JEFFRIES, P., YOUNG, T.W.K.: Mycologia 75, 509-517 (1983). - JOHRI, B.N.: Curr. Sci. 51, 521-524 (1982). - JONES, E.B.G.: Recent Advances in Aquatic Mycology, 770 pp. London: Elek 1983. - JONES, E.B.G., MOSS, S.T.: Bot. Mar. 23, 483-500 (1980a); - Trans. Br. Mycol. Soc. 74, 625-

631 (1980b). - JONES, E.B.G., MOSS, S.T., CUOMO, V.: Trans. Br. Mycol. Soc. 80, 193-200 (1983).

KAPLAN, J.D., GOOS, R.D.: Mycologia 74, 684-686 (1982). - KITZ, D.J., EMBREE, R.W., CAZIN, J., Jr.: Mycologia 74, 110-116 (1982). - KNIGHTS, J.K., DAVEY, M.R., LUCAS, J.A.: Protoplasma 113, 57-68 (1982). - KOCH, J., JONES, E.B.G., MOSS, S.T.: Bot. Mar. 26, 265-270 (1983). - KÜNKEL, W., RÖMER, W.: Z. Allg. Mikrobiol. 20, 195-207 (1980).

LANGE, L., OLSON, L.W.: Dansk Bot. Arkiv 33, 1-95 (1979); - Protoplasma 106, 69-82 (1981); - The fungal zoospore - Its structure and biological significance, 1-42. In: Zoosporic Plant Pathogens, ed. S.T. BUCZACKI, London: Academic Press 1983. - LATGÉ, J.P., PRÉVOST, M.C., PERRY, D.F., REISINGER, O.: Can. J. Bot. 60, 413-431 (1982). - LITTLEFIELD, L.J., HEATH, M.C.: Ultrastructure of Rust Fungi, 277 pp. New York: Academic Press 1979. - LU, B.C.: J. Cell. Biol. 76, 761-766 (1978). - LUCAROTTI, C.J.: Can. J. Bot. 59, 137-148 (1981).

MACE, M.E., BELL, A.A., BECKMAN, C.H.: Fungal Wilt Diseases of Plants, 608 pp. New York: Academic Press 1981. - MADELIN, M.F.: An appraisal of the taxonomic significance of some different modes of producing blastic conidia, 63-80. In: The Whole Fungus, ed. B. KENDRICK. Ottawa: National Museums of Canada 1979. - MARMIROLI, N., FERRARI, C., TEDESCHI, F., PUGLISI, P.P., BRUSCHI, C.: Biol. Cell 41, 79-84 (1981). - McDANIEL, L.L., HINDAL, D.F.: Mycologia 74, 271-274 (1982). - McLaughlin, D.J.: Am. J. Bot. 64, 1-16 (1977); - ibid. 67, 1225-1235 (1980); - Ultrastructure and Cytochemistry of Basidial and Basidiospore Development, 37-74. In: Basidium and Basidiocarp, eds. K. WELLS. Berlin, Heidelberg, New York: Springer 1982. - MICHEA-HAMZEHPOUR, M., GRANGE, F., TON-THAT, T., TURIAN, G.: Arch. Microbiol. 125, 53-58 (1980). - MILLS, G.L., CANTINO, E.C.: Arch. Microbiol. 130, 72-77 (1981). - MIMS, C.W.: Can. J. Bot. 59, 1041-1049 (1981). - MINTER, D.W., KIRK, P.M., SUTTON, B.C.: Trans. Br. Mycol. Soc. 79, 75-93 (1982); - ibid. 80, 39-66 (1983. - MIYATA, Y.: Sci. Rep. Kyoto Prefect. Univ. Agric. 33, 16-31 (1981). - MIYATA, Y., OSUGI, T., MASAGO, H.: Sci. Rep. Kyoto Prefect. Univ. Agric. 32, 39-50 (1980). - MÜLLER, E., LOEFFLER, W.: Mykologie, 366 pp., 4. Aufl. Stuttgart, New York: Thieme 1982.

NELSON, R.K., ORLOWSKI, M.: Arch. Microbiol. 130, 189-194 (1981).

OBERWINKLER, F., BANDONI, R.J., BLANZ, P., DEML, G., KISIMOVA-HOROVITZ, L.: Plant Syst. Evol. 140, 251-277 (1982). - O'DONNELL, K.L., McLAUGHLIN, D.J.: Protoplasma 108, 225-244 (1981a); - ibid. 108, 245-263 (1981b); - ibid. 108, 265-288 (1981c). - OLAH, G.M., REISINGER, O.R.: Ontogenesis and ultrastructure of spore walls in higher Basidiomycetes, 131-149. In: The Fungal Spore: Morphogenetic Controls, eds. G. TURIAN, H.R. HOHL. London, New York, Toronto, Sydney, San Francisco: Academic Press 1981. - OLSON, L.W., EDEN, U.M., LANGE, L.: Zoosporogenesis model systems - Problems - possible approaches 43-70. In: Fungal Spore: Morphogenetic Controls, eds. G. TURIAN, H.R. HOHL, London, New York, Toronto, Sydney, San Francisco: Academic Press 1981. - ORLOWSKI, M.: Arch Microbiol. 126, 133-140 (1980).

PEZET, R., PONT, V., GIRADET, F.: Can. J. Microbiol. 29, 179-184 (1983). - PORTER, D.: Mycologia 74, 363-375 (1982). - POWELL, M.J.: BioSystems 10, 167-180 (1978); - Can. J. Bot. 59, 2049-2061 (1981).

RAMBERG, J.E., McLAUGHLIN, D.J.: Can. J. Bot. 58, 1548-1561 (1980). - RAST, D.M., STÜSSI, H., HEGNAUER, H., NYHLEN, L.E.: Melanins, 507-531. In: The Fungal Spore: Morphogenetic Controls, eds. G. TURIAN, H.R.

HOHL. London, New York, Toronto, Sydney, San Francisco: Academic Press 1981. - ROSING, W.C.: Mycologia 74, 960-974 (1982).

SALAKO, E.A.: Mycopathol. 76, 3-12 (1981). - SAUER, H.W.: Developmental Biology of Physarum, 237 pp. Cambridge, London, New York, New Rochelle, Melbourne, Syndney: Cambridge Univers. Press 1982. - SCOTT, K.J., CHAKRAVORTY, A.K.: The Rust Fungi, 288 pp. London, New York, Paris, San Diego, San Francisco, Sao Paulo, Sydney, Tokyo, Toronto: Academic Press 1982. - SENN, H., GISI, U.: Phytophatol. Z. 105, 230-247 (1982). - SEVIOUR, R.J.: FEMS Microbiol. Lett. 12, 287-293 (1981). - SMITH, J.E.: Asexual sporulation in filamentous fungi, 214-239. In: The Filamentous Fungi, Vol. 3, eds. J.E. SMITH, D.R. BERRY, London: Arnold 1978. - SMITH, J.E., ANDERSON, J.G., KRISTIANSEN, B., ALRAWI, A., YAHYA, A.G.: Microcycle conidiation, 627-650. In: The Fungal Spore: Morphogenetic Controls, eds. G. TURIAN, H.R. HOHL. London, New York, Toronto, Sydney, San Francisco: Academic Press 1981. - SUBRAMA-NIAN, C.V.: Hyphomycetes, 496 pp. London: Academic Press 1983. - SUNDBERG, W.J.: Hymenial cytodifferentiation in Basidiomycetes, 298-314. In: The Filamentous Fungi, Vol. III, eds. J.E. SMITH, D.R. BERRY, London: Arnold 1978.

TANAKA, K., HIRATA, A.: J. Cell Sci. 56, 263-279 (1982). - TAYLOR, J. W., FULLER, M.S.: Exp. Mycol. 5, 35-59 (1981). - THIELKE, CH.: Z. Pilzk. 42, 57-66 (1976); - Meiotic Divisions in the basidium, 75-91. In: Basidium and Basidiocarp. eds. K. WELLS, E.K. WELLS. Berlin, Heidelberg, New York: Springer 1982. - THOMAS D. Des S.: Cytochalasin effects in plant and eukaryotic microbial systems. 257-276. In: Cytochalasins - Biochemical and Cell Biological Aspects, ed. S.W. TANENBAUM. Amsterdam: North-Holland 1978. - TON-THAT, T., MICHEA-HAMZEHPOUR, M., TURIA, G.: Protoplasma 116, 149-154 (1983). - TSUNEDA, A. HIRATSUKA, Y.: Can. J. Bot. 59, 1192-1195 (1981). - TURIAN, G.: Ber. Schweiz. Bot. Ges. 90, 203-212 (1980).

VAN GESTEL, J.F.E.: Exp. Mycol. 7, 287-291 (1983).

WEBSTER, J.: Pilze, 641 pp. Berlin, Heidelberg, New York: Springer 1983. - WELLS, K., WELLS, E.K.: Basidium and Basidiocarp, 187 pp. Berlin, Heidelberg, New York: Springer 1982. - WENZLER, H., BRAMBL, R.: J. Biol. Chem. 256, 7166-7172 (1981).

ZICKLER, D., SIMONET, J.M.: Exp. Mycol. 4, 191-206 (1980).

Dr. ECKART JELKE
Zentralinstitut für Mikrobiologie und experimentelle Therapie der Akademie der Wissenschaften
Beutenbergstr. 11
DDR 69 Jena

II. Morphologie und Histologie der höheren Pflanzen:

a) Vegetative Anatomie der Spermatophyta

Von HELMUT UHLARZ

1. Allgemeines

In einem Brief LEEUWENHOEK's, vom 1. Juli 1674, an die Royal Society in London, fand man vor kurzem einige beigelegte, originale Handschnittpräparate von Korkgewebe *(Quercus suber)*. Ein identischer Ausschnitt eines solchen Präparates wurde mit Hilfe von LEEUWENHOEK's Utrechter Mikroskop, einem modernen Lichtmikroskop sowie einem Raster-Elektronenmikroskop fotografiert (FORD 1982). Ein Vergleich der so unterschiedlich entstandenen Abbildungen führt vor Augen, in welchem Maße der Fortschritt der Pflanzenanatomie auf die Verbesserung der optischen (und präparativen) Techniken zurückzuführen ist. Der Weg, den die Pflanzenanatomie ging, nämlich über die Grenzen der Lichtoptik hinaus, wurde ihr schon in den Anfängen gewiesen. Jedenfalls betrifft eine briefliche Mitteilung LEEUWENHOEK's, vom 1. Mai 1695, nahezu eine Ultrastruktur. Sie kann als die erste Beschreibung und zeichnerische Darstellung des fibrillären Aufbaus einer verholzten Zellwand gelten (BAAS 1982a).

Der eigentliche Fortschritt der Pflanzenanatomie spiegelt sich in der Verbesserung ihrer Methode und ihrer Begriffe. Die Methode hat sich allerdings seit den Anfängen nicht grundsätzlich geändert. Sie besteht nach wie vor im Vergleichen und Ordnen von Strukturen nach Form- und Funktionskriterien, wobei letztere schwerer überprüfbar sind als die Erstgenannten. Verbessert wurde und wird vor allem der Begriffsapparat der Pflanzenanatomie mit der wachsenden Anzahl beobachteter Strukturen.

Heute verzeichnen wir eine Flut von Publikationen anatomischer Befunde, die mit Hilfe lichtoptischer und zunehmend auch elektronenoptischer Techniken erhoben wurden. Den Hauptstrom in ihr bilden die zahlreichen Arbeiten, in welchen anatomische Daten die empirische Grundlage für taxonomische Schlüsse und merkmalsphylogenetische Interpretationen sind. Sie tragen vielfach zu einer Verbesserung der Systematik der Spermatophyta bei, in jedem Falle zu einer Erweiterung unseres Wissens auf dem Gebiete der inzwischen klassischen Systematischen Pflanzenanatomie.

Daneben gibt es eine wachsende Anzahl von Publikationen, die an erster Stelle bloße Beobachtungsprotokolle sind. Oft zeichnen sie sich durch eine besonders ausgeprägte Akribie aus. Auch sie tragen mit bewährten Techniken zu einer Erweiterung der empirischen Basis, zu einer Vermehrung dokumentierter anatomischer Daten bei. Sehr viele von ihnen dienen jedoch lediglich der Sicherung von bereits Bekanntem, der Verifikation an einem weiteren Beispiel. Zum Fortschritt der Pflanzenanatomie im Sinne einer Verbesserung von Methode und Begriffen tragen sie in der Regel nicht bei.

Progress in Botany, Vol. 46

Von ihnen sind jene Beobachtungsprotokolle zu unterscheiden, welche über Strukturen berichten, die mit Hilfe neuer Untersuchungstechniken genauer als bisher oder überhaupt erst erfaßt werden konnten. Hier erweist sich vor allem die Raster-Elektronenmikroskopie, oft verbunden mit energiedispersiver Röntgenspektrometrie, als eine datenfördernde Technik. Es bleibt zu hoffen, daß die hier sich anhäufenden Daten einmal im Zusammenhang besser verstanden und genutzt werden können.

Ein anwachsender Strom von Publikationen enthält schließlich jene Arbeiten, welche sich um ein Verständnis des Zusammenhanges von Struktur und Funktion bemühen. Es sind vor allem Untersuchungen an Leitelementen und Leitgeweben sowie an epidermalen Strukturen. Vielfach liegt ihnen ein Komplex phylogenetisch-systematischer, ökologischer und anatomischer Fragestellungen zugrunde. In manchen dieser Studien können Korrelationen von anatomischen Strukturen und ökologischen Bedingungen des Habitats wahrscheinlich gemacht werden. Es zeichnet sich jedoch ab, daß Untersuchungen dieser Art vielleicht schon bald durch eine Synthese anatomischer und physiologischer Forschung überholt sein werden.

Der vorliegende Bericht erfaßt Publikationen der Jahre 1982 und 1983.

2. Standardwerke

Der jetzt vorliegende 2. Band der *Anatomy of the Dicotyledons* von METCALFE and CHALK (1983) schließt den Allgemeinen Teil der 2. Edition des monumentalen Werkes ab. Damit ist ein breites Fundament für die nachfolgenden, systematischen Bände gelegt, deren Gliederung einem revidierten System der Dicotyledoneae nach TAKH-TAJAN entsprechen wird. Noch bevor diese erschienen sind, werden sich die beiden ersten Bände als Quellen der Information und der Inspiration sicherlich schon vielfach bewährt haben. Nicht minder zu begrüßen ist die Fortsetzung der *Anatomy of the Monocotyledons* durch den Band VII: Helobiae (Alismatidae) von TOMLINSON (1982). Mit dem Band *Anatomie des Galles* von MEYER und MARESQUELLE (1983) wird das Handbuch der Pflanzenanatomie *(Encyclopedia of Plant Anatomy)* um einen Band ergänzt, dessen Inhalt weit über die reine Pflanzenanatomie hinausgeht.

Unter den Maßstäbe setzenden Werken muß auch ein Lehrbuch genannt werden: *Plant Anatomy*, 3. Edition, von FAHN (1982). Gegenüber seiner 2. Auflage ist das Werk in Text und Abbildungen drastisch überarbeitet. Zahlreiche elektronenoptische Abbildungen (SEM und TEM) steigern seinen Informationsgehalt und erleichtern das räumliche Verständnis histologischer, zellulärer und subzellulärer Strukturen. Daß der Autor zahlreiche Beispiele aus der weniger bekannten, xerischen und mediterranen Flora heranzieht, darf als eine Bereicherung gewertet werden. Ältere große Lehrbücher der Pflanzenanatomie werden unentbehrlich bleiben. Das Lehrbuch von FAHN ist das derzeit umfassendste.

3. Der Stamm

Zu den Fundamenten der Systematik und Phylogenetik von Gefäßpflanzen gehören die Stelen- bzw. die Knoten-Typen. Obwohl gerade die Knotentypen seit ca. 30 Jahren zum Teil recht kontrovers phylogenetisch interpretiert werden, gelten sie letztlich doch als unangefochtene Paradigmen, in ungezählten knotenanatomischen Studien. Neuere, sehr detaillierte Untersuchungen der Stelenkonstruktion von Spermatophyta stellen allerdings den Wert dieser bislang erfolgreichen Problemlösungsmodelle zunehmend in Frage. Vor allem sind es Studien, die sich nicht auf die maturen Knotenstrukturen beschränken. Sie berücksichtigen vielmehr primäre und sekundäre Systeme von Leitgeweben differenzierter als bisher, und vor allem den Stelenbau im ganzen.

Eine äußerst präzise Beschreibung des Xylemanschlusses eines Seitenzweiges an das Xylem der Abstammungsachse bei *Populus deltoides* geben LARSON und FISHER (1983; s. auch HAGEMAN in Progr. Bot. 45, 42). Den Knotenbau einiger Rubiaceae, insbesondere die Ausbildung des kommissuralen Leitgeweberinges und den anatomischen Anschluß der Seitenachsen an die Hauptachse, beschreibt NEUBAUER (1981) sehr ausführlich. Bemerkenswert sind auch die Beschreibungen des Leitbündelverlaufes bei einigen Apiaceae (PILLAI und KUMAR 1981) und bei *Orobus vernus* (GULENKOVA 1981). In ihnen wird vor allem die mit dem Erstarkungswachstum in Zusammenhang stehende Variabilität der Leitgewebeanordnung und der Knotenstruktur demonstriert.

Untersuchungen des Leitbündelverlaufes in Achsen von Monocotyledonen schließen immer mehr eine Lücke unseres Wissens. Möglicherweise hat in der Vergangenheit schon der für die "typische" Leitbündelanordnung geprägte Begriff "Atactostele" den einen oder anderen Forscher davon abgehalten, nach Gesetzmäßigkeiten auch dort zu suchen, wo anscheinend keine zu erwarten waren. Heute zeigt sich immer deutlicher, daß sich in der sogenannten Atactostele diskrete Muster und Zonen miteinander verknüpfter Leitbündelsysteme unterscheiden lassen, so u.a. bei Araceae (FRENCH und TOMLINSON 1981a,b,c) und bei Poaceae (PATEL et al. 1981). Hingegen lassen die Stelenmerkmale der unterirdischen rübenförmigen und zylindrischen Speicherkörper von *Dioscorea cotinifolia* noch nicht einmal eine Entscheidung darüber zu, ob es sich bei diesen "Organen" um Sprosse oder um Wurzeln handelt. Der auch von anderen *Dioscorea*-Arten seit langem bekannte "intermediäre Charakter" dieser "Organe" bleibt weiterhin rätselhaft (von TEICHMA und ROBBERTSE 1981).

Fundierte Kritik an den herkömmlichen Stelärtheorien und an der Überbewertung knotenanatomischer Merkmale in der Systematik und Phylogenetik von Spermatophyta üben BECK et al. (1982). In ihrer umfassenden Abhandlung ist die in über einem Jahrhundert angesammelte relevante Literatur verarbeitet. Den zahlreichen offengelegten Fragwürdigkeiten herkömmlicher Theorien und Interpretationen einzelner Merkmale stellen sie ihre Konzeption gegenüber. Im wesentlichen übertragen sie das schon ältere "protostelic dissection concept", das insbesondere auf BECK's Untersuchungen an Progymnospermen fußt, auf die rezenten Spermatophyta insgesamt. Danach soll sich die Eustele der Samenpflanzen nicht durch Reduktion über die Solenostele und Dictyostele der Farne entwickelt haben. Sie soll vielmehr unmittelbar aus der Protostele der gemeinsamen Vorfahren (Rhyniophyta) entstanden sein, eben durch eine longitudinale Dissektion der Protostele in sogenannten Scheinachsen. Mit "Scheinachsen" sind letztlich axiale Spurstränge gemeint, die "reparativ" akropetal fortgesetzt werden. "Die ursprüngliche Eustele wird als ein offenes Primärgefäßsystem mit schraubenförmigem Spurabgang charakterisiert, welches aus fünf Scheinachsen besteht." Im Laufe der Evolution soll sich bei den Gymnospermen die Stele durch Vermehrung der "Scheinachsen" adaptativ erweitert haben. Bei den Angiospermen soll vor allem die Anzahl der Spurabgänge bzw. deren Verzweigungsgrad erhöht worden sein.

Da bei den rezenten Samenpflanzen das System der primären Leitgewebe vielfach durch sekundäre Gewebe überlagert wird, weisen die Autoren zurecht darauf hin, daß in stelenanatomischen Untersuchungen allein die primären und nicht die maturen Leitgewebesysteme zu vergleichen seien. Aufgrund aller verfügbaren Daten gelangen sie u.a. zu dem Schluß, daß das System der primären Leitgewebe von Monocotyledonen eine zwar in hohem Maße modifizierte, aber eben doch eine Eustele ist. Sie soll sich also nicht grundsätzlich von der Eustele der Dicotyledonen unterscheiden. Den Schluß der Abhandlung bilden beherzigenswerte Empfehlungen für künftige Beschreibungen von Stelärstrukturen. Sie sollen dazu beitragen, überhaupt zu vergleichbaren Daten zu gelangen.

Die durchaus legitime und logische Verknüpfung paläobotanischer Daten (wir haben keine vollständigeren und besseren) mit rezent-anatomischen unter einem Gesichtspunkt, insbesondere die Konsistenz der Argumentation, läßt die Konzeption von BECK, SCHMID und ROTHWELL (l.c.) als eine "abgeschlossene Theorie" erscheinen. Ihr Axiom ist entwicklungsgeschichtlich leider nur an rezenten Pflanzen weiter überprüfbar. Es besteht die Annahme, daß der Sproßkörper (der Samenpflanzen) ein achseneigenes primäres Leitgewebesystem besitzt, von welchem die Blattspurstränge abzweigen, wenngleich der Augenschein doch eher dafür spricht, daß das Leitgewebesystem des Sprosses von Blattspursträngen aufgebaut wird (s. auch HAGEMANN in Progr. Bot. 45, 42). Das Axiom der Theorie bleibt im besten Sinne des Wortes fragwürdig.

Ergänzt wird die umfangreiche Abhandlung durch eine Klassifikation der Stelen und einen Überblick über das Vorkommen der einzelnen Stelentypen im System der Gefäßpflanzen (SCHMID 1982). Beide Arbeiten bilden - auch im gemeinsamen Literaturverzeichnis - eine abgerundete Einheit. Die von SCHMID geleistete Kompilation der bislang eingeführten Stelenbezeichnungen kann nicht genug dankbar anerkannt werden. Der riesige Katalog von Synonymen zeigt, in welchem Ausmaß sich die Stelenforschung mangels eines umfassenden Konzepts immer mehr in der Schaffung bloßer Termini für einzelne Fallbeschreibungen erschöpfte. In der vorliegenden Klassifikation werden die Protostele, die Solenostele (Farne) und die Eustele (Samenpflanzen) als übergeordnete Typen angenommen, welchen jeweils mehrere Untertypen untergeordnet sind.

4. Xylem

Im Berichtszeitraum sind zahlreiche xylotomische Studien unter systematischen Aspekten erschienen. Von ihnen soll hier nur eine erwähnt werden, da es sich um eine holzanatomische Erstbeschreibung für ein Taxon höheren Ranges handelt. Das Xylem von Nepenthaceae wurde zum ersten Male beschrieben (CARLQUIST 1981).

Die Suche nach Adaptationen und nach "ökologischen Trends" in Xylemstrukturen hält an. Wegen ihrer meisterlichen Darstellung der komplexen Problematik seien hier die Arbeiten von CARLQUIST an erster Stelle erwähnt. In ihnen werden neben systematischen und phylogenetischen auch funktionell-ökologische Aspekte angesprochen.

Untersuchungen liegen vor für Dipsacaceae (CARLQUIST 1982a), Buxaceae (CARLQUIST 1982b), Daphniphyllaceae (CARLQUIST 1982c), Illiciaceae (CARLQUIST 1982d) und Winteraceae (CARLQUIST 1983). Vergleichbare Studien wurden veröffentlicht über Balanitaceae (PARAMESWARAN und CONRAD 1982), Scrophulariaceae (MICHENER 1981) und die Anacardiacee

Pachycormus discolor (GIBSON 1981). In einem floristischen Ansatz vergleichen BAAS et al. (1983) ökologische Trends in den Gefäßmerkmalen von Pflanzen Israels und angrenzender Gebiete und solchen tropischer sowie gemäßigt kühler Zonen.

In den zitierten Arbeiten können "ökologische Trends" vor allem in den Gefäßmerkmalen z.T. deutlich nachgewiesen werden. So klar dabei eine Korrelation bestimmter Strukturen und - zumeist nur vage bestimmter - ökologischer Daten hervortreten mag, so wenig ist damit der Zusammenhang der Strukturen und ihrer Funktionen auch schon verstanden. Dies heißt nicht, daß derartige Studien letztlich überflüssig seien. Sie bilden vielmehr ein Fundament für Untersuchungen, in welchen Anatomie und Physiologie (und Physik) fruchtbar zusammenwirken.

Ein Wegweiser zu solchen Untersuchungen ist das Buch von ZIMMERMANN (1983): *Xylem Structure and the Ascent of Sap.* Der Autor gibt darin zunächst einen Überblick über die leitenden Elemente (Tracheiden, Tracheen) und ihre dreidimensionale Verknüpfung im Holzkörper und diskutiert anschließend die Kohäsionstheorie. Die hydraulische Architektur der Pflanzen und andere strukturelle Anpassungen werden von ihm ebenso ausführlich angesprochen wie der Verfall der Xylemfunktionen und pathologische Erscheinungen. Der besondere Wert des Buches besteht darin, daß der renommierte Autor absichtlich nicht nur gesicherte Daten im Zusammenhang präsentiert, sondern - vor allem zu Stress-Phänomenen - viele originale Ideen und ungeprüfte Hypothesen. Sie fordern zu einer Überprüfung auf und machen das Buch in besonderer Weise anregend. Wer über die Grenzen einer in erster Linie strukturforschenden Holzanatomie hinausblicken möchte, muß dieses Buch in die Hand nehmen.

In einer knappen und sehr klaren Darstellung ergänzt BRAUN (1983) die Kohäsionstheorie durch die aufgrund von Meßdaten triftige Hypothese, "daß neben dem Transpirationsstrom, der nur in bestimmten täglichen Zeitspannen fließt, als weiteres Prinzip des Wassersteigens eine osmotisch bedingte Wasserverschiebung angesehen werden muß."

Im Zusammenhang mit den rheologischen und hydraulischen Problemen in der Holzanatomie gewinnen nun auch Arbeiten an Bedeutung, die in der Vergangenheit vielfach nur als Meßprotokolle verstanden werden konnten. Es sind dies die Untersuchungen der Größen-Variation von Gefäßgliedern und Gefäßen und deren Verteilung innerhalb der einzelnen Pflanze.

FUKAZAWA und OHTANI (1982) untersuchen die Variation aller Xylem-Elemente einer *Tilia japonica* in axialer und radialer Richtung. Die Verteilung der Gefäßlängen und Gefäßdurchmesser im Vegetationskörper von *Acer rubrum* analysieren ZIMMERMANN und POTTER (1982). Eine vergleichende Analyse liegt für ringporige Hölzer *(Quercus rubra, Fraxinus americana)* und das Holz einer Kletterpflanze *(Vitis labrusca)* vor (ZIMMERMANN und JEJE 1981). Darin sind auch ausführlichere Angaben zu Technik und Methode enthalten. BOSSHARD et al. (1982) beschreiben die radiale Verdrängung einzelner Gefäße bei der Entstehung der Ringporigkeit von *Fraxinus excelsior.*

Abschließend sei auf einen von BAAS (ed. 1982) herausgegebenen Sammelband hingewiesen, der Symposiumsbeiträge vom 13. Internationalen Botanischen Kongress in Sydney, 1981, enthält. Sein Titel *New Perspectives in Wood Anatomy* ist dem Inhalt durchaus angemessen. Mehrere der darin enthaltenen Beiträge dürfen als zukunftsweisend für die Holzanatomie angesehen werden, so u.a. der Beitrag von BURLEY über *Genetic Variation in Wood Properties;* von BURLEY und MILLER über *The Application of Statistics and Computing in Wood Anatomy* und von ZIMMERMANN über *Functional Xylem Anatomy of Angiosperm Trees.*

a) Einzelne Xylemstrukturen unter verschiedenen Aspekten

Die Kenntnis der Gefäßglied-Typen von Monocotyledoneae (s. Progr. Bot. 44, 49) wird durch eine die Commelinales umfassende Untersuchung erweitert (CHEADLE und KOSAKAI 1980). Ebenfalls unter phylogenetisch-systematischen Aspekten bemerkenswert ist die Gefäßstruktur von *Gnetum*-Arten, die von MUHAMMAD und SATTLER (1982) beschrieben wird. Entgegen der gängigen Meinung können die Autoren nachweisen, daß die Tüpfel und Perforationen bei *Gnetum* eben nicht völlig andersartig sind als bei primitiven Angiospermen. Vielmehr ist die Variationsbreite der Gefäßstrukturen bei *Gnetum* so groß, daß sie die typischen Hoftüpfel der Coniferen ebenso umfaßt wie Tüpfelungs- und Perforationsmuster von Angiospermen. Eine sehr detaillierte Beschreibung der Tüpfelmembranen in den gefäßlosen Hölzern von Winteraceae geben MEYLAN und BUTTERFIELD (1982). Die intertracheidalen Tüpfelmembranen haben danach einen geringeren Ligningehalt als solche zwischen Parenchymzellen bzw. zwischen Tracheiden und Parenchymzellen. Torusähnliche Strukturen sind nicht vorhanden. Hingegen ist die Tüpfelmembran der Hoftüpfelpaare von Fasern bei *Pyrus*- und *Prunus*- Arten mit torusähnlichen Verdickungen ausgestattet (PARAMESWARAN und LIESE 1981).

Bei den vermeintlichen Verzierungen der Tüpfel mancher *Nothofagus*-Arten handelt es sich um bloße Inkrustationen, möglicherweise um Reste des Cytoplasmas. GALE (1982) weist auf die Notwendigkeit hin, solche "Pseudo-Verzierungen" von echten Tüpfel-Verzierungen zu unterscheiden, da diese vielfach taxonomisch relevant sind. Eine Unterscheidung ist aufgrund des unterschiedlichen Verhaltens gegenüber Solventien (Hypochlorid) möglich. THOMPSON und JAGELS (1983) weisen nach, daß es sich bei den amorphen "Pfropfen" in den Tüpfelhöfen an der Grenze zwischen Naßkern und Splint von *Ulmus americana* um Präzipitate handelt, die Calcium und Kalium (als Carbonat?) enthalten. An ihrer Bildung sollen möglicherweise Bakterien beteiligt sein.

Im Druckholz von *Pinus thunbergii* und *Cryptomeria japonica* wurden von MIO und MATSUMOTO (1982) zarte Lamellen entdeckt, die möglicherweise Reste der cambialen Mutterzellwände sind.

Nach einer Untersuchung von FINK (1982) steht die Ausbildung abnorm breiter Holzstrahlen in Nadel- wie in Laubhölzern anscheinend immer in Zusammenhang mit latenten Anlagen von Adventivwurzeln im Bast. In den Holzstrahlen einiger Annonaceae konnten BOTOSSO und GOMES (1982) zum ersten Male für die Familie radiale Gefäße und Reihen perforater Zellen nachweisen.

Transferzellen im primären Xylem von *Avicennia schaueriana* besitzen im Wandlabyrinth eine mittellamellen-ähnliche Zone, die bei anderen Transferzellen bislang nicht beobachtet wurde. Aus der zunehmenden Verdikkung und Lignifizierung der Wandproliferationen sowie aus der Rückbildung des Protoplasten schließt PARAMESWARAN (1982), daß Transferzellen nur in der primären Phase der Sproßentwicklung dem Stofftransport dienen. Auf die unterschiedliche cytologische Struktur der Kontaktzellen im primären Xylem von *Dianthus caryophyllus* weisen CATESSON et al. (1982) hin. Während die Kontaktzellen des Protoxylems stärker vakuolisiert sind, zeigen jene des Metaxylems eine feinere Vakuolisierung und mehr entwickelte Mitochondrien. Welche Bedeutung diese Differenzierung der Kontaktzellen im Stoffwechsel hat, ist noch unklar.

5. Phloem, Bast und Borke

Auch auf dem Gebiete der Phloem-Anatomie dürfen solche Arbeiten als richtungsweisend angesehen werden, welche zur Aufdeckung der kausalen Zusammenhänge von Strukturen und Funktionen beitragen. Die komplexeste Problematik liegt dabei wohl jenen Untersuchungen zugrunde, welche den Kausalkonnex von Form und Funktion bei der Differenzierung von Leitelementen zu entflechten versuchen. Wie wenig wir diesen Konnex im Grunde verstehen, wird in einer Arbeit von KOLLMANN et al. (1983) betont. Zugleich zeigt diese Arbeit möglicherweise einen Weg zum Verständnis der funktionellen Differenzierung der Assimilatleitbahnen auf. Verglichen wurden Siebelemente, die durch Umdifferenzierung aus parenchymatischen Zellen entstanden waren. Solche Umdifferenzierungen lassen sich experimentell durch Verwundungen von Leitbahnen bzw. durch Propfungen erzielen. Zudem können sie u.a. im Grundgewebe von Haustorialorganen parasitischer Pflanzen beobachtet werden. Feinstrukturelle Untersuchungen an den Siebelementen, die durch Umdifferenzierung aus parenchymatischen Zellen entstanden sind, zeigen "daß ein grundsätzlicher Unterschied zur Differenzierung von Leitbündelsiebelementen nicht besteht." Damit erweist sich der von KOLLMANN et al. vorgestellte experimentelle Ansatz als erfolgversprechend für die Gewinnung allgemeingültiger Erkenntnisse. Dies um so mehr, als bei der Umdifferenzierung von Grundgewebszellen zu Siebelementen "manche Differenzierungsschritte deutlicher hervortreten, als wir dies bislang bei Leitelementen procambialer bzw. cambialer Herkunft kannten."

In diesem Zusammenhang gewinnt der Nachweis von Siebelementen in den Haustorien einzelner parasitischer Spermatophyta an Gewicht, da er u.U. weitere mögliche Objekte für Untersuchungen der Siebelement-Differenzierung erschließt. BA (1983) konnte Siebröhrenelemente im Haustorium von *Striga gesnerioides* (Scrophulariaceae) nachweisen, die bei *Striga hermonthica* fehlen.

Unter einem merkmalsphylogenetischen Aspekt ist bemerkenswert, daß die Siebelemente aus dem Blattstiel von *Liriodendron tulipifera* (Magnoliaceae) nicht mehr als Siebzellen, sondern schon als Siebröhrenglieder anzusehen sind (DUTE 1983). Eine vergleichende Studie über das sekundäre Phloem von 444 holzigen Dicotyledonen aus 24 Familien (der Elfenbeinküste) führte DEN OUTER (1983) durch. Sie läßt eine Assoziation bestimmter Siebröhrenglied-Typen und Geleitzell-Typen sowie deren Korrelation mit bestimmten Axialparenchym- und Sklerenchym-Typen erkennen.

Siebelemente mit Nacré-Wand sind innerhalb der sogenannten Seegräser sehr unterschiedlich verbreitet. Nach KUO (1983) ist eine Nacré-Wand bei Zosteraceae vorhanden, bei Cymodoceaceae nur in der Gattung *Halodule*, während sie den "Seegräsern" aus den Hydrocharitaceae und den *Posidonia*-Arten fehlt. Am bemerkenswertesten ist, daß bei den untersuchten Pflanzen Siebelemente mit Nacré-Wand in Blättern und/oder Sproßachsen festgestellt werden konnten, niemals aber in Wurzeln.

Die Aktivität auf dem Gebiete der Borkenanatomie zeigt sich zunächst in einer größeren Anzahl von borkenanatomischen Beschreibungen für einzelne oder einige wenige verwandte Arten zu rein diagnostischen Zwecken. Eine große Ausnahme bilden hier die zusammenhängenden Untersuchungen von RICHTER (1981a,b,c) über die Borkenanatomie (und Holzanatomie) der Lauraceae, die als beispielhaft gründliche Studien gelten dürfen.

Besonders hervorzuheben ist jedoch ein Werk von ROTH (1982), mit welchem das Handbuch der Pflanzenanatomie *(Encyclopedia of Plant Anatomy)* eine wertvolle Erweiterung erfährt: *Structural Patterns of Tropical Barks*. Die einführenden Kapitel behandeln Definitionen und Beschreibungen der verschiedenen Borken-Elemente, Gewebe- und Zelltypen sowie Einschlüsse; den Zusammenhang von innerer Struktur und äußerer Erscheinung; Funktion und Struktur; Veränderungen im nichtleitenden Phloem; die taxonomische Bedeutung borkenanatomischer Merkmale; ökologische und phylogenetische Aspekte; Grundmuster in der Borkenbildung, und schließlich auch die wirtschaftliche Nutzung von Borken. Im Hauptteil des Werkes werden die Borken von ungefähr 280 Baumarten Venezuelas, aus 48 Familien, nach Labor- und Felduntersuchungen ausführlich beschrieben. Ein gewisser Mangel ist darin zu erblicken, daß bei den Angaben zu den einzelnen Familien die ältere Literatur, die vor allem für die "offizinellen Rinden" ja vorliegt, kaum berücksichtigt werden konnte. Dieser Mangel findet seine Erklärung wohl im Habitat der Autorin, mit seinem erschwerten Zugang gerade zur älteren Literatur. Im ganzen jedoch bietet das Werk ROTH's eine Fülle von Informationen, wenngleich es nur einen Ausschnitt aus der Mannigfaltigkeit tropischer Borken präsentieren kann. Die Autorin, die nicht zuletzt mit ihren "feldanatomischen" Untersuchungen an Frischmaterial wahre Pionierleistungen erbracht hat, wird am besten wissen, daß auf dem Gebiet der Borkenanatomie noch viel zu tun bleibt.

6. Sekretionsgewebe

Eine ausführliche Dokumentation liegt nun für die seit langem bekannten "Milchschläuche" von *Cyclanthus bipartitus* (Cyclanthaceae) vor. Danach handelt es sich um ungegliederte, wenig verzweigte, dünnwandige Milchröhren (WILDER und HARRIS 1982). Für den raschen und einfachen Nachweis von Milchröhren in Callus-Kulturen von *Euphorbia marginata* haben BRUNI et al. (1981) eine Färbetechnik entwickelt.

Die Verteilungsmuster der "Harztaschen" in den Blättern von *Hymenaea*-Arten *(Caesalpinioideae)* wurden von LANGENHEIM et al. (1982) eingehend auf ihre systematische Relevanz untersucht. Zum Teil sind sie, zusammen mit ihren Inhaltsstoffen, wahrscheinlich auf den Selektionsdruck herbivorer Insekten zurückzuführen.

Mehrere Untersuchungen befassen sich mit der taxonomischen Bedeutung extrafloraler Nektarien in einzelnen Sippen. Neu ist die Entdeckung extrafloraler Nektarien bei Punicaceae durch TURNER und LERSTEN (1983). Sie beschreiben die an den Blattspitzen entwickelten Nektarien ausführlich und analysieren chromatografisch die Zucker (Glucose, Fructose, Saccharose) der ausgeschiedenen Nektartropfen. Daß in den Areolen von Cactaceae einzelne Blätter nicht als Dornen, sondern als extraflorale Nektarien entwickelt sein können, ist seit langem bekannt. MAUSETH (1982) beschreibt die kaum bekannte Entwicklung eines solchen Nektariums bei *Ancistrocactus scheerii*, einschließlich der cytologischen Merkmale in der Sekretionsphase.

Hydathoden von Spiraeoideae (Rosaceae) wurden zum ersten Male untersucht. LERSTEN und CURTIS (1982) weisen nach, daß sie an den Blattzähnen von *Physocarpus opulifolius* strukturell vollständig, aber funktionslos sind. Bei anderen *Physocarpus*- Arten zeigen die "Hydathoden" auch strukturelle Reduktionserscheinungen, oder sie fehlen ganz.

7. Ergastische Körper

Eine erste grobe Übersicht über das Vorkommen und die Formen pflanzlicher Opale in Blättern dicotyler Bäume geben KONDO und PEASON (1981). Danach lassen sich vorläufig acht Gruppen von Opalen nach ihrer Form bzw. nach ihrer Herkunft aus den einzelnen Geweben unterscheiden. Von den in Gräsern, Palmen und Coniferen gefundenen Opalen scheinen sie sich durchwegs durch ihre Form zu unterscheiden. Es ist abzusehen, daß künftige Untersuchungen die Klassifikation der bislang weniger bekannten Opale dicotyler Pflanzen verfeinern werden und ihre systematische Bedeutung stärker hervortreten lassen.

Unter den organischen ergastischen Körpern erweisen sich die Stärkekörner immer wieder als taxonomisch wertvolle Merkmale. MAHLBERG (1982) kann aus seinen Untersuchungen an diploiden und tetraploiden Varietäten von *Euphorbia pulcherrima (Poinsettia)* schließen, daß die Ausbildung der besonders geformten Stärkekörner in den Milchröhren einem eigenen genetischen Kontrollmechanismus unterliegt. Für die Tribus der Euphorbieae lassen sich vier Typen von Stärkekörnern in Milchröhren aufstellen: stäbchenförmig, spindelförmig, osteoid und discoid. Die Verbreitung der einzelnen Typen in den verschiedenen Untergattungen von *Euphorbia* und in näher verwandten Gattungen unterstreicht manche der bisherigen phylogenetischen Interpretationen (BISBOER und MAHLBERG 1981).

Die im Mesophyll der Cotyledonen einiger Convolvulaceae und bei *Cuscuta japonica* entdeckten "sphärischen Körper" konnten als ölspeichernde Riesenzellen von ca. 100 µm Durchmesser identifiziert werden. Ihre Verbreitung innerhalb des Verwandtschaftskreises ist bislang nicht näher bekannt (WADA et al. 1981).

8. Das Blatt

Die meisten blattanatomischen Studien standen wieder in Zusammenhang mit systematischen Fragestellungen. Mit der wachsenden Anzahl blattanatomischer Untersuchungen unter ökophysiologischen Aspekten zeigt sich die Dringlichkeit einer intensiveren Zusammenarbeit von Anatomie und Physiologie immer deutlicher. Immer mehr Strukturen werden bekannt, die man als spezielle Anpassungen an besondere ökologische Gegebenheiten des Habitats interpretieren kann. Ihr funktioneller Beitrag zum Überleben der Pflanze bleibt allerdings vielfach, wenn nicht zumeist, im einzelnen unklar.

Sehr merkwürdig sind die äquifazialen Blätter von *Ixorhea tschudian* (Boraginaceae) dem ausgesprochen trockenen natürlichen Standort angepaßt. Sie sind breit, weich und mit einem sehr reduzierten Schwammparenchym ausgestattet. Erstaunlich ist, daß die Stomata der amphistomatischen Blätter - wie bei hygromorphen! Blättern - über die Blattoberfläche emporgehoben sind. Wahrscheinlich kann *Ixorhea tschudian* (u.a.) darum unter xerischen Bedingungen gedeihen, weil ihre luftexponierten Teile von einem Harz überzogen sind, das von Haaren sezerniert wird (DI FULVIO 1982). Die Blätter von Bäumen der Miombo-Gehölze Zaires zeigen, trotz unterschiedlicher systematischer Zugehörigkeit, eine weitgehend übereinstimmende, im ganzen xeromorphe Struktur. Für die Blätter anderer zambesischer Gehölzformationen konnte eine entsprechend einheitliche "Anpassung" hingegen nicht nachgewiesen werden (MALAISSE et al. 1983). Eine sehr detaillierte blattanatomische Analyse an Bäumen der hochmontanen Regenwälder Jamaicas ergab, daß sich die Blätter in durchaus verschiedenen Waldtypen strukturell nicht signifikant unterscheiden. Von den Baumblättern der tropischen Tiefland-Regenwälder Jamaicas differieren sie im ganzen deutlich, wenngleich nur in wenigen Merkmalen. Neben diesen statistisch abgesicherten Erkenntnissen erbrachte die Analyse auch den Nachweis von Korrelationen bestimmter Strukturen im einzelnen Blatt (TANNER und KAPOS 1982). Von sehr klar genannten Voraussetzungen geht eine Studie über die Evolution der Blattarchitektur von *Fremontodendron californicum* ssp. *californicum* (Sterculiaceae) aus. BISSING (1982) nimmt an, daß der im Chaparral heimische *Fremontodendron californicum* und der Nebelwaldbewohner *Cheirostemon platanoides* gemeinsamen phylogenetischen Ursprungs sind. Das Blatt von *Cheirostemon platanoides* verwendet er als Modell für das Blatt der gemeinsamen Ahnenform. Es soll also die konservativen Merkmale enthalten. Von dieser "strukturellen Grundlinie" ausgehend analysiert BISSING Art und Umfang der Veränderungen, die im Laufe der Evolution des Blattes von *Fremontodendron* wahrscheinlich erfolgt sind, für eine ganze Reihe einzelner Strukturen. Nur einzelne dieser Veränderungen können unmittelbar als Anpassungen an die xerischen Bedingungen des Chaparrals verstanden werden.

Die Suche nach dem Kranzzellen-Syndrom wird in verschiedenen Sippen fortgesetzt: Chenopodiaceae (CAROLIN et al. 1982); Amaranthaceae (FISHER und EVERT 1982); Poaceae (SHAW und SMEINS 1981). Dabei zeigt sich, daß anatomische Daten für die Identifikation von C_3- und C_4-Pflanzen innerhalb einer Sippe verschiedentlich nicht genügen, so z.B. bei manchen Poaceae (HATTERSLEY et al. 1982).

Welche Bedeutung andererseits die möglichst exakte, quantitative Erfassung blattanatomischer Strukturen für die Interpretation physiologischer Daten hat, geht aus einer Untersuchung von JELLINGS und LEECH (1982) and *Triticum, Hordeum* und *Avena* (Poaceae) hervor. Zwei unabhängige Methoden zur Messung der inneren Oberfläche des Mesophylls beschreiben MORRIS und THAIN (1983). Eine quantitative Analyse der Leitbündelanordnung und Leitbündelstruktur im Verhältnis zur Blattbreite und zum Querschnitt leitender Elemente liegt für das Blatt von *Saccharum officinarum* vor (COLBERT und EVERT 1982).

a) Trichome und Stomata

SAHU (1982a) weist zum ersten Male für eine Asteraceae *(Parthenium hysterophorus)* zwei neue Haarformen, multiseriate und zottige Haare, nach. Seine Studien über die Haarformen der Asteraceae (SAHU 1982b) werden ergänzt durch eine Untersuchung von KELSEY und SHAFIZADEH (1980), in welcher die Autoren Sesquiterpen-Lactone glandulärer Trichome durch eine Farbreaktion lichtoptisch sichtbar machen. Ausführlich beschreiben ARORA et al. (1982) die Blattepidermis-Strukturen von *Heracleum mantegazzianum* und *H. sphondylium* sowie deren Hybride, bei welcher sie vor allem in den Trichomen intermediäre, aber auch völlig aberrante Formen finden.

POZGAJ (1981) analysiert die Variabilität der Blattbehaarung der 9 autochthonen *Quercus*-Arten der Slowakei in ihrer Beziehung zu den ökologischen Daten der Standorte. LORENZONI et al. (1981) können für (französische, italienische, israelische) Sippen von *Quercus coccifera* s.l. einen von West nach Ost gerichteten Gradienten zunehmender Blattbehaarung nachweisen.

Die merkmalsphylogenetische Beziehung der Epidermistypen von Mesembryanthemaceae diskutieren IHLENFELDT und JUERGENS (1982). Danach soll die mesomorphe Epidermis, die durch auffällige blasenartige Idioblasten ausgezeichnet ist, die ursprüngliche sein. Aus ihr soll der xeromorphe Epidermis-Typ durch ontogenetische Abbreviation der Protodermentwicklung (mit dem Verlust der Idioblastenbildung) evoluiert sein.

Die Entwicklung und Organisation der Stomata einer ganzen Reihe von Orchidaceae beschreibt SINGH (1981) und beobachtet drei verschiedene Entwicklungstypen, die auf dem einzelnen Blatt auch gemeinsam vorkommen können: perigen, mesoperigen und mesogen. Bemerkenswert ist seine Beobachtung, daß sich Stomata terrestrischer und epiphytischer Orchideen durch ihre Größe sowie die Zahl und Anordnung von Neben- und Nachbarzellen unterscheiden. Auf die Diversität der Stoma-Entwicklung innerhalb der Orchidoideae weist RASMUSSEN (1981) hin. In der sehr sorgfältigen Studie werden zum ersten Male für die Monocotyledoneae die Entwicklungsschritte des hemimesogenen Stoma abgebildet. Für einen bei südafrikanischen *Eugenia*-Arten entdeckten, herkömmlich nicht klassifizierbaren Stoma-Typ schlagen VAN WYK et al. (1982) den Terminus "anomostaurocytisch" vor.

9. Die Wurzel

Besondere Beachtung verdient das kleine, illustrierte Buch *Tree Roots and Buildings*, von CUTLER und RICHARDSON (1981). In Text und Abbildungen ist es von unmittelbarem, praktischem Nutzen, nicht zuletzt für die forensische Botanik. Es enthält eine Fülle von Daten über Bauschäden und über die Wurzeln, durch welche sie verursacht wurden. Häufiger und seltener angepflanzte Laubbäume, Coniferen und Sträucher sind berücksichtigt. Angaben über beobachtete Entfernungen von Pflanzort und Bauschaden sind für die einzelenen Arten ebenso gemacht wie über Art und Häufigkeit der "Invasion". Die Identifizierung von "Delinquenten" anhand wurzelholz-anatomischer Strukturen wird durch den Vergleich mit anderen Spezies abgesichert und erleichtert. Dazu ist angegeben, ob sich nahe verwandte Arten anhand der Wurzelholzstrukturen überhaupt unterscheiden lassen. Im ganzen ist es das Buch, das sich mancher Botaniker wünschte, der zur Beurteilung von Bauschäden durch Wurzeln als Fachmann herangezogen wurde. Obwohl es allein auf Daten basiert, die in England erhoben wurden, wird es sicherlich auch außerhalb Nordwest-Europas ein äußerst hilfreiches Nachschlagewerk sein.

Weiterhin bleibt die Anzahl wurzelanatomischer Untersuchungen vergleichsweise gering. Bemerkenswert ist die Entdeckung eines Caspary-Streifens in der Hypodermis von Zwiebel- und Getreidewurzeln (PETERSON et al. 1982). Die Entwicklung der Leitgewebeverbindung zwischen Haupt- und Seitenwurzel bei *Lycopersicum esculentum* beschreiben BYRNE und EMMITT (1982) sehr detailliert. Dabei ist im Hinblick auf die funktionelle Differenzierung von Leitgewebe bemerkenswert, daß die Xylembrücke zwischen dem Protoxylem der Seitenwurzel und dem Metaxylem der Abstammungswurzel durch Umdifferenzierung parenchymatischer Zellen hergestellt wird.

Literatur

ARORA, K., GRACE, J., STEWART, F.: Bot. J. Linn. Soc. 85, 169-178 (1982).

BA, A.T.: C.R. Seances Acad. Sci, Ser. III, Sci. Vie. 296, 375-380 (1983). - BAAS, P.: IAWA-Bull. 3, 3-6 (1982a); - (ed.): New Perspectives in Wood Anatomy. The Hague-Boston-London: Nijhoff/Junk 1982b. - BAAS, P., WERKER, E., FAHN, A.: IAWA-Bull. 4, 141-159 (1983). - BECK, CH.B., SCHMID, R., ROTHWELL, G.W.: Bot. Rev. 48, 691-816, (1982). - BISBOER, D.D., MAHLBERG, P.G.: Nord. J. Bot. 1, 447-457 (1981). - BISSING, D.R.: Am. J. Bot. 69, 957-972 (1982). - BOSSHARD, H.H., KUCERA, L.J., STOCKER, U.: Vierteljahrsschr. Naturforsch. Ges. Zürich 127, 29-48 (1982). - BOTOSSO, P.C., GOMES, A.V.: IAWA-Bull. 3, 39-44 (1982). - BRAUN, H.J.: Ber. Dtsch. Bot. Ges. 96, 29-47 (1983). - BRUNI, A., VIANNINI, G.L., DALL'OLIO, G.: Z. Pflanzenphysiol 103, 373-378 (1981). - BYRNE, J.M., EMMITT, D.P.: Am. J. Bot. 69, 287-297 (1982).

CARLQUIST, S.: Bull. Torrey Bot. Club. 108, 324-330 (1981); - Taxon 31, 443-450 (1982a); - Flora (Jena) 172, 463-491 (1982b); - Brittonia 34, 252-266 (1982c); - Am. J. Bot. 69, 1587-1598 (1982d); - Am. J. Bot. 70, 578-590 (1983). - CAROLIN, R.C., JACOBS, S.W.L., VESK, M.: Aust. J. Bot. 30, 387-392 (1982). - CATESSON, A.M., MOREAU, M., DUVAL, J.C.: IAWA-Bull. 3, 11-14 (1982). - CHEADLE, V.I., KOSAKAI, H.: Phytomorphology 30, 98-117 (1980). - COLBERT, J.T., EVERT, R.F.: Planta (Berl.) 156, 136-151 (1982). - CUTLER, D.F., RICHARDSON, I.B.K.: Tree Roots and Buildings, New York, London: Construction Press 1981.

DEN OUTER, R.W.: Acta Bot. Neerl. 32, 29-38 (1983). - DI FULVIO, E.T.: Kurtziana 15, 9-18 (1982). - DUTE, R.R.: Am. J. Bot. 70, 64-73 (1983).

FAHN, A.: Plant Anatomy, 3. ed. Oxford: Pergamon Press 1982. - FINK, S.: IAWA-Bull. 3, 31-38 (1982). - FISHER, D.G., EVERT, R.F.: Am. J. Bot. 69, 1133-1147 (1982). - FORD, B.: IAWA-Bull. 3, 7-10 (1982). - FRENCH, J.C., TOMLINSON, P.B.: Am. J. Bot. 68, 1115-1129 (1981a). - Bot. Gaz. 142, 366-381 (1981b); - Bot. Gaz. 142, 550-563 (1981c). - FUKAZAWA, K., OHTANI, J.: IAWA-Bull. 3, 201-206 (1982).

GALE, R.: IAWA-Bull. 3, 179-184 (1982). - GIBSON, A.C.: Bot. J. Linn. Soc. 83, 273-284 (1981). - GULENKOVA, M.A.: Bot. Zh. (Leningr.) 66, 1014-1021 (1981).

HATTERSLEY, P.W., WATSON, L., JOHNSTON, C.R.: Bot. J. Linn. Soc. 84, 265-272 (1982).

IHLENFELDT, H.-D., JUERGENS, N.: Mitt. Inst. Allg. Bot. Hamburg 18, 103-116 (1982).

JELLINGS, A.J., LEECH, R.M.: New Phytol. 92, 39-48 (1982).

KELSEY, R.G., SHAFIZADEH, F.: Biochem. Syst. Ecol. 8, 371-378 (1980). - KOLLMANN, R., DÖRR, I., SCHULZ, A., BEHNKE, H.D.: Ber. Dtsch. Bot. Ges. 96, 117-132 (1983). - KONDO, R., PEASON, T.: Res. Bull. Obihiro Univ., Ser. I, 12, 217-230 (1981). - KUO, J.: Am. J. Bot. 70, 159-164 (1983).

LANGENHEIM, J.H., LINCOLN, D.E., STUBBLEBINE, W.H.: Am. J. Bot. 69, 595-607 (1982). - LARSON, P.R., FISHER, D.G.: Can. J. Bot. 61, 1040-1051 (1983). - LERSTEN, N.R., CURTIS, J.D.: Can. J. Bot. 60, 850-855 (1982). - LORENZONI, F.C., LORENZONI, G.G., COLOMBO, P.M.: Studi Trentini Sci. Nat. Acta Biol. 58, 141-152 (1981).

MAHLBERG, P.G.: Bot. Gaz. 143, 206-209 (1982). - MALAISSE,F., MPALA, K., COLONVAL-ELENKOV, E.: Bull. Soc. R. Bot. Belg. 115, 357-371 (1982). - MAUSETH, J.D.: Bot. Gaz. 143, 273-277 (1982). - METCALFE, C.R., CHALK, L.: Anatomy of the Dicotyledons, 2nd ed., Vol. II, Wood Structure and Conclusion of the General Introduction. Oxford: University Press 1983. - MEYER, J., MARESQUELLE, H.J.: Anatomie des Galles, Handbuch der Pflanzenanatomie (Encyclopedia of Plant Anatomy), Vol. XIII/1. Berlin-Stuttgart: Borntraeger 1983. - MEYLAN, B.A., BUTTERFIELD, B.G.: IAWA-Bull. 3, 167-176 (1982). - MICHENER, D.: Aliso 10, 39-58 (1981). - MIO, S., MATSUMOTO, T.: IAWA-Bull. 3, 56-58 (1982). - MORRIS, P., THAIN, J.F.: J. Exp. Bot. 34, 95-98 (1983). - MUHAMMAD, A.F., SATTLER, R.: Am. J. Bot. 69, 10004-1021 (1982).

NEUBAUER, H.F.: Plant Syst. Evol. 139, 103-112 (1981).

PARAMESWARAN, N.: Ber. Dtsch. Bot. Ges. 95, 291-296 (1982). - PARAMESWARAN, N., CONRAD, H.: IAWA-Bull. 3, 75-88 (1982). - PARAMESWARAN, N., LIESE, W.: IAWA-Bull. 2, 89-93 (1981). - PATEL, J.D., BHAT, V., DEVI, G.S., KOTHARI, I.L.: Flora (Jena) 171, 410-418 (1981). - PETERSON, C.A., EMANUEL, M.E., WILSON, C.: Can. J. Bot. 60, 1529-1535 (1982). - PILLAI, S.K., KUMAR, K.B.: Proc. Indian Acad. Sci., Plant Sci. 90, 457-464 (1981). - POZGAJ, J.: Preslia (Prague) 53, 339-344 (1981).

RASMUSSEN, H.: Bot. J. Linn. Soc. 82, 381-393 (1981). - RICHTER, H.G.: Anatomie des sekundären Xylems und der Rinde der Lauraceae. Hamburg, Berlin: Parey 1981a; - Sonderb. Naturwiss. Ver. Hamburg 0, 1-148 (1981b); - IAWA-Bull. 2, 79-87 (1981c). - ROTH, I.: Structural Patterns of Tropical Barks, Handbuch der Pflanzenanatomie (Encyclopedia of Plant Anatomy), Vol. IX/3. Berlin, Stuttgart: Borntraeger 1982.

SAHU, T.R.: Feddes Repert. 93, 437-441 (1982a); - J. Econ. Taxon. Bot. 3, 517-522 (1982b). - SCHMID, R.: Biol. Rev. 48, 817-931 (1982). - SHAW, R.B., SMEINS, F.E.: Bot. Gaz. 142, 534-544 (1981). - SINGH, H.: Acta Bot. Indica 9, 94-100 (1981).

TANNER, E.V.J., KAPOS, V.: Biotropica 14, 16-24 (1982). - THOMPSON, T., JAGELS, R.: IAWA-Bull. 4, 27-31 (1983). - TOMLINSON, P.B.: Anatomy of the Monocotyledons, Vol. VII: Helobiae (Alismatidae). Oxford: University Press 1982. - TURNER, G.W., LERSTEN, N.R.: Am. J. Bot. 70, 475-480 (1983).

VAN WYK, A.E., ROBBERTSE, P.J., KOK, P.D.F.: Bot. J. Linn. Soc. 84, 41-56 (1982). - VON TEICHMA, I., ROBERTSE, P.J.: J.S. Afr. Bot. 47, 637-652 (1981).

WADA, K., KADOTA, A., TANIHIRA, H., SUZUKI, Y.: Bot. Mag. Tokyo 94, 239-248 (1981). - WILDER, G.J., HARRIS, D.H.: Bot. Gaz. 143, 84-93 (1982).

ZIMMERMANN, M.H.: Xylem Structure and the Ascent of Sap. (Springer Series in Wood Science, Vol. 1) Berlin, Heidelberg, New York, Tokyo: Springer 1983. - ZIMMERMANN, M.H., JEJE, A.A.: Can. J. Bot. 59, 1882-1892 (1981). - ZIMMERMANN, M.H., POTTER, D.: IAWA-Bull. 3, 103-109 (1982).

Professor Dr. HELMUT UHLARZ
Abteilung Spezielle Botanik
(Biologie V)
Universität Ulm
D-7900 Ulm, FRG

B. Physiology

I. Plant Water Relations

By RAINER LÖSCH

Less than 20% of the about 1700 articles on plant water relations which appeared between 1981 and 1983 will be quoted here. For a more complete enumeration the reader is referred to the *Water-in-Plants Bibliography* (POSPISILOVA and SOLAROVA 1982, Vol. 7; 1983, Vol. 8) with its extensive indices. Also books entirely on plant water metabolism or with larger chapters on the topic were published, e.g., PALEG and ASPINALL 1981, EVANS and THAMES 1981, UNGER and STÖCKER 1981, MONTEITH and WEBB 1981, SPANSWICK et al. 1980, KOZLOWSKI 1981, JOHNSON 1981, SIMPSON 1981. The reader is also referred to the four volumes about ecological physiology of the *Encyclopedia of Plant Physiology, New Series* (LANGE et al. 1981, 1982a, 1983a,b).

1. Water Relations of Cells and Tissues

A period of consolidation has apparently been reached with respect to theory and methods of study of plant water relation parameters. MEIDNER (1983) summarizes concisely the present state of the understanding of plant water relations at the cellular and whole-plant level, and the dynamics of water flow through the plant to the atmosphere with emphasis on the morphological and physiological interrelations of these processes. McCLENDON (1981) analyzes the balance of forces resulting in the matric potential of cell walls. For the symplast a relationship is drawn between the cell volume due to different water contents and the potentials, pressure and stresses within the wall in a manner similar to the Höfler-diagram.

Technically, attention is paid to reductions of error in the application of the instruments for water potential measurements (temperature correction in thermocouple psychrometry: SAVAGE et al. 1982, avoidance of water loss from tissue during pressure bomb measurements: LEACH et al. 1982) and on the calibration of both methods against each other (e.g., ROY and BERGER 1983). A third instrument for such measurements, the hydraulic press, has now been incluced into such comparisons (e.g., GRANT et al. 1981, RADULOVICH et al. 1982). As yet its use is not completely accepted, but it might prove to be a suitable tool for routine monitoring of stress thresholds in the irrigation area. The evaluation of water potential/water content data by pressure-volume analyses is applied in many studies. By combining an exponential turgor function with a hyperbolic osmotic function JANE and GREEN (1983) have been able to describe mathematically the PV-curves. By this method it is possible to determine more exactly the turgor loss point and bulk elastic modulus values than by visual estimation of these parameters from the shape of the PV-curve. These techniques and evaluation procedures facilitate comparative studies of stress load and resistance (CLAYTON-GREENE 1983).

Progress in Botany, Vol. 46

In particular, more information has become available from PV-techniques about the bulk elastic modulus ε and its dependence on turgor (ψ_P). In *Triticum* ε increases by drought conditioning (MELKONIAN et al. 1982). Measured in forest tree species through a season, ε shows various patterns of dependence on increasing turgor (ROBERTS et al. 1981): a linear relationship, a linear increase up to a constant value, an increase up to a maximum at intermediate turgor with following decrease, and an exponential increase. Such species- and growth-stage-dependent differences could account for different ways by the species in coping with habitat water conditions. SOVONICK-DUNFORD et al. (1983) measured water relation parameters of the phloem by observation of changes in tissue thickness at different osmotic pressures of the bathing solutions. Similarly, by relating cell volume to (external osmotic pressure)$^{-1}$ MUNNS et al. (1983) determined ε and ψ_P values for the unicellular alga *Chlorella emersonii* grown in solutions with different NaCl concentrations. ε and direct turgor measurements for single cells in tissues were done with the pressure probe technique. In epidermal cells of *Elodea densa* ε was linearly related to turgor (STEUDLE et al. 1982); the link between hydraulic conductivity (L_p) and ψ_P is discussed with respect to pressure-dependent membrane structures. (About mobility of charges in algal membranes depending on turgor: ZIMMERMANN et al. 1982). In epidermal cells of *Tradescantia virginiana* L_p increased by a factor of 2 - 4 when temperature was increased by 10 K (TOMOS et al. 1981). If the pressure probe technique is applied to growing pea epicotyl segments (COSGROVE and STEUDLE 1981), very high L_p values can be found with very short half times of water exchange of individual cells (L_p in the magnitude $2 \cdot 10^{-1} - 10^{-6}$ cm s^{-1} bar^{-1} compared with $2 \cdot 10^{-6} - 10^{-7}$ cm s^{-1} bar^{-1} in mature epidermis cells of *Tradescantia virginiana*: TOMOS et al. 1981, TYERMAN and STEUDLE 1982). The light-induced leaf expansion of bean is not mediated, however by drastic changes in L_p, cell osmotic potential or wall-yield stress, but rather by increasing wall extensibility (VAN VOLKENBURGH and CLELAND 1981). COSGROVE (1981) showed that both the process of growth and the cell turgor depend on these peculiar wall extensibilities and also on hydraulic conductances. The analysis of these dependencies probably will become an important aspect of future research in cell and tissue water relationships. They will help to describe quantitatively the dynamic processes of growth and also of morphogenesis. In *Callitriche heterophylla*, for instance, the heterophylly is governed by turgor and wall extensibility that depend on environmental conditions: DESCHAMP and COOKE (1983).

Contrasting water-stress sensitivity of young and old leaves at comparable total water potentials may result just from differences in turgor potentials (KARLIC and RICHTER 1983). Periods of high and low stress sensitivity during ontogenesis (HOCHMAN 1982) could find an explanation in such less obvious cell water properties. (Susceptibility to water deficits during crop development is often modified from mineral metabolism: K^+ supply can effectively modify drought stress: MENGEL and ARNEKE 1982, JENSEN 1982.)

Ontogenetic stress adaptation has been studied very intensively with respect to osmotic adjustment. In addition to the growth- and maturation-dependent long-term changes of osmotic properties (ZUR et al. 1981) water stress brings about increases in the amount of osmotica and also changes in the cell water parameters (e.g., decrease of ε: IKE and THURTELL 1981). Osmotic adjustment is often confined to distinct parts of a plant (tomato roots, more than shoots: TAYLOR et al. 1982, maize leaf elongation region, not the exposed blade: MICHELENA and BOYER 1982). Sometimes, however, the beneficial effect of osmoregulation on leaf growth is limited: MICHELENA and BOYER 1982, WILSON and LUDLOW (1983). HANSON and HITZ (1982) summarize the knowledge about the biochemical events that increase the amounts of betaine and/or proline in the course of osmoregulation.

In halophytes three groups can be distinguished according to the osmotica used in osmoregulation: species producing soluble carbohydrates only, species accumulating both carbohydrates and nitrogenous compounds, and species producing more nitrogenous solutes than soluble carbohydrates (BRIENS and LARHER 1982). In NaCl-treated *Spartina alterniflora* glycinebetaine contents are ten times higher than proline contents (CAVALIERI 1983). Sea grasses possess a Cl^--extrusion system for osmoregulation (JAGELS 1983); their elastic moduli are different for swelling and shrinking cells (hysteresis in the relationship between volume and turgor pressure): TYERMAN (1982).

2. Root Water Uptake and Water Movement Through the Plant

Attempts continue to get more insight into the nature and magnitude of resistances against water uptake from soil into the roots. Salt accumulation immediately at the root surface (SCHLEIFF 1982) causes low soil-water potentials particularly in this transition zone (see also ZUR et al. 1982). Different soil porosities (SHEIN 1982) can offset or increase this effect. It is conceivable that soil/root resistance increases when plants are stressed by a strong evaporation and a contraction of the roots occurs, resulting in vapor gaps between root and soil (FAIZ and WEATHERLEY 1982). In dry soils an equilibrium between soil and *Triticum* root water potential did not occur even overnight, the potential drop across the border zone of both compartments being twice that along the stalks (WALLACE et al. 1983). In order to measure root water potentials directly in situ, ADEOYE and RAWLINS (1981) used a split-root technique with one branch of the root system enclosed together with a thermocouple psychrometer in a teflon tube in the soil. Vesicular-arbuscular mycorrhizae increase the water-absorbing surface area, reducing the resistances and allowing higher transpiration rates (which sometimes, however, can bring about a too rapid exploitation of limited soil-water reserves): ALLEN (1982b) – *Bouteloua*, LEVY et al. (1983) – *Citrus*. SANDS et al. (1982) report a modification of the hydraulic conductivity of pine and bean roots by different suberization. Chilling can reduce root hydraulic conductivities and impede stomatal regulation thus giving rise to wilting in chilling-sensitive plants (McWILLIAM et al. 1982). Changes in the resistance of roots meeting higher transpiration demands may occur preferentially in older regions of the root where the production of lateral roots contributes effectively to the water input (SANDERSON 1983). Species-specific strong or weak adaptation of the root system to soil structure and water content can influence crop productivity (GEISLER 1982). Root growth itself is influenced by soil-water potential and temperature. Low soil-water potentials and temperatures favor dense root growth and an increase of the number of growing roots. At high water potentials, however, root elongation rate reaches a maximum rate (TESKEY and HINCKLEY 1981). Such growth changes occur by the response of the root meristem cell cycles to ambient moisture conditions (THOMAS and DAVIDSON 1982, YEE and ROST 1982).

Longitudinal water-flow resistances in roots and stems are comparatively small. However, they can restrict water transport in particular cases, as is inferred from comparisons of root and intracoleoptile internode vessel numbers and radii in wheat genotypes (CORNISH 1981). In this case a proper combination of genotype and planting depth can improve coleoptile water relations and crop performance. Large differences in whole-plant conductances (volume flow/$\Delta\psi$ root-leaves) depending on xylem anatomy can be found, when more primitive species (WOODHOUSE and NOBEL 1982: ecologically different ferns) are compared with more advanced ones. While it is certain that this whole-plant conductance undergoes real changes during ontogenesis and is different even within populations of the same species (ROY and MOONEY 1982), its short term changes are questionable. They cannot be inferred from the ratio (transpiration/$\Delta\psi$ root-leaves) without taking account of transpirational water expenditure from storage reservoirs in the plant. Calculations based on flow measurements by the heat pulse technique (COHEN et al. 1983) result in constant resistance values. Constant resistances result also from calculations that make allowance for use of stored water (KATERJI et al. 1983). Time constants and resistances for the exchange between such reservoirs and the functional xylem can be calculated (EDWARDS and JARVIS 1982). Complex models combine time courses in changes of transpiration rate, shoot-water potential and stem radius (MILNE et al. 1983). By measurement of the attenuation of γ-radiation a nondestructive determination of tree-stem water contents is possible (EDWARDS and JARVIS 1983). With the cross-sectional area of conducting sapwood known it is possible according to the pipe model theory to predict also the canopy leaf area that is serviced by the sapwood vessels (WARING et al. 1982). AMIN (1982) dissociates himself from the conventional view of the ascent of sap and suggests the formation of a double-layer capacitor within the vessels balancing the gravitational forces. Real negative pressures would not be required by this model. This stresses the question of the physical state of water in the xylem vessels (see Fortschr. Bot. 34, 93).

3. Stomatal Behavior and Transpirational Water Loss

During this abstracting period some reviews on guard-cell physiology and their significance for gas exchange have appeared (ZEIGER 1983, OUTLAW 1982, 1983, MACROBBIE 1982, FARQUHAR and SHARKEY 1982) which summarize the present knowledge and give access to primary literature for detailed information.

A comprehensive review of the metabolism of the stomatal guard cells is given by ZEIGER (1983) with the proposal of a unifying chemiosmotic mechanism of stomatal function based on a primary H^+ gradient as the driving force for all the ion fluxes associated with stomatal movements (see also SHARPE and ZEIGER 1981).

WEYERS et al. (1983) outline methods to obtain the new tool of stomatal physiologists, guard-cell protoplasts (GCP), and the facilities of research with them. Corroborated by findings at the levels of epidermal strips and whole leaves, results from GCP investigations can offer insight into the guard cell metabolism. GCP show shrinking and swelling, respond to light, CO_2 concentration, ABA, and fusicoccin (e.g., GOTOW et al. 1982, BROWN and OUTLAW 1982, further references see WEYERS et al. 1983). GCP have been used to provide evidence for an ABA production by guard cells themselves (WEILER et al. 1982) and to answer questions about the photosynthetic properties of guard cells. Some recent findings (OUTLAW et al. 1981, VAUGHN and OUTLAW 1983, FITZSIMONS and WEYERS 1983) tend again to reaffirm the existence of PS II in guard cells, but generalizations must be avoided. GCP have been used extensively to assess enzymes activities of the various metabolic pathways (e.g., SCHNABL 1981).

SCHNABL et al. (1982) analyzed, in GCP, increases and decreases in activity of enzymes involved in the starch-malate balance; OUTLAW et al. (1981) and DONKIN et al. (1982) studied enzymatic activities of this metabolism also in epidermal extracts. HAMPP et al. (1982) assayed key enzymes in guard cells of the major C-pathways. They failed to detect RubP-carboxylase and hydroxypyruvate reductase, but found high levels of specific activity of enzymes of the catabolic pathways (glycolysis, the oxidative pentose-P-pathway, and the Krebs-cycle). For the direct influence of light on these enzymatic activities see RAO et al. (1983a, b); for a review devoted particularly to guard-cell carbon metabolism see OUTLAW (1982).

Updated information about the Cl^--metabolism of guard cells is given by MACROBBIE (1982), about K^+ contents and fluxes by OUTLAW (1983). Both reviews also pay due attention to the electrical peculiarities of the guard-cell membranes.

LAFFRAY et al. (1982) contribute measurements to the yet patchy existing information that under certain circumstances K^+-shifts can be delayed compared with the progress of the pore movement (which implies that in these cases K^+ changes would not be the primary cause for aperture changes). Contradictory reports about the involvement of K^+-changes in movements of the chloroplast-free *Paphiopedilum* stomata (OUTLAW et al. 1982, WILLMER et al. 1983) continue the deadlock between the conflicting findings of meanwhile four different working groups.

Studies about the involvement of phytohormones continue to be focused on abscisic acid, the effects of which can be modified however by other substances. Thus IAA interferes with the stomata-closing effect produced by the combined action of CO_2 and ABA (SNAITH and MANSFIELD 1982). Stomata of the CAM plant *Kalanchoe daigremontiana* respond to lower ABA concentrations than is known for any other species (JEWER and INCOLL 1981). Generally, ABA accelerates the passive K^+ efflux out of the cells with a consequent depolarization of the membrane potential (KASAMO 1981). EM-detectable membrane alterations following stomatal closure by PMA or farnesol treatment can be produced, however, by these substances themselves and need not be ABA-mediated (SCHULT and DÖRFFLING 1981). In the causal relationships of stomatal responses to the environment progress can be reported particularly with light and humidity responses. Light response is reviewed in detail by ZEIGER (1983). It can now be accepted as a fact that as well as the light effect on stomatal apertures mediated by leaf photosynthesis – thus depending on the PAR wavelengths and working via leaf internal CO_2 concentrations – another, independent blue-light effect gives rise to stomatal opening. (References quoted by ZEIGER 1983, more recent papers are, e.g., MORISON and JARVIS 1983a,b: blue light more than three times as effective than red light and analysis of the interrelations between the effects of irradiance quality and quantity and the leaf internal CO_2 concentration in Scots pine, KARLSSON et al. 1983: blue light responses of stomata in chlorophyll-deficient wheat seedlings.) ZEIGER and FIELD (1982) assume a membrane-bound flavin as the blue-light photoreceptor. This system, which is most effective at very low light intensities, could offer the plant ecological advantages at dawn (ZEIGER et al. 1981). Light effects are further complicated by the probable involvement of the phytochrome system in stomatal movements which could influence membrane permeability and energy turnover (ROTH-BEJERANO and ITAI 1981, ROTH-BEJERANO et al. 1982, NEJIDAT et al. 1983).

Humidity (VPD) responses of stomata and their interrelations with other factors were observed in many field and controlled environment studies not itemized here. They can start within seconds after change in VPD (FANJUL and JONES 1982). Stomata of CAM plants can respond sensitively in the dark to humidity gradients, thus impeding nocturnal CO_2 fixation (MEDINA 1982), whereas the effect of this response on photosynthetic rates of C_4 plants is less drastic due to their high degree of CO_2 utilization (BUNCE 1982, 1983). Stomatal humidity responses are quite different in different populations of the same species growing in microclimatically different environments (ROY and MOONEY 1982, MOONEY and CHU 1983, MOONEY et al. 1983). This is certainly based on differences in guard-cell cuticle development, a feature that turns out more and more to be distinctive in the determination of the degree of this ecologically so important stomatal behavior. More indirect hints on this relationship come from comparative measurements of ad- and abaxial stomatal humidity responses (SILCOCK and WILSON 1982) and from the comparison of these responses in current- and previous-year needles of Douglas fir (MEINZER 1982a). In the latter case the decline of VPD responses of older needles could be restored by hexan washes of the needles. Light microscope and TEM studies of stomata from broad-leaved trees and from *Pinus silvestris* showed water-permeable and even cuticle-free areas at the outer pore walls that remain exposed to the external air even with closing stomata (APPLEBY and DAVIES 1983a,b). Their partial exposure to the increased VPD allows a very sensitive adaptation of the pore area to the prevailing humidity conditions. Water-loss differences between the stomatal apparatus and neighboring cells (peristomatal transpiration) due to a peculiar fibrillation of the guard-cell wall structures were found even in the primitive stomata of *Funaria hygrometrica* (SACK and PAOLILLO 1983). While the structures facilitate the appropriate movements of the guard cells, the hydraulic properties of them are likewise fitted to their functions. Thus MEIDNER (1982) was able to measure differences in the guard-cell elastic moduli during various phases of movement. They are somewhat reduced during the "Spannungsphase" and increase during the "Motorphase". SHERIFF (1982) discusses the magnitudes of hydraulic conductances and their functional operation in plants with humidity-sensitive and -insensitive stomata.

Model calculations relate the transpirational water loss from the substomatal cavity to its wall area and volume. These dimensions fit better in terms of mesophyll CO_2 uptake than in terms of water-loss avoidance (PICKARD 1981, 1982a,b; calculations of the gas transfer through stomatas: CARY 1981, LEUNING 1983). Often stomatal regulation keeps leaf-internal CO_2 concentration (c_i) rather constant (e.g., MEINZER 1982b; stronger c_i control in C_4 than in C_3 plants: RAMOS and HALL 1982). Fruiting reduces stomatal resistances drastically by increased photosynthetic rates, i.e., by lower plant internal CO_2 concentrations: HERBORN and LENZ (1981).

In addition to the complicated interaction of factors influencing stomatal apertures long-term and long-distance effects complicate the analysis of the natural stomatal behavior (e.g., dependence of stomatal conductances on previous night temperatures: DREW and BAZZAZ 1982, influences of day-to-day changes in root temperature on conductance: KÜPPERS et al. 1982). Nevertheless, attempts continue in modeling stomatal responses to the environmental conditions for predictive purposes (e.g., KAUFMANN 1982a,b, UPADHAYA et al. 1983).

Evidence increases that different behavior of ab- and adaxial stomata is inherent and not only dependent on the microenvironments (TRAVIS and MANSFIELD 1981). PEMADASA investigated this problem in a series of papers (1981, 1982a,b, 1983a,b) with the result that disparity in behavior is based on different K^+-permeabilities of the membranes in ab- and adaxial stomata, the K^+-fluxes normally being more restricted in the latter ones.

The analysis of the relationships between stomatal conductance, transpiration, and photosynthesis under various environmental conditions and treatments is improved by the development of portable devices (based on the concept of steady state porometers and IRGA CO_2 monitoring) (e.g., GRIFFITHS and JARVIS 1981, BELL and INCOLL 1981a,b, SCHULZE et al. 1982, FIELD et al. 1982).

The interrelations between stomatal conductance and photosynthesis and the hypothesis about optimization of water use (example for a partial confirmation of the hypothesis: WILLIAMS 1983) are reviewed by FARQUHAR and SHARKEY (1982). The authors conclude from their analysis that stomata limit water loss more substantially than carbon assimilation. This dynamic regulation of water expenditure is measured in many basic and applied studies. The midday depression of transpiration – as an obvious effect of stomatal responses mainly to atmospheric conditions and only modified by plant water status – provokes attention in investigations of its seasonal variations (e.g., ANDERSON 1982, LÖSCH et al. 1982, LANGE et al. 1982b, TENHUNEN et al. 1982). Leaf size and leaf orientation influence transpiration rates primarily by altered leaf temperatures (GELLER and SMITH 1982); toothed leaf margins increase convective heat transfer in comparison with entire leaves of the same size (GOTTSCHLICH and SMITH 1982). Leaf temperature regulation by transpiration cooling, on the other hand, may enable geophyllous plants to maintain tolerable temperatures despite being adpressed to a heat-laden soil surface (ELLER and GROBBELAAR 1982).

4. Water-Stress Effects on Growth and Metabolism

Many papers deal with the effects of reduced water supply on plant productivity. Since they are mostly dealing with specific problems in crop or forest plants, the examples quoted in previous reports (e.g., Progr. Bot. 41, 25; 44, 75) may suffice showing the broadness of research and will not be enlarged upon this time. Rather, emphasis will be paid to several investigations of relationships between water shortage and metabolic processes. A review about metabolic consequences of water deficits in mesophytes is given by HANSON and HITZ (1982).

Osmoregulation, as an active adaptation of the plant to reduced water potentials, has already been mentioned in Sect. 1. Proline (and also glycinebetaine) accumulation (PALEG and ASPINALL 1981) within the cytoplasm often accompanies solute accumulation in the vacuole and acts to warrant osmotic balance between the compartments and for membrane protection: Thus, K^+-increase in the course of osmoregulation of *Sorghum bicolor* is paralleled by a proline increase when a certain threshold concentration of K^+ is passed (WEIMBERG et al. 1982). Species-specifically, however, K^+ can also directly control the proline biosynthesis by synergistic interaction with arginine that apparently acts as a precursor for proline in *Eleusine coracana* (finger millet), but not so in other plants (investigated, e.g., groundnut: RAO et al. 1981a,b). According to the categories of HANSON and HITZ (1982), in other cases glutamate could be the precursor (unlikely to be so in *Phaseolus vulgaris*: PAHLICH et al. 1982), or a lowered rate of proline oxidation (SELLS and KOEPPE 1981), or a slower incorporation of proline into protein. FUKUTOKU and YAMADA (1981, similarly DUNGEY and DAVIES 1982) assume some de novo synthesis of proline at the cost of protein degradation. Interesting differences in drought response between the succulent stems and the leaves of *Euphorbia trigona*, where proline and glycinebetaine accumulate only in the latter, were reported by HUBER and EDER (1982).

Despite the increases in pool sizes of small molecules, which are assumed to be useful adaptations to lowered water potentials, all other metabolism is also affected by water shortage. Attention has been paid, in this respect, particularly to the photosynthetic processes.

Water shortage reduces chlorophyll levels in the leaves (SANCHEZ et al. 1983). Under water stress a reduction occurs in chloroplast volume, membranes (amount of phospholipids), and grana thylakoids (that swell simultaneously); starch grains increase and become fused, plastoglobuli accumulate. All these processes are reversed after rehydration but much slower than the rate of degradation (SAMIEV and MARFINA 1982, SILAEVA and TKACHUK 1982, CHETAL et al. 1983). Interaction between water molecules and (proto-)chlorophyll layers during their aggregation were spectroscopically followed by BODDI et al. (1983) and CHAPADOS and LEBLANC (1983). In order to reduce the photosynthetic rate by 50% hygrophytic mesophyll tissue was rinsed in solutions with an osmotic potential of about -2 MPa, mesophyll from xerophytes in such of ca. -5 MPa. These differences originate mostly from different osmotic potentials of the cell sap. The highest tolerable reduction in total protoplast volume (ca. 55% of the maximum) was the same for all tested tissues (KAISER 1982). In comparative investigations with stroma enzymes, intact chloroplasts, protoplasts, and leaf slices only the enzymes recovered completely from the osmotic stress. Dark reactions and ATP synthesis were more affected by stress than electron transport processes (KAISER et al. 1981a,b). Apparently, stromal acidification can ameliorate the effects of low osmotic potentials on the enzyme systems (BERKOWITZ and GIBBS 1983). In a manner similar to KAISER using sorbitol stressed chloroplasts BERKOWITZ and GIBBS (1982) defined the most stress-sensitive steps of the Calvin cycle (the step Fructose-1,6-bis-P to Fructose-6-P and some step between Ribose-5-P and Glycerate-3-P). Further papers on water-stress effects on Calvin cycle enzymes include SHARKEY and BADGER (1982), VAPAAVUORI and VALANNE (1982), MAYORAL et al. (1981) – all: RubP-carboxylase. When the facultative CAM plant *Mesembryanthemum crystallinum* changes from C_3 to CAM under NaCl-stress the activities of the acid-metabolism enzymes become increased while those of the Calvin cycle enzymes remain relatively constant (HOLTUM and WINTER 1982).

Although the stress sensitivity of the dark reactions primarily controls the inhibition of photosynthesis at low water potentials, nevertheless the light reactions can be affected by water stress too. In dehydrated *Pisum* chloroplasts MATORIN et al. (1982) found blocked the electron transfer between the Hill reaction point and PS II, and between plastoquinone and P 700. From fluorescence spectra FORK and OQUIST (1981), OQUIST and FORK (1982a) derived energy distribution patterns from PS II to PS I, which occur under desiccation stress in the intertidal alga *Porphyra*. By such energy shifts damage to the more photolabile PS II is avoided when photosynthesis is inhibited at low tide, under light-exposed conditions. By contrast, no increased energy transfer from PS II to PS I occurs in the deep-shade liverwort *Porella* and in *Trebouxia* isolates: OQUIST and FORK (1982b).

The soluble fraction of mitochondrial enzymes is, as a rule, only moderately sensitive to dehydration; whereas the activity of their membrane-bound enzymes could be reduced sharply by stress (PRIESTLEY and BRUINSMA 1982).

Ethylene production in wheat leaves, triggered by water deficits, increases 30-fold and, after some time, declines again. APELBAUM and YANG (1981) and HOFFMAN et al. (1983) analyze the biochemical pathway of ethylene production and recommend the residual metabolite N-malonyl-aminocyclopropan-carboxylic acid as an indicator for the stress history of leaves.

Similarly, the glucosyl ester of ABA can serve as accumulative indicator of the water-stress experience of a leaf. Unlike the major metabolite of ABA turnover, phaseic acid, it does not decline after release from the stress. Dynamic time courses of turgor-dependent increases and decreases of ABA and phaseic acid, their separation within the leaf, the complication of these courses by repeated stress recovery, and the possible enhancement of ABA methods by ethylene in the darkness (PIERCE and RASCHKE 1981, ACKERSON 1982, ZEEVAART 1983) can very easily veil dose-effect relationships. Nevertheless, one can often determine clear correlations between leaf water potential depressions and ABA contents (e.g., KANNANGARA et al. 1983), and such correlations are used in screening for drought-adapted genotypes (e.g., HENSON et al. 1981).

Other methods of screening for drought-tolerant genotypes include comparative porometer measurements, IR-measurements of canopy temperature, and assessment of the drying rate of excised leaves (CLARKE and McCAIG 1982).

Freeze-Dehydration. Frost injury to plants largely occurs by intracellular ice nucleation which must be distinguished from intercellular ice formation (BAKRADZE et al. 1981, PASICHNYI et al. 1980). Specific frost tolerance depends largely on the amount of cellular dehydration tolerance (KRASAVTSEV and KHVALIN 1982, SAKAI 1982) which in turn reflects pecularities of membrane permeability (MANUIL'SKII 1982). Freezing tolerance can be shifted by temperature effects on membrane properties (GUSTA et al. 1982, HONG and SUCOFF 1982). Deep supercooling to below -40°C, which prevents ice damage by the remaining intracellular water (QUAMME et al. 1982), causes the high winter hardiness of timberline flora (BECWAR et al. 1981). Bacteria as ice-nucleating agents, however, can bring about frost damage to crop plants even at moderately low temperatures (LINDOW et al. 1982a,b).

Frost-drought, with low temperature based plant dehydration as the actual damage effect, decreases gas exchange by lowering plant-water potentials drastically. Frost-drought is mitigated by high water storage capacities of the plant; fertilizing can alleviate its effect only in its very initial phase (PESCHL 1982a,b, 1983).

5. Aspects of Applied Water Biology

It might astonish that drought-induced losses of productivity occur even in areas with high rainfall like the N.- and NW.-European countries. Even in the per-humid west of Ireland pasture-grass losses in dry years may amount to 1.5 t/ha (BRERETON and KEANE 1982). To assess the regional hygric conditions many attempts are made to calculate (e.g., ASLYNG and HANSEN 1982, GUNSTON and BATCHELOR 1983) and to map (e.g., ALLERUP et al. 1982: suboceanic Denmark, ROCHE 1982: tropical Amazonian area) evapotranspiration data, to compare them with precipitation (e.g., NIEUWOLT 1982: agricultural rainfall index against precipitation irregularities in the tropics), and to give advice for farming improvement (e.g., BLASSE 1982). In the agriculturally highest developed countries, however, economic costs of yield-maximizing irrigation and fertilizing more and more can exceed any economic gain (CABELGUENNE et al. 1982).

Irrigation with saline water should be carefully planned to avoid salt accumulation at the soil/root interface that prevents further water uptake in Cl^--excluding plants (SCHLEIFF 1981, 1983). Large differences in root conductivities to water and to NaCl exist in halotolerant and -intolerant species (OWNBEY and MAHALL 1983). In species, where Cl^- can pass the roots, the soil/root water potential gradient becomes shifted to a lower level, but in these cases the salt toxicity itself may stress the plants (SPIRIG 1981: for urban trees under the impact of de-icing salts). Saline irrigation influences the pool sizes of various amino acids (LABANAUSKAS et al. 1981, DREIER 1983), ground nut (seedling) lipid synthesis is inhibited and the ratio saturated/unsaturated fatty acids decreases (GOPAL and RAO 1982).

Appropriate substrate moisture conditions are needed for fungal infections. Once a phytapathogen has invaded the host, internal plant water relations may become changed. The diseases can modify root water uptake (e.g., *Phymatotrichum* in cotton plants: OLSEN et al. 1983, *Erysiphe* in barley: WALTERS and AYRES 1982), they can hamper water transport by xylem vessel occlusion (e.g., *Verticillium* in tulip poplar: MOREHART and MELCHIOR 1982), or they upset stomatal functioning (e.g., *Erysiphe* in sugar beet: GORDON and DUNIWAY 1982).

6. Habitat Water Relations and Plant Performance

Several studies of the habitat assortment in dioecious species coincide in showing the male plant always to be associated with more xeric habitats (SHEREMET'EV 1981, FOX and HARRISON 1981, ZIMMERMANN and LECHOWICZ 1982). Differences in drought tolerance and stomatal sensitivity split populations into ecological races and initiate the evolutionary speciation (examples: *Hypochoeris radicata* : VAN DE DIJK 1981, *Heliotropium curassavicum*: ROY and MOONEY 1982, *Diplacus aurantiacus*: MOONEY and CHU 1983). Niche seperation between co-occurring species is often based on their habitat water relations: Reciprocal superiority exists in the dry-to-wet soil gradient between the tropcial grasses *Dichanthium annulatum* and *Botriochloa pertusa* (TRIPATHI and GUPTA 1981), between the annual *Salsola kali* and the perennial grasses *Agropyron smithii* and *Bouteloua gracilis* (ALLEN 1982a). The balance between the latter two species depends on a high (favoring *Bouteloua*) or a low variability of water supply (SALA et al. 1982). Distribution differences between *Solidago juncea* and *S. canadensis* along soil moisture gradients are based, on the other hand, more on morphological and phenological peculiarities than on differences in their water-use patterns (POTVIN and WERNER 1983).

Plant water relation parameters with respect to habitat microclimate and soil moisture conditions were determined in the various vegetation zones of the world.

Species distribution and dominance in the N.-American desert and semi-desert areas depend on the specific capabilities of drought adjustment (MONSON and SMITH 1982), of sufficient carbon gain in cacti suffering from water stress and adverse thermoperiods (GREEN and WILLIAMS 1982), or of sufficient seedling establishment, if rainfall occurs (JORDAN and NOBEL 1982). In desert annuals, summer- and winter-growing species can be distinguished, the dormancy in the latter being genetically fixed and its break being environmentally controlled in all cases (FREAS and KEMP 1983). The autecology of representative examples for both types was studied by FORSETH and EHLERINGER (1982) and EHLERINGER (1983). Plant communities of the KaraKum differ from each other by storage and turnover of water, the reserves being stored mainly in the underground phytomass (DEDKOV and GUNIN 1982). According to CO_2-exchange measurements the desert gymnosperm *Welwitschia mirabilis* possesses a C_3 gas-exchange with two-peaked daily courses (at least in younger leaft parts) and considerably high transpiration rates (up to 1.9 m mol $m^{-2}s^{-1}$). The δ^{13} values and the organic acid contents in the tissues (no diurnal oscillations) are, however, in the order of magnitude of CAM plants (WILLERT et al. 1982).

Soil and plant water balances were measured in *Acacia harpophylla* and *Eucalyptus marginata* woodlands in Australia (TUNSTALL and CONNOR 1981, CARBON et al. 1981). In both habitats plants exploit water stored deep in the soil by long-lasting low water potentials.

Measurements of stomatal behavior, transpiration, and water potential in dependence on microclimatic and edaphic conditions in Mediterranean-type areas are given, e.g., by LÖSCH et al. (1982: macchia), CONRAD and RADOSEVICH (1981: chaparral), MILLER et al. (1983: fynbos), AUSSENAC and VALETTE (1982: *Quercus ilex/Qu. pubescens* forests). The presented order is one of decreasing environmental water stress with plant response patterns being quite species specific.

The water-use efficiency of *Taxodium distichum* in its wetland ecosystem is rather good; the stands respond to the local water availability by differences in their leaf area index (0.5 - 3.4 at drier sites, 8.5 in floodplain forests): BROWN (1981).

Special attention has been given to plant water relations at the ecologically very specialized sites of coastal salt marshes and alpine timberline. In the first case the water biology is involved with the salt effects, in the latter with influences of low temperatures. The papers about the salt marshes (USTIN et al. 1982: California, GIURGEVICH and DUNN 1982, ANTLFINGER and DUNN 1983: Georgia, COOPER 1982: North Ireland) equally emphasize the small-scale habitat pattern in the marsh ecotones, where salinity and more or less waterlogged zones determine species distribution. The patterns of alpine herbaceous vegetation along topographic-moisture gradients were studied by OBERBAUER and BILLINGS (1981). Low water potentials were found in plants of ridgetops, high potentials combined with rather high leaf conductances were found in wet meadows at the foots of the slopes. Runoff from snowbanks offers favorable moisture conditions to many alpine forbs and grasses; hence they are not too well adapted to drought conditions (PETERSON and BILLINGS 1982). Gas exchange of timberline conifers is influenced greatly by slope exposure (HASLER 1982). Cuticle development, wax deposits and exposure to wind strongly affect needle life spans and water relations of timberline trees (GÜNTHARDT and WANNER 1982, HADLEY and SMITH 1983).

KÖRNER et al. (1983) investigated the dependence of ecomorphology and leaf conductance on microclimate along an altitudinal transect in montane New Guinea.

In the host-hemiparasite system of mistletoes and trees the stomatal control systems of both species are basically independent, but become linked when transpiration demand exceeds supply. K^+ which is delivered by the xylem sap is accumulated in the hemiparasite, since due to lack of phloem connections to the host a redistribution (that normally is effected via photosynthate transport) cannot occur (GLATZEL 1983).

Waterlogging and Flooding. Flooding-tolerant plants coincide in having an extensive aerenchyma that allows O_2 supply to those plant parts which grow in an O_2-poor environment (SMIRNOFF and CRAWFORD 1983, HALDEMAN and BRÄNDLE 1983, submersed *Lobelia*: SAND-JENSEN and PRAHL 1982). When bulrush rhizomes are brought under anoxia artificially, they produce and release ethanol and lactic acid (DUSS and BRÄNDLE 1982). Production of ethylene is a well-known feature in flooding-intolerant plants under anaerobic conditions (e.g., BRADFORD et al. 1982).

Plants respond to flooding by formation of adventitious roots and hypertrophied lenticels, by turgor loss and stomatal closure in spite of increased water potentials (TANG and KOZLOWSKI 1982, BRADFORD and HSIAO 1982). Proline content is transiently increased (ALONI and ROSENSHTEIN 1982). Roots undergo changes in ultrastructure (CAMPBELL and DREW 1983). With emphasis on the anoxia sensitivity of root mitochondrial membranes, VARTAPETYAN (1982) attempts a unified theory about waterlogging tolerance. However, the discussions on causes of the long-distance effects of root anaerobiosis on shoot water relations (JACKSON and KOWALEWSKA 1983) and the real role that ethanol plays in flooding injury (JACKSON et al. 1982) still remain at the level of speculative hypotheses.

7. Water Relations of Poikilohydric Plants

Cyanobacteria and Algae. There are again several contributions on osmoregulation in unicellular algae. Marine blue-green algae as a group use glucosylglycerol as the major osmoregulatory compound and can be cultivated in seawater media of up to 6% - 11% NaCl (MACKAY et al. 1983). RAVEN (1982) compares the energy costs of cell volume regulation of freshwater algae which have either a wall mechanically resisting the turgor pressure (numerically equivalent to intracellular osmolarity) or a contractile vacuole for osmoregulation. If growth rates are low, wall synthesis will be energetically superior to the contractile vacuole mechanism. Zonation of intertidal macroalgae mirrors their different capabilities in recovering photosynthesis after desiccation (DRING and BROWN 1982).

Fungi and Lichens. Droplet formation at the hyphal tips of *Serpula lacrimans* is a metabolically governed process; a pressure-driven water flow exists along the hyphae (BROWNLEE and JENNINGS 1981). Investigations on lichen ecology have further concentrated on the relations between thallus water content and gas exchange (assessment of CO_2 transport and carboxylation resistances within foliose lichens: GREEN and SNELGAR 1981, SNELGAR et al. 1981; species- and habitat-specifity of optimal thallus hydration: COXSON et al. 1983, TÜRK 1983, data compilation about this matter: LANGE and MATTHES 1981). KAPPEN (1983) measured lichen microclimate in arid Antarctic habitats; he demonstrates a niche convergence where microclimate is similar under different macroclimates in arid regions, and where water availability is decisive (KAPPEN and FRIEDMANN 1983). LANGE and REDON (1983) characterized water relations and CO_2 gas exchange of epiphytic lichens of the Chilean "fog oasis" Fray Jorge.

Poikilohydric Cormophytes. While solute leakage due to membrane lesions occurs from homoiohydric plants under desiccation, this is not very pronounced in *Selaginella lepidophylla* (LEOPOLD et al. 1981). It occurs in this spikemoss (the "false rose of Jericho") only if drying proceeds rapidly, whereas if water loss from this plant is too slow then a de novo protein synthesis is impaired (EICKMEIER 1983). The course of rehydration of *Selaginella* can be divided into three steps (BERGTROM et al. 1982): First of all condensation and density changes can be seen in cytoplasma and chloroplasts, this is follwed by polysome formation and structure changes in chloroplasts, and RubP-carboxylase activity and photosynthesis increase. Finally de novo protein synthesis, so far cytoplasma-directed, becomes organelle-directed (EICKMEIER 1982), resulting in increased grana formation. Photosynthesis increases further and starch deposition in chloroplasts can be seen. The "true rose of Jericho" *(Anastatica hierochuntia)* is desiccation-tolerant to a high degree particularly during the seedling stage (FRIEDMAN et al. 1981). In *Xerophyta villosa* polyribosome content and protein production double transiently during dehydration (TYMMS et al. 1982). When humidity-sensitive bleaching occurs in leaves of *Borya nitida*, PS II activity is initially stimulated, and PS I-mediated electron transfer is reduced. Finally both are lost, chloroplasts swell and thylakoid structure changes (HETHERINGTON and SMILLIE 1982a, HETHERINGTON et al. 1982). The species is temperature-resistent between 0°C and 55°C (chlorophyll fluorescence as test for vitality: HETHERINGTON and SMILLIE 1982b).

Acknowledgement. The critical revision and linguistic improvement of the manuscript by Prof. Dr. L. KAPPEN and Dr. T.G.A. GREEN is gratefully acknowledged.

References

ACKERSON, R.C.: Plant Physiol. 69, 609-613 (1982). - ADEOYE, K.B.; RAWLINS, S.L.: Plant Physiol. 68, 44-47 (1981). - ALLEN, E.B.: Ecology 63, 732-741 (1982a). - ALLEN, M.F.: New Phytol. 91, 191-196 (1982b). - ALLERUP, B., MADSEN, H., RIIS, J.: Nord. Hydrol. 13, 263-278 (1982). - ALONI, B., ROSENSHTEIN, G.: Physiol. Plant. 56, 513-517 (1982). - AMIN, M.: J. Biol. Phys. 10, 103-109 (1982). - ANDERSON, J.E.: Ecology 63, 48-56 (1982). - ANTLFINGER, A.E., DUNN, E.L.: Am. J. Bot. 70, 561-567 (1983). - APELBAUM, A., YANG, S.F.: Plant Physiol. 68, 594-596 (1981). - APPLEBY, R.F., DAVIES, W.J.: Ann. Bot. 52, 459-468 (1983a); - Oecologia 56, 30-40 (1983b). - ASLYNG, H.C., HANSEN, S.: Water Balance and Crop Production Simulation. Copenhagen: Roy. Vet. Agric. Univ. Publ. 1982. - AUSSENAC, G., VALETTE, J.C.: Ann. Sci. For. 39, 41-62 (1982).

BAKRADZE, N.G., MOISTRAPISHVILI, K.M., KESHELASHVILI, L.V.: Biofizika 26, 119-123 (1981). - BECWAR, M.R., RAJASHEKAR, C., HANSEN-BRISTOW, K.E., BURKE, M.J.: Plant Physiol. 68, 111-114 (1981). - BELL, C.J., INCOLL, L.D.: J. Exp. Bot. 32, 1125-1134 (1981a); - ibid. 32, 1135-1142 (1981b). - BERGTROM, G., SCHALLER, M., EICKMEIER, W.G.: J. Ultrastruct. Res. 78, 269-282 (1982). - BERKOWITZ, G.A., GIBBS, M.: Plant Physiol. 70, 1535-1540 (1982); - ibid. 71, 905-911 (1983). - BODDI, B., RAKASZ, E., LANG, F.: Photobiochem. Photobiophys. 5, 27-34 (1983). - BLASSE, W.: Wiss. Z. Humboldt-Univ. Berl. Math.-Naturwiss. Reihe 31, 45-52 (1982). - BRADFORD, K.J., HSIAO, T.C.: Plant Physiol. 70, 1508-1513 (1982). - BRADFORD, K.J., HSIAO, T.C., YANG, Sh.F.: Plant Physiol. 70, 1503-1507 (1982). - BRERETON, A.J., KEANE, T.: Ir. J. Agric. Res. 21, 227-248 (1982). - BRIENS, M., LARHER, F.: Plant Cell Environ. 5, 287-292 (1982). - BROWN, P.H., OUTLAW, W.H.: Plant Physiol. 70, 1700-1703 (1982). - BROWN, S.: Ecol. Monogr. 51, 403-428 (1981). - BROWNLEE, C., JENNINGS, D.H.: Trans. Br. Mycol. Soc. 77, 33-40 (1981). - BUNCE, J.A.: Oecologia 54, 233-235 (1982); - ibid. 57, 262-265 (1983).

CABELGUENNE, M., MARTY, J.R., HILAIRE, A.: Agronomie 2, 567-576 (1982). - CAMPBELL, R., DREW, M.C.: Planta 157, 350-357 (1983). - CARBON, B.A., BARTLE, G.A., MURRAY, A.M.: Austr. For. Res. 11, 191-200 (1981). - CARY, J.W.: Photosynth. Res. 2, 185-194 (1981). - CAVALIERI, A.J.: Oecologia 57, 20-24 (1983). - CHAPADOS, C., LEBLANC, R.M.: Biophys. Chem. 17, 211-244 (1983). - CHETAL, S., WAGLE, D.S., NAINAWATEE, H.S.: Plant Sci. Lett. 29, 273-278 (1983). - CLARKE, J.M., McCAIG, T.N.: Crop Sci. 22, 503-506 (1982). - CLAYTON-GREENE, K.A.: Oecologia 57, 368-373 (1983). - COHEN, Y., FUCHS, M., COHEN, S.: J. Exp. Bot. 34, 451-460 (1983). - CONRAD, S.G., RADOSEVICH, S.R.: For. Sci. 27, 627-639 (1981). - COOPER, A.: New Phytol. 90, 263-276 (1982). - CORNISH, P.S.: Plant Soil 59, 119-126 (1981). - COSGROVE, D.J.: Plant Physiol. 68, 1439-1446 (1981). - COSGROVE, D.J., STEUDLE, E.: Planta 153, 343-350 (1981). - COXSON, D.S., BROWN, D., KERSHAW, K.A.: New Phytol. 93, 247-260 (1983).

DEDKOV, V.P., GUNIN, P.D.: Bot. Zh. 67, 1256-1264 (1982). - DESCHAMP, P.A., COOKE, T.J.: Science 219, 505-507 (1983). - DONKIN, M.E., TAFFS, J., MARTIN, E.S.: Planta 155, 416-422 (1982). - DREIER, W.: Biol. Plant. 25, 81-87 (1983). - DREW, A.P., BAZZAZ, F.A.: Oecologia 54, 76-79 (1982). - DRING, M.J., BROWN, F.A.: Mar. Ecol. Progr. Ser. 8, 301-308 (1982). - DUNGEY, N.O., DAVIES, D.D.: J. Exp. Bot. 33, 12-20 (1982). - DUSS, F., BRÄNDLE, R.: Flora 172, 217-222 (1982).

EDWARDS, W.R.N., JARVIS, P.G.: Plant Cell Environ. 5, 271-278 (1982); - ibid. 6, 255-260 (1983). - EHLERINGER, J.: Oecologia 57, 107-112 (1983). - EICKMEIER, W.G.: Oecologia 58, 115-120 (1983); - Plant Physiol. 69, 135-138 (1982). - ELLER, B.M., GROBBELAAR, N.: J. Exp. Bot. 33, 366-375 (1982). - EVANS, D.D., THAMES, J.L. (ed.): Water in Desert Ecosystems. Stroudsburg: Dowden, Hutchinson & Ross 1981.

FAIZ, S.M.A., WEATHERLEY, P.E.: New Phytol. 92, 333-344 (1982). - FANJUL, L., JONES, H.G.: Planta 154, 135-138 (1982). - FARQUHAR, G.D., SHARKEY, T.D.: Annu. Rev. Plant Physiol. 33, 317-345 (1982). - FIELD, C., BERRY, J.A., MOONEY, H.A.: Plant Cell Environ. 5, 179-186 (1982). - FITZSIMONS, P.J., WEYERS, J.B.D.: J. Exp. Bot. 34, 55-66 (1983). - FORK, D.C., OQUIST, G.: Z. Pflanzenphysiol. 104, 385-394 (1981). - FORSETH, I.N., EHLERINGER, J.R.: Oecologia 54, 41-49 (1982). - FOX, J.F., HARRISON, T.: Oecologia 49, 233-235 (1981). - FREAS, K.E., KEMP, P.R.: J. Ecol. 71, 211-218 (1983). - FRIEDMAN, J., STEIN, Z., RUSHKIN, E.: Oecologia 51, 400-403 (1981). - FUKUTOKU, Y., YAMADA, Y.: Plant Cell Physiol. 22, 1397-1404 (1981).

GEISLER, G.: Z. Acker-Pflanzenb. 151, 99-117 (1982). - GELLER, G.N., SMITH, W.K.: Oecologia 53, 227-234 (1982). - GIURGEVICH, J.R., DUNN, E.L.: Oecologia 52, 404-410 (1982). - GLATZEL, G.: Oecologia 56, 193-201 (1983). - GOPAL, G.R., RAO, G.R.: Z. Pflanzenphysiol. 106, 1-14 (1982). - GORDON, T.R., DUNIWAY, J.M.: Phytopathology 72, 723-726 (1982). - GOTOW, K., KONDO, N., SYONO, K.: Plant Cell Physiol. 23, 1063-1070 (1982). - GOTTSCHLICH, D.E., SMITH, A.P.: Oecologia 53, 418-420 (1982). - GRANT, R.F., SAVAGE, M.J., LEA, J.D.: S. Afr. J. Sci. 77, 398-400 (1981). - GREEN, J.M., WILLIAMS, G.J.: Oecologia 52, 43-48 (1982). - GREEN, T.G.A., SNELGAR, W.P.: Physiol. Plant. 52, 411-416 (1981). - GRIFFITHS, J.H., JARVIS, P.G.: J. Exp. Bot. 32, 1157-1168 (1981). - GUNSTON, H., BATCHELOR, C.H.: Agric. Water Manage. 6, 65-78 (1983). - GÜNTHARDT, M.S., WANNER, H.: Bot. Helv. 92, 47-60 (1982). - GUSTA, L.V., RAJASHEKAR, C., CHEN, P.M., BURKE, M.J.: Cryo. Lett. 3, 27-34 (1982).

HADLEY, J.L., SMITH, W.K.: Arct. Alp. Res. 15, 127-136 (1983). - HALDEMANN, C., BRÄNDLE, R.: Physiol. Veg. 21, 109-114 (1983). - HAMPP, R., OUTLAW, W.H., TARCYNSKI, M.C.: Plant Physiol. 70, 1582-1585 (1982). - HANSON, A.D., HITZ, W.D.: Annu. Rev. Plant Physiol. 33, 163-203 (1982). - HASLER, R.: Oecologia 54, 14-22 (1982). - HENSON, I.E., MAHALAKSHI, V., BIDINGER, F.R., ALAGARSWAMY, G.: J. Exp. Bot. 32, 899-910 (1981). - HERBORN, A., LENZ, F.: Gartenbauwissenschaft 46, 181-186 (1981). - HETHERINGTON, S.E., SMILLIE, R.M.: Aust. J. Plant Physiol. 9, 587-600 (1982a); - Planta 155, 76-81 (1982b). - HETHERINGTON, S.E., HALLAM, N.D., SMILLIE, R.M.: Aust. J. Plant Physiol. 9, 601-610 (1982). - HOCHMAN, Z.: Field Crops Res. 5, 55-68 (1982). - HOFFMAN, N.E., LIU, Y., YANG, S.F.: Planta 157, 518-523 (1983). - HOLTUM, J., WINTER, K.: Planta 155, 8-16 (1982). - HONG, S.G., SUCOFF, E.: Plant Physiol. 69, 697-700 (1982). - HUBER, W., EDER, A.: Biochem. Physiol. Pflanz. 177, 184-191 (1982).

IKE, I.F., THURTELL, G.W.: Physiol. Plant. 52, 257-262 (1981).

JACKSON, M.B., KOWALEWSKA, A.K.B.: J. Exp. Bot. 34, 493-506 (1983). - JACKSON, M.B., HERMAN, B., GOODENOUGH, A.: Plant Cell Environ. 5, 163-172 (1982). - JAGELS, R.: Am. J. Bot. 70, 327-333 (1983). - JANE, G.T., GREEN, T.G.A.: Oecologia 57, 380-390 (1983). - JENSEN, C.R.: Irrig. Sci. 3, 111-121 (1982). - JEWER, P.C., INCOLL, L.D.: Planta 153, 317-318 (1981). - JOHNSON, C.B. (ed.): Physiological Processes Limiting Plant Productivity. London, Boston, Sidney, Wellington, Durban, Toronto: Butterworths 1981. - JORDAN, P.W., NOBEL, P.S.: Bot. Gaz. 143, 511-517 (1982).

KAISER, W.M.: Planta 154, 538-545 (1982). - KAISER, W.M., KAISER, G., PRACHUAB, P.K., WILDMAN, S.G., HEBER, U.: Planta 153, 416-422 (1981a). - KAISER, W.M., KAISER, G., SCHÖNER, S., NEIMANIS, S.: Planta 153, 430-435 (1981b). - KANNANGARA, T., SEETHARMA, N., DURLEY, R.C., SIMPSON, G.M.: Can. J. Plant Sci. 63, 147-156 (1983). - KAPPEN, L.: Ber. Dtsch. Bot. Ges. 96, 87-101 (1983). - KAPPEN, L., FRIEDMAN, I.: Verh. Ges. Ökol. 10, 517-519 (1983). - KARLIC, H., RICHTER, H.: Flora 173, 143-150 (1983). - KARLSSON, P.E., HOEGLUND, H.O., KLOCKARE, R.: Physiol. Plant. 57, 417-421 (1983). - KASAMO, K.: Plant Cell Physiol. 22, 1257-1268 (1981). - KATERJI, N., HALLAIRE, M., PERRIER, A., DURRAND, R.: Acta Oecol. Oecol. Plant 4, 11-26 (1983). - KAUFMANN, M.R.: Plant Physiol. 69, 1018-1022 (1982); - ibid. 69, 1023-1026 (1982). - KÖRNER, C., ALLISON, A., HILSCHER, H.: Flora 174, 91-135 (1983). - KOZLOWSKI, T.T.: Water Deficits and Plant Growth. VI. Woody Plant Communities. New York, San Francisco, London: Academic Press 1981. - KRASAVTSEV, O.A., KHVALIN, N.N.: Fiziol. Rast. 29, 437-446 (1982). - KÜPPERS, M., HALL, A.E., SCHULZE, E.D.: Oecologia 52, 116-120 (1982).

LABANAUSKAS, C.K., STOLZY, L.H., HANDY, M.F.: Plant Soil 59, 299-316 (1981). - LAFFRAY, D., LOUGUET, P., GARREC, J.P.: J. Exp. Bot. 35, 771-782 (1982). - LANGE, O.L., MATTHES, U.: Photosynthetica 15, 555-574 (1981). - LANGE, O.L., REDON, J.: Flora 174, 245-284 (1983). - LANGE, O.L., NOBEL, P.S., OSMOND, C.B., ZIEGLER, H. (ed.): Encyclopedia of Plant Physiology, N.S. Vol. 12 A,B,C,D, Physiological Plant Ecology. Berlin, Heidelberg, New York: Springer 1981/1982a/1983a/1983b. - LANGE, O.L., TENHUNEN, J.D., BRAUN, M.: Flora 172, 563-579 (1982b). - LEACH, J.E.T., WOODHEAD, T., DAY, W.M.: Agric. Meteorol. 27, 257-264 (1982). - LEOPOLD, A.C., MUSGRAVE, M.E., WILLIAMS, K.M.: Plant Physiol. 68, 1122-1225 (1981). - LEUNING, R.: Plant Cell Environ. 6, 181-194 (1983). - LEVY, Y., SYVERTSEN, J.P., NEMEC, S.: New Phytol. 93, 61-66 (1983). - LINDOW, S.E., ARNY, D.C., UPPER, C.D.: Plant Physiol. 70, 1084-1089 (1982a). - LINDOW, S.E., HIRANO, S.S., BARCHET, W.R., ARNY, D.C., UPPER, C.D.: Plant Physiol. 70, 1090-1093 (1982b). - LÖSCH, R., TENHUNEN, J.D., PEREIRA, J.S., LANGE, O.L.: Flora 172, 138-160 (1982).

MACKAY, M.A., NORTON, R.S., BOROWITZKA, L.J.: Mar. Biol. 73, 301-308 (1983). - MACROBBIE, E.A.C.: Philos. Trans. R. Soc. Lond. B. Biol. Sci. 299, 469-482 (1982). - MANUIL'SKII, V.D.: Dokl. Akad. Nauk Ukr. SSR. Ser. B. Geol. Khim. Biol. Nauki 1982, 62-65 (1982). - MATORIN, D.N., ORTOIDZE, T.V., NIKOLAEV, G.M., VENEDIKTOV, P.S., RUBIN, A.B.: Photosynthetica 16, 226-233 (1982). - MAYORAL, M.L., ATSMON, D., SHIMSHI, D., GROMET-ELHANAN, Z.: Austr. J. Plant Physiol. 8, 385-394 (1981). - McCLENDON, J.H.: Am. J. Bot. 68, 1263-1268 (1981). - McWILLIAM, J.R., KRAMER, P.J., MUSSER, R.L.: Aust. J. Plant Physiol. 9, 343-352 (1982). - MEDINA, E.: Z. Pflanzenphysiol. 107, 251-258 (1982). - MEIDNER, H.: J. Exp. Bot. 33, 355-359 (1982); - ibid. 34, 1606-1618 (1983). - MEINZER, F.C.: Oecologia 54, 236-242 (1982a); - ibid. 54, 270-274 (1982b). - MELKONIAN, J.J., WOLFE, J., STEPONKUS, P.L.: Crop Sci. 22, 116-123 (1982). - MENGEL, K., ARNEKE, W.W.: Physiol. Plant. 54, 402-408 (1982). - MICHELENA, V.A., BOYER, J.S.: Plant Physiol. 69, 1145-1149 (1982). - MILLER, P.C., MILLER, J.M., MILLER, P.M.: Oecologia 56, 392-396 (1983). - MILNE, R., FORD, E.D., DEANS, J.D.: For. Ecol. Manage. 5, 1-26 (1983). - MONSON, R.K., SMITH, S.D.: Ecology 63, 113-123 (1982). - MONTEITH, J., WEBB, C. (ed.): Soil Water and Nitrogen in Mediterranean-type Environments. Development in Plant and Soil Science, Vol. 1. The Hague, Boston, London: Nijhoff/Junk 1981. - MOONEY, H.A., CHU, C.: Oecologia 57, 148-150 (1983). - MOONEY, H.A., FIELD, C., WILLIAMS, W.E., BERRY, J.A., BJÖRKMAN, O.: Oecologia 57, 38-42 (1983). - MOREHART, A.L., MELCHIOR, G.L.: Can. J. Bot. 60, 201-209 (1982). - MORISON, J.I.L. JARVIS, P.G.: Plant Cell Environ. 6, 95-102 (1983a); - ibid. 6, 103-110 (1983b). - MUNNS, R., GREENWAY, H., SETTER, T.L., KUO, J.: J. Exp. Bot. 34, 144-155 (1983).

NEJIDAT, A., ITAI, C., ROTH-BEJERANO, N.: Physiol. Plant. 57, 367-370 (1983). - NIEUWOLT, S.: Agric. Environ. 7, 135-148 (1982).

OBERBAUER, S.F., BILLINGS, W.D.: Oecologia 50, 325-331 (1981). - OLSEN, M.W., MISAGHI, I.J., GOLDSTEIN, D., HINE, R.B.: Phytopathology 73, 213-216 (1983). - OQUIST, G., FORK, D.C.: Physiol. Plant. 56, 56-62 (1982a); - ibid. 56, 63-68 (1982b). - OUTLAW, W.H., Jr.: Recent Adv. Phytochem. 16, 185-222 (1982); - Physiol. Plant. 59, 302-311 (1983). - OUTLAW, W.H., Jr., MANCHESTER, J., BROWN, P.H.: Plant Physiol. 68, 1047-1051 (1981). - OUTLAW, W.H., Jr., MANCHESTER, J., ZENGER, V.E.: Can. J. Bot. 60, 240-244 (1982). - OWNBEY, R.S., MAHALL, B.E.: Physiol. Plant. 57, 189-195 (1983).

PAHLICH, E., JÄGER, H.-J., HORZ, M.: Z. Pflanzenphysiol. 105, 475-478 (1982). - PALEG, L.G., ASPINALL, D. (ed.): The Physiology and Biochemistry of Drought Resistance in Plants. Sidney, New York, London, Toronto, S.Francisco: Academic Press 1981. - PASICHNYI, A.P., PONOMAREVA, I.D., TSEPKOW, G.V.: Fiziol. Biokhim. Kul't. Rast. 12, 548-553 (1980). - PEMADASA, M.A.: New Phytol. 89, 373-384 (1981); - ibid. 92, 21-30 (1982a); - J. Exp. Bot. 33, 92-99 (1982b); - New Phytol. 93, 13-24 (1983a); - ibid. 94, 201-210 (1983b). - PESCHL, A.: Flora 172, 427-447 (1982a); - ibid. 172, 449-461 (1982b); - ibid. 173, 33-47 (1983). - PETERSON, K.M., BILLINGS, W.D.: Arct. Alp. Res. 14, 189-194 (1982). - PICKARD, W.F.: Math. Biosci. 56, 111-128 (1981); - ibid. 60, 73-88 (1982a); - Plant Physiol. 69, 971-974 (1982b). - PIERCE, M., RASCHKE, K.: Planta 153, 156-165 (1981). - POSPISILOVA, J., SOLAROVA, J.: Water in Plants Bibliography, Vol. 7 (1981). The Hague, Boston, London: Junk 1982; - Water in Plants Bibliography, Vol. 8 (1982). The Hague, Boston, Lancaster: Junk 1983. - POTVIN, M.A., WERNER, P.A.: Oecologia 56, 148-152 (1983). - PRIESTLEY, D.A., BRUINSMA, J.: Physiol. Plant. 56, 303-311 (1982).

QUAMME, H.A., CHEN, P.M., GUSTA, L.V.: J. Am. Soc. Hortic. Sci. 107, 299-304 (1982).

RADULOVICH, R.A., PHENE, C.J., DAVIS, K.R., BROWNELL, J.R.: Agron J. 74, 383-385 (1982). - RAMOS, C., HALL, A.E.: Photosynthetica 16, 343-355 (1982). - RAO, I.M., ANDERSON, L.E.: Plant Physiol. 71, 451-455 (1983a); - ibid. 71, 456-459 (1983b). - RAO, R.C.N., KRISHNASASTRY, K.S., UDAYAKUMAR, M.: Plant Sci. Lett. 23, 327-334 (1981a); - ibid. 23, 335-340 (1981b). - RAVEN, J.A.: New Phytol. 92, 1-20 (1982). - ROBERTS, S.W., STRAIN, B.R., KNOERR, K.R.: Physiol. Plant. 52, 245-250 (1981). - ROCHE, M.A.: Cah. ORSTOM Ser. Hydrol. 19, 37-44 (1982). - ROTH-BEJERANO, N., ITAI, C.: Physiol. Plant. 52, 201-206 (1981). - ROTH-BEJERANO, N., NEJIDAT, A., ITAI, C.: Physiol. Plant. 56, 80-83 (1982). - ROY, J., BERGER, A.: New Phytol. 93, 43-52 (1983). - ROY, J., MOONEY, H.A.: Oecologia 52, 370-375 (1982).

SACK, F.D., PAOLILLO, D.J., Jr.: Protoplasma 116, 1-13 (1983). - SAKAI, A.: Plant Cell Physiol. 23, 1219-1228 (1982). - SALA, O.E., LAUENROTH, W.K., REID, C.P.P.: J. Appl. Ecol. 19, 647-658 (1982). - SAMIEV, Kh.S., MARFINA, K.G.: Fiziol. Biokhim. Kul't. Rast. 14, 346-349 (1982). - SANCHEZ, R.A., HALL, A.J., TRAPANI, N., COHEN DE HUNAU, R.: Photosynth. Res. 4, 35-48 (1983). - SAND-JENSEN, K., PRAHL, C.: New Phytol. 91, 103-120 (1982). - SANDERSON, J.: J. Exp. Bot. 34, 240-253 (1983). - SANDS, R., FISCUS, E.L., REID, C.P.P.: Aust. J. Plant Physiol. 9, 559-570 (1982). - SAVAGE, M.J., CASS, A., DE JAEGER, J.M.: Plant Physiol. 69, 526-530 (1982). - SCHLEIFF, U.: Agric. Water Manage. 3, 317-325 (1981); - Z. Pflanzenernähr. Bodenk. 145, 436-447 (1982); - Z. Acker- Pflanzenb. 152, 9-18 (1983). - SCHNABL, H.: Planta 152, 307-313 (1981). - SCHNABL, H., ELBERT, C., KRÄMER, G.: J. Exp. Bot. 33, 996-1003 (1982). - SCHULT, S., DÖRFFLING, K.: Physiol. Plant. 53, 487-490 (1981). - SCHULZE, E.-D.,

HALL, A.E., LANGE, O.L., WALZ, H.: Oecologia 53, 141-145 (1982). - SELLS, G.D., KOEPPE, D.E.: Plant Physiol. 68, 1058-1063 (1981). - SHARKEY, T.D., BADGER, M.R.: Planta 156, 199-206 (1982). - SHARPE, P.J.H., ZEIGER, E.: Chemiosmotic hypothesis of ion transport in guard cells, 47-64. In: Integrated View of Guard Cells, ed. C. ROGERS. Symp. South. Sect. Soc. Plant Physiol. Rockville, MD: Am. Soc. of Plant Physiologists 1981. - SHEIN, E.V.: Biol. Nauki 1982, 90-94 (1982). - SHEREMET'EV, S.N.: Bot. Zh. 66, 565-572 (1981). - SHERIFF, D.W.: Ann. Bot. 50, 535-548 (1982). - SILAEVA, A.M., TKACHUK, E.S.: Fiziol. Biokhim. Kul't. Rast. 14, 148-152 (1982). - SILCOCK, R.G., WILSON, D.: New Phytol. 90, 201-208 (1982). - SIMPSON, G.M. (ed.): Water Stress in Plants. New York: Praeger 1981. - SMIRNOFF, N., CRAWFORD, R.M.M.: Ann. Bot. 51, 237-250 (1983). - SNAITH, P.J., MANSFIELD, T.A.: J. Exp. Bot. 33, 360-365 (1982). - SNELGAR, W.P., GREEN, T.G.A., BELTZ, C.K.: Physiol. Plant. 52, 417-422 (1981). - SOVONICK-DUNFORD, S., FERRIER, J.M., DAINTY, J.: Ann. Bot. 51, 27-38 (1983). - SPANSWICK, R.M., LUCAS, W.J., DAINTY, J. (ed.): Plant Membrane Transport. Current conceptual issue. Amsterdam, New York, Oxford: Elsevier/North-Holland Biomedical Press 1980. - SPIRIG, A.: Veröff. Geobot. Inst. ETH, Stiftg. Rübel, Zürich 74, 1-61 (1981). - STEUDLE, E., ZIMMERMANN, U., ZILLIKENS, J.: Planta 154, 371-380 (1982).

TANG, Z.C., KOZLOWSKI, T.T.: Can. J. For. Res. 12, 196-202 (1982). - TAYLOR, A.G., MOTES, J.E., KIRKHAM, M.B.: J. Am. Hortic. Sci. 107, 387-390 (1982). - TENHUNEN, J.D., LANGE, O.L., JAHNER, D.: Oecologia 55, 165-169 (1982). - TESKEY, R.O., HINCKLEY, T.M.: Physiol. Plant. 52, 363-369 (1981). - THOMAS, J.E., DAVIDSON, D.: Caryologia 35, 191-204 (1982). - TOMOS, A.D., STEUDLE, E., ZIMMERMANN, U., SCHULZE, E.-D.: Plant Physiol. 68, 1135-1143 (1981). - TRAVIS, A.J., MANSFIELD, T.A.: J. Exp. Bot. 32, 1169-1180 (1981). - TRIPATHI, R.S., GUPTA, G.P.: Trop. Ecol. 22, 1-9 (1981). - TÜRK, R.: Phyton 23, 1-18 (1983). - TUNSTALL, B.R., CONNOR, D.J.: Austr. J. Bot. 29, 311-320 (1981). - TYERMAN, S.D.: Plant Physiol. 69, 957-965 (1982). - TYERMAN, S.D., STEUDLE, E.: Aust. J. Plant. Physiol. 9, 461-480 (1982). - TYMMS, M.J., GAFF, D.F., HALLAM, N.D.: J. Exp. Bot. 33, 332-343 (1982).

UNGER, K., STÖCKER, G. (ed.): Biophysikalische Ökologie und Ökosystemforschung. Berlin: Akademie-Verlag 1981. - UPADHYAYA, S.K., RAND, R.H., COOKE, J.R.: J. Theor. Biol. 101, 425-440 (1983). - USTIN, S.L., PEARCY, R.W., BAYER, D.E.: Bot. Gaz. 143, 368-373 (1982).

VAN DE DIJK, S.J.: Plant Soil 63, 149-164 (1981). - VAN VOLKENBURGH, E., CLELAND, R.E.: Planta 153, 572-577 (1981). - VAPAAVUORI, E.M., VALANNE, N.S.: Photosynthetica 16, 1-6 (1982). - VARTAPETYAN, B.B.: Fiziol. Rast. 29, 985-994 (1982). - VAUGHN, K.C., OUTLAW, W.H.: Plant Physiol. 71, 420-424 (1983).

WALLACE, J.S., CLARK, J.A., McGOWAN, M.: J. Agric. Sci. 100, 581-590 (1983). - WALTERS, D.R., AYRES, P.G.: Physiol. Plant Pathol. 20, 275-284 (1982). - WARING, R.H., SCHROEDER, P.E., OREN, R.: Can. J. For. Res. 12, 556-560 (1982). - WEILER, E.W., SCHNABL, H., HORNBERG, C.: Planta 154, 24-28 (1982). - WEIMBERG, R., LERNER, H.R., POLJAKOFF-MAYBER, A.: Physiol. Plant. 55, 5-10 (1982). - WEYERS, J.B.D., FITZ-SIMONS, P.J., MANSEY, G.M., MARTIN, E.S.: Physiol. Plant. 58, 331-339 (1983). - WILLERT, D.J. von, ELLER, B.M., BRINCKMANN, E., BAASCH, R.: Oecologia 55, 21-29 (1982). - WILLIAMS, W.E.: Plant Cell Environ. 6, 145-152 (1983). - WILLMER, C.M., RUTTER, J.C., MEIDNER, H.: J. Exp. Bot. 34, 507-513 (1983). - WILSON, J.R., LUDLOW, M.M.: Austr. J. Plant Physiol. 10, 15-24 (1983). - WOODHOUSE, R.M., NOBEL, P.S.: Am. J. Bot. 69, 135-140 (1982).

YEE, V.F., ROST, T.L.: Cytologia 47, 615-624 (1982).

ZEEVAART, J.A.D.: Plant Physiol. 71, 477-481 (1983). - ZEIGER, E.: Annu. Rev. Plant Physiol. 34, 441-475 (1983). - ZEIGER, E., FIELD, C.: Plant Physiol. 70, 370-375 (1982). - ZEIGER, E., FIELD, C., MOONEY, H.A.: Stomatal opening at dawn: Possible roles of the blue light response in nature, 391-407. In: Plants and the Daylight Spectrum, ed. H. SMITH. New York: Academic Press 1981. - ZIMMERMANN, J.K., LECHOWICZ, M.J.: Oecologia 53, 305-309 (1982). - ZIMMERMANN, U., BÜCHNER, K.-H., BENZ, R.: J. Membr. Biol. 67, 183-198 (1982). - ZUR, B., BOOTE, K.J., JONES, J.W.: J. Exp. Bot. 32, 1181-1192 (1981). - ZUR, B., JONES, J.W., BOOTE, K.J., HAMMOND, L.C.: Agron. J. 74, 99-105 (1982).

Dr. RAINER LÖSCH
Botanisches Institut
Olshausenstr. 40-60
D-2300 Kiel, FRG

II. Mineral Nutrition: Vacuoles and Tonoplasts

By DAVID T. CLARKSON and ULRICH LÜTTGE

1. Introduction

In a much respected review published in 1976, MATILE and WIEMKEN (1976) wrote: "Although a number of biochemical properties of vacuoles have been investigated, the mechanisms of interaction between cytoplasm and vacuole remain largely to be elucidated."

Since that time a number of technical advances have been made, especially in understanding the coupling of tonoplast functions to metabolic energy (see Sect. 2.f). This characterization of tonoplast ATPases and the establishment of proton-electrochemical gradients across the tonoplast associated with their activity has yet rarely been linked directly to physiological processes in vivo.

Nevertheless, the quantitative aspects of the interactions between cytoplasm and vacuole begin to attract increasing interest. The compartmentation of ions and metabolites in relation to their regulatory functions in cell tugor maintenance, intermediary metabolism and membrane transport has become a major research preoccupation in recent years. In higher plant cells the large volume fraction and the relatively inert metabolic activity of vacuoles suggest that they are storage compartments. From this it follows that they should provide a means whereby cytoplasm is buffered against damaging changes in the concentration of ions and metabolites. Before these assumptions can be validated, reliable techniques have to be developed for measuring the distribution of materials in vacuole and cytoplasm, estimating the fluxes of materials between them and understanding how these fluxes are regulated.

2. Experimental Approaches

a) Direct Access

Direct separation and analysis of the vacuolar and cytoplasmic compartments have long been practiced with giant coenocytes of algae. While in exceptional cases of giant idioblastic cells in higher plants it may be possible to obtain vacuolar sap of individual cells for direct analysis (e.g., RYGOL and LÜTTGE 1983), algal coenocytes remain the only objects readily allowing direct access to the vacuolar compartment. The preparation of tonoplast-free internodal cells of characean algae in conjunction with perfusion techniques is used to modify the cytoplasm and to understand plasmalemma functions (TAZAWA et al. 1976, SHIMMEN and TAZAWA 1977, 1980a, KIKUYAMA et al. 1979, TAZAWA and SHIMMEN 1982, MIMURA et al. 1983). Its potential for understanding the

Progress in Botany, Vol. 46

functions of vacuoles and tonoplasts, however, is also evident (KIYOSAWA and TAZAWA 1977, SHIMMEN and TAZAWA 1980b).

b) Washout or Efflux Analysis

This is perhaps the oldest and most conventional approach to compartmental analysis. It depends on the rate of exchange between labeled and unlabeled forms of a substance across barriers of differing conductivity between separate compartments (PITMAN 1963). The approach was originally developed with giant coenocytes of algae, where it could be checked by direct analysis of the vacuolar and the cytoplasmic compartment (MAC ROBBIE and DAINTY 1958).

Ideally the tissues analyzed should be at flux equilibrium, but the technique has been applied widely to root and leaf tissues where net accumulation and growth are occurring. Workers have ignored the constraints of theory in many cases: recently THAINE (1984) has produced a modified approach to the analysis of tracer experiments, which is valid where cells are not in a steady state. The method involved the concept of a unidirectional rate constant. Further refinements of analysis allow fluxes of tracer into the xylem to be accounted for (WALKER and PITMAN 1976, JESCHKE 1982, JESCHKE and JAMBOR 1981, JESCHKE and STELTER 1983, STELTER and JESCHKE 1983). Earlier flux analyses were concerned largely with nonmetabolized ions, but recently further analyses have been made with SO_4^{2-} (THOIRON et al. 1981, CRAM 1983a,b), and $H_2PO_4^-$ (LEFEBVRE and CLARKSON 1984a,b).

c) Microphobe Analysis

The undoubted attraction of this technique is that (in theory at least) it can provide direct chemical analysis on specific areas of sections of plant material, thus adjacent volumes of vacuole and cytoplasm might be compared in the same cell. Practical difficulties abound to prevent the achievement of this objective. Some procedures involved in specimen preparation introduce great uncertainties about redistribution of cell contents. Freeze-substition of water for some nonpolar, resin solubilizing solvent (ether or acetone) has been developed by HARVEY et al. (1976); impetus to this work has been given by the requirements for precise geometrical relationships between the specimen surface and the X-ray detector in quantitative work. Resin-embedded thin sections meet these requirements and have the advantage that the morphology of the cytoplasm is easily recognized. Against such advantages must be set the possible dislocation of polar solutes from vacuoles as ice becomes replaced by the substituting solvent. To avoid this problem, several groups prefer to examine freeze-fractured surfaces directly by holding the tissue at or near liquid N_2 temperature on a cold stage (PITMAN et al. 1981, VAN STEVENINCK et al. 1982a,b). Such surfaces have irregular topography and density, and structures are very poorly resolved. The results are therefore only semiquantitative but redistribution is unlikely.

Compartmentation of K^+, Na^+, and Cl^- was studied in the red alga *Porphyra umbilicalis* (WIENCKE et al. 1983). In cells with large vacuoles the area of cytoplasm is so small that the measuring beam at its minimal practical diameter is frequently wider than the cytoplasm (PITMAN et al. 1981). Vacuolar concentrations can, however, be measured readily and such studies are already showing up very interesting differences in vacuolar contents of different cell types in roots and leaves (VAN STEVENINCK et al. 1982a,b, GORHAM and WYN JONES 1983, STOREY et al.

1983a,b), for instance outer cortical cell vacuoles contained 30-fold higher Cl^- concentrations as those in xylem parenchyma in roots of *Lupinus luteus* (VAN STENVENINCK et al. 1982b).

Attempts to immobilize ions by precipitation techniques appear to have been largely abandoned because of the unspecific nature of the reagents or redistribution (see VAN IREN et al. 1979, HARVEY et al. 1979).

d) Differential Extraction

A number of methods have been devised for creating pores or instability in plasma membranes so that cytosol contents can be quickly leached out or extracted. The contents of the vacuole can be determined after a second extraction or osmotic shock. Most of these methods were developed for microorganisms and their application to multicellular tissues is problematical (see LERNER et al. 1983a). The plasma membranes of yeast cells become extremely leaky when treated with polybasic agents, such as cytochrome c (SCHLENK et al. 1970), diethylaminoethyl-dextran (HUBER-WÄLCHLI and WIEMKEN 1979) and poly-L-lysine (LERNER and REUVENI 1982, LERNER et al. 1983a). These large molecules are likely to be excluded from the pores in walls which comprise free space of roots (see CARPITA et al. 1979).

Smaller pore-forming antibiotic molecules do seem to penetrate the free space of callus and root tissues and may have more general application if their effects on plasma membrane and tonoplast can be separated. Polymyxin B and Polymyxin E and Fillipin were found to depolarize partially the membrane potential of cortical cells of excised *Sorghum* roots by about 40 mV (LERNER et al. 1983a) and created pores which released adenylates into the outer medium. Polymyxin-B-treated tobacco released K^+ into the medium in two quite distinct phases; initially about 10% of cell K^+ was released followed by a much more rapid release of the bulk cell K^+. These two phases may correspond to cytoplasmic and vacuolar compartments (WEIMBERG et al. 1983). In general, such methods, where both plasma membrane and tonoplast are affected by the poreforming agent, have little to recommend them for ion-compartmentation study in comparison with conventional tracer exchange techniques. Their value may be in release of metabolites which do not normally cross the plasma membrane.

Smaller pore-forming agents such as dimethylsulfoxide (DELMER 1979), toluene (WEIMBERG et al. 1981, GORHAM and WYN JONES 1983), triton X (ADAMOWICZ 1980) and EDTA (WEIMBERG et al. 1983) have also been used in attempts to distinguish cytoplasmic and vacuolar (storage) pools of ions and metabolites.

e) Nuclear Magnetic Resonance Spectroscopy (NMR)

Using root tips and cultured cells for measurements it has been possible to distinguish two pools of inorganic phosphate in vivo using $[^{31}P]$-NMR. The detection of these pools depends largely on the difference in pH in the cytoplasm and vacuole, which gives two peaks for P_i which are distinctly separated on the chemical shift axis (ROBERTS et al. 1980). This useful observation can be used to determine either the P_i-concentration in the two compartments or their pH (ROBERTS et al. 1981a,b, 1982, ADUCCI et al. 1982, MARTIN et al. 1982, KIME et al. 1982a,b). In addition, the level of hexose phosphates, adenylate P and some other P compounds in the cytoplasm can be measured on living cells. A number of reviews on this topic have appeared recently (ROBERTS and JARDETZKI 1981, MOORE, RATCLIFFE and WILLIAMS 1983).

There are prospects at least in theory for the use of NMR in the study of NO_3^- and K^+ distribution between cytoplasm and vacuole. In these cases no difference in chemical shift due to the ΔpH between the compartments is measured, but model systems show that the relaxation time of the resonance in cytoplasm and protein solutions (BSA) is very much faster than in the vacuole. Kinetic analysis of resonance relaxation reveals two distinct phases for $[^{14}N]O_3^-$ which have time constants of 30 ms and 100 ms (LEE and RATCLIFFE 1984). The latter relates to $[^{14}N]O_3^-$ in the vacuolar sap, while the more rapid relaxation occurs in the protein environment. Spectra from intact roots, however, have not given a convincing demonstration of this rapid phase and it seems likely that cytoplasmic NO_3^- is at a low concentration. A similar method of analysis may be possible for K^+ or Rb^+ and may be more fruitful because of the relatively high K^+ in the cytoplasm (LEE, pers. comm.).

At the moment NMR studies have an important limitation because they are restricted to immature cells or those which have a high proportion of cytoplasm in the cell volume. Thus, fully expanded cells in plant roots, for instance, do not give sufficiently large peaks for atomic nuclei in the cytoplasm for quantitation. A general comment is that the spectra obtained are an average of all the compartments of a given designation in a tissue or organ. Microanalysis reveals that there may be very considerable differences in vacuolar P_i and other ions in cells at the periphery and at the inner cortex of a root (VAN STEVENINCK et al. 1982b). The histological compartmentation of P_i in *Vicia faba* leaves has established concentration differences up to 50-fold with palisade parenchyma cells 4 mM, spongy parenchyma cells 5 mM, epidermal cells 60 mM, stomatal guard cells 100 to 200 mM (OUTLAW et al. 1984). Microfluorimetric measurements using 9-aminoacridine have also revealed heterogeneity of cell populations in respect to vacuolar pH (MANIGAULT et al. 1983). Thus, the NMR technique is a very important advance in that it is nonintrusive and gives information on living cells, but in tissues, in common with all physical averaging techniques, it lacks the resolution to measure meaningful cell to cell variation.

f) Isolation of Vacuoles and Tonoplast Vesicles

Basically, two different options exist in the attempt to prepare purified fractions of tonoplast material.

First, separation of membrane fractions from homogenates of tissues or isolated protoplasts by gradient centrifugation. In this case the unambiguous identification of tonoplast fractions generally causes the greatest problem. The situation rapidly improves, however, as reliable markers become available from studies with isolated vacuoles.

Second, preparation of isolated vacuoles. In this case stability and yield cause the greatest problem. Vacuoles can be either obtained mechanically from thinly sliced tissue (LEIGH and BRANTON 1976, LEIGH and TOMOS 1983) or by lysis of protoplasts. The latter is achieved osmotically by transfer to mildly hypertonic buffer (WAGNER and SIEGELMAN 1975, METTLER and LEONARD 1979). The lytic properties of polybases (see Sect. 2.d) have also been employed in releasing vacuoles (DÜRR et al. 1975). SCHMIDT and POOLE (1980) found 97% of red beetroot protoplasts ruptured to release intact vacuoles after 1-min treatment by 50 g ml^{-1} DEAE-dextran. Discontinuous gradients with a DEAE-dextran layer to lyse the protoplasts passing through them can be readily used for preparing and purifying isolated vacuoles in one centrifugation step at about 1200 g (BOUDET et al. 1981, ALIBERT and BOUDET 1982, SMITH et al. 1984, JOCHEM et al. 1984). Protoplasts may also be disrupted mechanically by passing a stream of them through appropriately sized syringe needles (MARTINOIA et al. 1981, LILLEY et al. 1982).

The properties of isolated vacuoles or tonoplast vesicles are investigated biochemically, e.g., for ATPase activity and its characteristics. The electrochemical H^+ gradient across their membranes is measured via the distribution of weak acids and bases, lipophilic cations and pH-sensitive or fluorescent dyes (for references see report in the previous volume pp. 79 - 80: LÜTTGE 1983, and CHURCHILL et al. 1983, MILLER et al. 1984), or by using microelectrodes (RONA et al. 1980, JOCHEM et al. 1984). The tonoplast ATPase of higher plant cells is insensitive to molybdate and vanadate, has a high specificity for Mg-ATP as substrate, is insensitive to univalent cations, is stimulated by Cl^- and inhibited by NO_3^- and tends to have a slightly alkaline pH optimum (see pp. 79 - 80 in LÜTTGE 1983, BENNET and SPANSWICK 1983, O'NEILL et al. 1983, THOM and KOMOR 1984).

MARTINOIA et al. (1981) recovered 99% of the free nitrate of protoplasts from barley leaves in the isolated vacuoles, but only 52% of the amino N. In view of the fact that NO_3^- efflux across plasma membranes is frequently very much greater than that of other ions (DEANE-DRUMMOND and GLASS 1983) and that the plasmalemma of protoplasts may become "leaky" to anions during plasmolysis or isolation (RONA and CORNEL 1979, RONA et al. 1980, CORNEL et al. 1983, LEFEBVRE and CLARKSON 1984a,b), these estimates of vacuolar compartmentation may be excessive. There is also evidence that vacuoles may leak relatively large amounts of K^+ (LEIGH and TOMOS 1983) and NO_3^- (GRANSTEDT and HUFFAKER 1982) at some stage during their isolation. Conversely, vacuoles isolated from leaves of CAM plants appear to retain most of their malate (BUSER and MATILE 1977, KENYON et al. 1978). Vacuoles obtained from sugarcane suspension cultures also appeared to loose little sugar to the isolation medium (THOM et al. 1982).

The proton permeability of tonoplasts in situ must be rather low, e.g., regarding the considerable intravacuolar acidification which is possible down to pH 3.3 in leaves of CAM plants at the end of the dark period (LÜTTGE and SMITH 1984) or the extreme case of the H_2SO_4-accumulating brown alga *Desmarestia* with a pH of 0.5 to 0.8 (Mc CLINTOCK et al. 1982). Isolated vacuoles seem to maintain a low tonoplast permeability for protons but become highly permeable when ionophores and KCl are present in the external medium (WEIGEL and WEIS 1984). Furthermore, protons may leak from vacuoles indirectly in the form of protonated molecules, e.g., as the nondissociated malic acid during the deacidification phase in CAM plants (LÜTTGE and SMITH 1984). The proton electrochemical gradient between the vacuole and its surroundings is not maintained in isolation from the cytoplasm and its energy source for proton pumping (MARIN et al. 1981, CHRESTIN 1984, JOCHEM et al. 1984).

Such observations cast doubt on the use of isolated vacuoles for accurate quantitation of the ionic contents.

g) Transcellular Electrical Profiles

The establishment of transcellular electrochemical profiles has a long history with giant algal coenocytes (MAC ROBBIE 1970). Separate measurements of electrical properties of plasmalemma and tonoplast using microelectrodes in cells of higher plants were much more difficult and tricks like centrifugation were used (LÜTTGE and ZIRKE 1974).

Recently, comparative estimates of electrical membrane resistances have much improved the situation (BATES et al. 1982a,b, 1983). The technique of gently moving electrodes across cells to obtain transcellular profiles has been perfected (RONA and CORNEL 1979, RONA et al. 1977, 1980, 1982), and this comprises the use of ion-specific electrodes to measure ion activities in the cytoplasmic and vacuolar compartments (RONA et al. 1982, KURKDJIAN and BARBIER-BRYGDOO 1983).

New prospects for estimation of membrane resistances are offered by the charge-pulse technique. With microelectrodes in the vacuole, cells are charged with very short pulses (200 ns) and the decay of the obtained membrane voltage is studied. The existence of mobile charges is demonstrated and specific membrane resistances can be assessed (BENZ and ZIMMERMANN 1983). The interpretations in terms of separating plasmalemma and tonoplast are not yet unambiguous though.

3. Relationship Between the Vacuole and the Cytoplasm

Evidence from microorganisms shows quite clearly that conditions which cause marked changes in the ionic or metabolite contents of cells usually result in large fluctuations in vacuolar concentrations with little change in the cytosol. In *Saccharomyces carlsbergensis* external changes in K^+, Mg^{2+}, and P_i had scarcely any effect on cytosolic concentrations, while those in the vacuole varied widely (LICHKO and OKOROKOV 1976, LICHKO et al. 1982). In *Candida utilis* the cytosolic concentrations of arginine differed little in cells provided with either arginine or ornithine or citrulline as nitrogen source, but in the former instance there was a very great accumulation of arginine in the vacuole (HUBER-WÄLCHLI and WIEMKEN 1979). These investigations used polybasic molecules to render cell membranes leaky.

In higher plant cells NMR observations indicate homeostasis within the cytoplasm with respect to its P_i concentration (FOYER et al. 1982, LEE and RATCLIFFE 1983, REBEILLE et al. 1983). In root tips of pea *(Pisum sativum)* variations in total P_i concentration, brought about by P-starvation for various times, were perfectly correlated with changes in the magnitude of the vacuolar P_i peak, while the cytosolic P_i peak was unchanged (LEE and RATCLIFFE 1983). Prolonged starvation or growth in the absence of an adequate P_i supply eventually exhausts the vacuolar store and cytoplasmic P_i concentration then decreases. When P_i-starved cells of *Acer pseudoplatanus* were transferred to a medium containing 300 μM P_i, P_i entered the cells rapidly and accumulated in the cytoplasm. Only when the cytoplasm was "full" was there any significant P_i accumulation in the vacuole (REBEILLE et al. 1983). By contrast, when cells with adequate P_i were transferred to a medium containing a high concentration of phosphate, P_i entered the cells slowly and accumulated in the vacuole, not in the cytoplasm.

The fluxes of many ions across the tonoplast are appreciably slower than those across the plasmalemma. Recent observations of sulfate movements confirm this view (THOIRON et al. 1981, CRAM 1983a,b). The slow fluxes may serve to buffer the cytoplasm against long-term changes in its ionic concentration, but in the short-term rapid uptake or release of ions by organelles may alter cytosolic concentrations, thus allowing some regulation of processes within the organelles by inorganic ions. A recent analysis of compartmentation and fluxes of P_i in isolated cladophyll cells from *Asparagus officinalis* suggests that cycling of P_i between the cytosol and the plastids may cause transient fluctuations in cytosolic P_i concentration. The half-time for these fluctuations is short by comparison with fluxes of P_i between cytosol and vacuole. Thus transient depletion of P_i during the operation of the phosphate translocator in the chloroplast envelope would not cause a net input of P_i into the cytoplasm from the vacuole (WOODROW et al. 1984). One may envisage, therefore, other short-term fluctuations which could interact with ion transport mechanisms at the plasmalemma, augmenting or repressing their activities by allosteric mechanisms (GLASS 1983). In general, however, the concept that regulation of ion transport in the longer term is based on variation of cytosolic ionic concentrations seems to contradict the few hard facts which have so far emerged from compartmentation analysis.

Kinetics of the initial rate of ion transport by root cells probably measure the plasmalemma fluxes and these have been found to vary with the ionic status of whole cells under consideration. Thus GLASS (1976) found good correlations between K^+ levels in roots and both the K_m and V_{max} of the high affinity K^+ uptake. The weakness of this work is that no assessment was made of cytosolic K^+ concentration. CRAM (1983a,b) showed that sulfate accumulation in *Daucus* storage root tissue is determined by the flux across the tonoplast and that the kinetics respond to vacuolar SO_4^{2-} concentration. The correlation in the work of LEE and RATCLIFFE (1983) between vacuolar P_i-concentration and initial uptake rate of P_i also casts doubt on direct involvement of cytosolic ionic concentration as the controller of uptake rate. Much needs to be done to clarify these matters.

The factors which regulate tonoplast fluxes are almost completely unknown. The proton-electrochemical gradient established across the tonoplast by the vacuolar ATPase could drive a wealth of solute fluxes via secondary flux coupling. However, this most promising concept so far received only little attention experimentally. A correlation between the proton-electrochemical gradient and citrate uptake was demonstrated in *Hevea* latex vacuoles (MARIN et al. 1981). The vacuolar H^+-translocating ATPase was related to vacuolar sugar accumulation in sugar cane (KOMOR et al. 1982) and malate accumulation in CAM (SMITH et al. 1984).

Although Na transport is such a widely described activity of the tonoplast, there has been no demonstration of Na^+ transport by ATPase in vesicles or isolated vacuoles. The lack of modern investigations on this important aspect has already been marked in the previous review (p.80: LÜTTGE 1983). LERNER et al.(1983b) demonstrated differences in sensitivities of membrane ATPases to salt between a glycophyte and a halophyte. Crude preparations were used, however, leaving uncertainties about the intracellular localization of the ATPases in vivo. The morphology of the tonoplast in the alga *Porphyra umbilicalis* changes when the salinity of its medium is increased (WIENCKE and LÄUCHLI 1983, KNOTH and WIENCKE 1984); the number of intercalated membrane particles on the cytoplasmic side of the tonoplast increased three- to four-fold as salt burden rose. These particles are regarded as ion-transporting complexes.

The release of ions stored in the vacuole may be regulated by metabolic demand and subject to genetic control (see Sect. 4). Physiological studies of NO_3^- reduction indicated, many years ago (BURSTRÖM 1943), the presence of metabolic and storage pools of NO_3^- in leaves and roots. Depletion of the former does not necessarily promote release from the latter. In leaves of barley it seems that only in the light is there appreciable NO_3^--flow from the vacuole to the metabolic pool in the cytoplasm (ASLAM et al. 1976, 1979). The release of an ion from the vacuole probably involves counter movement of another ion. Proton pumping into the vacuole may provide a proton motive force, directed toward the cytoplasm that could be coupled to anion symport systems, indeed the counterpart of anion transport at the plasmalemma. The release of vacuolar K^+ can be stimulated by the uptake of Na^+ by barley roots; the quantitative aspects of this exchange have been studied intensively (HOOYMANS 1974, JESCHKE 1977a,b, 1979, JESCHKE et al. 1983). In barley K^+ released from the vacuole by Na^+ seems to become selectively channeled into symplast movement and is largely exported from the root in the xylem (JESCHKE 1979).

4. The Vacuole and "Efficiency" of Mineral Nutrition

The efficiency of utilization of mineral nutrients may be the crucial factor in determining overall efficient nutrition, especially when nutrient supply from the soil is suboptimal (GABELMAN and GERLOFF 1983, MOLL et al. 1982, FAWOLE et al. 1982). Evidence showing that efficiency of utilization varies with genotypes of various crop plants has been summarized by GERLOFF and GABELMAN (1983) and GABELMAN and GERLOFF (1983). It is clear that efficiency of this kind is an inheritable trait, but it is not clear how this efficiency is achieved. In only one instance has it been suggested to have a biochemical basis; C_4 plants are inherently more efficient in N utilization because their mode of photosynthesis economizes in the synthesis of RUDP carboxylase/oxygenase – this is the principle protein in C_3 species accounting for more than 40% of leaf protein-N (BROWN 1978). In general, however, it seems unlikely that plants which utilize nutrients with high efficiency use less inorganic nutrients in the assimilatory or structure-building processes than less efficient plants. Some curious results of EPSTEIN (1978) suggest that differences in efficiency may depend on compartmentation within cells. Using a K^+-inefficient tomato mutant, he showed that gross tissue K^+-concentration in poorly growing leaves of the mutant was as high or higher than that in the wild-type tomato; only very high inputs of exogenous K^+ kept the plants alive. The genetic lesion had no effect on either initial uptake or long distance transport of K^+ and this left EPSTEIN to suggest that, in the mutant, abnormally large amounts of K^+ were sequestered in the vacuole. This might be due to some derangement of the tonoplast mechanism which normally keeps cytosolic Na^+ at a low level – some change in orientation or binding site specificity of a pump or channel might do this. This extreme example has more restrained counterparts in experiments of MAKMUR et al. (1978) using 156 selection lines of tomato grown with limiting supplies of K^+. Under K^+ stress, efficient lines produced nearly twice as much dry weight per unit K^+ absorbed as less efficient ones, and tissue concentrations, especially in young leaves, were lower. The performance of less efficient genotypes could be improved by providing Na^+ to the growth medium. By analogy with barley this could be envisaged as releasing K^+ sequestered in the vacuole (JESCHKE 1977a,b). The sequestration of K^+ in root vacuoles can be strongly influenced by the nutrient "demand" created by shoot growth. Tonoplast K^+ fluxes were decreased in barley where a high shoot:root ratio was created by partial root excision; in such circumstances recently absorbed K^+ becomes selectively channeled into the symplast (JESCHKE 1982).

In barley varieties supplied with adequate K^+ there was evidence of variation of utilization efficiency (PETTERSSON and JENSEN 1983). In these results efficiency was not associated with low tissue concentrations of K^+, neither was it related to measurements of uptake efficiency as judged by $[^{86}Rb]$ influx; however, these influx results did not correlate with net K^+ uptake. The authors suggest that $[^{86}Rb]$ is not a good tracer for K^+ – a similar conclusion has been reached in other experiments, most recently by BEHL and JESCHKE (1982). It should be stressed that where subtleties of compartmentation are being studied chemical analogs should not be used as tracers; the use of $[^{86}Rb]$ for K^+, $[^{85}Sr]$ for Ca^{2+}, $[^{36}Cl]O_4$ for NO_3^-, should be avoided.

Despite the importance of this matter there does not seem to have been any systematic attempt to measure the compartmentation of nutrients in cells or leaves or roots in genotypes differing in efficiency of utilization. This may be because of reluctance to use the rather imperfect analytical methods available, but a start might be made on simple tracer exchange studies. These have been applied successfully to leaves as well as to roots (see WALKER and PITMAN 1976).

5. Outlook

The vacuoles are not only important for steady-state conditioning (e.g., turgor, storage, detoxification) but are dynamically involved in plant cell metabolism. It is important that the rapidly increasing knowledge of biochemical activities of tonoplasts soon merges with attempts to understand the role of vacuoles in nutrition and in ecological adaptations. Phosphorus, nitrogen and sulfur nutrition and intracellular Na^+, K^+, and Ca^{2+} balance at present seem to be problems which can be approached using the techniques reviewed above.

References

ADAMOWICZ, S.: Physiol. Vég. 18, 453-461 (1980). - ADUCCI, P., FEDERICO, R., CARPINELLI, G., PODO, F.: Planta 156, 579-582 (1982). - ALIBERT, G., BOUDET, A.: Physiol. Vég. 20, 289-309 (1982). - ASLAM, M., DAKS, A., HUFFAKER, R.C.: Plant Physiol. 58, 588-591 (1976). - ASLAM, M., HUFFAKER, R.C., RAINS, D.W., RAD, K.P.: Plant Physiol. 63, 1205-1209 (1979).

BATES, G.W., GOLDSMITH, M.H.M., GOLDSMITH, T.H.: Origins and measurement of the membrane potentials in *Avena* coleoptiles, 241-246. In: Plasmalemma and Tonoplast: Their Functions in the Plant Cell, eds. D. MARMÉ, E. MARRÈ, R. HERTEL. Amsterdam: Elsevier, Biomedical Press 1982a; - J. Membr. Biol. 66, 15-23 (1982b); - Plant Sci. Lett. 30, 279-284 (1983). - BEHL, R., JESCHKE, W.D.: J. Exp. Bot. 33, 584-600 (1982). - BENNETT, A.B., SPANSWICK, R.M.: J. Membr. Biol. 71, 95-107 (1983). - BENZ, R., ZIMMERMANN, U.: Biophys. J. 43, 13-26 (1983). - BOUDET, A.M., CANUT, H., ALIBERT, G.: Plant Physiol. 68, 1354-1358 (1981). - BROWN, R.H.: Crop Sci. 18, 93-98 (1978). - BURSTRÖM, H.: Ann. Agric. Coll. Sweden 2, 1-50 (1943). - BUSER, C., MATILE, P.: Z. Pflanzenphysiol. 82, 462-466 (1977).

CARPITA, N., SABULASE, D., MONTEZINOS, D., DELMER, D.P.: Science 205, 1144-1147 (1979). - CHRESTIN, H.: Le Compartiment Vacuo-Lysosomal (les Lutoides) du Latex d'*Hevea brasiliensis*. Son Rôle dans le Maintien de l' Homeostasie et dans les Processus de Senescence des Cellules Laticifères. Montpellier: Thèse de Docteur ès Sciences Naturelles, Université des Sciences et Techniques du Languedoc 1984. - CHURCHILL, K.A., HOLAWAY, B., SZE, H.: Plant Physiol. 73, 921-928 (1983). - CORNEL, D., GRIGNON, C., RONA, J.P., HELLER, R.: Physiol. Plant. 57, 203-209 (1983). - CRAM, W.J.: Plant Sci. Lett. 31, 329-338 (1983a); - Plant Physiol. 72, 204-211 (1983b).

DEANE-DRUMMOND, C.E., GLASS, A.D.M.: Plant Physiol. 73, 100-104 (1983). - DELMER, D.P.: Plant Physiol. 64, 623-629 (1979). - DÜRR, M., BOLLER, T., WIEMKEN, A.: Arch. Microbiol. 105, 319-327 (1975).

EPSTEIN, E.: Plant Physiol. 62, 582-585 (1978).

FAWOLE, I., GABELMAN, W.H., GERLOFF, G.C., NORDHEIM, E.V.: J. Am. Soc. Hort. Sci. 107, 94-97 (1982). - FOYER, C.H., WALKER, D.A., SPENCER, C., MANN, B.: Biochem. J. 202, 429-434 (1982).

GABELMAN, W.H., GERLOFF, G.C.: Plant Soil 72, 335-350 (1983). - GERLOFF, G.C., GABELMAN, W.H.: Genetic basis of inorganic plant nutrition, 453-480. In: Inorganic Plant Nutrition, eds. A. LÄUCHLI, R.L. BIELESKI. Encyclopedia of Plant Physiology, Vol. 15B. Berlin, Heidelberg, New York, Tokyo: Springer 1983. - GLASS, A.D.M.: Plant Physiol. 58, 33-37 (1976); - Annu. Rev. Plant Physiol. 34, 311-326 (1983). - GORHAM, J., WYN JONES, R.G.: Planta 157, 344-349 (1983). - GRANSTEDT, R.C., HUFFAKER, R.C.: Plant Physiol. 70, 410-413 (1982).

HARVEY, D.M.R., HALL, J.L., FLOWERS, T.J.: J. Microsc. 107, 189-197 (1976). - HARVEY, D.M.R., FLOWERS, T.J., HALL, J.L.: J. Microsc. 116, 213-226 (1979). - HOOYMANS, J.J.M.: Z. Pflanzenphysiol. 73, 234-242 (1974). - HUBER-WÄLCHLI, V., WIEMKEN, A.: Arch. Microbiol. 120, 141-149 (1979).

JESCHKE, W.D.: J. Exp. Bot. 28, 1289-1305 (1977a); - Z. Pflanzenphysiol. 84, 247-264 (1977b); - Univalent cation selectivity and compartmentation in cereals, 37-61. In: Recent Advances in the Biochemistry of Cereals, eds. D.L. LAIDMAN, R.G. WYN JONES. London, New York, San Francisco: Academic Press 1979; - J. Exp. Bot. 33, 601-618 (1982). - JESCHKE, W.D., JAMBOR, W.: J. Exp. Bot. 32, 1257-1272 (1981). - JESCHKE, W.D., STELTER, W.: Physiol. Plant. 58, 505-514 (1983). - JESCHKE, W.D., STELTER, W., REISING, B., BEHL, R.: J. Exp. Bot. 34, 964-979 (1983). - JOCHEM, P., RONA, J.-P., SMITH, J.A.C., LÜTTGE, U.: Physiol. Plant. submitted (1984).

KENYON, W.H., KRINGSTAD, R., BLACK, C.C.: FEBS Lett. 94, 281-283 (1978). - KIKUYAMA, M., HAYAMA, T., FUJII, S., TAZAWA, M.: Plant Cell Physiol. 20, 993-1002 (1979). - KIME, M.J., RATCLIFFE, R.G., LOUGHMAN, B.C.: J. Exp. Bot. 33, 670-681 (1982a). - KIME, M.J., RATCLIFFE, R.G., WILLIAMS, R.J.P., LOUGHMAN, B.C.: J. Exp. Bot. 33, 656-669 (1982b). - KIYOSAWA, K., TAZAWA, M.: J. Membr. Biol. 37, 157-166 (1977). - KNOTH, A., WIENCKE, C.: Plant Cell Environ. 7, 113-119 (1984). - KOMOR, E., THOM, M., MARETZKI, A.: Plant Physiol. 69, 1326-1330 (1982). - KURKDJIAN, A.C., BARBIER-BRYGDOO, H.: Anal. Biochem. 132, 96-104 (1983).

LEE, R.B., RATCLIFFE, R.G.: J. Exp. Bot. 34, 1222-1224 (1983); - J. Exp. Bot. in press (1984). - LEFEBVRE, D.D., CLARKSON, D.T.: Can. J. Bot. in press (1984a); - J. Exp. Bot. in press (1984b). - LEIGH, R.A., BRANTON, D.: Plant Physiol. 58, 656-662 (1976). - LEIGH, R.A., TOMOS, D.: Planta 159, 469-475 (1983). - LERNER, H.R., REUVENI, M.: Induction of pore formation selectively in the plasmalemma of plant cells by poly-L-lysine treatment: a method for the direct measurement of cytosol solutes in plant cells, 49-52. In: Plasmalemma and Tonoplast: Their

Functions in the Plant Cell, eds. D. MARMÉ, E. MARRÈ, R. HERTEL. Amsterdam: Elsevier, Biomedical Press 1982. - LERNER, H.R., BALLARIN DENTI, A., PRADET, A.: Plant Physiol. 57, 90-94 (1983a). - LERNER, H. R., REINHOLD, L., GUY, R., BRAUN, Y., HASIDIM, M., POLJAKOFF-MAYBER, A.: Plant Cell Environ. 6, 501-506 (1983b). - LICHKO, L.P., OKOROKOV, L.A.: Dokl. Akad. Nauk. USSR 227, 756-758 (1976). - LICHKO, L.P., OKOROKOV, L.A., KULAEV, I.S.: Arch. Microbiol. 132, 289-293 (1982). - LILLEY, R. Mc C., STITT, M., MADER, G., HELDT, H.W.: Plant Physiol. 70, 965-970 (1982). - LÜTTGE, U.: Progr. Bot. 45, 76-88 (1983). - LÜTTGE, U., SMITH, J.A.C.: J. Membr. Biol. in press (1984). - LÜTTGE, U., ZIRKE, G.: J. Membr. Biol. 18, 305-314 (1974).

MAC ROBBIE, E.A.C.: Quart. Rev. Biophys. 3, 251-294 (1970). - MAC ROBBIE, E.A.C., DAINTY, J.: J. Gen. Physiol. 42, 335-353 (1958). - MAKMUR, A., GERLOFF, G.C., GABELMAN, W.H.: J. Am. Soc. Hort. Sci. 103, 545-549 (1978). - MANIGAULT, P., MANIGAULT, J., KURKDJIAN, A.: Physiol. Vég. 21, 129-136 (1983). - MARIN, B., SMITH, J.A.C., LÜTTGE, U.: Planta 153, 486-493 (1981). - MARTIN, J.B., BLIGNY, R., REBEILLE, F., DOUCE, R., LEGUAY, J.J., MATHIEU, Y., GUERN, J.: Plant Physiol. 70, 1156-1161 (1982). - MARTINOIA, E., HECK, U., WIEMKEN, A.: Nature 289, 292-294 (1981). - MATILE, Ph., WIEMKEN, A.: Interactions between cytoplasm and vacuole, 255-287. In: Transport in Plants III., eds. C.R. STOCKING, U. HEBER. Encyclopedia of Plant Physiology, Vol. 3. Berlin, Heidelberg, New York: Springer 1976. - McCLINTOCK, M., HIGINBOTHAM, N., URIBE, E.G., CLELAND, R.E.: Plant Physiol. 70, 771-774 (1982). - METTLER, I.J., LEONARD, R.T.: Plant Physiol. 63, 183-190 (1979). - MILLER, A.J., BRIMELOW, J.J., JOHN, P.: Planta 160, 59-65 (1984). - MIMURA, T., SHIMMEN, T., TAZAWA, M.: Planta 157, 97-104 (1983). - MOLL, R.H., KAMPRATH, E.J., JACKSON, W.A.: Agron. J. 74, 562-564 (1982). - MOORE, G.R., RATCLIFFE, R.G., WILLIAMS, R.J.P.: NMR and the biochemist, 142-195. In: Essays in Biochemistry, eds. P.N. CAMPBELL, R.D. MARSHALL, Vol. 19. London: Academic Press 1983.

O'NEILL, S.D., BENNETT, A.B., SPANSWICK, R.M.: Plant Physiol. 72, 837-846 (1983). - OUTLAW, W.H., TARCZYNSKI, M.C., MILLER, W.I.: Plant Physiol. 74, 430-433 (1984).

PETTERSSON, S., JENSEN, P.: Plant Soil 72, 231-237 (1983). - PITMAN, M.G.: Aust. J. Biol. Sci. 16, 647-668 (1963). - PITMAN, M.G., LÄUCHLI, A., STELZER, R.: Plant Physiol. 67, 673-679 (1981).

REBEILLE, F., BLIGNY, R., MARTIN, J.-B., DOUCE, R.: Arch. Biochem. Biophys. 225, 143-148 (1983). - ROBERTS, J.K.M., JARDETZKY, O.: Biochim. Biophys. Acta 639, 53-76 (1981). - ROBERTS, J.K.M., RAY, P.M., WADE-JARDETZKY, N., JARDETZKY, O.: Nature 283, 870-872 (1980). - ROBERTS, J.K.M., RAY, P.M., WADE-JARDETZKY, N., JARDETZKY, O.: Planta 152, 74-78 (1981a). - ROBERTS, J.K.M., WADE-JARDETZKY, N., JARDETZKY, O.: Biochem. 20, 5389-5394 (1981b). - ROBERTS, J.K.M., WEMMER, D., RAY, P.M., JARDETZKY, O.: Plant Physiol. 69, 1344-1347 (1982). - RONA, J.-P., CORNEL, D.: Physiol. Vég. 17, 1-11 (1979). - RONA, J.-P., CORNEL, D., HELLER, R.: Bioelectrochem. Bioenerg. 4, 185-194 (1977). - RONA, J.-P., VAN DE SYPE, G., CORNEL, D., GRIGNON, C., HELLER, R.: Bioelectrochem. Bioenerg. 7, 377-391 (1980). - RONA, J.-P., CORNEL, D., GRIGNON, C., HELLER, R.: Physiol. Vég. 20, 459-463 (1982). - RYGOL, J., LÜTTGE, U.: Plant Cell Environ. 6, 545-553 (1983).

SCHLENK, F., DAINKO, J.L., SVIHLA, G.: Arch. Biochem. Biophys. 140, 228-236 (1970). - SCHMIDT, R., POOLE, R.J.: Plant Physiol. 66, 25-28 (1980). - SHIMMEN, T., TAZAWA, M.: J. Membr. Biol. 37, 167-192 (1977); - Plant Cell Physiol. 21, 1007-1013 (1980a); - J. Membr. Biol. 55, 223-232 (1980b). - SMITH, J.A.C., URIBE, E.G., BALL, E., HEUER, S.,

LÜTTGE, U.: Eur. J. Biochem. in press (1984). - STELTER, W., JESCHKE, W.D.: Physiol. Plant. 58, 505-514 (1983). - STOREY, R., PITMAN, M.G., STELZER, R., CARTER, C.: J. Exp. Bot. 34, 778-794 (1983a). - STOREY, R., PITMAN, M.G., STELZER, R.: J. Exp. Bot. 34, 1196-1206 (1983b).

TAZAWA, M., SHIMMEN, T.: Bot. Mag. Tokyo 95, 147-154 (1982). - TAZAWA, M., KIKUYAMA, M., SHIMMEN, T.: Cell Struct. Funct. 1, 165-176 (1976). - THAINE, J.F.: J. Exp. Bot. 35, 444-453 (1984). - THOIRON, A., THOIRON, B., DEMARTY, M., THELLIER, M.: Biochim. Biophys. Acta 644, 24-35 (1981). - THOM, M., KOMOR, E.: Eur. J. Biochem. 138, 93-99 (1984). - THOM, M., MARETZKI, A., KOMOR, E.: Plant Physiol. 69, 1315-1319 (1982).

VAN IREN, F., ESSEN-JOOLEN, L., VAN DER DUIJN SCHOUTEN, P., BOERS-VAN DER SLUIJS, P., DE BRUIJN, W.C.: Histochemistry 63, 273-294 (1979). - VAN STEVENINCK, R.F.M., VAN STEVENINCK, M.E., STELZER, R., LÄUCHLI, A.: Z. Pflanzenphysiol. 107, 91-95 (1982a). - VAN STEVENINCK, R.F.M., VAN STEVENINCK, M.E., LÄUCHLI, A.: Z. Pflanzenphysiol. 108, 215-222 (1982b).

WAGNER, G.J., SIEGELMAN, H.W.: Science 190, 1298-1299 (1975). - WALKER, N.A., PITMAN, M.G.: Measurement of fluxes across membranes, 93-126. In: Transport in Plants II. A., eds. U. LÜTTGE, M.G. PITMAN. Encyclopedia of Plant Physiology, Vol. 2A. Berlin, Heidelberg, New York: Springer 1976. - WEIGEL, H.J., WEIS, E.: Plant Sci. Lett. 33, 163-175 (1984). - WEIMBERG, R., LERNER, H.R., POLJAKOFF-MAYBER, A.: Plant Physiol. 68, 1433-1438 (1981). - WEIMBERG, R., LERNER, H.R., POLJAKOFF-MAYBER, A.: J. Exp. Bot. 34, 1333-1346 (1983). - WIENCKE, C., STELZER, R., LÄUCHLI, A.: Planta 159, 336-341 (1983). - WIENCKE, C., LÄUCHLI, A.: Planta 159, 342-346 (1983). - WOODROW, I.E., ELLIS, J.R., JENNINGS, A., FOYER, C.H.: Planta in Press (1984).

Dr. DAVID T. CLARKSON
Agricultural Research Council
Letcombe Laboratory
Wantage Oxon OX13 9JT
United Kingdom

Professor Dr. ULRICH LÜTTGE
Institut für Botanik der
Technischen Hochschule Darmstadt
Schnittspahnstr. 3-5
D-6100 Darmstadt, FRG

III. Photosynthesis. Carbon Metabolism: On Land and at Sea[1]

By GRAHAME J. KELLY and ERWIN LATZKO

"If you were to row out from the coast on a summer's day and haul in a bucket of water, it might surprise you to learn that it contained several million plants."

E. STEMANN NIELSEN in
The Galathea Deep Sea Expedition
(1950-52)

1. Introduction

This contribution celebrates, for our part, a decade of reviewing photosynthetic carbon metabolism in *Progress in Botany*. Each 2 years an attempt has been made to highlight a particular aspect of this topic, and one would expect that by now the supply of aspects would be exhausted. Indeed, this may have appeared likely, had not one of us moved to a new Institute. This move gave birth to a realization that we have, to a considerable extent, ignored about 30% of the world's CO_2 fixation over the past decade of reviewing. The greater portion of this 30% belongs to the layer of unicellular algae (commonly termed marine phytoplankton) that occurs in the world's oceans. Suspended at a depth where light intensity is low, light quality is blue-greenish, and in a medium of pH about 8 and containing 2 mM HCO_3^-, these marine photosynthetic CO_2 fixers face problems quite unlike those that have made CAM, the C_4 pathway, and photorespiration so predominant in their terrestrial counterparts. The remainder of the 30% is attributed to macrophytic algae and sea grasses, mainly in shallow waters near land masses. Carbon metabolism by the macrophytic algae has been recently thoroughly reviewed by KREMER (1981a).

Consequently, this 1984 review endeavors to look into the ocean, as well as to consider investigations with terrestrial plants reported since our last review in Vol. 44. The great majority of publications relate to the latter, but a glance at the available reports on marine algae should at least promote a more balanced outlook on global photosynthetic carbon metabolism.

[1]Abbreviations: CAM, crassulacean acid metabolism; CoA, coenzyme A; FBP, fructose-1,6-P_2; GAP, glyceraldehyde-3-P; OAA, oxaloacetate; PEP, P-enolpyruvate; P_i, inorganic phosphate; RuBP, ribulose-1,5-P_2; SBP, sedoheptulose-1,7-P_2

Progress in Botany, Vol. 46

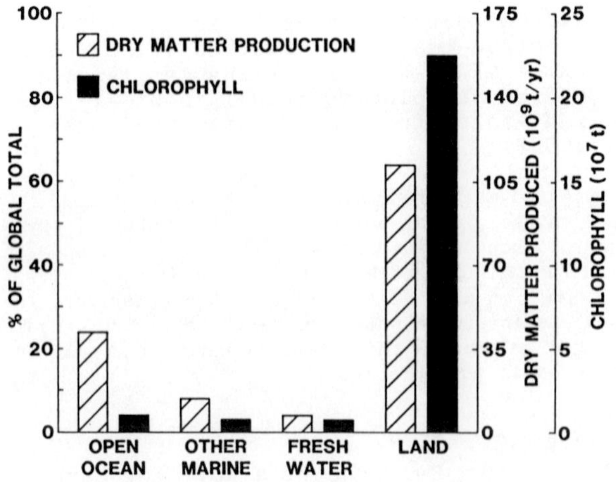

Fig. 1. Distribution of dry matter production and chlorophyll between land and sea. Adapted from WHITTAKER and LIKENS (1975)

2. The Superior Performance of Marine Phytoplankton in Photosynthetic Carbon Assimilation

It is estimated that the open ocean phytoplankton carry out 24% of the world's carbon assimilation, using only 4% of its chlorophyll (WHITTAKER and LIKENS 1975, Fig. 1). Therefore, the marine algae appear to be eight times more effective than terrestrial plants as far as CO_2-fixation expressed on a chlorophyll basis is concerned: the respective rates, derived from Figure 1 on the assumption that dry matter is 45% carbon, are (in µmol CO_2-fixed/mg chlorophyll·h) 36 for marine algae and 4.3 for terrestrial plants. The superior performance by the marine algae is all the more impressive when it is noted that the irradiance at the average depth where they occur is below that required to saturate their photosynthesis. Productivity experiments with samples of seawater exposed to saturating or near-saturating irradiance have yielded rates of 33 to over 1000 µmol/mg chl·h, with a mean value of about 350 µmol/mg chl·h (RYTHER and YENTSCH 1957, STRICKLAND et al. 1969, VENRICK et al. 1977, FURNAS 1983, WILLIAMS et al. 1983). A similarly high mean value was obtained for 12 unialgal cultures in the laboratory, although the distribution of values was not so great, the highest being 480 µmol/mg chl·h (KREMER and BERKS 1978, GLOVER 1980). One exception noted by GLOVER (1980) was *Coccolithus pelagicus* for which a rate of 1730 µmol/mg chl·h was obtained, and attributed to a low content of chlorophyll. However, the high CO_2 fixation rates for marine algae cannot generally be explained by low chlorophyll concentrations (and the "missing" chlorophyll being replaced by other light-harvesting accessory pigments), since, in fact, most measurements point to chlorophyll contents similar to those of green algae and higher plant cells (PRÉZELIN and ALBERTE 1978, FALKOWSKI 1981). One tends to conclude, then, that the open ocean phytoplankton consists mostly of healthy algal cells capable of photosynthesizing at maximal rates, and suffering neither water nor CO_2 limitation (seawater contains 2 mM HCO_3^-). The main factor that reduces their in situ CO_2 fixation from the potential 350 to actual 36 µmol/mg chl·h is the low irradiance, and this is a consequence of the depth at which the algae choose to grow.

The viewpoint that the phytoplankton in seawater samples photosynthesize at maximum possible rates when exposed to saturating light is supported by the fact that the value of 350 μmol/mg ch·h is equal to the best rates of photosynthetic electron transport measured (albeit with higher plants; DELANEY and WALKER 1978). Electron transport (specifically, the rate of oxidation of reduced plastoquinone) is widely believed to limit the rate of light-saturated and CO_2-saturated photosynthesis (RICHARDSON et al. 1983).

The above discussion does not pretend that the photosynthetic cells of land plants are not capable of photosynthesizing at a rate approaching that of phytoplankton. Indeed, suspensions of such cells isolated from tobacco leaves fixed 120 to 150 μmol CO_2/mg chl·h (CATALDO and BERLYN 1974), and intact alfalfa leaflets can fix CO_2 at 90 μmol/mg chl·h (PLATT et al. 1976). However, the rigors of land-based photosynthesis (periodic water and/or CO_2 deficits, shading, aged cells) result in actual rates well below the potential rates, and only one-eighth of the actual rates for oceanic phytoplankton.

3. The Uptake of Inorganic Carbon, and Evidence of a Role for Carbonic Anhydrase

Terrestrial plants obtain all their inorganic carbon as CO_2 from the air, and although some of this CO_2 may be converted to HCO_3^- (see below) in an aqueous film surrounding the plasmalemma, earlier evidence that HCO_3^- may be transported across the plasmalemma has been reinvestigated and it now seems that CO_2 is the principal permeating species of inorganic carbon (ESPIE and COLMAN 1982, VOLOKITA et al. 1983).

Aquatic plants, however, are exposed to both CO_2 and HCO_3^-, because CO_2 from the air dissolves in water to a concentration of 10 μM, and this dissolved CO_2 can be converted to HCO_3^- [Eq. (1)], the equilibrium concentration of which is dependent on the H^+ concentration, i.e., on the pH

$$CO_2 + H_2O \rightleftharpoons H_2CO_3 \rightleftharpoons H^+ + HCO_3^-. \quad (1)$$

It appears that algae and cyanobacteria, both freshwater and marine, utilize the option of taking up HCO_3^- (MARCUS et al. 1982, MERRETT and ARMITAGE 1982, COLMAN and GEHL 1983, IMAMURA et al. 1983). The uptake by marine algae is of particular interest because the seawater pH of 8.2 results in a HCO_3^- concentration 200 times that of CO_2. Uptake of HCO_3^- appears to be an active process, involving an enzyme-mediated transport of HCO_3^- across the plasmalemma, driven by ATP hydrolysis and capable of concentrating HCO_3^- inside the cell (BADGER and ANDREWS 1982, KAPLAN et al. 1982, OGAWA and INOUE 1983, SPALDING and OGREN 1983, SPALDING et al. 1983a). Once inside the cell CO_2 can be obtained from the HCO_3^- by a reversal of Eq. (1), and may be held in the cell as protein carbamate (YOKOTA et al. 1983a). The reaction producing CO_2 by H_2CO_3 dehydration is catalyzed by carbonic anhydrase. Studies supporting the importance of this enzyme (NILSEN and JOHNSEN 1982, SPALDING and OGREN 1982, SHIRAIWA and MIYACHI 1983) have been reinforced by a description from SPALDING et al. (1983b) of a *Chlamydomonas reinhardii* mutant with low carbonic anhydrase activity. The mutant grew poorly, yet it accumulated HCO_3^- to an internal concentration six times that of wild-type cells. The authors concluded that the accumulated HCO_3^-, in the absence of sufficient carbonic anhydrase activity, was not dehydrated to CO_2 rapidly enough to support normal rates of photosynthesis.

Nevertheless, there is still much to be learnt about carbonic anhydrase. Reports from Tokyo indicate that some algae have carbonic anhydrase on the cell surface, outside the plasmalemma, and also that certain algae take up CO_2 instead of, or as well as, HCO_3^- (KIMPEL et al. 1983, MIYACHI et al. 1983, TSUZUKI 1983). In higher plants, where the enzyme makes up 1% - 2% of all protein (OKABE et al. 1984), its function is unclear since HCO_3^- uptake seems not to be important (see above). Thus it is interesting that PORTER and GRODZINSKI (1984) quite recently observed a much greater decline in carbonic anhydrase activity in bean leaves maturing under 1200 µl/l CO_2 than under air-level CO_2 (300 µl/l). This response is reminiscent of the very low enzyme activity in algae that are grown under a high level of CO_2 (SPALDING and OGREN 1982, SHIRAIWA and MIYACHI 1983).

4. Chloroplast Carbon Metabolism

A comprehensive and thorough review of chloroplast metabolism, emphasizing not only enzyme properties, but also metabolite concentrations and the considerably artificial environments in which isolated enzymes and chloroplasts are usually studied, has been prepared recently by LILLEY (1983).

a) RuBP Carboxylase

The aerial roots of leafy orchids have become the latest addition to the list of photosynthetic tissues with the demonstration by KWOK-KI et al. (1983) that they contain substantial levels of RuBP carboxylase activity. Studies with rice leaves have provided further evidence that this enzyme serves an additional purpose to that of being the Calvin cycle's CO_2 fixer, viz. it is a store of nitrogen which is remobilized for the growth of young tissues (MAE et al. 1983). Its level in rice leaves can be a limiting factor to photosynthesis by these leaves (MAKINO et al. 1983).

The reaction catalyzed by RuBP carboxylase has received special attention during the last 2 years. Students being introduced to photosynthesis in introductory biochemistry courses have inevitably been told that "carboxylation of RuBP produces an unstable six-carbon intermediate that breaks down to two molecules of glycerate-3-P". It now appears that this six-carbon intermediate finally has been trapped and identified by SCHLOSS and LORIMER (1982) as 2-carboxy-3-keto-D-arabinitol-1,5-P_2. Combined with direct evidence for the intermediacy of the enediol between carbons 2 and 3 (SAVER and KNOWLES 1982), details

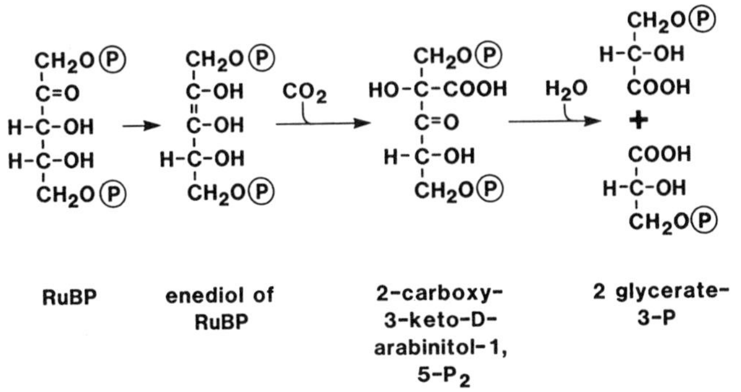

Fig. 2. Details of the reaction catalyzed by RuBP carboxylase. Based on SCHLOSS and LORIMER (1982) and SAVER and KNOWLES (1982)

of the CO_2 fixation reaction can now be written with some considerable degree of confidence (Fig. 2).

The true CO_2-acceptor substrate of the higher plant enzyme also has been investigated, and found to be $RuBP^{4-}$ (rather than $Mg \cdot RuBP^{2-}$; ROACH and McFADDEN 1983). The high affinity reported for this substrate (Km = 2.2 μM) concurs with the conclusion of BOURQUIN and FOCK (1983) that its concentration will normally not restrict CO_2 assimilation.

Activation of RuBP carboxylase via formation of the enzyme$\cdot CO_2 \cdot Mg^{2+}$ complex has been studied further. It is claimed that Ca^{2+} can replace Mg^{2+} in this complex (BÁRCENA 1983, PARRY et al. 1983). The activation may be more involved than previously thought, in that GUTTERIDGE et al. (1982) have found two forms of wheat RuBP carboxylase that differ in their rates of activation by CO_2 and Mg^{2+}, the half-times for activation being 8 s and 120 min, respectively. These authors, and also JORDAN and CHOLLET (1983), JORDAN et al. (1983), and VATER et al. (1983) have attributed the action of RuBP carboxylase effectors, such as gluconate-6-P, FBP, NADPH, and RuBP to somewhat complex influences on the CO_2 + Mg^{2+} activation process. There is also renewed interest in the effect of light on RuBP carboxylase activity in vivo, and recent strong evidence that the activity doubles upon illumination (SICHER 1982, VU et al. 1983) has been followed by a suggestion that light leads to conditions more favorable for activation of the carboxylase by CO_2 and Mg^{2+} (PERCHOROWICZ and JENSEN 1983). It seems that more attention should be devoted to the in vivo enzyme (STUMPF and JENSEN 1982), the properties of which may be altered upon purification (BRAUN et al. 1983).

The RuBP carboxylase enzyme is composed of eight large and eight small subunits. Each large subunit contains one active site, but the function of the small subunit has been something of a mystery. Working with the enzyme from blue-green algae, ANDREWS and BALLMENT (1983) and TAKABE et al. (1984) have now concluded that large subunits are not catalytically competent unless they have small subunits attached to them. It will be important to ascertain the extent to which this conclusion applies to the enzyme from other organisms.

RuBP carboxylase has been identified in a number of marine algae, but reported levels of activity usually have been below observed rates of photosynthesis. Best agreement has been obtained with the diatom *Phaeodactylum tricornutum* (BEARDALL and MORRIS 1976, HOLDSWORTH and COLBECK 1976), but with several other species (GLOVER and MORRIS 1979, RIVKIN et al. 1982) and with natural populations of phytoplankton (PRISCU and GOLDMAN 1983, SMITH et al. 1983) photosynthesis was more than three times the measured RuBP carboxylase activity. Some authors are tempted to attribute the balance of carbon fixation to enzymes using PEP as CO_2 acceptor, but it should be remembered that none of these is connected to an autocatalytic sequence, such as the Calvin cycle, that permits growth. The demonstration by GLOVER and MORRIS (1979) of a positive correlation between RuBP carboxylase activity (but not PEP carboxylase activity) and rate of photosynthesis attests to the requirement for more effort being directed toward demonstrating RuBP carboxylase activities in marine algae commensurate with their capabilities for CO_2 fixation.

b) Other Calvin Cycle Enzymes

The struggle to understand chloroplast GAP dehydrogenase is continuing. Much of the confusion appears to be due to the presence of homotetrameric (identical subunits) and heterotetrameric (two different subunits) forms of the enzyme, the ability of light in vivo and NAD^+ (during extraction) to influence the ratio of these two forms, and the proclivity toward aggregation (IADAROLA et al. 1983, LOOZE and WAGNER 1983a,b). Sophisticated experiments by CERFF and KLOPPSTECH (1982), using messenger RNA translation by wheat germ ribosomes, have confirmed that the chloroplast enzyme is distinct from its cytosolic counterpart.

Chloroplast and cytosolic forms of triose-P isomerase (PICHERSKY and GOTTLIEB 1984) and FBP-aldolase (KRÜGER and SCHNARRENBERGER 1983) have been purified. In the latter case, a separate experiment with corn leaves indicated only one form of aldolase, although GASPERINI and PUPILLO (1982) did find two. Wheat, spinach, pea and rye leaves all have been shown to contain transketolase in both chloroplasts and the cytosol (MURPHY and WALKER 1982a, FEIERABEND and GRINGEL 1983).

The pivotal regulatory function of the Calvin cycle's FBPase has continued to attract attention. The light-mediated reductive activation of this enzyme (LEEGOOD and WALKER 1982) has been shown to be very dependent on pH and the presence of Mg^{2+} and FBP (LEEGOOD et al. 1982, MINOT et al. 1982, HERTIG and WOLOSIUK 1983). An unusual observation is that certain polyamines interfere with this activation (CORLEY et al. 1983). Chloroplast FBPase now has been purified using affinity chromatography (PLÁ et al. 1981). In the prokaryote *Anacystis nidulans* two FBPases have been identified (GERBLING et al. 1984). One of these was found to have properties similar to those of the higher plant chloroplast enzyme (UDVARDY et al. 1982). As with chloroplast FBPase, light-mediated activation of SBPase also benefits from the presence of the substrate (SBP) and Mg^{2+} (WOODROW and WALKER 1982, WOLOSIUK et al. 1982). Specifically, SBP increases the rate of reductive activation (WOODROW and WALKER 1983), while Mg^{2+} influences the dissociation constant for protonation, and thus modulates the effect of pH on the activity (WOODROW et al. 1984). A continuous spectrophotometric assay for SBPase has been developed by WOODROW and WALKER (1982).

Intensive efforts in Göttingen (GARDEMANN et al. 1982, 1983, FLÜGGE et al. 1982) have produced indications that the rapid cessation of CO_2 fixation upon darkening of leaves is largely due to ribulose-5-P kinase being rapidly inactivated by the effects of rapid changes in stromal metabolite levels superimposed on dark-mediated oxidation. Simulation experiments suggest that the enzyme achieves only 4% of its potential activity during illumination, and is completely inactive in darkened leaves. An affinity label that binds only to the reduced form of the enzyme was discovered recently (ASHTON 1984).

Other experiments on light-mediated enzyme regulation have been reported. WIRTZ et al. (1982) managed to measure the activities of enzymes from illuminated pea leaves, within 5 to 10 s after extraction, by developing a violent squashing technique. They detected light-mediated activations of FBPase, SBPase, and ribulose-5-P kinase, but not of RuBP carboxylase or chloroplast GAP dehydrogenase. The presence of thioredoxins that can mediate light activation, and activations of FBPase and SBPase, have been confirmed in a CAM plant (HUTCHESON and BUCHANAN 1979). Ferralterin, an alternative to the ferrodoxin/thioredoxin system, has been discovered (DE LA TORRE et al. 1982).

Certain enzymes outside the Calvin cycle are also light-regulated. Recent information indicates that chloroplast NADP-linked malate dehydrogenase activation is inhibited by $NADP^+$ (SCHEIBE and JACQUOT 1983), that the light-mediated regulation of glucose-6-P dehydrogenase is not mechanistically comparable to reduction of the enzyme by dithiothreitol

(SRIVASTAVA and ANDERSON 1983), and that even a pig-heart enzyme (in this case aconitase) can be activated by the chloroplast's light-activation machinery (MOHAMED and ANDERSON 1983). Evidence for some light-mediated activation of PEP carboxylase also has appeared (KARABOURNIOTIS et al. 1983, RAO and ANDERSON 1983).

c) Starch Metabolism

A transitory reserve of reduced carbon, in the form of starch, is deposited in chloroplasts during the day, and dissimilated to soluble sugar-Ps at night (KRUGER et al. 1983a). The evidence is as strong as ever (CAM plants and stomatal guard cells included) that the ratio of the concentrations of glycerate-3-P to P_i regulates leaf starch synthesis through its effect on ADP-glucose pyrophosphorylase activity (PREISS 1982, PORTIS 1982, OUTLAW and TARCZYNSKI 1984, SINGH et al. 1984). In *Chlorella*, synthesis was concentrated into the first half of the cell cycle (HIROKAWA et al. 1982) and was affected by temperature: growth at 38°C (rather than 20°C) produced a different quality starch, more amylose-like (NAKAMURA and MIYACHI 1982, NAKAMURA and IMAMURA 1983).

Concerted efforts continue to probe the relative contributions of hydrolysis (via amylases) and phosphorolysis (via phosphorylase) to starch degradation. Leaf phosphorylases, including the enigmatic extrachloroplast enzyme, have been further investigated (SHIMOMURA et al. 1982, HAMMOND and PREISS 1983, KRUGER and AP REES 1983a). The properties of amylase are less well understood, perhaps because so many forms have been identified in leaves (STEUP and GERBLING 1983). A chloroplast debranching enzyme has been recently purified (LUDWIG et al. 1984). Two reports imply that hydrolysis is the primary degradation reaction (CHANG 1982, STEUP et al. 1983) and the suggestion has been made that phosphorolysis is most likely restricted to a pool of soluble glucan intermediates (STEUP et al. 1983). However, other reports give phosphorolysis at least an equal status to hydrolysis (HAMMOND and BURTON 1983, HAMPP and GOLLER 1983, LEVI and GIBBS 1984). Interestingly, maltose (and higher maltodextrins) are formed in chloroplasts both in the light and in darkness, but enzymes other than amylases are involved (SCHILLING 1982, KRUGER and AP REES 1983b, LINDEN and SCHILLING 1984).

d) "Chloroplast Respiration"

Conversion of the immediate products of starch degradation to triose-P may be regulated at the reaction catalyzed by chloroplast phosphofructokinase [now also reported in a green alga (KOMBRINK and WÖBER 1982)]. The leaf enzyme is reportedly inhibited by NADPH (CSÉKE et al. 1982a) and reduced in activity by a light-mediated mechanism (HEUER et al. 1982). In the dark, ATP for the phosphofructokinase could be generated if one of the two triose-P (formed from the product FBP) is converted further to glycerate-3-P. However, this also produces an NADPH, and KOW et al. (1982a) have coined the novel term "chloroplast respiration" to accommodate evidence for the oxidation of this NADPH, with oxygen as the final electron acceptor, to produce H_2O_2 (Fig. 3). The ascorbate-mediated reduction of this H_2O_2 to water is quite possibly linked to the oxidation of another NADPH through glutathione (LATZKO and KELLY 1983a). A similar type of chloroplast respiration has been described by BENNOUN (1982).

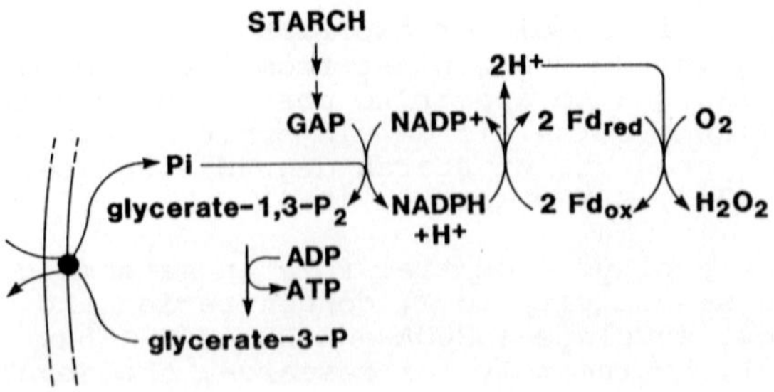

Fig. 3. "Chloroplast respiration". Adapted from KOW et al. (1982a). *Fd* ferredoxin (oxidized or reduced); ● phosphate translocator

e) Potassium, Calcium, and pH

Several new regulatory aspects of chloroplast carbon metabolism have appeared. An intriguing study by DEMMIG and GIMMLER (1983) showed that CO_2 fixation by isolated spinach chloroplasts could be increased from 10 to 110 µmol/mg chl·h by boosting the external KCl concentration from 10 to 100 mM. Calcium may also regulate carbon metabolism, possibly by activating NAD kinase. It does this in the form of a complex with the small regulatory protein calmodulin (MUTO 1983), which is receiving much attention in the world of mammalian biochemistry. A small but definite percentage (0.6% to 2%) of the cell's calmodulin was found to be in chloroplasts (JARRETT et al. 1982, MUTO 1982), and uptake of Ca^{2+} by chloroplasts has been demonstrated (MUTO et al. 1982). On the other hand, NAD kinase has been found in both the stroma and envelope of pea chloroplasts, and in the outer mitochondrial membrane of dark-grown corn seedlings, but only the membrane-bound enzymes underwent Ca^{2+}-calmodulin activation (DIETER and MARMÉ 1984, SIMON et al. 1984). It will be interesting to find out whether Ca^{2+}-calmodulin-dependent protein kinases, recently found in wheat germ and dark-grown corn coleoptiles (POLYA and MICUCCI 1984, VELUTHAMBI and POOVAIAH 1984), also occur in photosynthetic cells.

Experiments with salt (NaCl)-stressed plants have revealed a remarkable ability of chloroplasts to maintain their NaCl content (at about 70 µM) when the total leaf content varied 60-fold (KAISER et al. 1983). The chloroplasts in water-stressed plants have carbon fixation inhibited by stromal acidification which is somehow induced by the increased osmolarity, and results in reduced FBPase activity (BERKOWITZ and GIBBS 1983, BERKOWITZ et al. 1983).

5. Cytosolic Synthesis of Sucrose from Carbon Exported from the Chloroplast

a) Movement of Metabolites Across the Chloroplast Envelope

The outer membrane of the chloroplast envelope poses no problem for metabolite movement: it has pores of 2.5 - 3 nm diameter, and thus acts as a molecular sieve, excluding only those molecules with molecular weight over 7000 (FLÜGGE and BENZ 1984). It is the inner membrane that controls the import/export of most small molecules, usually via specific transporting proteins. Reduced carbon is exported as triose-P or glycerate-3-P via the well-established phosphate translocator. Evidence for the presence of this translocator in chloroplasts from two algae, *Chlamydomonas reinhardii* (BELKNAP 1983, KLEIN et al. 1983) and *Codium fragile* (RUTTER and COBB 1983), has been obtained, but SCHÜRMANN and ORTIZ (1982) could not detect evidence for a translocator in *Euglena gracilis* chloroplasts. The phosphate translocator of C_4-plant mesophyll chloroplasts has acquired the principal role of exporting PEP (HALLBERG and LARSSON 1983) generated in the chloroplast from imported pyruvate. Chloroplasts from CAM plants also appear to possess phosphate translocators, but somehow seem able to shield Calvin cycle intermediates from export, so that only products of starch degradation move into the cytosol (PIAZZA et al. 1982, MONSON et al. 1983).

The phosphate translocator clearly plays a central role in metabolic regulation, as evidenced by its sensitivity to P_i concentration and to the ΔpH across the chloroplast envelope (USUDA and EDWARDS 1982, FLÜGGE et al. 1983, PORTIS 1983). Fortunately for research, the development by KOW and GIBBS (1982) of a particulate reconstituted chloroplast system allows the sensitive translocator to be circumvented where desired during investigations.

Other chloroplast envelope translocators have also received attention. The capacity of the equally well-established dicarboxylate translocator to indirectly export reducing equivalents to the cytosol has been seriously questioned (GIERSCH 1982, SOMERVILLE and OGREN 1983), and the probability that there is more than one type of dicarboxylate translocator has been raised (DRY and WISKICH 1983, SOMERVILLE and OGREN 1983, WOO 1983, PROUDLOVE et al. 1984). In addition, a light-stimulated glycerate carrier has been identified (ROBINSON 1982a,b), and a translocator specific to C_4 and CAM plants has been proposed from experiments with isolated envelope membranes (FOSTER and EDWARDS 1983).

b) The Regulation of Sucrose Biosynthesis

Until recently (see following paragraph), the regulation of sucrose biosynthesis was elusive to the point where its nonexistence was contemplated. Feedback inhibition of any enzyme by sucrose was effectively ruled out by FOYER et al. (1983a), a conclusion consistent with the partitioning of sucrose into the vacuole in isolated protoplasts (KAISER et al. 1982). However, the possibility of some regulatory phenomena was kept alive by the persistent efforts of HUBER (1983) and co-workers who established positive correlations between the synthesis and/or content of sucrose, and the activity of sucrose-P synthase (HUBER and ISRAEL 1982, RUFTY and HUBER 1983, RUFTY et al. 1983). Supporting results were obtained with *Solanum melongena* (CLAUSSEN and LENZ 1983) and with wheat leaf protoplasts (STITT et al. 1983a). However, great sensitivity of the synthase to known effector molecules has not been found (AMIR and PREISS 1982, DOEHLERT and HUBER 1983), and this no doubt caused some workers to consider an alternative enzyme for regulation, i.e., the cytosolic FBPase (FOYER et al. 1982, STITT et al. 1983a). Again, however, apart from the interesting property of cold lability (WEEDEN and BUCHANAN 1983), great sensitivity to effectors had not been recorded for this enzyme.

This changed, and a completely new perspective to the regulation of sucrose biosynthesis was introduced, when CSÉKE et al. (1982b) and STITT et al. (1982a) reported that a sugar bisphosphate recently discovered in animal tissues, fructose-2,6-P_2, also occurs in the cytosol of leaf cells and is a potent inhibitor of the cytosolic FBPase. Subsequent work correlated decreased sucrose formation with increased fructose-2,6-P_2 levels (STITT et al. 1983b, HUBER and BICKETT 1984). However, this neat correlation overshadows the obvious complexity of the regulatory system involved. Fructose-2,6-P_2 is not a metabolic intermediate; it occurs in (and is effective at) nanomolar concentrations. [In fact, it markedly influenced experiments when present as a contaminant of commercial fructose-6-P at a level of only 0.0002% (KRUGER et al. 1983b)!] A special kinase exists for its synthesis

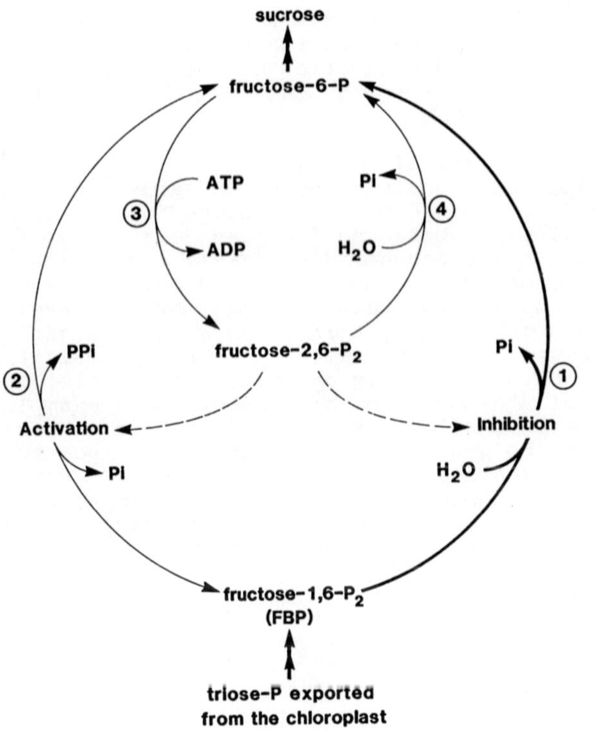

Fig. 4. Regulation of cytosolic sucrose biosynthesis at the stage of FBP hydrolysis (*reaction 1*; enzyme FBPase). An apparently small amount of the product of FBP hydrolysis, fructose-6-P, is used to generate (*reaction 3*; enzyme fructose-6-P,2 kinase) a different fructose bisphosphate, i.e., fructose-2,6-P_2, that inhibitis FBPase and thus slows down the hydrolysis of further FBP. The fructose-2,6-P_2 can be eliminated subsequently by hydrolysis (*reaction 4*; enzyme fructose-2,6-bisphosphatase). The significance of *reaction 2* (enzyme: PP_i-linked P-fructokinase), which is activated by fructose-2,6-P_2, is not yet clear. Adapted from several references (see text)

(CSÉKE and BUCHANAN 1983) and a special phosphatase hydrolyzes it (SMYTH et al. 1984). Obviously, regulation of the activities of these two enzymes will also have to be considered. In addition, fructose-2,6-P_2 is a powerful effector (in this case, activator) of a second cytosolic enzyme, the PP_i-linked P-fructokinase that has been found so far only in plants and certain microorganisms (PFLEIDERER and KLEMME 1980, SABULARSE and ANDERSON 1981, CSÉKE et al. 1982b, STITT et al. 1982a, CARNAL and BLACK 1983, SMYTH et al. 1984). Since the reaction catalyzed by this unusual P-fructokinase is reversible, its role vis-a-vis that of FBPase remains in question. The current status of these new discoveries is outlined in Figure 4. Interestingly, there is essentially no evidence for fructose-2,6-P_2, or for enzymes regulated by it, in the chloroplast (CSÉKE et al. 1982b).

c) Soluble Low-Molecular-Weight Carbohydrates Other than Sucrose

While it must be acknowledged that sucrose is the major soluble carbohydrate in a wide variety of plants, it should be emphasized that other soluble carbohydrates occur at comparable concentrations in many species. In higher plants alditols, such as mannitol and sorbitol, are particularly prevalent. RUMPHO et al. (1983) have recently demonstrated that celery leaves, which form as much mannitol as sucrose, contain in the cytosol activities of the three enzymes required for mannitol synthesis. They propose that the reduction of mannose-6-P to mannitol-1-P is driven by photosynthetically generated NADPH indirectly transferred from the chloroplast by the triose-P/glycerate-3-P shuttle that utilises nonreversible GAP dehydrogenase in the cytosol. In the leaves of apple, and most probably other members of the Rosaceae, NADPH is used to reduce glucose-6-P to sorbitol-6-P for the production of sorbitol, which is the major translocated form of carbon (LOESCHER et al. 1982), and if this reaction is also cytosolic, then operation of the same shuttle could be envisaged here.

Marine algae are noted for the variety of soluble carbohydrates accumulated (KREMER 1981a). An extensive analysis by KREMER and KIRST (1982) led to the identification of four main groups: (1) those that accumulate free monosaccharides, mainly glucose and fructose (e.g., cryptomonads), (2) those that accumulate disaccharides (mainly sucrose) as well as some glucose and fructose (e.g., *Ulva lactuca* and certain other green algae), (3) those that accumulate alditols (e.g., brown algae, renowned for mannitol accumulation, and certain green algae – a variety of C_3 to C_7 alditols are encountered), and (4) those that accumulate heterosides, i.e., a monosaccharide (usually galactose or mannose) attached to an aglycone (typically glycerol or glycerate); many examples belong to the red algae. In almost all cases it would seem feasible that these carbohydrates could be synthesized from triose-P exported from chloroplasts, in a manner analogous to sucrose synthesis. Even volemitol, a C_7 alditol derived from sedoheptulose-7-P, could be generated by partial operation of the oxidative pentose-P pathway with a loss (as CO_2) of only one-sixth of the entering carbon.

6. Nitrogen Assimilation, and Lipid Biosynthesis

a) Nitrogen Assimilation

Plants acquire most nitrogen either as NH_4^+, NO_3^- (which is then reduced to NH_4^+), or through nitrogen fixation (which produces NH_4^+). The NH_4^+ is then incorporated into the amino acid pool by glutamine synthetase. Higher plant leaves (McNALLY et al. 1983) and green algae (FLORENCIO and VEGA 1983, SUMAR et al. 1984) contain two forms of this enzyme. With higher plants one of these is in the chloroplast while the other is cytosolic, but while many species have both forms, some lack the cytosolic enzyme (HIREL et al. 1982a, 1983, McNALLY et al. 1983). Activity of the chloroplast enzyme markedly increases when dark-grown leaves are illuminated (HIREL et al. 1982b, NISHIMURA et al. 1982a). The next enzyme of NH_4^+ assimilation, glutamate synthase, has been purified from rice leaves (SUZUKI and GADAL 1982). Evidence for light-mediated activation of this enzyme in *Chlorella sorokiniana* has been reported by TISCHNER and SCHMIDT (1982).

Carbon skeletons for the net synthesis of amino acids must be derived ultimately from the Calvin cycle. The conversion of triose-P, exported from the chloroplast, to pyruvate, oxaloacetate, and α-oxoglutarate must therefore be an integral component of nitrogen assimilation (LATZKO and KELLY 1983b). This conversion involves PEP as an intermediate, which then may be either carboxylated by PEP carboxylase or converted to pyruvate by pyruvate kinase. There are conflicting reports as to whether NH_4^+ uptake leads to increased PEP carboxylase activity (KANAZAWA et al. 1983, MARQUES et al. 1983). Two forms of pyruvate kinase have been found in leaves, and the regulatory properties of one of these seemed particularly suited to NH_4^+ assimilation (RUYTERS and MIYACHI 1983, BAYSDORFER and BASSHAM 1984). Growth under blue light, which favors protein synthesis, appears to stimulate pyruvate kinase activity (RUYTERS 1982, RUYTERS and MIYACHI 1983).

Marine phytoplankton obtain nitrogen as NH_4^+, NO_3^- or urea (REES and BEKHEET 1982, ROSSLENBROICH and DÖHLER 1982). In the latter case, Ni becomes an essential nutrient where urease, a Ni-containing enzyme, is used to hydrolyze the urea. Recent evidence suggests that marine phytoplankton take up NH_4^+, using an active transport system with very high affinity, presumably to compensate for the low background concentration (less than 30 mM) of NH_4^+ in the ocean (KOIKE et al. 1983).

b) Lipid Biosynthesis

The active research into lipid biosynthesis mentioned in our last review has continued unabated. Controversy has emerged concerning the source of acetate for fatty acid synthesis in the chloroplast. The pathway proposed for spinach leaves, i.e., hydrolysis of acetyl-CoA in the mitochondrion, and movement of the acetate to the chloroplast where resynthesis of acetyl-CoA occurs (LIEDVOGEL and STUMPF 1982, MURPHY and WALKER 1982b), does not seem applicable to pea leaves, in which direct generation of acetyl-CoA by pyruvate dehydrogenase appears more likely (GIVAN 1983, GIVAN and HODGSON 1983). The distribution of these alternative approaches among other plants remains to be established.

The synthesis of fatty acids from acetyl-CoA proceeds in the stroma of plastids, catalyzed by a fatty acid synthetase complex of the non-associated type, i.e., similar to that in the prokaryote *Escherichia coli* (SHIMAKATA and STUMPF 1982). In chloroplasts, synthesis appears to be favored during illumination (SAUER and HEISE 1983). The synthetase produces palmitate, which may be incorporated into lipids, or used for the production of other fatty acids. In the plastid stroma it may be elongated to stearate, which may in turn be desaturated to oleate (in both reactions the fatty acid is attached to acyl carrier protein), but further desaturation of oleate occurs only after the oleate is exported from plastids and esterified into a lipid, probably phosphatidylcholine in the endoplasmic reticulum (ROUGHAN and SLACK 1982). Enzymes for the esterification of oleate into phosphatidylcholine and for its desaturation to linoleate have been demonstrated in microsomal preparations from pea leaves (MURPHY et al. 1983a,b), and evidence for the channeling of newly synthesized dioleoyl-phosphatidylcholine directly to the desaturase enzyme has been obtained by MURPHY et al. (1984). Details of the further desaturation of linoleate to linolenate have not been worked out. Current indications are that it occurs on the chloroplast envelope, where galactose units are also added to produce the envelope's galactolipids (DORNE et al. 1982, OHNISHI and YAMADA 1982, ROUGHAN and SLACK 1982, BLOCK et al. 1983, CLINE and KEEGSTRA 1983, FRENTZEN et al. 1983), but desaturation on the endoplasmic reticulum has not been ruled out (DUBACQ et al. 1983).

Oceanic algae have some unique features to their fatty acid composition. Linoleic (18 : 2, i.e., 18 carbons, 2 double bonds) and linolenic (18 : 3) acids are less prevalent than in land plants and, in fact, linoleic acid is essentially absent from diatoms and dinoflagellates, two principal algal groups in marine phytoplankton. On the other hand, longer chain polyunsaturated fatty acids, especially eicosapentaenoic (20 : 5) and docosahexaenoic (22 : 6), seldom encountered in land plants, are particularly predominant in marine algae (KATES and VOLCANI 1966, ACKMAN et al. 1968, VOLKMAN et al. 1981, JONES et al. 1983). Nothing is known about the biosynthesis of these marine fatty acids, but the incentive for research will no doubt be stimulated by burgeoning evidence that these fatty acids, which persist through the oceanic food web into fish, are particularly effective at reducing the incidence of cardiovascular disease in man (SIESS et al. 1980, GOODNIGHT et al. 1982, COREY et al. 1983).

7. C_4 Metabolism

a) C_4 Plants

Some species of *Cymbopogon* and *Vetiveria*, which are aromatic tropical grasses, and *Microstegium vimineum*, a shade-adapted grass, have been added to the list of plants with the C_4 pathway of photosynthesis (RAJENDRUDU and DAS 1981, WINTER et al. 1982a). The pathway can make use of aspartate as a reservoir of dicarboxylic acid in some species (CHAPMAN and HATCH 1981, SHIEH et al. 1982). Decarboxylation of malate in the bundle-sheath cells is catalyzed, in one subgroup of C_4 plants, by a mitochondrial NAD-linked malic enzyme. GARDESTRÖM and EDWARDS (1983) have now reported the interesting observation that the mitochondria from these cells (but not those from mesophyll cells) oxidize malate without respiratory control, apparently via the alternative oxidase (cyanide-resistant) pathway, thus ensuring that the C_4 pathway in these plants may operate without the constraints that coupled electron flow might impose.

The remaining research with C_4 plants has concentrated on the enzymes involved. A report by SUGIYAMA and HIRAYAMA (1983) that the biomass of maize seedlings correlated with the PEP carboxylase (but not RuBP carboxylase) activity agreed with figures from BASSI and PASSERA (1982), showing that higher light intensity during growth was associated with a much higher level of PEP carboxylase, but only slightly higher level of RuBP carboxylase. There are indications for more than one form of PEP carboxylase in leaves (PAN and TAN 1981, STIBOROVÁ and LEBLOVÁ 1983, VIDAL and GADAL 1983), and also for light-mediated activation (KARABOURNIOTIS et al. 1983). Immunochemical techniques have confirmed that the main form is cytosolic, while the subsequent C_4 pathway enzyme, NADP-linked malate dehydrogenase, is in the chloroplast (PERROT-RECHENMANN et al. 1982, 1983). This latter enzyme is synthesized de novo during greening of leaves (VIDAL and GADAL 1981), and undergoes reductive light-mediated activation, the extent of which depends on the redox state of ferredoxin and associated electron carriers (LEEGOOD and WALKER 1983). An affinity label that binds only to the reduced, active enzyme has been reported (ASHTON 1983). The location of PEP carboxykinase in those C_4 plants that use this enzyme for decarboxylation in bundle-sheath cells has been finally established as the cytosol (CHAPMAN and HATCH 1983, WATANABE et al. 1984).

The unique mechanism of light/dark-mediated regulation of pyruvate P_i dikinase has been consistently probed (NAKAMOTO and SUGIYAMA 1982, NAKAMOTO and EDWARDS 1983a,b,c, HATCH and BURNELL 1983). Inactivation was finally traced to a protein factor (enzyme?)-mediated phosphorylation of the enzyme, with ADP as the phosphate donor (ASHTON and HATCH 1983). This is the first known protein phosphorylation by ADP. It only occurred with enzyme already phosphorylated "conventionally", i.e., by ATP (BURNELL and HATCH 1984). Activation by P_i appears to involve the same protein factor (BURNELL and HATCH 1983). A further report that pyruvate P_i dikinase occurs in C_3 plants, at levels lower than in C_4 plants, has appeared (AOYAGI and BASSHAM 1983).

b) C_3-C_4 Intermediate Plants

Evidence for the existence of C_3-C_4 intermediate plants has become tenuous. Properties of the enzymes PEP carboxylase and glutamine synthetase from species claimed to be C_3-C_4 were in some respects intermediate between those of the enzymes from closely related C_3 and C_4 plants (NAKAMOTO et al. 1983, HIREL et al. 1983). However, from the three main examples of C_3-C_4 intermediates, viz. *Panicum milioides*, *Moricandia arvensis* (HOLADAY et al. 1982), and several species of *Flaveria* (APEL and MAASS 1981, KU et al. 1983), only the latter remains unchallenged. A consensus has emerged that *M. arvensis* is a C_3 plant (WINTER et al. 1982b, HOLADAY and CHOLLET 1983), but more interesting is the erosion of evidence for *P. milioides* being C_3-C_4, since its reputation has been more long-standing. In particular, previous claims that *P. milioides* leaves contain pyruvate P_i dikinase activity have been squelched (MEYER et al. 1982, EDWARDS et al. 1982, HOLADAY and CHOLLET 1983). The report by KU et al. (1983) that this enzyme is in C_3-C_4 *Flaveria* species will therefore bear further investigation.

c) CAM

Interest in CAM has not waned. Rapid publication of the proceedings of a symposium on CAM (TING and GIBBS 1982) attests to this. The conclusions by these editors that "CAM is an adaptation to dry environments" is reflected in an odd way by the findings of WINTER et al. (1983) that Australian CAM plants, although found mainly in rain forests, are epiphytic species (e.g., many Orchidaceae) and therefore liable to short but frequent periods of water stress. Among more conventional CAM plants, a range of types is emerging, including C_3 plants with some CAM characteristics (MARTIN and ZEE 1983, TING and BURK 1983), *Cissus quadrangularis* which has C_3 leaves but CAM stems (TING et al. 1983), and further examples of species with CAM when well watered, but CAM-idling when water-stressed (RAYDER and TING 1983a,b).

Malate formed from nocturnal PEP carboxylation during CAM is thought to be actively transported into vacuoles (SMITH et al. 1982) in which the pH falls (STIDHAM et al. 1983), although a permease capable of mediating an exchange diffusion of malate across the vacuole membrane also exists (BUSER-SUTER et al. 1982). Cytosolic pH in CAM cells is held relatively constant during CAM (MARIGO et al. 1983). Malate decarboxylation the following day proceeds more rapidly than photosynthesis, thus CO_2 accumulates behind closed stomata (BARROW and COCKBURN 1982).

With respect to the enzymology of CAM, NAD-linked malic enzyme has been purified and shown to be hysteretic (WEDDING and BLACK 1983), and pyruvate P_i dikinase has been shown to have a wide distribution among CAM plants (SANADA and NISHIDA 1982). Adenylate kinase was found to be 4 - 20 times more active in CAM plants utilizing malic enzyme for decarboxylation, compared to those utilizing PEP carboxykinase (MANETAS et al. 1983). Most other studies have centered on PEP carboxylase. This enzyme was more prominent in older leaves (AMAGASA 1982, WINTER et al. 1982c). Its regulatory properties appear to change diurnally (BRULFERT and QUEIROZ 1982, BRULFERT et al. 1982a, MANETAS 1982, WINTER 1982), and are affected by temperature in vitro (BUCHANAN-BOLLIG et al. 1984). Discussion continues as to whether more than one form of enzyme is present: in most cases only one form is seen (MÜLLER et al. 1982, NOTT and OSMOND 1982), but BRULFERT et al. (1982b) claim that a second form appeared when CAM was induced in *Kalanchoe blossfeldiana*. Marked increase in activity due to synthesis (but not necessarily of a new enzyme form) also occurred with another species in which CAM can be induced, viz. *Mesembryanthemum crystallinum* (FOSTER et al. 1982, HOLTUM and WINTER 1982). Chloroplasts have been obtained from protoplasts of this plant and, rather remarkably, showed distinctly different properties depending on whether the parent tissue was performing CAM or C_3 photosynthesis (WINTER et al. 1982d, DEMMIG and WINTER 1983).

d) Freshwater Macrophytes, and Stomata

C_4 metabolism is a notable feature in submersed freshwater plants. It is most conspicuous under summer-like growth conditions, and is distinguished by high PEP carboxylase activity (BEER and WETZEL 1982, ASCENCIO and BOWES 1983). Low photorespiration was also observed, and attributed to CO_2-concentration via either a C_4-like mechanism or active uptake of inorganic carbon (SALVUCCI and BOWES 1983a,b).

Stomatal guard cells accumulate malate via carboxylation of PEP (THORPE 1983) generated from starch. Activation of P-fructokinase (SCHNABL et al. 1982) and a high capacity for glycolysis (HAMPP et al. 1982a) must benefit this process. These cells also possess a high potential for respiration (SHIMAZAKI et al. 1983). However, the absence of RuBP carboxylase and a Calvin cycle has been confirmed (HAMPP et al. 1982a, OUTLAW et al. 1982), except perhaps for CAM plants (MADHAVAN and SMITH 1982).

e) Marine Algae

C_4 metabolism appears to have assumed great importance in the ocean, particularly for the unicellular algae. This is because high growth rates, coupled to high rates of cell division, result in these plants concentrating on protein synthesis, rather than synthesis of carbohydrates. Protein synthesis requires, in turn, a supply of amino acids, the majority of which have carbon skeletons derived from the tricarboxylic acid cycle. Replenishment of intermediates taken from this cycle is a principal function of PEP carboxylation (see LATZKO and KELLY 1983b), hence it is not surprising that short-term feeding of $[^{14}C]O_2$ to marine algae has resulted in labeling of C_4 compounds approaching, or even surpassing, that of C_3 compounds (BEARDALL et al. 1976, MUKERJI et al. 1978, ROSSLENBROICH and DÖHLER 1982, KULKARNI and NIMBALKAR 1983). Labeling becomes almost exclusively C_4 type in cultures approaching or at stationary phase (MUKERJI et al. 1978) when PEP carboxylation persists after RuBP carboxylation has significantly decreased (GLOVER and MORRIS 1979). Initially, results such as these led to speculation that, in some algae, PEP carboxylase had achieved an importance comparable to that of RuBP carboxylase, and although such thoughts still appear (PRISCU and GOLDMAN 1983), they have in general been checked by reminders from KREMER (1981a,b) that all the PEP provided is derived from the Calvin cycle, either directly from glycerate-3-P in the light, or (in the dark) from stored carbohydrates synthesized from Calvin cycle intermediates during a preceding light period. Figure 5 shows that such metabolic activity can lead to a spectrum of C_3/C_4 labeling ratios.

An interesting aspect of marine PEP carboxylation is that it is most predominant in brown algae (KREMER 1981a), including diatoms, in which the carboxylating enzyme is PEP carboxykinase; PEP carboxylase appears to be absent (AKAGAWA et al. 1972, HOLDSWORTH and BRUCK 1977, ROSSLENBROICH and DÖHLER 1982, KERBY and EVANS 1983).

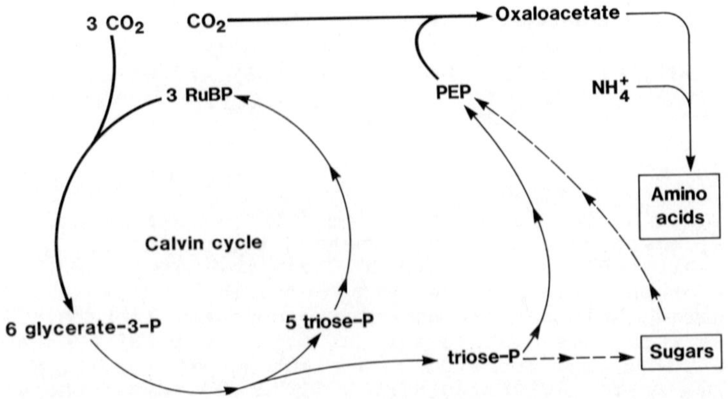

Fig. 5. The net fixation of three CO_2 by the Calvin cycle produces one triose-P. If each triose-P is used to synthesize a C_4 amino acid (in a metabolic sequence that includes fixation of a fourth CO_2), a short-term labeling pattern of "75% C_3/ 25% C_4" will ensue *(heavy lines)*. The proportion of label in C_4 compounds would, however, be expected to be less since the triose-P is also required for other biosyntheses, e.g., sugars *(broken lines)*. On the other hand, at later growth stages these sugars may be remobilized to generate additional PEP *(broken lines)*, and if this occurs when the Calvin cycle activity is low, label may appear predominantly in C_4 compounds

8. Respiration

a) Mitochondrial (Dark) Respiration

The popular belief that mitochondrial respiration in photosynthetic cells is largely inhibited during photosynthesis (presumably due to a lack of ADP which is fully utilized by the chloroplast) has received support from some quarters, but has been challenged in others. In green and blue-green algae, inhibition seemed to vary according to growth rate and species (SCHERER and BÖGER 1982, HARRIS and PICCININ 1983). In higher plants, illumination led to a marked increase in the cytosolic dihydroxyacetone-P/glycerate-3-P ratio, which could act through the adenylate system to inhibit respiration (HEBER et al. 1982). However, two groups of workers, using new techniques for the rapid fractionation of protoplasts, could find little or no change in the cytosolic and mitochondrial ATP/ADP ratios when dark-kept protoplasts were illuminated (GOLLER et al. 1982, HAMPP et al. 1982b, LILLEY et al. 1982, STITT et al. 1982b) and it has been concluded that any inhibition of mitochondrial respiration in the light cannot be due to changes in the cytosolic ATP/ADP ratio (DRY and WISKICH 1982, STITT et al. 1982b). Adenylate pools are equilibrated in the chloroplasts (by a stromal adenylate kinase) and cytosol (by an adenylate kinase in the mitochondrial intermembrane space), but not in the mitochondrial matrix (BIRKENHEAD et al. 1982, HAMPP et al. 1982b, STITT et al. 1982b).

Studies with intact wheat leaves have produced data to support the inhibition of dark respiration by light in one case (PEISKER and APEL 1981), but not in another (AZCÔN-BIETO and OSMOND 1983). If any respiration proceeds in the light, a new question will arise, viz. what proportion can be attributed to the alternative (cyanide-resistant) pathway recently demonstrated to be present in leaves, particularly when the supply of respiratory substrate is abundant (AZCÔN-BIETO et al. 1983a,b). These and other questions should receive answers soon, now that high-quality preparations of leaf mitochondria are available (NEUBURGER et al. 1982, NISHIMURA et al. 1982b, NASH and WISKICH 1983).

b) Photorespiration

Oxygenation of the Calvin cycle's CO_2 acceptor (RuBP), catalyzed by RuBP carboxylase/oxygenase, and subsequent metabolism of one of the products (glycolate-P) to glycerate-P and CO_2, via the glycolate pathway, form the basis of photorespiration. From one viewpoint it is a consequence of an unfavorable ratio: that of the concentrations of CO_2 and O_2 in air. These two gases compete for the active site of the enzyme, but only when CO_2 is incorporated does plant growth result. By increasing the CO_2/O_2 ratio, WULFF and STRAIN (1982) increased the dry matter production of *Desmodium paniculatum* twofold, despite a decline in photosynthesis/unit leaf area, and KAMIŃSKA and MALESZEWSKI (1982) and SOMERVILLE and SOMERVILLE (1983) found a decreased flow of assimilated carbon through the glycolate pathway in bean and *Arabidopsis* leaves when the CO_2/O_2 ratio was similarly increased. The complementary dream of researchers, i.e., to develop a mutant RuBP carboxylase with lowered affinity for O_2 (i.e., low oxygenase activity in air) is still being pursued (SPREITZER et al. 1982).

Improved methods of estimating photorespiration have been developed (PETERSON 1983, VINES et al. 1983). The apparent increase of photorespiration with temperature (MONSON et al. 1982, PETERSON 1983) has been explained by HALL and KEYS (1983) to be due to changes in the relative solubilities of CO_2 and O_2 with temperature. In C_4 plant leaves, O_2 uptake by RuBP carboxylase/oxygenase seemed to be insensitive to CO_2 levels, but some evidence indicated that increased CO_2 did, as expected, decrease this O_2 uptake, but that the decrease was masked by a concomitant increase in O_2 uptake by the Mehler reaction (i.e., oxygen reduction leading to H_2O_2 formation) (FURBANK and BADGER 1982, 1983, JOLIVET-TOURNIER and GERSTER 1984).

Extensive investigation of the glycolate pathway has occurred. Six enzymes of the pathway have received detailed attention (BIEKMANN and FEIERABEND 1982, WINKLER et al. 1982, KLECZKOWSKI and RANDALL 1983, NISHIMURA et al. 1983a,b, SAROJINI and OLIVER 1983, TITUS et al. 1983, HUSIC and TOLBERT 1984). A transporter that facilitates glycolate export from the chloroplast has been identified (HOWITZ and McCARTY 1983); the glycolate can then easily enter peroxisomes (LIANG and HUANG 1983). Quality preparations of peroxisomes can now be obtained from leaves (LIANG et al. 1982, SCHMITT and EDWARDS 1982). Peroxisomal oxidation of glycolate produces glyoxylate, for which four fates have been considered: (1) reduction back to glycolate (PETERSON 1982), (2) condensation with succinate to produce isocitrate (SINGH and NAIK 1983), (3) nonenzymic decarboxylation by H_2O_2 [no longer considered physiologically important in leaves (WALTON 1982, WALTON and WOOLHOUSE 1983, KENDALL et al. 1983), but is significant in the unusual case of *Euglena* (YOKOTA and KITAOKA 1982, YOKOTA et al. 1983b)], and (4) amination to form glycine (the most favored possibility). New evidence suggests that alanine may be an important amine donor to glyoxylate (BETSCHE 1983, NAKAMURA and TOLBERT 1983, WALTON 1983). Glycine moves into mitochondria, either by diffusion or via an active uptake system (PROUDLOVE and MOORE 1982, WALKER et al. 1982a, YU et al. 1983). In the mitochondrion, glycine is oxidized in preference to tricarboxylic acid cycle intermediates (WALKER et al. 1982b, BERGMAN and ERICSON 1983, DRY et al. 1983). An indication that glycine is generated by a glycolate pathway in C_4 plant leaves (MAREK and STEWART 1983) has been supplemented by the observation that only bundle-sheath mitochondria are capable of oxidizing glycine; mesophyll mitochondria have lost this ability (OHNISHI and KANAI 1983). Glycine oxidation produces CO_2, NH_3, and a one-carbon unit that condenses with a second glycine to form serine which returns to the peroxisome and is deaminated to hydroxypyruvate. The latter is reduced to glycerate, but the source of reducing power is unclear now that a malate/aspartate shuttle has been effectively ruled out (SCHMITT and EDWARDS 1983). Finally, glycerate returns to the chloroplast via a light-stimulated chloroplast transporter (ROBINSON 1982a,b).

Reassimilation of the NH_3 produced during glycine oxidation occurs in the chloroplast, utilizing the enzymes normally employed for nitrogen assimilation, viz. glutamine synthetase and glutamate synthase. The first of these enzymes is absent from mitochondria (NISHIMURA et al. 1982b). The chloroplast system has been demonstrated, using inhibitors, in C_3 and C_4 plant leaves and green algae (LARSSON et al. 1982, PLATT and RAND 1982, MARTIN et al. 1983, PELTIER and THIBAULT 1983), and further confirmed using $[^{15}N]$ (WOO et al. 1982, BERGER and FOCK 1983). It has been shown operative with isolated chloroplasts (WOO and OSMOND 1982) and a reconstituted chloroplast system (ANDERSON and WALKER 1983). The principal role of the chloroplast envelope's dicarboxylate transporter is now considered to be the import of α-oxoglutarate and export of glutamate during this reassimilation of NH_3 (SOMERVILLE and OGREN 1983).

The role of photorespiration in the disposal of excess photosynthetically generated energy (10 ATP and 6 NADPH are utilized for each CO_2 released and subsequently refixed: see LATZKO and KELLY 1970) has been discussed less vigorously in the last 2 years. It had been shown that intact leaves suffered photoinhibition (i.e., damage to photosystem II) in the light if not allowed to regenerate $NADP^+$ through either photorespiration or photosynthesis, but it was not possible to imitate this situation with isolated chloroplasts (CORNIC et al. 1982). This result means that, while the prevention of photoinhibition by photorespiration is still a real possibility, the mechanism is not completely clear. Photoinhibition can be potentiated by other stresses, e.g., low temperature (POWLES et al. 1983). It is thought to be initiated by the Mehler reaction, i.e., electrons being passed from reduced ferredoxin to oxygen (in the absence of $NADP^+$, the "usual" electron acceptor), with consequent accumulation of toxic products such as superoxide ions and H_2O_2. However, it is clear that the Mehler reaction proceeds to some extent even under more-or-less normal conditions (BEHRENS et al. 1982, FURBANK et al. 1982), concomitantly with $NADP^+$ reduction (ROBINSON and GIBBS 1982), and is coupled to ATP synthesis (i.e., "pseudocyclic" photophosphorylation) in both C_3 and C_4 plant leaves, thus enabling chloroplasts to balance their ATP to NADPH requirements (FURBANK et al. 1982, 1983). The H_2O_2 produced by this Mehler reaction is presumably reduced to H_2O by NADPH through the ascorbate/glutathione system shown in Fig. $7B_1$ of our last review (KELLY and LATZKO 1982). This system has been demonstrated with both intact and ruptured chloroplasts (JABLONSKI and ANDERSON 1982, ANDERSON et al. 1983a,b), and another of the constituent enzymes characterized (HOSSAIN and ASADA 1984). Uptake of ascorbate by spinach chloroplasts has been observed (ANDERSON et al. 1983b, BECK et al. 1983), and considerable concentrations of both ascorbate and glutathione, each maintained predominantly in the reduced form in both light and darkness, have been demonstrated in chloroplasts (ANDERSON et al. 1983a, FOYER et al. 1983b, LAW et al. 1983). In darkness, NADPH generated during mobilization of starch to glycerate-3-P is believed to be oxidized by a similar ascorbate-based system (KOW et al. 1982b).

Further clues to the mechanism of photoinhibition, and the proposed protective role of photorespiration, may be expected from studies with phytoplankton algae. These algae, especially those in the ocean, grow at depths where the light intensity is between 1% and 10% of full sunlight. Many species become photoinhibited at higher light intensities, even in their natural environment where photosynthesis and photorespiration would seem possible. For example, photoinhibition of the dinoflagellate *Amphidinium carterae* occurs at about 1/20th of full sunlight (SAMUELSSON and RICHARDSON 1982). The reason for such sensitivity remains to be explained, but two hints are available. Firstly, in brown algae the Mehler reaction appears to be quite active (POPOVIC et al. 1983). Secondly, it is likely that photorespiration does not occur in the ocean (KELLY 1984). Although there is evidence for the production of glycolate by phytoplankton (FOGG 1983), and for enzymes of the glycolate pathway in the diatom *Cylindrotheca fusiformis* (PAUL and VOLCANI 1976), studies with the diatom *Navicula pelliculosa* and the red unicellular alga *Porphyridium cruentum* suggest that photorespiration is largely repressed by a high internal CO_2 concentration consequent upon the ocean's abundance of inorganic carbon (as 2 mM HCO_3^-) and the proclivity for these algae to actively transport HCO_3^- into the cell (BIRMINGHAM et al. 1982, COLMAN and GEHL 1983). Similar conclusions may apply to freshwater algae, since MERRETT and ARMITAGE (1982) found that the CO_2-concentrating mechanism of *Chlamydomonas reinhardii* apparently repressed glycolate formation, and VINCENT et al. (1984) found that the bright light experienced by phytoplankton in the high-altitude Lake Titicaca (Peru) led to marked photoinhibition, an indication that the level (if any) of photorespiration in these algae was too low to prevent them from experiencing this photoinhibition.

References

ACKMAN, R.G., TOCHER, C.S., McLACHLAN, J.: J. Fish. Res. Board Can. 25, 1603-1620 (1968). - AKAGAWA, H., IKAWA, T., NISIZAWA, K.: Plant Cell Physiol. 13, 999-1016 (1972). - AMAGASA, T.: Plant Cell Physiol. 23, 1471-1474 (1982). - AMIR, F., PREISS, J.: Plant Physiol. 69, 1027-1030 (1982). - ANDERSON, J.W., WALKER, D.A.: Planta 159, 247-253 (1983) - ANDERSON, J.W., FOYER, C.H., WALKER, D.A.: Biochim. Biophys. Acta 724, 69-74 (1983a); - Planta 158, 442-450 (1983b). - ANDREWS, T.J., BALLMENT, B.: J. Biol. Chem. 258, 7514-7518 (1983). - AOYAGI, K., BASSHAM, J.A.: Plant Physiol. 73, 853-854 (1983). - APEL, P., MAASS, I. Biochem. Physiol. Pflanzen 197, 396-399 (1981). - ASCENCIO, J., BOWES, G.: Photosynth. Res. 4, 151-170 (1983). - ASHTON, A.R.: Biochim. Biophys. Acta 749, 24-31 (1983); - Biochem. J. 217, 79-84 (1984). - ASHTON, A.R., HATCH, M.D.: Biochem. Biophys. Res. Commun. 115, 53-60 (1983). - AZCÓN-BIETO, J., OSMOND, C.B.: Plant Physiol. 71, 574-581 (1983). - AZCÓN-BIETO, J., LAMBERS, H., DAY, D.A.: Aust. J. Plant Physiol. 10, 237-245 (1983a); - Plant Physiol. 72, 598-603 (1983b).

BADGER, M.R., ANDREWS, T.J.: Plant Physiol. 70, 517-523 (1982). - BÁRCENA, J.A.: Biochem. Int. 7, 755-760 (1983). - BARROW, S.R., COCKBURN, W.: Plant Physiol. 69, 568-571 (1982). - BASSI, R., PASSERA, C.: Photosynth. Res. 3, 53-58 (1982). - BAYSDORFER, C., BASSHAM, J.A.: Plant Physiol. 74, 374-379 (1984). - BEARDALL, J., MORRIS, I.: Mar. Biol. 37, 377-387 (1976). - BEARDALL, J., MUKERJI, D., GLOVER, H.E., MORRIS, I.: J. Phycol. 12, 409-417 (1976). - BECK, E., BURKERT, A., HOFMANN, M.: Plant Physiol. 73, 41-45 (1983). - BEER, S., WETZEL, R.G.: Plant Physiol. 70, 488-492 (1982). - BEHRENS, P.W., MARSHO, T.V., RADMER, R.J.: Plant Physiol. 70, 179-185 (1982). - BELKNAP, W.R.:

Plant Physiol. 72, 1130-1132 (1983). - BENNOUN, P.: Proc. Natl. Acad. Sci. USA 79, 4352-4356 (1982). - BERGER, M.G., FOCK, H.P.: Photosynth. Res. 4, 3-7 (1983). - BERGMAN, A., ERICSON, I.: Physiol. Plant. 59, 421-427 (1983). - BERKOWITZ, G.A., GIBBS, M.: Plant Physiol. 72, 1100-1109 (1983). - BERKOWITZ, G.A., CHEN, C., GIBBS, M.: Plant Physiol. 72, 1123-1126 (1983). - BETSCHE, T.: Plant Physiol. 71, 961-965 (1983). - BIEKMANN, S., FEIERABEND, J.: Biochim. Biophys. Acta 721, 268-279 (1982). - BIRKENHEAD, K., WALKER, D., FOYER, C.: Planta 156, 171-175 (1982). - BIRMINGHAM, B.C., COLEMAN, J.R., COLMAN, B.: Plant Physiol. 69, 259-262 (1982). - BLOCK, M.A., DORNE, A.-J., JOYARD, J., DOUCE, R.: J. Biol. Chem. 258, 13281-13286 (1983). - BOURQUIN, P.W., FOCK, H.P.: Photosynthetica 17, 182-188 (1983). - BRAUN, P., BODE, J., WILD, A.: Z. Naturforsch. 38c, 243-246 (1983). - BRULFERT, J., QUEIROZ, O.: Planta 154, 339-343 (1982). - BRULFERT, J., VIDAL, J., GADAL, P., QUEIROZ, O.: Planta 156, 92-94 (1982a). - BRULFERT, J., MÜLLER, D., KLUGE, M., QUEIROZ, O.: Planta 154, 326-331 (1982b). - BUCHANAN-BOLLIG, I.C., KLUGE, M., MÜLLER, D.: Plant Cell Environ. 7, 63-70 (1984). - BURNELL, J.N., HATCH, M.D.: Biochem. Biophys. Res. Commun. 111, 288-293 (1983); - ibid. 118, 65-72 (1984). - BUSER-SUTER, C., WIEMKEN, A., MATILE, P.: Plant Physiol. 69, 456-459 (1982).

CARNAL, N.W., BLACK, C.C.: Plant Physiol. 71, 150-155 (1983). - CATALDO, D.A., BERLYN, G.P.: Am. J. Bot. 61, 957-963 (1974). - CERFF, R., KLOPP-STECH, K.: Proc. Natl. Acad. Sci. USA 79, 7624-7628 (1982). - CHANG, C.W.: Phytochemistry 21, 1263-1269 (1982). - CHAPMAN, K.S.R., HATCH, M.D.: Aust. J. Plant Physiol. 8, 237-248 (1981); - Plant Sci. Lett. 29, 145-154 (1983). - CLAUSSEN, W., LENZ, F.: Z. Pflanzenphysiol. 109, 459-468 (1983). - CLINE, K., KEEGSTRA, K.: Plant Physiol. 71, 366-372 (1983). - COLMAN, B., GEHL, K.A.: J. Phycol. 19, 216-219 (1983). - COREY, E.J., SHIH, C., CASHMAN, J.R.: Proc. Natl. Acad. Sci. USA 80, 3581-3584 (1983). - CORLEY, E., WOLOSIUK, R.A., HERTIG, C.M.: Biochem. Biophys. Res. Commun. 115, 707-714 (1983). - CORNIC, G., WOO, K.C., OSMOND, C.B.: Plant Physiol. 70, 1310-1315 (1982). - CSÉKE, C., BUCHANAN, B.B.: FEBS Lett. 155, 139-142 (1983). - CSÉKE, C., NISHIZAWA, A.N., BUCHANAN, B.B.: Plant Physiol. 70, 658-661 (1982a). - CSÉKE, C., WEEDEN, N.F., BUCHANAN, B.B., UYEDA, K.: Proc. Natl. Acad. Sci. USA 79, 4322-4326 (1982b).

DELANEY, M.E., WALKER, D.A.: Biochem. J. 171, 477-482 (1978). - DE LA TORRE, A., LARA, C., YEE, B.C., MALKIN, R., BUCHANAN, B.B.: Arch. Biochem. Biophys. 213, 545-550 (1982). - DEMMIG, B., GIMMLER, H.: Plant Physiol. 73, 169-174 (1983). - DEMMIG, B., WINTER, K.: Planta 159, 66-76 (1983). - DIETER, P., MARMÉ, D.: J. Biol. Chem. 259, 184-189 (1984). - DOEHLERT, D.C., HUBER, S.C.: Plant Physiol. 73, 989-994 (1983). - DORNE, A.-J., BLOCK, M.A., JOYARD, J., DOUCE, R.: FEBS Lett. 145, 30-34 (1982). - DRY, I.B., WISKICH, J.T.: Arch. Biochem. Biophys. 217, 72-79 (1982); - Plant Physiol. 72, 291-296 (1983). - DRY, I.B., DAY, D.A., WISKICH, J.T.: FEBS Lett. 158, 154-158 (1983). - DUBACQ, J.-P., DRAPIER, D., TRÉMOLIÈRES, A.: Plant Cell Physiol. 24, 1-9 (1983).

EDWARDS, G.E., KU, M.S.B., HATCH, M.D.: Plant Cell Physiol. 23, 1185-1195 (1982). - ESPIE, G.S., COLMAN, B.: Plant Physiol. 70, 649-654 (1982).

FALKOWSKI, P.G.: J. Plankton Res. 3, 203-216 (1981). - FEIERABEND, J., GRINGEL, G.: Z. Pflanzenphysiol. 110, 247-258 (1983). - FLORENCIO, F.J., VEGA, J.M.: Z. Naturforsch. 38c, 531-538 (1983). - FLÜGGE, U.I., BENZ, R.: FEBS Lett. 169, 85-89 (1984). - FLÜGGE, U.I., STITT, M., FREISL, M., HELDT, H.W.: Plant Physiol. 69, 263-267 (1982). - FLÜGGE, U.I., GERBER, J., HELDT, H.W.: Biochim. Biophys. Acta 725, 229-237 (1983). - FOGG, G.E.: Bot. Mar. 26, 3-14 (1983). - FOSTER, J.G., EDWARDS, G.E.: Plant Cell Physiol. 24, 1191-1202 (1983). - FOSTER, J.G., EDWARDS, G.E.,

WINTER, K.: Plant Cell Physiol. 23, 585-594 (1982). - FOYER, C., WALKER, D., LATZKO, E.: Z. Pflanzenphysiol. 107, 457-465 (1982). - FOYER, C., ROWELL, J., WALKER, D.: Arch. Biochem. Biophys. 220, 232-238 (1983a); - Planta 157, 239-244 (1983b). - FRENTZEN, M., HEINZ, E., McKEON, T.A., STUMPF, P.K.: Eur. J. Biochem. 129, 629-636 (1983). - FURBANK, R.T., BADGER, M.R.: Aust. J. Plant Physiol. 9, 553-558 (1982); - ibid. 10, 451-458 (1983). - FURBANK, R.T., BADGER, M.R., OSMOND, C.B.: Plant Physiol. 70, 927-931 (1982); - ibid. 73, 1038-1041 (1983). - FURNAS, M.J.: J. Plankton Res. 5, 637-655 (1983).

GARDEMANN, A., STITT, M., HELDT, H.W.: FEBS Lett. 137, 213-216 (1982); - Biochim. Biophys. Acta 722, 51-60 (1983). - GARDESTRÖM, P., EDWARDS, G.E.: Plant Physiol. 71, 24-29 (1983). - GASPERINI, C., PUPILLO, P.: Plant Sci. Lett. 28, 163-171 (1982). - GERBLING, K.-P., STEUP, M., LATZKO, E.: Arch. Microbiol. 137, 109-114 (1984). - GIERSCH, C.: Arch. Biochem. Biophys. 219, 379-387 (1982). - GIVAN, C.V.: Physiol. Plant. 57, 311-316 (1983). - GIVAN, C.V., HODGSON, J.M.: Plant Sci. Lett. 32, 233-242 (1983). - GLOVER, H.E.: J. Plankton Res. 2, 69-79 (1980). - GLOVER, H.E., MORRIS, I.: Limnol. Oceanogr. 24, 510-519 (1979). - GOLLER, M., HAMPP, R., ZIEGLER, H.: Planta 156, 255-263 (1982). - GOODNIGHT, S.H., HARRIS, W.S., CONNER, W.E., ILLINGWORTH, D.R.: Arteriosclerosis 2, 87-113 (1982). - GUTTERIDGE, S., PARRY, M.A.J., SCHMIDT, C.N.G.: Eur. J. Biochem. 126, 597-602 (1982).

HALL, N.P., KEYS, A.J.: Plant Physiol. 72, 945-948 (1983). - HALLBERG, M., LARSSON, C.: Physiol. Plant. 57, 330-338 (1983). - HAMMOND, J.B.W., BURTON, K.S.: Plant Physiol. 73, 61-65 (1983). - HAMMOND, J.B.W., PREISS, J.: Plant Physiol. 73, 709-712 (1983). - HAMPP, R., GOLLER, M.: Planta 159, 314-321 (1983). - HAMPP, R., OUTLAW, W.H., TARCZYNSKI, M.C.: Plant Physiol. 70, 1582-1585 (1982a). - HAMPP, R., GOLLER, M., ZIEGLER, H.: Plant Physiol. 69, 448-455 (1982b). - HARRIS, G.P., PICCININ, B.B.: J. Phycol. 19, 185-192 (1983). - HATCH, M.D., BURNELL, J.N.: Aust. J. Plant Physiol. 10, 179-186 (1983). - HEBER, U., TAKAHAMA, U., NEIMANIS, S., SHIMIZU-TAKAHAMA, M.: Biochim. Biophys. Acta 679, 287-299 (1982). - HERTIG, C.M., WOLOSIUK, R.A.: J. Biol. Chem. 258, 984-989 (1983). - HEUER, B., HANSEN, M.J., ANDERSON, L.E.: Plant Physiol. 69, 1404-1406 (1982). - HIREL, B., PERROT-RECHENMANN, C., SUZUKI, A., VIDAL, J., GADAL, P.: Plant Physiol. 69, 983-987 (1982a). - HIREL, B., VIDAL, J., GADAL, P.: Planta 155, 17-23 (1982b). - HIREL, B., LAYZELL, D.B., McCASHIN, B., McNALLY, S.F., CANVIN, D.T.: Can. J. Bot. 61, 2257-2259 (1983). - HIROKAWA, T., HATA, M., TAKEDA, H.: Plant Cell Physiol. 23, 813-820 (1982). - HOLADAY, A.S., CHOLLET, R.: Plant Physiol. 73, 740-745 (1983). - HOLADAY, A.S., HARRISON, A.T., CHOLLET, R.: Plant Sci. Lett. 27, 181-189 (1982). - HOLDSWORTH, E.S., BRUCK, K.: Arch. Biochem. Biophys. 182, 87-94 (1977). - HOLDSWORTH, E.S., COLBECK, J.: Mar. Biol. 38, 189-199 (1976). - HOLTUM, J.A.M., WINTER, K.: Planta 155, 8-16 (1982). - HOSSAIN, M.A., ASADA, K.: Plant Cell Physiol. 25, 85-92 (1984). - HOWITZ, K.T., McCARTY, R.E.: FEBS Lett. 154, 339-342 (1983). - HUBER, S.C.: Plant Physiol. 71, 818-821 (1983). - HUBER, S.C., BICKETT, D.M.: Plant Physiol. 74, 445-447 (1984). - HUBER, S.C., ISRAEL, D.W.: Plant Physiol. 69, 691-696 (1982). - HUSIC, H.D., TOLBERT, N.E.: Arch. Biochem. Biophys. 229, 64-72 (1984). - HUTCHESON, S.W., BUCHANAN, B.B.: Plant Physiol. 72, 877-885 (1979).

IADAROLA, P., ZAPPONI, M.C., FERRI, G.: Experentia 39, 50-52 (1983). - IMAMURA, M., TSUZUKI, M., SHIRAIWA, Y., MIYACHI, S.: Plant Cell Physiol. 24, 533-540 (1983).

JABLONSKI, P.P., ANDERSON, J.W.: Plant Physiol. 69, 1407-1413 (1982). - JARRETT, H.W., BROWN, C.J., BLACK, C.C., CORMIER, M.J.: J. Biol. Chem. 257, 13795-13804 (1982). - JOLIVET-TOURNIER, P., GERSTER, R.:

Plant Physiol. 74, 108-111 (1984). - JONES, G.J., NICHOLS, P.D., JOHNS, R.B.: J. Phycol. 19, 416-420 (1983). - JORDAN, D.B., CHOLLET, R.: J. Biol. Chem. 258, 13752-13758 (1983). - JORDAN, D.B., CHOLLET, R., OGREN, W.L.: Biochemistry 22, 3410-3418 (1983).

KAISER, G., MARTINOIA, E., WIEMKEN, A.: Z. Pflanzenphysiol. 107, 103-113 (1982). - KAISER, W.M., WEBER, H., SAUER, M.: Z. Pflanzenphysiol. 113, 15-27 (1983). - KAMIŃSKA, Z., MALESZEWSKI, S.: Z. Pflanzenphysiol. 108, 201-206 (1982). - KANAZAWA, T., DISTEFANO, M., BASSHAM, J.A.: Plant Cell Physiol. 24, 979-986 (1983). - KAPLAN, A., ZENVIRTH, D., REINHOLD, L., BERRY, J.A.: Plant Physiol. 69, 978-982 (1982). - KARABOURNIOTIS, G., MANETAS, Y., GAVALAS, N.A.: Plant Physiol. 73, 735-739 (1983). - KATES, M., VOLCANI, B.E.: Biochim. Biophys. Acta 116, 264-278 (1966). - KELLY, G.J.: Trends Biochem. Sci. 9, 255-256 (1984). - KELLY, G.J., LATZKO, E.: Prog. Bot. 44, 103-131 (1982). - KENDALL, A.C., KEYS, A.J., TURNER, J.C., LEA, P.J., MIFLIN, B.J.: Planta 159, 505-511 (1983). - KERBY, N.W., EVANS, L.V.: J. Phycol. 19, 1-3 (1983). - KIMPEL, D.L., TOGASAKI, R.K., MIYACHI, S.: Plant Cell Physiol. 24, 255-259 (1983). - KLECZKOWSKI, L.A., RANDALL, D.D.: FEBS Lett. 158, 313-316 (1983). - KLEIN, U., CHEN, C., GIBBS, M.: Plant Physiol. 72, 488-491 (1983). - KOIKE, I., REDALJE, D.G., AMMERMAN, J.W., HOLM-HANSEN, O.: Mar. Biol. 74, 161-168 (1983). - KOMBRINK, E., WÖBER, G.: Arch. Biochem. Biophys. 213, 602-619 (1982). - KOW, Y.W., GIBBS, M.: Plant Physiol. 69, 179-186 (1982). - KOW, Y.W., ERBES, D.L., GIBBS, M.: Plant Physiol. 69, 442-447 (1982a). - KOW, Y.W., SMYTH, D.A., GIBBS, M.: Plant Physiol. 69, 740-741 (1982b). - KREMER, B.P.: Oceanogr. Mar. Biol. Annu. Rev. 19, 41-94 (1981a); - Phycologia 20, 242-250 (1981b). - KREMER, B.P., BERKS, R.: Z. Pflanzenphysiol. 87, 149-165 (1978). - KREMER, B.P., KIRST, G.: Z. Naturforsch. 37c, 761-771 (1982). - KRÜGER, I., SCHNARRENBERGER, C.: Eur. J. Biochem. 136, 101-106 (1983). - KRUGER, N.J., ap REES, T.: Phytochemistry 22, 1891-1898 (1983a); - Planta 158, 179-184 (1983b). - KRUGER, N.J., BULPIN, P.V., ap REES, T.: Planta 157, 271-273 (1983a). - KRUGER, N.J., KOMBRINK, E., BEEVERS, H.: Biochem. Biophys. Res. Commun. 117, 37-42 (1983b). - KU, M.S.B., MONSON, R.K., LITTLEJOHN, R.O., NAKAMOTO, H., FISHER, D.B., EDWARDS, G.E.: Plant Physiol. 71, 944-948 (1983). - KULKARNI, A.H., NIMBALKAR, J.D.: Photosynthetica 17, 189-192 (1983). - KWOK-KI, H., HOCK-HIN, Y., CHOY-SIN, H.: Plant Cell Physiol. 24, 1317-1321 (1983).

LARSSON, M., LARSSON, C.-M., ULLRICH, W.R.: Plant Physiol. 70, 1637-1640 (1982). - LATZKO, E., KELLY, G.J.: Prog. Bot. 40, 99-118 (1978); - pp. 59-69 in: The New Frontiers in Plant Biochemistry, ed. T. AKAZAWA, T. ASAHI, H. IMASEKI. Tokyo: Japan Scientific Societies Press 1983a; - Physiol. Vég. 21, 805-815 (1983b). - LAW, M.Y., CHARLES, S.A., HALLIWELL, B.: Biochem. J. 210, 899-903 (1983). - LEEGOOD, R.C., WALKER, D.A.: Planta 156, 449-456 (1982); - Plant Physiol. 71, 513-518 (1983). - LEEGOOD, R.C., KOBAYASHI, Y., NEIMANIS, S., WALKER, D.A., HEBER, U.: Biochim. Biophys. Acta 682, 168-178 (1982). - LEVI, C., GIBBS, M.: Plant Physiol. 74, 459-463 (1984). - LIANG, Z., HUANG, A.H.C.: Plant Physiol. 73, 147-152 (1983). - LIANG, Z., YU, C., HUANG, A.H.C.: Plant Physiol. 70, 1210-1212 (1982). - LIEDVOGEL, B., STUMPF, P.K.: Plant Physiol. 69, 897-903 (1982). - LILLEY, R.McC.: Plant Cell Environ. 6, 329-343 (1983). - LILLEY, R.McC., STITT, M., MADER, G., HELDT, H.W.: Plant Physiol. 70, 965-970 (1982). - LINDEN, J.C., SCHILLING, N.: Plant Physiol. 74, 795-799 (1984). - LOESCHER, W.H., MARLOW, G.C., KENNEDY, R.A.: Plant Physiol. 70, 335-339 (1982). - LOOZE, S. de, WAGNER, E.: Physiol. Plant. 57, 231-237 (1983a); - ibid. 57, 238-242 (1983b). - LUDWIG, I., ZIEGLER, P., BECK, E.: Plant Physiol. 74, 856-861 (1984).

MADHAVAN, S., SMITH, B.N.: Plant Physiol. 69, 273-277 (1982). - MAE, T., MAKINO, A., OHIRA, K.: Plant Cell Physiol. 24, 1079-1086 (1983). -

MAKINO, A., MAE, T., OHIRA, K.: Plant Physiol. 73, 1002-1007 (1983). - MANETAS, Y.: Photosynth. Res. 3, 321-333 (1982). - MANETAS, Y., SAMARAS, Y., STAMATAKIS, K.: Z. Pflanzenphysiol. 111, 365-368 (1983). - MARCUS, Y., ZENVIRTH, D., HAREL, E., KAPLAN, A.: Plant Physiol. 69, 1008-1012 (1982). - MAREK, L.F., STEWART, C.R.: Plant Physiol. 73, 118-120 (1983). - MARIGO, G., LÜTTGE, U., SMITH, J.A.C.: Z. Pflanzenphysiol. 109, 405-413 (1983). - MARQUES, I.A., OBERHOLZER, M.J., ERISMANN, K.H.: Plant Physiol. 71, 555-561 (1983). - MARTIN, C.E., ZEE, A.K.: Plant Physiol. 73, 718-723 (1983). - MARTIN, F., WINSPEAR, M.J., MacFARLANE, J.D., OAKS, A.: Plant Physiol. 71, 177-181 (1983). - McNALLY, S.F., HIREL, B., GADAL, P., MANN, A.F., STEWART, G.R.: Plant Physiol. 72, 22-25 (1983). - MERRETT, M.J., ARMITAGE, T.L.: Planta 155, 95-96 (1982). - MEYER, A.O., KELLY, G.J., LATZKO, E.: Plant Physiol. 69, 7-10 (1982). - MINOT, R., MEUNIER, J.-C., BUC, L., RICARD, J.: FEBS Lett. 142, 118-120 (1982). - MIYACHI, S., TSUZUKI, M., AVRAMOVA, S.T.: Plant Cell Physiol. 24, 441-451 (1983). - MOHAMED, A.H., ANDERSON, L.E.: Plant Physiol. 71, 248-250 (1983). - MONSON, R.K., STIDHAM, M.A., WILLIAMS, G.J., EDWARDS, G.E., URIBE, E.G.: Plant Physiol. 69, 921-928 (1982). - MONSON, R.K., RUMPHO, M.E., EDWARDS, G.E.: Planta 159, 97-104 (1983). - MUKERJI, D., GLOVER, H.E., MORRIS, I.: J. Phycol. 14, 137-142 (1978). - MÜLLER, D., KLUGE, M., GRÖSCHEL-STEWART, U.: Plant Cell Environ. 5, 223-230 (1982). - MURPHY, D.J., WALKER, D.A.: Planta 155, 316-320 (1982a); - ibid. 156, 84-88 (1982b). - MURPHY, D.J., MUKHERJEE, K.D., LATZKO, E.: Biochem. J. 213, 249-252 (1983a). - MURPHY, D.J., WOODROW, I.E., LATZKO, E., MUKHERJEE, K.D.: FEBS Lett. 162, 442-446 (1983b). - MURPHY, D.J., MUKHERJEE, K.D., WOODROW, I.E.: Eur. J. Biochem. 139, 373-379 (1984). - MUTO, S.: FEBS Lett. 147, 161-164 (1982); - Z. Pflanzenphysiol. 109, 385-393 (1983). - MUTO, S., IZAWA, S., MIYACHI, S.: FEBS Lett. 139, 250-254 (1982).

NAKAMOTO, H., EDWARDS, G.E.: Plant Physiol. 71, 568-573 (1983a); - Aust. J. Plant Physiol. 10, 279-289 (1983b); - Biochem. Biophys. Res. Commun. 115, 673-679 (1983c). - NAKAMOTO, H., SUGIYAMA, T.: Plant Physiol. 69, 749-753 (1982). - NAKAMOTO, H., KU, M.S.B., EDWARDS, G.E.: Plant Cell Physiol. 24, 1387-1393 (1983). - NAKAMURA, Y., IMAMURA, M.: Plant Sci. Lett. 31, 123-131 (1983). - NAKAMURA, Y., MIYACHI, S.: Plant Sci. Lett. 27, 1-6 (1982). - NAKAMURA, Y., TOLBERT, N.E.: J. Biol. Chem. 258, 7631-7638 (1983). - NASH, D., WISKICH, J.T.: Plant Physiol. 71, 627-634 (1983). - NEUBURGER, M., JOURNET, E.-P., BLIGNY, R., CARDE, J.-P., DOUCE, R.: Arch. Biochem. Biophys. 217, 312-323 (1982). - NILSEN, S., JOHNSON, Ø.: Physiol. Plant. 56, 273-280 (1982). - NISHIMURA, M., BHUSAWANG, P., STRZALKA, K., AKAZAWA, T.: Plant Physiol. 70, 353-356 (1982a). - NISHIMURA, M., DOUCE, R., AKAZAWA, T.: Plant Physiol. 69, 916-920 (1982b). - NISHIMURA, M., AKHMEDOV, Y.D., STRZALKA K., AKAZAWA, T.: Arch. Biochem. Biophys. 222, 397-402 (1983a). - NISHIMURA, M., AKHMEDOV, Y.D., AKAZAWA, T.: Photosynth. Res. 4, 99-109 (1983b). - NOTT, D.L., OSMOND, C.B.: Aust. J. Plant Physiol. 9, 409-422 (1982).

OGAWA, T., INOUE, Y.: Biochim. Biophys. Acta 724, 490-493 (1983). - OHNISHI, J., KANAI, R.: Plant Cell Physiol. 24, 1411-1420 (1983). - OHNISHI, J., YAMADA, M.: Plant Cell Physiol. 23, 767-773 (1982). - OKABE, K., YANG, S.-Y., TSUZUKI, M., MIYACHI, S.: Plant Sci. Lett. 33, 145-153 (1984). - OUTLAW, W.H., TARCZYNSKI, M.C.: Plant Physiol. 74, 424-429 (1984). - OUTLAW, W.H., TARCZYNSKI, M.C., ANDERSON, L.C.: Plant Physiol. 70, 1218-1220 (1982).

PAN, D., TAN, K.H.: Photosynthetica 15, 511-517 (1981). - PARRY, M.A.J. SCHMIDT, C.N.G., KEYS, A.J., GUTTERIDGE, S.: FEBS Lett. 159, 107-111 (1983). - PAUL, J.S., VOLCANI, B.E.: Arch. Microbiol. 110, 247-252 (1976). - PEISKER, M., APEL, P.: Photosynthetica 15, 435-441 (1981). -

PELTIER, G., THIBAULT, P.: Plant Physiol. 71, 888-892 (1983). - PERCHOROWICZ, J.T., JENSEN, R.G.: Plant Physiol. 71, 955-960 (1983). - PERROT-RECHENMANN, C., VIDAL, J., BRULFERT, J., BURLET, A., GADAL, P.: Planta 155, 24-30 (1982). - PERROT-RECHENMANN, C., JACQUOT, J.P., GADAL, P., WEEDEN, N.F., CSÉKE, C., BUCHANAN, B.B.: Plant Sci. Lett. 30, 219-226 (1983). - PETERSON, R.B.: Plant Physiol. 69, 192-197 (1982); - ibid. 73, 983-988 (1983). - PFLEIDERER, C., KLEMME, J.-H.: Z. Naturforsch. 35c, 229-238 (1980). - PIAZZA, G.J., SMITH, M.G., GIBBS, M.: Plant Physiol. 70, 1748-1758 (1982). - PICHERSKY, E., GOTTLIEB, L.D.: Plant Physiol. 74, 340-347 (1984). - PLÁ, A., CHUECA, A., LÓPEZ-GORGÉ, J.: Photosynth. Res. 2, 291-296 (1981). - PLATT, S.G., RAND, L.: Plant Cell Physiol. 23, 917-921 (1982). - PLATT, S.G., PLAUT, Z., BASSHAM, J.A.: Plant Physiol. 57, 69-73 (1976). - POLYA, G.M., MICUCCI, V.: Biochim. Biophys. Acta 785, 68-74 (1984). - POPOVIC, R., COLBOW, K., VIDAVER, W., BRUCE, D.: Plant Physiol. 73, 889-892 (1983). - PORTER, M.A., GRODZINSKI, B.: Plant Physiol. 74, 413-416 (1984). - PORTIS, A.R.: Plant Physiol. 70, 393-396 (1982); - ibid. 71, 936-943 (1983). - POWLES, S.B., BERRY, J.A., BJÖRKMAN, O.: Plant Cell Environ. 6, 117-123 (1983). - PREISS, J.: Annu. Rev. Plant Physiol. 33, 431-454 (1982). - PRÉZELIN, B.B., ALBERTE, R.S.: Proc. Natl. Acad. Sci. USA 75, 1801-1804 (1978). - PRISCU, J.C., GOLDMAN, C.R.: Limnol. Oceanogr. 28, 1168-1181 (1983). - PROUDLOVE, M.O., MOORE, A.L.: FEBS Lett. 147, 26-30 (1982). - PROUDLOVE, M.O., THURMAN, D.A., SALISBURY, J.: New Phytol. 96, 1-5 (1984).

RAJENDRUDU, G., DAS, V.S.R.: Photosynth. Res. 2, 225-233 (1981). - RAO, I.M., ANDERSON, L.E.: Plant Physiol. 71, 451-455 (1983). - RAYDER, L., TING, I.P.: Photosynth. Res. 4, 203-211 (1983a); - Plant Physiol. 72, 606-610 (1983b). - REES, T.A.V., BEKHEET, I.A.: Planta 156, 385-387 (1982). - RICHARDSON, K., BEARDALL, J., RAVEN, J.A.: New Phytol. 93, 157-191 (1983). - RIVKIN, R.B., SELIGER, H.H., SWIFT, E., BIGGLEY, W.H.: Mar. Biol. 68, 181-191 (1982). - ROACH, D.J.W., McFADDEN, B.A.: Photosynth. Res. 4, 111-117 (1983). - ROBINSON, J.M., GIBBS, M.: Plant Physiol. 70, 1249-1254 (1982). - ROBINSON, S.P.: Biochem. Biophys. Res. Commun. 106, 1027-1034 (1982a); - Plant Physiol. 70, 1032-1038 (1982b). - ROSSLENBROICH, H.-J., DÖHLER, G.: Biochem. Physiol. Pflanzen 177, 363-373 (1982). - ROUGHAN, P.G., SLACK, C.R.: Annu. Rev. Plant Physiol. 33, 97-132 (1982). - RUFTY, T.W., HUBER, S.C.: Plant Physiol. 72, 474-480 (1983). - RUFTY, T.W., KERR, P.S., HUBER, S.C.: Plant Physiol. 73, 428-433 (1983). - RUMPHO, M.E., EDWARDS, G.E., LOESCHER, W.H.: Plant Physiol. 73, 869-873 (1983). - RUTTER, J.C., COBB, A.H.: New Phytol. 95, 559-568 (1983). - RUYTERS, G.: Z. Pflanzenphysiol. 108, 207-214 (1982). - RUYTERS, G., MIYACHI, S.: Plant Cell Physiol. 24, 863-871 (1983). - RYTHER, J.H., YENTSCH, C.S.: Limnol. Oceanogr. 2, 281-286 (1957).

SABULARSE, D.C., ANDERSON, R.L.: Biochem. Biophys. Res. Commun. 103, 848-855 (1981). - SALVUCCI, M.E., BOWES, G.: Plant Physiol. 73, 488-496 (1983a): - Planta 158, 27-34 (1983b). - SAMUELSSON, G., RICHARDSON, K.: Mar. Biol. 70, 21-26 (1982). - SANADA, Y., NISHIDA, K.: Z. Pflanzenphysiol. 105, 189-192 (1982). - SAROJINI, G., OLIVER, D.J.: Plant Physiol. 72, 194-199 (1983). - SAUER, A., HEISE, K.-P.: Plant Physiol. 73, 11-15 (1983). - SAVER, B.G., KNOWLES, J.R.: Biochemistry 21, 5398-5403 (1982). - SCHEIBE, R., JACQUOT, J.-P.: Planta 157, 548-553 (1983). - SCHERER, S., BÖGER, P.: Arch. Microbiol. 132, 329-332 (1982). - SCHILLING, N.: Planta 154, 87-93 (1982). - SCHLOSS, J.V., LORIMER, G.H.: J. Biol. Chem. 257, 4691-4694 (1982). - SCHMITT, M.R., EDWARDS, G.E.: Plant Physiol. 70, 1213-1217 (1982); - ibid. 72, 728-734 (1983). - SCHNABL, H., ELBERT, C., KRÄMER, G.: J. Exp. Bot. 33, 996-1003 (1982). - SCHÜRMANN, P., ORTIZ, W.: Planta 154, 70-75 (1982). - SHIEH, Y.-J., KU, M.S.B., BLACK, C.C.: Plant Physiol. 69, 776-780 (1982). - SHIMAKATA, T., STUMPF, P.K.: Plant Physiol. 69, 1257-1262 (1982). - SHIMAZAKI, K.,

GOTOW, K., SAKAKI, T., KONDO, N.: Plant Cell Physiol. 24, 1049-1056 (1983). - SHIMOMURA, S., NAGAI, M., FUKUI, T.: J. Biochem. 91, 703-717 (1982). - SHIRAIWA, Y., MIYACHI, S.: Plant Cell Physiol. 24, 919-923 (1983). - SICHER, R.C.: Plant Physiol. 70, 366-369 (1982). - SIESS, W., SCHERER, B., BÖHLIG, B., ROTH, P., KURZMANN, I., WEBER, P.C.: Lancet 1, 441-444 (1980). - SIMON, P., BONZON, M., GREPPIN, H., MARME, D.: FEBS Lett. 167, 332-338 (1984). - SINGH, B.K., GREENBERG, E, PREISS, J.: Plant Physiol. 74, 711-716 (1984). - SINGH, P., NAIK, M.S.: Plant Sci. Lett. 30, 9-16 (1983). - SMITH, J.A.C., MARIGO, G., LÜTTGE, U., BALL, E.: Plant Sci. Lett. 26, 13-21 (1982). - SMITH, J.C., PLATT, T., HARRISON, W.G.: Prog. Oceanogr. 12, 425-549 (1983). - SMYTH, D.A., WU, M.-X., BLACK, C.C.: Plant Sci. Lett. 33, 61-70 (1984). - SOMERVILLE, S.C., OGREN, W.L.: Proc. Natl. Acad. Sci. USA 80, 1290-1294 (1983). - SOMERVILLE, S.C., SOMERVILLE, C.R.: J. Exp. Bot. 34, 415-424 (1983). - SPALDING, M.H., OGREN, W.L.: FEBS Lett. 145, 41-44 (1982); - ibid. 154, 335-338 (1983). - SPALDING, M.H., SPREITZER, R.J., OGREN, W.L.: Plant Physiol. 73, 273-276 (1983a); - ibid. 73, 268-272 (1983b). - SPREITZER, R.J., JORDAN, D.B., OGREN, W.L.: FEBS Lett. 148, 117-121 (1982). - SRIVASTAVA, D.K., ANDERSON, L.E.: Biochim. Biophys. Acta 724, 359-369 (1983). - STEUP, M., GERBLING, K.-P.: Anal. Biochem. 134, 96-100 (1983). - STEUP, M., ROBENEK, H., MELKONIAN, M.: Planta 158, 428-436 (1983). - STIBOROVÁ, M., LEBLOVÁ, S.: Photosynthetica 17, 379-385 (1983). - STIDHAM, M.A., MORELAND, D.E., SIEDOW, J.N.: Plant Physiol. 73, 517-520 (1983). - STITT, M., MIESKES, G., SOLING, H.-D., HELDT, H.W.: FEBS Lett. 145, 217-222 (1982a). - STITT, M., LILLEY, R.McC., HELDT, H.W.: Plant Physiol. 70, 971-977 (1982b). - STITT, M., WIRTZ, W., HELDT, H.W.: Plant Physiol. 72, 767-774 (1983a). - STITT, M., GERHARDT, R., KÜRZEL, B., HELDT, H.W.: Plant Physiol. 72, 1139-1141 (1983b). - STRICKLAND, J.D.H., EPPLEY, R.W., ROJAS de MENDIOLA, B.: Bol. Inst. Mar Peru (Callao) 2, 1-45 (1969). - STUMPF, D.K., JENSEN, R.G.: Plant Physiol. 69, 1263-1267 (1982). - SUGIYAMA, T., HIRAYAMA, Y.: Plant Cell Physiol. 24, 783-787 (1983). - SUMAR, N., CASSELTON, P.J., McNALLY, S.F., STEWART, G.R.: Plant Physiol. 74, 204-207 (1984). - SUZUKI, A., GADAL, P.: Plant Physiol. 69, 848-852 (1982).

TAKABE, T., RAI, A.K., AKAZAWA, T.: Arch. Biochem. Biophys. 229, 202-211 (1984). - THORPE, N.: Plant Sci. Lett. 30, 331-338 (1983). - TING, I.P., BURK, J.H.: Plant Sci. Lett. 32, 279-285 (1983). - TING, I.P., GIBBS, M. (eds.). pp. 316 in: Crassulacean Acid Metabolism. Baltimore: Waverly Press 1982. - TING, I.P., STERNBERG, L.O., DeNIRO, M.J.: Plant Physiol. 71, 677-679 (1983). - TISCHNER, R., SCHMIDT, A.: Plant Physiol. 70, 113-116 (1982). - TITUS, D.E., HONDRED, D., BECKER, W.M.: Plant Physiol. 72, 402-408 (1983). - TSUZUKI, M.: Z. Pflanzenphysiol. 110, 29-37 (1983).

UDVARDY, J., GODEH, M.M., FARKAS, G.L.: J. Bacteriol. 151, 203-208 (1982). - USUDA, H., EDWARDS, G.E.: Plant Physiol. 69, 469-473 (1982).

VATER, J., GAUDSZUN, T., LANGE, B., ERDIN, N., SALNIKOW, J.: Z. Naturforsch. 38c, 418-427 (1983). - VELUTHAMBI, K., POOVAIAH, B.W.: Science 223, 167-169 (1984). - VENRICK, E.L., BEERS, J.R., HEINBOKEL, J.F.: J. Exp. Mar. Biol. Ecol. 26, 55-76 (1977). - VIDAL, J., GADAL, P.: Physiol. Vég. 19, 483-489 (1981); - Physiol. Plant. 57, 119-123 (1983). - VINCENT, W.F., NEALE, P.J., RICHERSON, P.J.: J. Phycol. 20, 201-211 (1984). - VINES, H.M., TU, Z.-P., ARMITAGE, A.M., CHEN, S.-S., BLACK, C.C.: Plant Physiol. 73, 25-30 (1983). - VOLKMAN, J.K., SMITH, D.J., EGLINTON, G., FORSBERG, T.E.V., CORNER, E.D.S.: J. Mar. Biol. Ass. U.K. 61, 509-527 (1981). - VOLOKITA, M., KAPLAN, A., REINHOLD, L.: Plant Physiol. 72, 886-890 (1983). - VU, C.V., ALLEN, L.H., BOWES, G.: Plant Physiol. 73, 729-734 (1983).

WALKER, G.H., SAROJINI, G., OLIVER, D.J.: Biochem. Biophys. Res. Commun. 107, 856-861 (1982a). - WALKER, G.H., OLIVER, D.J., SAROJINI. G.: Plant Physiol. 70, 1465-1469 (1982b). - WALTON, N.J.: Planta 155, 218-224 (1982); - Plant Sci. Lett. 30, 203-209 (1983). - WALTON, N.J., WOOLHOUSE, H.W.: Planta 158, 469-471 (1983). - WATANABE, M., OHNISHI, J., KANAI, R.: Plant Cell Physiol. 25, 69-76 (1984). - WEDDING, R.T., BLACK, M.K.: Plant Physiol. 72, 1021-1028 (1983). - WEEDEN, N.F., BUCHANAN, B.B.: Plant Physiol. 72, 259-261 (1983). - WHITTAKER, R.H., LIKENS, G.E.: pp. 305-328 in: Primary Productivity of the Biosphere, ed. H. LIETH, R.H. WHITTAKER. Berlin,Heidelberg, New York: Springer 1975. - WILLIAMS, P.J.leB., HEINEMANN, K.R., MARRA, J., PURDIE, D.A.: Nature 305, 49-50 (1983). - WINKLER, U., SÄFTEL, W., STABENAU, H.: Plant Physiol. 70, 340-343 (1982). - WINTER, K.: Planta 154, 298-308 (1982). - WINTER, K., SCHMITT, M.R., EDWARDS, G.E.: Plant Sci. Lett. 24, 311-318 (1982a). - WINTER, K., USUDA, H., TSUZUKI, M., SCHMITT, M., EDWARDS, G.E., THOMAS, R.J., EVERT, R.F.: Plant Physiol. 70, 616-625 (1982b). - WINTER, K., FOSTER, J.G., SCHMITT, M.R., EDWARDS, G.E.: Planta 154, 309-317 (1982c). - WINTER, K., FOSTER, J.G., EDWARDS, G.E., HOLTUM, J.A.M.: Plant Physiol. 69, 300-307 (1982d). - WINTER, K., WALLACE, B.J., STOCKER, G.C., ROKSANDIC, Z.: Oecologia 57, 129-141 (1983). - WIRTZ, W., STITT, M., HELDT, H.W.: FEBS Lett. 142, 223-226 (1982). - WOLOSIUK, R.A., HERTIG, C.M., NISHIZAWA, A.N., BUCHANAN, B.B.: FEBS Lett. 140, 31-35 (1982). - WOO, K.C.: Plant Physiol. 71, 112-117 (1983). - WOO, K.C., OSMOND, C.B.: Plant Physiol. 69, 591-596 (1982). - WOO, K.C., MOROT-GAUDRY, J.F., SUMMONS, R.E., OSMOND, C.B.: Plant Physiol. 70, 1514-1517 (1982). - WOODROW, I.E., WALKER, D.A.: Arch. Biochem. Biophys. 216, 416-422 (1982); - Biochim. Biophys. Acta 722, 508-516 (1983). - WOODROW, I.E., MURPHY, D.J., LATZKO, E.: J. Biol. Chem. 259, 3791-3795 (1984). WULFF, R.D., STRAIN, B.R.: Can. J. Bot. 60, 1084-1091 (1982).

YOKOTA, A., KITAOKA, S.: Plant Physiol. 70, 760-764 (1982). - YOKOTA, A., KOMURA, H., KITAOKA, S.: Biochem. Biophys. Res. Commun. 111, 544-550 (1983a). - YOKOTA, A., KAWABATA, A., KITAOKA, S.: Plant Physiol. 71, 772-776 (1983b). - YU, C., CLAYBROOK, D.L., HUANG, A.H.C.: Arch. Biochem. Biophys. 227, 180-187 (1983).

Dr. GRAHAME J. KELLY
CSIRO
Marine Laboratories
7000 Hobart, Tasmania
Australia

Professor Dr. ERWIN LATZKO
Botanisches Institut der
Universität Münster
Schloßgarten 3
D-4400 Münster, FRG

IV. Metabolism of Organic Acids

By Manfred Kluge

The last review on the metabolism of organic acids (Progr. Bot. 43, 1981) was dedicated to malate. Since then, most of the publications on acid metabolism concerned malate. Thus, malate will, once again, be the subject of the present article.

Besides malate, some recent aspects of oxalate, malonate and shikimate metabolism will also be considered. It is not the aim of this review to provide a gapless bibliography on acid metabolism, rather, to outline some recent trends.

1. PEP-carboxylase as the Key Enzyme in Plant Metabolism

During recent years the view has emerged that the synthesis of C_4 dicarboxylic acids by β-carboxylation of phosphoenolpyruvate (PEP) is a key process in plant metabolism. Much interest has, therefore, been paid to the PEP-carboxylase (PEPC) which catalyzes this reaction. In order to compile the recent knowledge about the reaction mechanism, regulation, biological role and enzymology on PEPC, an international symposium dedicated to the enzyme was held in spring 1983 at Aussois (France). The papers (original papers and reviews) presented at this meeting are published in Physiologie Végétale 21, issue 5, pp. 801-1089. This publication therefore represents a kind of monography on PEPC and treats PEPC in the context of photosynthetic pathways in higher plants (C_3-, C_4-, CAM-photosynthesis), metabolic pathways in bacteria, intracellular pH-state mechanisms, ion balance, stomatal movements, and symbiotic nitrogen fixation. Also problems of regulation, reaction mechanisms, and carbon isotope effects are considered. All these contributions support the view that PEPC has to be considered as one of the most important carboxylating enzymes in plants.

2. Crassulacean Acid Metabolism (CAM)

a) Current Reviews on CAM

During the last two years the majority of the publications on organic acids in plants dealt with CAM. Recent developments in this field have been reviewed by KLUGE (1982), and HOLTUM and OSMOND (1981). Further reviews and original papers on many aspects of CAM can be found in proceedings of an international symposium on CAM held in 1982 at Riverside [*Crassulacean Acid Metabolism*, I.P. TING and M. GIBBS (eds.) Amer. Soc. of Plant Physiologists, Rockville, 1982].

Progress in Botany, Vol. 46

b) Carbon Flow in CAM

α) Metabolic Pattern of CO_2-fixation During Phases II and IV of the Diurnal CAM Cycle. According to OSMOND (1978), four phases can be distinguished during the diurnal CAM cycle. Phase I represents the nocturnal CAM where external CO_2 is fixed by PEP-carboxylase, followed by malic acid synthesis from OAA, produced by the carboxylation. The malic acid is stored overnight in the vacuoles. During the day malic acid is released from the vacuoles, decarboxylated, and the products of this step are photosynthetically converted to carbohydrates (phase III). It becomes, now, increasingly clear that phase II (i.e., the early light period) and phase IV (i.e., the late afternoon) represent merely transits between the nocturnal and day-time CAM processes (phase II) and vice versa (phase IV). This was shown by the finding that during the mentioned transitory phases the labeling pattern after fixation of external $[^{14}C]O_2$ shifts continuously. During the initial light period external carbon is still fixed mainly by PEPC, with malic acid being the end product, while the C_3-pathway of CO_2-fixation operates on a low level and serves C_3 skeletons to the PEPC pathway (MEYER 1979, KLUGE et al. 1982, FISCHER and KLUGE 1984, LITTLEJOHN and KU 1984) rather than to the synthesis of carbohydrates. This pattern changes during phase II until, toward its end, the PEPC-pathway is increasingly inhibited, while the C_3-pathway of CO_2-fixation becomes dominant. Later, during phase III, external CO_2 is excluded and photosynthesis, then, operates exclusively at the expense of carbon deriving from malic acid depletion. WINTER and TENHUNEN (1982) supported the view of the shift in the carbon flow pathways during phase II by their finding that initially the CO_2-fixation rate was not sensitive to oxygen concentrations, while later, during phase II, high oxygen concentrations were inhibitory. This indicates that CO_2 is initially mainly fixed by PEPC; later, however, by RubP carboxylase/oxygenase.

Phase IV appears to be a mirror image of phase II, i.e., during early phase IV the C_3-pathway is the main mechanism harvesting external CO_2, while during late phase IV products of the PEPC-pathway become increasingly important, suggesting that PEP now contributes substantially to the CO_2-fixation (KLUGE et al. 1982). However, HOLTUM et al. (1983) concluded, from the lack of significant CO_2-fixation at a CO_2 concentration of 100 μbar, that PEPC is not involved in harvesting external CO_2 during phase IV.

Recently, D. RITZ and M. KLUGE (unpublished) could show that malate is synthetized by CO_2-fixation during phase IV. By means of mass-spectroscopical analyses of $[^{13}C]$-malate labeled by $[^{13}C]O_2$-fixation during phase IV (compare POPP et al. 1982), they showed that about 60% of the labeled malate molecules carried the isotope only in one carboxyl group, whereas 40% of the labeled malate was labeled in both carboxyl groups. This allows the conclusion that the singly labeled malate is derived from β-carboxylation of PEP, which has been provided from (not-labeled) reserve carbohydrates. The existence of doubly labeled malate molecules, however, indicates that carbon-3 skeletons synthetized by RubPC-mediated $[^{13}C]O_2$-fixation (thus carrying label in C-1) are also used as CO_2-acceptors in the PEP reaction, which then transfers label into the C_4-carboxyl of malate.

It can be concluded from this experiment that during phase IV both the C_3-pathway and the PEPC-pathway of CO_2-fixation are active. However, it is conceivable that the relative contribution of the two pathways may depend on the plant species. With the above-mentioned finding in mind, that the C_3-pathway of photosynthetic CO_2-fixation can feed carbon into malic acid synthesis, it is interesting to observe that in some facultative or obligate CAM plants a malic acid rhythm may occur which is inverse to that of normal CAM, i.e., malic acid accumulates during the day and disappears during the night. Already WOLF (1960) reported on such cases. Recently, AMAGASA (1982) found such inversed malic acid rhythms in young leaves of *Kalanchoë daigremontiana* , BRULFERT (pers. comm.) in the facultative CAM plant *Crassula sieberina* , and v. WILLERT et al. (1984) in several plants growing in situ in the Namib desert. In the latter case, it was clearly shown that the daytime malic acid accumulation depended on net CO_2-fixation by photosynthesis.

β) Malate Synthesis Under Permanent Light. Under permanent light, CAM plants perform a circadian rhythm of net CO_2-exchange, with CO_2-uptake and malate accumulation during the prospective night, and CO_2-output together with malic depletion during the prospective day. BUCHANAN-BOLLIG (1984) showed that the duration of the rhythms depended on irradiance and temperature. With high irradiance and/or temperature the rhythm damps out rapidly. It is, however, maintained over many cycles at moderate irradiance and temperatures. KLUGE et al. (1982), BUCHANAN-BOLLIG et al. (1984) found that during the peaks of CO_2-uptake the carbon flow is very similar, if not identical, to the late phase IV of CAM, where malate is synthetized by PEPC-mediated CO_2-fixation, which can proceed parallel to and also in series with RubPC-mediated CO_2-fixation. The studies by BUCHANAN-BOLLIG et al. (1984) suggest also that not the malate synthetizing pathway itself, but the capability of the vacuoles to store malic acid temporarily is disturbed by the high irradiance and temperatures. This is consistent with the view that the oscillator controlling the rhythmicity of CAM under permanent light is identical with the vacuole (WILKINS 1983, NUNGESSER et al. 1984).

γ) Malic Acid Metabolism During Deacidification (Phase III). There is now further evidence that two types of CAM plants exist, which differ in the manner in which the previously stored malic acid is decarboxylated. One of the two types decarboxylates by NADP- and NAD-malic enzymes, yielding CO_2 and pyruvate, whereas in the other type malate is metabolized by PEP-carboxykinase. This latter pathway yields also CO_2, but PEP instead of pyruvate. HOLTUM and OSMOND (1981) suggest that in both types of CAM plants the C_3-residue remaining from the decarboxylation of malate can be used in gluconeogenesis. In the case of the malic enzyme type, the pyruvate has to be converted to PEP by pyruvate-Pi-dikinase. This enzyme is active only in the malic enzyme type plants; not, however, in the PEP-carboxykinase plants. The conversion of pyruvate to PEP requires, in addition to pyruvate-Pi-dikinase, adenylate kinase. According to MANETAS et al. (1983) this enzyme is highly active in the malic enzyme CAM plants; however not in the other group. The observed activities are more than sufficient to account for the required rates of AMP to ATP conversion. It was shown by HOLTUM and OSMOND (1981) that only the malic enzyme plants could substantially incorporate externally applied [3-^{14}C]-pyruvate into carbohydrates. In the PEP-carboxykinase type plants, such external pyruvate was oxidized via the Krebs cycle. This finding illustrates clearly the in vivo function of pyruvate P_i-dikinase of the malic enzyme plants, i.e. , to initiate the gluconeogenetic consumption of malate-derived pyruvate.

The rate of malic acid depletion during phase III of CAM is obviously controlled by photosynthesis. This can be concluded from the observation by BARROW and COCKBURN (1982) that the action spectrum of malate consumption is identical with the action spectrum of photosynthesis. Further evidence came from NISHIDA and HAYASHI (1980) who found that DCMU inhibits deacidification. NISHIDA (1982) argues that photosynthesis might be linked to malate depletion by the intercellular CO_2-concentrations. At low light intensities, the internal CO_2-concentration is assumed to be sufficiently high to push the reaction of the decarboxylating enzymes in the direction of carboxylation, or at least to inhibit decarboxylation. Effective decarboxylation, however, is required in order to keep the cytoplasmic malate concentration sufficiently low to permit passive export of malate from the vacuole. In fact, tracer studies by A. FISCHER and M. KLUGE (unpublished) showed that the photosynthetic carbon transfer from previously stored malate to carbohydrates could be inhibited by high CO_2-concentrations.

δ) Storage of Malic Acid in the Vacuoles. The reversible storage of malic acid in the vacuole is an essential feature of CAM. There are now several reports that intact vacuoles have been isolated and studied in CAM plants (KRINGSTAD et al. 1980, BUSER-SUTER et al. 1982, BLACK et al. 1982, LÜTTGE et al. 1981, 1982, 1984, SMITH et al. 1984a,b, AOKI and NISHIDA 1984). BUSER-SUTER et al. (1982) found that uptake of malate by isolated vacuoles was independent of ATP. On the other hand, AOKI and NISHIDA (1984) and, independently, SMITH et al. (1984a,b), established clearly the presence of a vacuolar ATP-ase.

It was concluded from inhibitor studies (AOKI and NISHIDA 1984, SMITH et al. 1984a,b) that the tonoplast ATP-ase in CAM plants transports H^+. The resulting $\Delta\mu H^+$ would then drive the transport of malate^{2-} into the vacuole. A malate-permease has been shown to be present in the tonoplast (BUSER-SUTER et al. 1982). JOCHEM et al. (1984) demonstrated the ATP-dependent H^+-pumping directly by measuring the transtonoplast electrical potentials and intravacuolar pH. These authors could show that the energetization of isolated vacuoles by Mg-ATP resulted in a positive polarization of 8.6 mV and drop of pH (1.1 pH unit) inside the vacuole.

The release of malic acid is by passive transport (LÜTTGE et al. 1982). LÜTTGE and SMITH (1984) provided an elegant hypothesis of how this release might be controlled. According to these authors, efflux occurs predominantly as H_2mal^o. The permeability coefficient for H_2mal^o would then determine the maximume concentration of malic acid which can be achieved in the vacuole during the night phase of CAM.

3. Malate Synthesis by PEPC-mediated CO_2-fixation in Nitrogen Fixing Root Nodules

The production of C_4 acid by PEPC plays, obviously, an important role in the symbiotic N_2-fixation in root nodules. This enzyme is highly effective in root nodules (see the review by GADAL et al. 1983). Within the nodule cells the enzyme is located in the cytoplasm (GADAL et al. 1983). In the nodules there may exist an isoenzyme of PEPC which is different from that of the roots and which is believed to play a specific role in the metabolism linked with N_2-fixation. GADAL (1983) suggests that the synthesis of C_4 acids by CO_2-fixation in root nodules performs several tasks. Firstly, since nitrogen fixation requires much ATP, thus high rates of respiration, the enzyme may help to avoid excessive loss of carbon by refixation of respiratory CO_2. Carbon skeletons produced by PEPC-mediated CO_2-fixation may also serve as acceptor of amino groups in aspartate and asparagine biosynthesis. This latter amide is then transported to the aerial part of the plant. Furthermore, it is assumed that malate, produced via PEPC, serves as a substrate for respiration, both in the host cells and in the symbiotic bacterioids. Finally, there is evidence that malate is important in the maintenance of charge balance in the xylem sap of nitrogen-fixing legumes (ISRAEL and JACKSON 1982). McCLURE et al. (1983) found that in *Alnus*-nodules, together with PEPC, carbonylphosphate synthase contribute to CO_2-fixation. The mentioned authors found in the cells of *Alnus*-nodules that at least two independent malate pools exist. A larger pool is physiologically inactive. Its role is unknown. The other pool is in fast exchange with amino acid synthesis.

PEPC-mediated CO_2-fixation is also supposed to allow recovery or detoxification of NH_3 released during protein turnover in roots (GIVAN 1979). POPP and SUMMONS (1983) and OSMOND and POPP (1983) found that in the presence of NH_4Cl the synthesis of amino acids has priority over the synthesis of malate.

4. Malate Metabolism and Proton Extrusion

Malate plays a central role in maintaining ionic equilibrium in the cells when cations are taken up. The contribution of PEPC-mediated malate synthesis in the context of ion balance has recently been reviewed (GUERN et al. 1983).

A major aspect of the involvement of malate and some other organic acids in the maintenance of ion balance during ion uptake concerns nitrate uptake. There are several reports that malate accumulates in the tissue when nitrogen is taken up by roots (KIRKBY and KNIGHT 1977, BLEVINS et al. 1978, FROST et al. 1978, SMITH 1978). The most likely explanation for this effect is that malic acid synthesis provides the protons which have to be released from the root cells when nitrate is taken up together with cations. DAHLBENDER and STRACK (1984) found recently that in *Raphanus sativus* not only malate accumulates upon nitrogen uptake, but also sinapoyl-bound malate. The enzyme involved in the synthesis of this substances, 1-sinapoyl-glucose: L-malate synapoyltransferase, appears to be controlled by the malate pool in the cells.

The metabolic pathway involved in ion-induced malate synthesis and its regulation were investigated in more detail by POPP et al. (1982). By means of an elegant mass-spectrometrical analysis of $[^{13}C]$-malate labeled in roots by $[^{13}C]O_2$-fixation, it could be shown that only singly labeled malate occurred. The label was found either in the C_4- or C_1-carbon of malate. This fits with the assumption that malate was synthetized via PEPC, and that the label was randomized afterward by a fumarase-like reaction, presumably in mitochondria.

In maize roots the release of H^+ and the accumulation of malate upon cation uptake is inhibited by Ni (MORGUTTI et al. 1984). A detailed analysis revealed that the primary effect oculd be the inhibition of PEPC by Ni. On the other hand, OSMOND and POPP (1983) found that in wheat-roots the actual malate level resulting from the demand in cation uptake is much more controlled by the rate of decarboxylation than by synthesis of malate.

Proton extrusion can compensate excess cation influx in cultured cells also. This was shown by KOMOR et al. (1981) in sugarcane cell suspensions. However, these authors found also that the regulation of ion balance in younger and older cells was different. While younger cultures compensate cation uptake by proton efflux, older cells show a balanced cation-anion uptake. The older cells synthetizes organic acid (mainly malate) more abundantly to obtain protons which are required in nitrate and sulfate assimilation.

Some of parasitic phanerogames studies so far have a high activity of PEPC, thus synthetizing malate via this enzyme. Although details are not yet known, it is conceivable that the substantial PEPC-activity found in the underground leaflets of *Lathrea* (M. KLUGE, unpublished, RENAUDIN et al. 1982) might be involved in the special ion (and water) extrusion mechanism shown by this root-xylem parasite. On the other hand, SOLDATINI et al. (1981) and RENAUDIN et al. (1982) interpret CO_2-fixation by PEPC found in *Orobanche*, *Viscum*, *Thesium* and *Osyris* as carbon-saving mechanism which might recover respiratory CO_2. It is conceivable that also the substantial malate synthesis and extrusion by leaf and shoot glands of *Cicer arietinum* is linked to the maintenance of ion equilibrium. In this context the studies of HERNANDEZ-NISTAL et al. (1983) have to be mentioned. They showed that in embryo axes of *Cicer* the PEPC-mediated CO_2-fixation and malate synthesis were positively correlated with K^+ uptake. SANTAKUMARI et al. (1979) believe that *Cicer* performs CAM. However, due to the recent investigations by KOUNDAL and SINHA (1981) and WINTER (1981) it is now unequivocally clear that CAM is lacking in *Cicer*.

5. Malate Oxidation by Mitochondria

During the last few years the view that malate in plant mitochondria can be oxidated both via NAD^+-malate dehydrogenase (yielding oxaloacetate) and NAD malic enzyme (yielding pyruvate) has emerged. Both enzymes together would allow full oxidation of malate via the Krebs cycle without the participation of glycolysis. This is because the prerequisites of the Krebs cycle, oxaloacetate and pyruvate, would be made available by MDH and NAD^+ malic enzyme. In a careful study on malate oxidation by *Arum* spadix mitochondria, WEDDING and WHATLEY (1984) showed that at pH 6.5 malate was oxidized primarily through NAD malic enzyme, while at pH 7.7 malate oxidation was mainly by MDH. In the spadix mitochondria NAD malic enzyme was found by the mentioned authors to be exceptionally high. The physiological significance of this phenomenon remains to be clarified.

RUSTIN et al. (1980) believe that malate oxidation via NAD malic enzyme is linked to the cyanide-insensitive electron transport. This idea has been critically evaluated in a review by LATIES (1982). Since this review further evidence was provided in the literature for malate oxidation via the "alternative", cyanide-insensitive pathway (CHAUVEAU et al. 1983, NEUBURGER and DOUCE 1980, PALMER et al. 1982, RUSTIN et al. 1983, WALKER and OLIVER 1983, WISKICH and DAY 1982). The physiological role of the alternative pathway is still unknown (see review by LATIES 1982). RUSTIN and ALIN (1984) and CLAUDINE QUEIROZ-CLARET (pers. comm.) have, however, outlined the hypothesis that in CAM plants mitochondrial NAD malic enzyme and the alternative pathway are of major importance in generating CO_2 from the nocturnally synthetized and stored malic acid during the light period.

6. Malic Acid in Grapes

Malic acids, together with tartaric acid, represent the dominant organic acids both in the leaves and berries of grapes *(Vitis vinifera)*. RUFFNER (1982a,b), RUFFNER et al. (1983) have shown that the malic acid synthesis proceeds via β-carboxylation of PEP, with the CO_2-acceptor deriving from the C_3-pathway of photosynthesis. Such malic acid synthesis is not unusual in C_3 plants and may have a many-faceted function (see review by LATZKO and KELLY 1983). MALIPIERO et al. (1981) found that environmental conditions which favor photorespiration enhanced the $[^{14}C]O_2$-malate. From these findings the author conclude that a precursor-product relationship must exist between metabolites of the photorespiratory pathway on the one hand, and malic acid on the other. However, studies by RUFFNER et al. (1983) revealed that this relationship, if it exists, plays only a minor role in malate synthesis by C_3 plants.

The degradation of malic acid in the berries of *Vitis vinifera* proceeds presumably via NADP malic enzyme. The latter enzyme has been isolated from grapes and biochemically characterized (POSSNER et al. 1981). RUFFNER et al. (1984) proposed that in accordance with demands of reduction equivalents in biosynthetic pathways, malate in the fruits serves as a storage pool for "reduction power". It is by NADP malic enzyme-mediated oxidative decarboxylation of malate that this pool can be used by $NADPH_2$ formation.

7. Metabolism of Organic Acids Other than Malic Acid

a) Ascorbate and Oxalate

In several oxalate-accumulating plants, ascorbate functions as precursor for oxalate (YANG and LOEWUS 1975, NUSS and LOEWUS 1978, LOEWUS 1980). L-ascorbic acid itself may be derived in a direct way from D-glucose (LOEWUS 1980). This pathway has now been demonstrated to occur also in single-cell algae (RENSTROM et al. 1982). RAVEN et al. (1982) showed clearly by means of elegant $[^{18}O]_2$ and $[^{13}C]O_2$ studies that oxalate in higher plants is not derived from a direct oxidation of the glyoxylate of the glycolate pathway. Rather, these authors suggest that PEPC-mediated CO_2-fixation may be involved in oxalate biosynthesis. A possible pathway would be the conversion of oxaloacetate deriving from the PEPC reaction by OAA-lyase. The above-mentioned results by RAVEN et al. (1982) implicate also that oxalate synthesis via lactate dehydrogenase-mediated dismutation of glyoxylate (DAVIES and ASKER 1983) should be of minor importance in photosynthetizing cells of higher plants.

WAGNER (1981) demonstrated the vacuolar localization of ascorbate-derived oxalic acid in barley seedlings.

Free oxalic acid was a major component of the organic acid spectra only in the salt-secreting species of Australian mangroves. Malate and citrate could be found in all mangrove species, quinic and shikimic acids occurred frequently. Altogether, organic anions seem to play only a minor role in maintaining ion balance in mangroves (POPP 1984).

A suitable object for future studies on oxalate metabolism may be the epidermal bladder cells of *Oxalis carnosa*, where the concentration of free oxalic acid is high enough to cause a pH value of ca. 1.0 (STEUDLE et al. 1983).

b) Malonate

The synthesis of malonate was shown in *Glycine max* to proceed via acetyl-CoA carboxylase (STUMPF and BURRIS 1981a). In the mentioned plant, malonate does not accumulate as a dead-end product of metabolism. Exogenously supplied [1-^{14}C]-malonate was easily metabolized in leaves of *Coffea arabica* (NUNES et al. 1983), with label appearing in respired CO_2, acids of the Krebs cycle, uronic acids, sugars and amino acids. Since labeled malonate was applied under illumination and is obviously respired, thus yielding [^{14}C]O_2, the appearance of label in sugars can be easily explained in terms of photosynthetic reassimilation of malonate-derived [^{14}C]O_2.

Drastic changes in concentrations of malonate can be shown within whole plants, plant organs, or with time (STUMPF and BURRIS 1981b). POPP and KINZEL (1981) found that in *Phaseolus* malonate plays a key role in balancing excess cations. SASSON and MONSELISE (1976) postulate that, at least in fruits, malonate is an indicator for senescence.

c) Trans-aconitic Acid

BRAUER and TEEL (1981) have studied the trans-aconitate synthesis via citrate dehydrogenase in *Zea mays*. The authors found two isoenzymes of the enzyme (named CD I and CD II). One isoform consists of four 80,000 MW subunits, the other isoform, however, of two 60,000 MW subunits. The two isoenzymes were found to be located in the cytosol and in the mitochondria, respectively. The isoenzymes showed different kinetics and regulatory properties (BRAUER and TEEL 1982). The isoenzyme CD I is regulated by K^+. This may explain why in certain grasses K^+ is positively correlated with the level of trans-aconitic acid.

d) Shikimic Acid

Shikimic acid is a precursor of sphagnorubin, a reddish-violet pigment extractable from the cell wall of several *Sphagnum* species (RUDOLPH et al. 1981). The key enzyme involved, shikimate dehydrogenase, showed an increase upon chilling-induced sphagnorubin synthesis. The enzyme obtained from tomato plants has been purified and kinetically characterized by LOURENCO and NEVES (1984). HOLLÄNDER-CZYTKO and AMRHEIN (1983) studied the subcellular compartmentation of shikimic acid in buckwheat cell suspension cultures grown in the presence of shikimate pathway inhibitors. Under such conditions the cells accumulate shikimate up to 8 µmol/g fresh weight. The acid is exclusively located in the vacuoles. Protoplasts as well as isolated vacuoles were able to take up shikimate-3-P and to hydrolyze the phosphate esters.

References

AMAGASA, T.: Z. Pflanzenphysiol. 108, 93-96 (1982). - AOKI, K., NISHIDA, K.: Physiol. Plant. 60, 21-26 (1984).

BARROW, S.R., COCKBURN, W.: Plant Physiol. 69, 568-571 (1982). - BLACK, C.C., CARNAL, N.W., KENYON, W.H.: pp. 51-68 in: Crassulacean Acid Metabolism, eds. I.P. TING, M. GIBBS. Am. Soc. Plant Physiol. Rockville 1982. - BLEVINS, D.G., BARNETT, N.M., FROST, W.B.: Plant Physiol. 62, 784-788 (1978). - BRAUER, D., TEEL, M.R.: Plant Physiol. 68, 1406-1408 (1981); - ibid. 70, 723-727 (1982). - BUCHANAN-BOLLIG, I.C.: Planta 160, 264-271 (1984). - BUCHANAN-BOLLIG, I.C., FISCHER, A., KLUGE, M.: Planta 161, 71-80 (1984). - BUSER-SUTER, C., WIEMKEN, A., MATILE, P.: Plant Physiol. 69, 456-459 (1982).

CHAUVEAU, M., DIZENGREMEL, P., ROUSSAUX, J.: Plant Physiol. 73, 945-948 (1983).

DAHLBENDER, B., STRACK, D.: Planta 161, 142-148 (1984). - DAVIES, D.D., ASKER, H.: Plant Physiol. 72, 134-138 (1983).

FISCHER, K., KLUGE, M.: Planta 160, 121-128 (1984). - FROST, W.B., BLEVINS, D.G., BARNETT, N.M.: Plant Physiol. 61, 323-326 (1978).

GADAL, P.: Physiol. Vég. 21, 1069-1074 (1983). - GADAL, P., PERROT-RECHENMANN, C., VIDAL, J.: Physiol. Vég. 21, 1055-1062 (1983). - GIVAN, C.V.: Phytochemistry 18, 375-382 (1979). - GUERN, J., MATHIEU, J., KURKDIJAN, A.: Physiol. Vég. 21, 855-866 (1983).

HERNANDEZ-NISTAL, J., ALDASORO, J., RODRIGUEZ, D., MATILLA, A., NICOLÁS, G.: Physiol. Plant. 57, 273-278 (1983). - HOLLÄNDER-CZYTKO, H., AMRHEIN, N.: Plant Sci. Lett. 29, 89-96 (1983). - HOLTUM, J.A.M., OSMOND, C.B.: Austr. J. Plant Physiol. 8, 31-44 (1981). - HOLTUM, J.A.M., O'LEARY, M.H., OSMOND, C.B.: Plant Physiol. 71, 602-609 (1983).

ISRAEL, D.W., JACKSON, W.A.: Plant Physiol. 69, 171-178 (1982).

JOCHEM, P., RONA, J.P., SMITH, J.A.C., LÜTTGE, U.: Physiol. Plant. (in press) (1984).

KIRKBY, E.A., KNIGHT, A.H.: Plant Physiol. 60, 349-353 (1977). - KLUGE, M.: pp. 231-262 in: Carbon Metabolism and Plant Productivity, ed. J. GOVINDJEE. New York: Academic Press 1982. - KLUGE, M., FISCHER, A., BUCHANAN-BOLLIG, I.C.: pp. 31-50 in: Crassulacean Acid Metabolism, eds. I.P. TING, M. GibbS. Rockville: Am. Soc. Plant Physiol. 1982. - KOMOR, E., THOM, M., MARETZKI, A.: Plant Cell Environ. 4, 359-365 (1981). - KOUNDAL, K., SINHA, S.: Phytochemistry 20, 1251-1252 (1981). - KRINGSTAD, R., KENYON, W., BLACK, C.: Plant Physiol. 66, 379-382 (1980).

LATIES, G.G.: Annu. Rev. Plant Physiol. 33, 519-555 (1982). - LATZKO, E., KELLY, G.J.: Physiol. Vég. 21, 805-815 (1983). - LITTLEJOHN, R.O., KU, M.S.B.: Plant Physiol. 74, 1050-1054 (1984). - LOEWUS, F.A., in: The Biochemistry of Plants, eds. P.K. STUMPF, E.E. COHN, Vol. 3. New York: Academic Press 1980. - LOURENCO, M.R., NEVES, V.A.: Phytochemistry 23, 497-400 (1984). - LÜTTGE, U., SMITH, J.A.C.: J. Membr. Biol. (in press) (1984). - LÜTTGE, U., SMITH, J.A.C., MARIGO, G., OSMOND, C.: FEBS Letters 126, 1981. - LÜTTGE, U., SMITH, J.A.C., MARIGO, G.: pp. 69-91 in: Crassulacean Acid Metabolism, eds. I.P. TING, M. GIBBS. Rockville: Am. Soc. Plant Physiol. 1982. - LÜTTGE, U., JOCHEM, P., SMITH, J.A.C., URIBE, E.G.: pp. 359-364 in: Membrane Transport in Plants, eds. W.J. CRAM, K. JANÂCEK, R. RYBOVÂ, K. SIEGLER. Prag: Academia Praha 1984.

MALIPIERO, U., RUFFNER, H.P., ROST, D.M.: Z. Pflanzenphysiol. 104, 243-251 (1981). - MANETAS, J., SAMARAS, J., STAMATAKIS, A.: Z. Pflanzenphysiol. 111, 365-368 (1983). - McCLURE, P.R., COKER, G.T., SCHUBERT, K.R.: Plant Physiol. 71, 652-657 (1983). - MEYER, C.P.: PhD Thesis, University of Melbourne 1979. - MORGUTTI, S., SACCHI, G.A., COCUCCI, S.M.: Physiol. Plant. 60, 70-74 (1984).

NEUBURGER, M., DOUCE, R.: Biochim. Biophys. Acta 589, 176-189 (1980). - NISHIDA, K.: Physiol. Plant. 54, 451-454 (1982). - NISHIDA, K., HAYASHI, Y.: Plant Sci. Lett. 19, 271-276 (1980). - NUNES, M.A., CONTREIRAS, J., da SILVA, D.F.: J. exp. Bot. 34, 1101-1107 (1983). - NUNGESSER, D., KLUGE, M., TOLLE, H., OPPELT, W.: Planta (in press) (1984). - NUSS, R.F., LOEWUS, F.A.: Plant Physiol. 61, 590-592 (1978).

OSMOND, C.B.: Annu. Rev. Plant Physiol. 29, 379-414 (1978). - OSMOND, C.B., POPP, M.: Plant Sci. Lett. PSL 1562-JC, Limerick: Elsevier 1983.

PALMER, J.M., SCHWITZGUEBEL, J.P., MOLLER, J.: Biochem. J. 208, 703-711 (1982). - POPP, M.: Z. Pflanzenphysiol. 113, 395-410 (1984). - POPP, M., KINZEL, H.: J. exp. Bot. 32, 1-8 (1981). - POPP, M., SUMMONS, R.E.: Physiol. Vég. 21, 1083-1089 (1983). - POPP, M., OSMOND, C.B., SUMMONS, R.E.: Plant Physiol. 69, 1289-1292 (1982). - POSSNER, D., RUFFNER, H.P., RAST, D.M.: Planta 151, 549-554 (1981).

RAVEN, J.A., GRIFFITHS, H., GLIDEWELL, S.M., PRESTON, T.: Proc. R. Soc. Lond. B 216, 87-101 (1982). - RENAUDIN, S., VIDAL, J., LARHER, F.: Z. Pflanzenphysiol. 106, 229-237 (1982). - RENSTROM, B., GRÜN, M., LOEWUS, F.A.: Plant Sci. Lett. 28, 299-305 (1982). - RUDOLPH, Ch., KRAUSE, H.-J., BLAICHER, M., HERMS, E.: Biochem. Physiol. Pflanzen 176, 728-736 (1981). - RUFFNER, H.P.: Vitis 21, 247-259 Part A (1982a); - ibid. 21, 346-358, Part B (1982b). - RUFFNER, H.P., BREM, S., RAST, D.M.: Plant Physiol. 73, 582-585 (1983). - RUFFNER, H.P., POSSNER, D., BREM, S., RAST, D.M.: Planta 160, 444-448 (1984). - RUSTIN, P., ALIN, M.F.: Physiol. Vég. 21, 22-30 (1984). - RUSTIN, P., MOREAU, F., LANCE, C.: Plant Physiol. 66, 457-463 (1980). - RUSTIN, P., DUPONT, J., LANCE, C.: Arch. Biochem. Biophys. 225, 630-639 (1983).

SANTAKUMARI, M., REDDY, C., REDDY, A., DAS, V.: Naturwissenschaften 66, 54-55 (1979). - SASSON, A., MONSELISE, S.P.: Experientia 32, 1116-1117 (1976). - SMITH, F.W.: Plant Soil 49, 367-379 (1978). - SMITH, J.A.C., URIBE, E.G., BALL, E., HEUER, S., LÜTTGE, U.: Eur. J. Biochem. (in press) (1984a). - SMITH, J.A.C., URIBE, E.G., BALL, E., LÜTTGE, U.: Planta (in press) (1984b). - SOLDATINI, G.F., LOTTI, G., PARADOSSI, C.: Biochem. Biophys. Pflanzen 176, 109-115 (1981). - STEUDLE, E., ZIEGLER, H., ZIMMERMANN, U.: Planta 159, 38-45 (1983). - STUMPF, D.K., BURRIS, R.H.: Plant Physiol. 68, 992-995 (1981a); - ibid. 68, 989-991 (1981b).

WAGNER, G.J.: Plant Physiol. 67, 591-593 (1981). - WALKER, G.H., OLIVER, D.J.: Arch. Biochem. Biophys. 225, 847-853 (1983). - WEDDING, R.T., WHATLEY, F.R.: New Phytol. 96, 505-517 (1984). - WILKINS, M.B.: Planta 157, 471-480 (1983). - WILLERT, D.J. v., BRINCKMANN, E., ELLER, B.M., SCHEITLER, B.: Oecologia 61, 393-397 (1984). - WINTER, K.: Z. Pflanzenphysiol. 101, 421-430 (1981). - WINTER, K., TENHUNEN, J.D.: Plant Physiol. 70, 1718-1722 (1982). - WISKICH, J.T., DAY, D.A.: Plant Physiol. 70, 959-964 (1982). - WOLF, J.: pp. 809-889 in: Encyclopedia of Plant Physiology, ed. W. RUHLAND, Vol. 12. Springer: Berlin, Heidelberg, New York 1960.

YANG, J.C., LOEWUS, F.A.: Plant Physiol. 56, 283-285 (1975).

Professor Dr. MANFRED KLUGE
Institut für Botanik der
Technischen Hochschule Darmstadt
Schnittspahnstr. 3-5
D-6100 Darmstadt, FRG

V. Metabolism of Organic N-Compounds Ammonium Assimilation and Amino Acid Metabolism[1]

By THOMAS HARTMANN and ADELHEID EHMKE

Plant nitrogen metabolism has been reviewed by DURZAN and STEWARD (1983) and with emphasis on ammonium assimilation and amino acid metabolism by MIFLIN and LEA (1982), LEA et al. (1982), WALLSGROVE et al. (1983a).

1. Assimilation and Reassimilation of NH_4^+

Ammonia is either assimilated from an external source (NO_3^-, N_2) or reassimilated after the release from an internal source (e.g., gly during photorespiration, ureides, arg, asn or the action of phenylalanine-ammonia-lyase). In the last years there has been growing awareness that the reassimilation can by far exceed the primary assimilation of ammonia. So there are many reports dealing with the recycling of NH_4^+ during photorespiration.

a) Physiology

All reports based upon inhibitor and labeling studies agree that the functioning of the GS/GOGAT cycle is necessary for the reassimilation of NH_4^+ in C_4 plants, assumed to have low photorespiratory activity, as well as in C_3 plants (PLATT and ANTHON 1981, FRANTZ et al. 1982, PLATT and RAND 1982, WOO et al. 1982, BERGER and FOCK 1983, MARTIN et al. 1983).

There are, however, some discrepancies concerning the amounts of NH_4^+, which accumulate in C_3 and C_4 plants. PLATT and coworkers using high concentrations (8 mmol/l) of the GS inhibitor MSO found similar NH_4^+ accumulation rates in both C_4 and C_3 leaves. In spinach, NH_4^+ accumulation was not relieved under nonphotorespiratory conditions, i.e., reduced O_2 concentration, suggesting an alternate source of NH_4^+ release (PLATT and ANTHON 1981, PLATT and RAND 1982). In the presence of lower MSO or tabtoxin concentrations ($\leq$ 2 mmol/l, noninhibitory for photosynthesis) there was a clear response to reduced O_2 concentrations in C_3 and C_4 plants (FRANTZ et al. 1982, MARTIN et al. 1983). So most of the NH_4^+ which accumulated during the inhibitor treatment came from photorespiration. The rates of NH_4^+ accumulation were higher in C_3 than in C_4 plants (MARTIN et al. 1983).

The GS inhibitors L-methionine sulfoximine and phosphophinotricin were shown to inhibit GS from pea leaves in a mixed competitive manner with K_i values of 0.16 mM and 0.073 mM respectively (LEASON et al. 1982). Tabtoxinine-β-lactam, a hy-

[1]Abbreviations: GS, glutamine synthetase; GOGAT, glutamate synthase; MSO, methionine sulfoximine; Fd, ferredoxin

Progress in Botany, Vol. 46

drolytic product of tabtoxin from *Pseudomonas syringae* was shown to be the effective metabolite, inhibiting plant GS (THOMAS et al. 1983).

[^{15}N] kinetic labeling studies with barley seedlings growing under steady-state conditions were fully consistent with a computer-simulated 3-compartment model in which NH_4^+ is assimilated solely via the GS/GOGAT pathway (FENTEM et al. 1983a). This result was found to be in agreement with inhibitor studies using MSO and azaserine (FENTEM et al. 1983b).

Solid-state [^{15}N]-NMR measurements were found to be a useful method to follow the utilization of gln and asn nitrogen for the synthesis of storage proteins in cultured developing soybean cotyledons (SKOKUT et al. 1982a). Whereas the amide-N and the amino-N of gln were metabolized equally, the two nitrogens of asn were used differentially. The amino-N donates approximately twice as much N to protein as does the amide-N, indicating the presence of an asparagine aminotransferase. In a similarly devised experiment the utilization of gly in the presence of gln was studied using solid-state [^{15}N]- and [^{13}C]-NMR (SKOKUT et al. 1982b). Addition of gly decreased the use of gln-N, and gly-N was found to be incorporated selectively into the protein, in amounts far exceeding what would be expected if it appeared solely in gly residues. This indicates transfer of gly-N by rapid transamination.

b) Glutamine Synthetase (GS)

Many reports deal with the isoforms of GS and their regulation (see McNALLY and HIREL 1983 for rev.) Earlier studies showed that in green leaves and shoots GS may occur in two isoforms. GS_1 is located in the cytosol, and GS_2 in the chloroplast (cf. Progr. Bot. 44, 155, HIREL and GADAL 1982, HIREL et al. 1982a). Investigating a wide range of higher plants, different patterns of GS isoforms could be observed (McNALLY et al. 1983a,b,c). According to the relative activities of GS_1 and GS_2 in shoots, four groups were characterized: (1) exhibits only GS_1 activity (achlorophyllous plant parasites), (2) possesses only GS_2 (e.g., spinach and other Chenopodiaceae and lupins), (3) contains GS_1 and GS_2, but GS_1 activity is less than 30% of total activity (e.g., many C_3 grasses and temperate legumes), (4) contains both isoforms but high GS_1 activity (many C_4 plants and tropical legumes). There is no positive correlation between photorespiratory activity and cytosolic GS_1. On the contrary, within closely related species of the genus *Panicum*, the C_3 plant lacks GS_1, whereas the C_4 plant has a relatively high GS_1 content (HIREL et al. 1983a). This raises the question about the physiological role of the isoforms. The chloroplastic form (GS_2) has been suggested to assimilate NH_4^+ derived from nitrate reduction. In green tissues the cytosolic form (GS_1) was originally thought to be responsible for the recycling of NH_4^+ produced by photorespiration and other NH_4^+-liberating reactions (KEYS et al. 1978). This view is supported by the results obtained with ungerminated seeds and greening cotyledons of peanuts (WINTER et al. 1982) and soybeans (DYVINIAK and DUNHAM 1983) adapted to remobilize storage nitrogen, as well as with the obligate achlorophyllous plant parasites (McNALLY et al. 1983c). In these systems only GS_1 is present. On the other hand, the general validity of this assumption has been doubted. Spinach chloroplasts can very effectively reassimilate NH_4^+ added together with 2-oxoglutarate (WOO and OSMOND 1982), or released by mitochondria during gly oxidation in a recombinated system (BERGMAN et al. 1981). As pointed out above, spinach possesses only the chloroplastic GS_2. One can assume that at least in this plant GS_2 has the function to deal with the reassimilation of NH_4^+ as well as with primary assimilation.

In *Phaseolus* root nodules three forms of GS could be separated on DEAE-Sephacel or by gel electrophoresis (CULLIMORE et al. 1982). One form is specific for the root tissue (GS_r), the other two are nodule-specific (GS_{n1}, GS_{n2}). As in most plants, the multiple GS forms differ in charge rather than in size. GS_r and GS_{n2} seem to be almost identical. During nodulation with the wild-type strain of *Rhizobium phaseoli* the GS activity increased considerably. This increase was solely due to the synthesis of the nodule specific isoform GS_{n1} (LARA et al. 1983). Messenger RNA for the GS from nodules and roots of *Phaseolus* were demonstrated, for the first time in higher plants, in preparations of poly$(A)^+$RNA by in vitro translation and immunoprecipitation (CULLIMORE and MIFLIN 1983). Probably the GS subunits are not processed before assembling the active oligomere, since the in-vitro translation products had the same molecular weight as the enzyme subunits isolated from the nodules. The amount of GS-mRNA is much greater in *Rhizobium*-infected nodules than in nonnodulated roots, confirming de novo synthesis of GS_{n1} during nodulation.

GS has been purified from many plant sources. All isoforms have a mass of 330,000 - 380,000 MW and are composed of eight identical subunits of about 45,000 KW (McCORMACK et al. 1982, GROAT and SCHRADER 1982, HIREL et al. 1982c, WINTER et al. 1982, CULLIMORE et al. 1983, FLORENCIO and VEGA 1983, PUSHKIN et al. 1983). Kinetic analysis indicates that the reaction mechanism is sequential with Mg-ATP binding first (McCORMACK et al. 1982). GS_1 and GS_2 differ in their heat stability (HIREL and GADAL 1982, AHMAD et al. 1982), their response to dithiothreitol (WALLSGROVE et al. 1983a), energy charge (HIREL et al. 1983b) and their susceptibility to proteolysis (STREIT and FELLER 1983).

The chloroplastic GS_2 is synthesized during greening of etiolated rice leaves (HIREL et al. 1982b), pumpkin cotyledons (NISHIMURA et al. 1982) and seedlings of peanut and soybean (WINTER et al. 1982, DYVINIAK and DUNHAM 1983). In lettuce seeds there was a phytochrome-mediated synthesis of GS (TAKEBA 1983). Thioredoxin activates GS (eigthfold) and GOGAT (fourfold) in synchronous *Chlorella sorokiniana* (TISCHNER and SCHMIDT 1982). Roots of sunflower plants display a 12-h rhythm of GS activity with one peak in the light and one in the dark phase (KNIGHT and WEISSMAN 1982).

c) Glutamate Synthase (GOGAT)

The two forms of GOGAT, one Fd-dependent (EC 1.4.1.7) the other NAD(P)H-dependent (EC 1.4.1.13/14), are located in the plastids. Fd-GOGAT is always present in shoots and roots. NAD(P)H-GOGAT formerly thought to be restricted to nonphotosynthetic tissues, was recently detected in pea seedlings (MATOH and TAKAHASHI 1981, 1982), greening barley and pea leaves (WALLSGROVE et al. 1982) and greening tobacco cell cultures (SUZUKI et al. 1984). Whereas Fd-GOGAT increases during seedling growth (MATOH and TAKAHASHI 1982) and, more rapidly, on greening of etiolated pea and barley leaves (WALLSGROVE et al. 1982), rice leaves (SUZUKI et al. 1982a) or tobacco cells (SUZUKI et al. 1984), NAD(P)H-GOGAT has a low constitutive level (WALLSGROVE et al. 1982, SUZUKI et al. 1984) or decreases during maturation (MATOH and TAKAHASHI 1982). In green leaves it comprises 2% - 3% (WALLSGROVE et al. 1982), in green tobacco cells 12% - 14% (SUZUKI et al. 1984) of total GOGAT activity. Fd-GOGAT from rice leaves was purified and shown to consist of two identical subunits. The dimer has a mass of about 224,000 MW (SUZUKI and GADAL 1982). This is higher than the molecular masses reported for the enzyme from other sources (155,000 - 180,000 MW). As all other known Fd-GOGATs, the enzyme is free of iron sulfide and flavin. In a chloroplast reconstituted system from rice leaves the dependence on photoreduced ferredoxin could be demonstrated (SUZUKI et al. 1982b). The formation of glu was strictly light-dependent. The K_m value for Fd was estimated to be 4.3 µmol/l. This is in accordance with the value of purified Fd-GOGAT (SUZUKI and GADAL 1982). Fd-GOGAT from green rice leaves was shown to be immunochemically identical to the Fd-GOGAT in etiolated leaves and related to the Fd-GOGAT in rice roots and from tobacco cells. There was no cross-reaction with the NAD(P)H-GOGAT from rice or tobacco cells (SUZUKI et al. 1982a, 1984). Thus, pyridine-nucleotide-dependent and Fd-dependent GOGATs are distinct proteins with different structural conformation.

2. Asparagine

Asn is the major transport and low-molecular storage form of N in many plants. The principle path of asn synthesis is thought to be gln-dependent amidation of asp catalyzed by asparagine synthetase (cf. Progr. Bot. 41, 86). Most enzyme preparations have been obtained from germinating seeds, little is known about asn synthesis in roots or shoot tissues. In fact, there are some difficulties in the general acceptance of this pathway: $[^{14}C]$-labeled asp is a poor precursor of asn in root and shoot tissues (cf. Progr. Bot. 34, 157), and in *Asparagus* (COONEY et al. 1980) the β-cyano-alanine pathway of asn synthesis has been detected, whereas asparagine synthetase appeared to be absent.

Important experimental arguments in favor of the asparagine synthetase pathway come from studies with pea leaves (JOY et al. 1983). Pea shoots metabolized $[^{14}C]$-asp predominantly to organic acids, however, addition of the aminotransferase inhibitor aminooxyacetate enhanced the flow of $[^{14}C]$ into asn, suggesting that asp is the normal precursor of asn. Only traces of asparagine synthetase could be detected in pea leaves or roots. Subsequently it was found that pea extracts contain an inhibitor which greatly reduced (over 95%) the activity of asparagine synthetase from lupin and mung bean cotyledons. The inhibitor was heat-stable with a low molecular weight. A similar inhibitor was present in *Asparagus* tissues. Thus the presence of an inhibitor may prevent detection of asparagine synthetase in pea and *Asparagus* tissues. Tracer experiments with $[^{14}C]O_2$ and application of inhibitors of NH_4^+ assimilation were consistent with a glutamine-dependent synthesis of asn from asp with oxalacetate as a precursor to asp in detached soybean root nodules (HUBER and STREETER 1984). A rather unstable asparagine synthetase could be isolated from the nodules; the highest enzyme activities were measured during the seed filling period.

The mobilization of N from asn translocated into a leaf may proceed via two enzymatic routes: asparaginase which produces asp and NH_4^+, and asparagine aminotransferase which produces 2-oxo-succinamate and an amino acid (cf. Progr. Bot. 44, 158). It has now been demonstrated that asparagine aminotransferase is localized exclusively in the peroxisomes of young and mature pea leaves. Asparaginase was found only in the cytosol of young pea leaves (IRELAND and JOY 1983a). Furthermore, asparagine aminotransferase purified from pea leaves was found to be the same enzyme as peroxisomal serine: glyoxylate aminotransferase (IRELAND and JOY 1983b). The role of asn transamination and its possible participation in photorespiration is still obscure.

3. 4-Methylene-L-Glutamine (4-Megln)

In peanuts 4-Megln is the principal nitrogen carrier from root to leaf (see WINTER et al. 1981). During germination 4-methylene-glutamic acid (4-Meglu) and 4-Megln are rapidly formed. GS was shown to be able to amidate 4-Meglu, but with low activity and low affinity (WINTER et al. 1982). Recently, a new amide synthetase could be detected in peanut seedlings (WINTER et al. 1983). The enzyme catalyzes the synthesis of 4-Megln from 4-Meglu and NH_4^+, it was named 4-methylene-glutamine synthetase [4-methyleneglutamate: ammonia ligase (AMP forming)]. It is distinct from GS and asparagine synthetase in chromatographic behavior and substrate specificity. A 4-methylene-L-glutamine amidohydrolase has been partially purified from peanut leaves (POWELL and DEKKER 1983). As in the case of 4-Megln-synthetase, the activity seems to be absent in dormant seeds, but appears during germination. The amidohydrolase activitiy is solely located in the soluble fraction of peanut leaves and cotyledons. The enzyme is relatively specific for 4-Megln (activity with gln 20%).

4. Prolin and Ornithine Degradation

Pro oxidation is a mitochondrial process that can be inhibited by water stress. It is known to yield glu via two enzymatic steps: Prolindehydrogenase and Δ^1-pyrroline-5-carboxylate (P5C) dehydrogenase. In corn mitochondria the two enzymes were shown to be bound to the matrix side of the inner membrane. In both reactions the electrons were shown to enter the respiratory chain directly. Free NAD^+ is not involved (ELTHON and STEWART 1981, 1982). Ornithine is metabolized within corn mitochondria via ornithine transaminase coupled with a NAD-dependent P5C-dehydrogenase, which is different from the respective enzyme involved in pro oxidation (ELTHON and STEWART 1982). A pro-requiring mutant of *Zea mays* was shown to possess a normal ability to synthesize pro. Evidence has been presented that pro deficiency is caused by enhanced pro catabolism which prevails over pro synthesis (DIERKS-VENTLING and TONELLI 1982).

5. Shikimate Pathway

Besides leading to the formation of the aromatic amino acids phe, tyr, and trp, the shikimate pathway represents the central route for the synthesis of a vast number of alicyclic and preferentially aromatic compounds (Fig. 1). Examples are quinic acid, gallic acid and other benzoic acid derivatives, the chorismate derived anthraquinoid pigments and the naphthoquinons, as well as the various phenylpropanoids derived from phe via cinnamic acid, including flavonoids, lignine precursors, simple phenols and coumarins. In addition, tyr and trp are precursors of important alkaloidal classes, such as the benzylisoquinolins and the indol alkaloids. Thus, the shikimate pathway with its many branches provides the precursors for the formation of many secondary constituents. In fact, the carbon flow via the shikimate sequence into lignine precursors or flavonoids may far exceed the demand for de novo synthesized proteinaceous amino acids. This situation has to be considered especially in respect of subcellular compartmentation of the pathway and its regulation. For details concerning the biosynthetic sequence branching off the main pathway the reader is referred to the volume in *Biochemistry of Plants*, edited by CONN (1981).

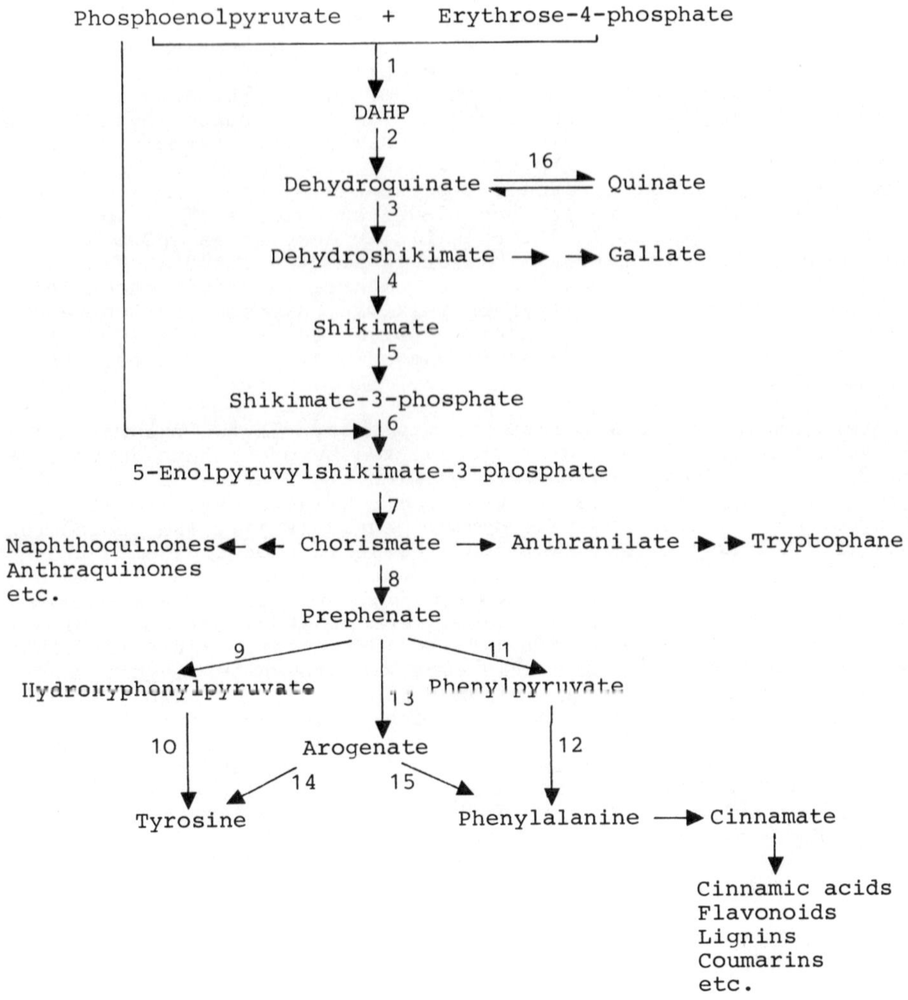

Fig. 1. The shikimate pathway. The numbers refer to the enzymes mentioned in the text

a) Pathway Enzymes

With the exception of chorismate synthase (Fig. 1, E7) all enzymes of the shikimate pathway have been demonstrated in plants. However, many enzymatic details and the regulation of the path need to be resolved. There are a few recent enzymatic studies concerning the pre-chorismate sequence. 3-Deoxy-D-arabino heptulosonate 7-phosphate synthase (DAHP synthase), the first enzyme of the sequence, has been partially characterized from shoots of *Zea mays* (GRAZIANA and BOUDET 1980). The enzyme is sensitive to inhibition by trp. This is the first report of a plant DAHP-synthase sensitive to feedback regulation by aromatic amino acids. Feedback sensitivity is easily lost during enzyme storage, and increased during seedling development. Maximal inhibition (ca. 60%) at 1 mM trp, was observed in extracts of 8-day-old shoots. The next enzymes are of considerable interest because they occur more or less aggregated in different groups of organisms. Until recently the multifunctional "arom multienzyme complex" comprising the enzymes 2 to 6 (Fig. 1) was considered a common feature of fungi. It has now been shown that at least yeasts can be devided in two groups, regarding the organization of the five enzymes. One group is characterized by the "arom multienzyme complex". Representatives are *Neurospora*, *Saccharomyces* and *Torulopsis* . In the other group, represented by *Hansenula*, *Candida* or *Pichia*, only two of the five enzymes, 5-dehydroquinase (E 3) and shikimate: NADP oxidoreductase (E 4) occur firmly associated (BODE and BIRNBAUM 1981a,b). This latter behavior is also realized in higher plants (KOSHIBA 1979), whereas in most bacteria the five sequential enzymes were separable. 5-Enolpyruvylshikimate 3-phosphate synthase (EPSP synthase) (E 6) has been purified from pea seedlings (MOUSDALE and COGGINS 1984), shikimate dehydrogenase (E 4) from tomatoes (LAURENCO and NEVES 1984).

The conversion of prephenate into tyr may occur via two pathways: (1) the 4-hydroxyphenylpyruvate route, catalyzed by prephenate dehydrogenase (E 9) and 4-hydroxyphenylpyruvate amino transferase (E 10); (2) the arogenate route catalyzed by prephenate aminotransferase (E 13) yielding arogenate and arogenate dehydrogenase (E 14). The arogenate pathway of tyr biosynthesis seems to prevail in photosynthetic prokaryotes (BYNG et al. 1982) and photosynthetic eukaryotes. The pathway has been established for *Euglena gracilis* (BYNG et al. 1981a), *Zea mays* (BYNG et al. 1981b), *Vigna radiata* (RUBIN and JENSEN 1979) and cell suspension cultures of *Nicotiana silvestris* (GAINES et al. 1982). In both *Zea mays* and tobacco cultures highly specific arogenate dehydrogenases could be isolated. They are either NAD^+-linked *(Zea)* or $NADP^+$-dependent *(Nicotiana)*. Both plants lack prephenate dehydrogenase acitivity. Tyr but not phe is an effective endproduct inhibitor of arogenate dehydrogenase (GAINES et al. 1982). In contrast to tyr biosynthesis, the respective branch leading to phe via arogenate dehydratase (E 15) has still to be established. To date *Euglena* is the only instance where this enzyme has been identified (BYNG et al. 1981a).

Evidence for a direct conversion of phe into tyr, generally thought not to take place in higher plants, could be demonstrated by tracer experiments in callus cultures of *Portulaca grandiflora* (ENDRESS 1981).

b) Inhibitor Studies

Continuing their search for inhibitors of the biosynthesis of aromatic compounds AMRHEIN and coworkers introduced glyphosate (N-[phosphonomethyl]glycine), a nonselective broadspectrum herbicide, as a potent and specific inhibitor of the shikimate pathway. Glyphosate was found to inhibit the light-induced formation of flavonoids and chlorogenic acid in buckwheat hypocotyls (HOLLÄNDER and AMRHEIN 1980) and of chorismate derived anthraquinoid pigments in cell suspension cultures of *Galium mollugo* (AMRHEIN et al. 1980). Glyphosate inhibition was reversed by phe (buckwheat) and chorismate *(Galium)* . 5-Enolpyruvylshikimate 3-phosphate (EPSP) synthase (E 6) from microbial and plant source was found to be the primary site of action of glyphosate (STEINRÜCKEN and AMRHEIN 1980, AMRHEIN et al. 1982, RUBIN et al. 1982, AMRHEIN et al. 1983, BOOCOCK and COGGINS 1983, MOUSDALE and COGGINS 1984). Glyphosate is a competitive inhibitor in respect of phosphoenolpyruvate. Glyphosate-treated plants or plant cell cultures accumulate considerable amounts of shikimate (BERLIN and WITTE 1981). In buckwheat cell cultures, shikimate which accumulated in response to the inhibitor was found to be exclusively located in the vacuole

(HOLLÄNDER-CZYTKO and AMRHEIN 1983). Furthermore, protoplasts as well as isolated vacuoles were able to take up and hydrolyze shikimate-3-phosphate. This vacuolar dephosphorylation explains why plant cells preferably accumulate shikimate instead of its 3-phosphate, the substrate of the inhibited enzyme. Plant cell cultures *(Corydalis sempervirens)* as well as bacteria *(Aerobacter aerogenes)* can adapt to increasing concentrations of glyphosate. This adaptation was found to be due to an increase in their EPSP-synthase activity, which is assumed to be enzyme overproduction (AMRHEIN et al. 1983). Furthermore, a glyphosate-resistant strain of *Aerobacter aerobenes* could be isolated (SCHULZ et al. 1984). This strain possesses a glyphosate-insensitive EPSP-synthase.

c) Regulation of Quinate: NAD^+-Oxidoreductase

The reversible conversion of quinate into dehydroquinate catalyzed by quinate: NAD^+ oxidoreductase (E 16) may be regarded as a side reaction of the shikimate pathway. The enzyme, which in corn seedlings occurs associated with a dehydroquinase isoenzyme (E 3), is assumed to be involved in a metabolic sequence connected or parallel the shikimate route (GRAZIANA et al. 1980). However, its metabolic role remains unclear.

Recently BOUDET and coworkers established this enzyme to be target of a fascinating mode of regulation. The activity of the reductase from carrot cell suspension cultures is modulated by reversible phosphorylation (REFENO et al. 1982a). The respective protein kinase and protein phosphatase could be isolated and partially characterized (REFENO et al. 1982b). The reductase is active in its phosphorylated form and inactive in its dephosphorylated form. The affinity for the substrate dehydroquinate decreased about 120-fold when the enzyme was deactivated, whereas its affinity for quinate was not modified. Further studies revealed that the protein kinase is dependent on Ca^{2+} and calmoduline (RANJEVA et al. 1983, GRAZIANA et al. 1983). Thus the regulation of the reductase can be described by a two-cycle cascade triggered by Ca^{2+}: (1) The activation/deactivation of calmoduline-dependent protein kinase. (2) The phosphorylation of the reductase.

d) Subcellular Compartmentation

The dual localization of the shikimate pathway enzymes in the chloroplast (cf. Progr. Bot. 41, 89, BUCHHOLZ and SCHULTZ 1980, SCHULTZ et al. 1981) and the cytosol has been further proved. Cytosolic and plastidal localization of shikimate oxidoreductase (E 4), used as marker enzyme of the prechorismate sequence, could be demonstrated in pea plants as well as some species with variegated leaves (ROTHE et al. 1983). Both activities are regulated differentially. The intraplastidic activity responded more sensitively to light intensity and nutritional conditions.

6. Aspartate-derived Amino Acids

Lys, thr, met, and ile derive from asp via a branched sequence of reactions. This pathway is well established in higher plants although some of the enzymatic steps especially in lys biosynthesis still await elucidation (cf. Prog. Bot. 41, 88). Recent work was focused mainly on the subcellular compartmentation of the synthesis of the aspartate family of amino acids and on further aspects of pathway regulation. In the latter case much effort is being devoted to the improvement of grain proteins in cereals, in respect of their dietary value. Furthermore, a great number of papers is concerned with the biosynthesis and metabolism of met and S-adenosylmethionine (SAM).

a) Subcellular Compartmentation

Isolated, intact pea chloroplasts synthesize lys, thr, ile and possibly met from [^{14}C]-asp (MILLS et al. 1980, BURDGE and WILSON 1981). This light-enhanced synthesis is regulated by exogenous lys and thr in a manner consistent with the properties of the isolated enzymes (see below). Several key enzymes of the pathway have been detected in the chloroplast: the initial enzyme aspartate kinase (WAHNBAECK-SPENCER et al. 1979, LEA et al. 1979), dihydrodipicolinate synthase and diaminopimelate decarboxylase, the first and the last enzymes unique to lys biosynthesis (WALLSGROVE and MAZELIS 1980). In many plants homoserine dehydrogenase exists in two or more molecular forms (BRYAN 1980). There appear to be two major classes, one of lower molecular mass (70,000 - 80,000 MW) and insensitive to thr inhibition, the other of higher molecular mass (150,000 - 190,000 MW) and generally sensitive to thr. SAINIS et al. (1981) could show that in pea and barley leaves the former isoenzyme is located in the cytosol, whereas the latter is in the chloroplast. A comprehensive analysis of the localization of the enzymes of thr and met biosynthesis in protoplasts prepared from pea and barley leaves revealed that aspartate kinase, homoserine kinase, threonine synthase, and cystathionine γ-synthase are confined to the chloroplast (WALLSGROVE et al. 1983b). About 50% of cystathionine β-lyase activity was found in the chloroplast fraction. By contrast, neither methionine synthase nor methionine adenosyltransferase were significantly associated with the chloroplasts. Thus, we can conclude that in leaves the biosynthetic pathway of the aspartate-derived amino acids, except the last step in met biosynthesis, is restricted to the chloroplast. It is interesting to note that the only enzymes of the aspartate family found in animal cells, methionine synthase and methionine adenosyltransferase, are precisely those that are not located in the chloroplast.

Lys is the common precursor of a number of alkaloids. In this context it is noteworthy that in lupin leaves basic steps of quinolizidine alkaloid biosynthesis take place in the chloroplast (WINK et al. 1980). The two initial alkaloid-specific enzymes, lysine decarboxylase (HARTMANN et al. 1980, SCHOOFS et al. 1983) and oxosparteine synthase (WINK and HARTMANN 1982) are located in the chloroplast.

b) Regulation

The synthesis of asp-derived amino acids is regulated in plants by a series of feedback loops (cf. Progr. Bot. 41, 88). The presence of lys- and thr-sensitive isoenzymes has been further proved for aspartate kinase from pea (LEA et al. 1979) and soybean (MATTHEWS and WIDHOLM 1979). The enzyme from barley can be separated into one thr-sensitive and two lys-sensitive isoenzymes (BRIGHT et al. 1982a, ROGNES et al. 1983). The lys-sensitive isoenzyme(s) of various plants are synergistically regulated by lys plus S-adenosylmethionine (SAM) at low levels of both effectors (ROGNES et al. 1980). Thus, in plants SAM functions as a feedback signal of met biosynthesis; met itself is completely ineffective. Since most cereal seed proteins are deficient in lys, thr and met, much effort has been devoted to their improvement. Various amino acid-overproducing mutant lines of maize (HIBBERD et al. 1980,

HIBBERD and GREEN 1982) and barley (BRIGHT et al. 1982b, ROGNES et al. 1983) have been selected. These mutants show alterations in the regulatory properties of only lys-sensitive aspartate kinase and overproduced thr, but not lys in tissue cultures (maize) and young plants and seeds (maize and barley). From this it can be assumed that dihydrodipicolinate synthase, the first enzyme unique to lys biosynthesis, is more tightly regulated by lys than homoserine dehydrogenase by thr. The relative amounts of thr-sensitive and thr-insensitive molecular forms of threonine dehydrogenase vary with age and plant tissue (BRYAN and LOCHNER 1981a,b). Dihydrodipicolinate synthase has been partially purified from spinach leaves and was found to be sensitively inhibited by lys ($I_{0.5}$ = 20 µM) (WALLSGROVE and MAZELIS 1981). Till now no lys-overproducing mutant line could be selected. Barley mutants resistant to inhibition by the lys analog S-[2-aminoethyl]-cysteine were shown to possess a greatly reduced ability for basic amino acid uptake (BRIGHT et al. 1979, 1983).

c) Degradation of Lysine

The enzyme lysine: 2-oxoglutarate reductase yielding saccharopine in a NADPH-dependent reaction was partially characterized from developing maize endosperm (ARRUDO et al. 1982). The main activity of the reductase coincides with optimal seed protein synthesis, where more than twice the lys required for proteine synthesis is imported into the seed (ARRUDO and DA SILVA 1983). This is the first indication of lys catabolism in plants via the saccharopine route well known from microorganisms and animals.

d) Biosynthesis of Methionine

In green plants O-phosphohomoserine is the branch point for de novo met biosynthesis. It is the physiological substrate for the transsulfuration which involves the transfer of sulfur from cys to homocysteine via the intermediate cystathionine. The enzymes involved are cystathionine γ-synthase and cystathionine β-lyase. The final step of met biosynthesis, the methylation of homocysteine is catalyzed by a tetrahydropteroyltriglutamate methyltransferase (for rev. see GIOVANELLI et al. 1980a,b).

Besides transsulfuration, direct sulfhydration of O-phosphohomoserine has been demonstrated as a possible alternative pathway of homocysteine formation in plants (DATKO et al. 1977). However, in-vivo labeling studies with $[^{35}S]O_4^{2-}$ have failed to detect any homocysteine formation via this shunt in *Lemna paucicostata* and have shown that at least 90% of homocysteine which is formed comes from cystathionine (MACNICOL et al. 1981). Furthermore, several lines of evidence indicate that O-phosphohomoserine sulfhydrylase and O-phosphohomoserine-dependent cystathionine γ-synthase are activities of one enzyme: (a) Propargylglycine, a suicide inhibitor of cystathionine synthase, inhibited the two activities to the same extent (THOMPSON et al. 1982a); (b) when *Lemna* was grown in the presence of 2 µM external met, the two activities are simultaneously regulated down to 15% of control (THOMPSON et al. 1982b). Thus, transsulfuration is the major, perhaps the sole, pathway of sulfur incorporation of de novo met biosynthesis.

"Reverse transsulfuration" and thus a direct flow of sulfur from met to cys is not likely to occur in plants, due to the absence of cystathionine γ-lyase. This has been proved by in-vivo experiments: Growth of *Lemna* is inhibited by molybdate known to interfere with sulfate uptake. This growth inhibition could be totally prevented by addition of

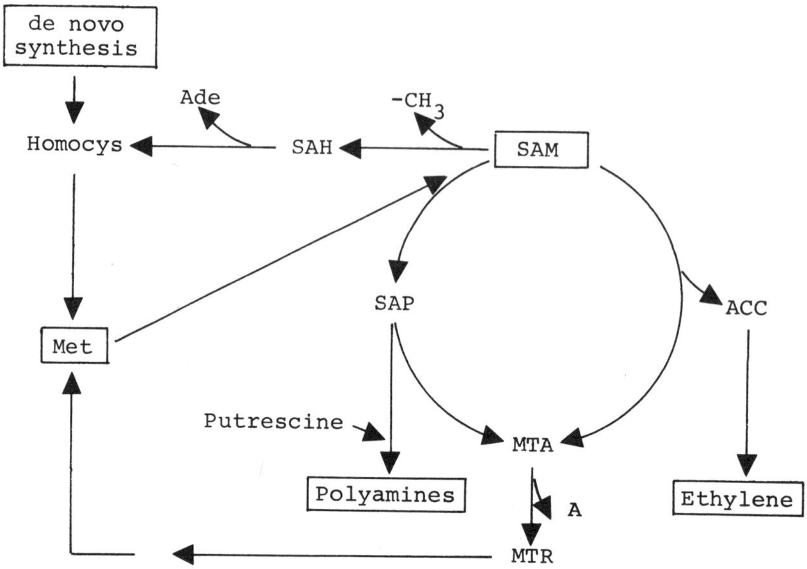

Fig. 2. Metabolism of methionine and S-adenosylmethionine. Methionine salvage during ethylene and polyamine biosynthesis. Abbreviations: *SAM* S-Adenosylmethionine; *SAP* S-Adenosyl-3-methylthiopropylamine (decarboxylated SAM); *SAH* S-Adenosylhomocysteine; *MTA* 5'-Methylthioadenosine; *MTR* 5-Methylthioribose; *Ade* Adenosine; *A* Adenine

external cystine but not by met, homocystine or cysthationine (DATKO and MUDD 1982). On the other hand, cell suspension cultures of *Catharanthus roseus*, although lacking cystathionine γ-lyase, were shown to grow on met as the sole source of sulfur. Met sulfur is supplied for cys biosynthesis as free H_2S via a series of dissimilatory enzymatic steps, including oxidative degradation giving rise to methanethiol (CH_3-SH), and replacement of homocysteine sulfane sulfur by CH_3-SH yielding met and free H_2S (SCHWENN et al. 1983).

e) Metabolism of S-Adenosylmethionine (SAM)

SAM has at least three important metabolic functions in plants (Fig. 2). (1) It is the predominant methyl donor in methyl transfer reactions. (2) It serves as precursor of ethylene biosynthesis in fruits and other plant tissues (see DÖRFFLING and TIETZ 1982). (3) It is the donor of the α-aminopropionyl moity in spermine and spermidine synthesis (see SMITH 1981).

Methylthioadenosine (MTA) arises as a product of both pathways, ethylene synthesis and polyamine formation (Fig. 2). Recent studies have established that this compound is effectively recycled into met. In actively ethylene producing tissues, such as ripening apples (YUNG et al. 1982) and tomatoes (WANG et al. 1982) labeled MTA is incorporated into met. A 5'-methylthio-adenosine nucleosidase producing 5-methylthioribose (MTR) from MTA has been characterized from *Lupinus luteus* seeds (GURANOWSKI et al. 1981). In growing *Lemna paucicostata*, which evolves little or no ethylene, the thiomethyl group of met is reincorporated into met, whereas the four-carbon moiety is not (GIOVANELLI et al. 1981). Subsequently, the operation of MTA recycling as a constitutive sequence could be demonstrated in *Lemna*. The adenine moiety of MTA is salvaged into adenine nucleotides, whereas four ribose carbons are incorporated into met (GIOVANELLI et al. 1983). This salvage pathway for met biosynthesis is of general importance in ethylene-producing plant tissues and in relation to polyamine synthesis. The same pathway also occurs in bacteria (SHAPIRO and BARRETT 1981), fungi (SHAPIRO and SCHLENK 1980, GURANOWSKI and PASZEWSKI 1982) and animals (BACKLUND and SMITH 1981, BACKLUND et al. 1982).

References

AHMAD, I., LAHRER, F., MANN, A.F., McNALLY, S.F., STEWART, G.R.: New Phytol. 91, 585-595 (1982). - AMRHEIN, N., DEUS, B., GEHRKE, P., STEINRÜCKEN, H.C.: Plant Physiol. 66, 830-834 (1980). - AMRHEIN, N., HOLLÄNDER-CZYTKO, H., LEIFELD, J., SCHULZ, A., STEINRÜCKEN, H.C., TOPP, H.: Groupe Polyphenols. Journées internationales d'études et assemblées générales, eds. A.M. BOUDET, R. RANJEVA. Bull. Liaison 11, 21-30 (1982). - AMRHEIN, N., JOHÄNNING, D., SCHAB, J., SCHULZ, A.: FEBS Lett. 157, 191-196 (1983). - ARRUDO, P., DA SILVA, W.J.: Phytochemistry 22, 2687-2689 (1983). - ARRUDO, P., SODEK, L., DA SILVA, W.J.: Plant Physiol. 69, 988-989 (1982).

BACKLUND, P.S., SMITH, R.A.: J. Biol. Chem. 256, 1533-1535 (1981). - BACKLUND, P.S., CHANG, C.P., SMITH, R.A.: J. Biol. Chem. 257, 4196-4202 (1982). - BERGER, M.G., FOCK, H.P.: Aust. J. Plant Physiol. 10, 187-194 (1983). - BERGMAN, A., GARDESTRÖM, P., ERICSON, I.: Physiol. Plant 53, 528-532 (1981). - BERLIN, J., WITTE, L.: Z. Naturforsch. 36c, 210-214 (1981). - BODE, R., BIRNBAUM, D.: Biochem. Physiol. Pflanzen 176, 329-341 (1981a); - Z. Allg. Mikrobiol. 21, 417-422 (1981b). - BOOCOCK, M.R., COGGINS, J.R.: FEBS Lett. 154, 127-133 (1983). - BRIGHT, S.W.J., FEATHERSTONE, L.C., MIFLIN, B.J.: Planta 146, 629-633 (1979). - BRIGHT, S.W.J., MIFLIN, B.J., ROGNES, S.E.: Biochem. Genet. 20, 229-243 (1982a). - BRIGHT, S.W.J., KUEH, J.S.H., FRANKLIN, J., ROGNES, S.E., MIFLIN, B.J.: Nature 299, 278-279 (1982b). - BRIGHT, S.W.J., KUEH, J.S.H., ROGNES, S.E.: Plant Physiol. 72, 821-824 (1983). - BRYAN, J.K.: 403-452 in: The Biochemistry of Plants, eds. P.K. STUMPF, E.E. CONN, Vol. 5. New York, London: Academic Press 1980. - BRYAN, J.K., LOCHNER, N.R.: Plant Physiol. 68, 1395-1399 (1981a); - 68, 1400-1405 (1981b). - BUCHHOLZ, B., SCHULTZ, G.: Z. Pflanzenphysiol. 100, 209-215 (1980). - BURDGE, E.L., WILSON, G.: FEBS Lett. 136, 322-324 (1981). - BYNG, G.S., WHITAKER, R.J., SHAPIRO, C.L., JENSEN, R.A.: Mol. Cell Biol. 1, 426-438 (1981a). - BYNG, G.S., WHITAKER, R.J., FLICK, C., JENSEN, R.A.: Phytochemistry 20, 1289-1292 (1981b). - BYNG, G.S., KANE, J.F., JENSEN, R.A.: Crit. Rev. Microbiol. 9, 227-252 (1982).

CONN, E.E. (ed.): Secondary Plant Products, 798 pp. In: The Biochemistry of Plants, eds. P.K. STUMPF, E.E. CONN, Vol. 7. New York: Academic Press 1981. - COONEY, D.A., JAYARAM, H.N., SWENGROS, S.G., ALTER, S.C., LEVINE, M.: Int. J. Biochem. 11, 69-83 (1980). - CULLIMORE, J.V., MIFLIN, B.J.: FEBS Lett. 158, 107-112 (1983). - CULLIMORE, J.V., LEA, P.J., MIFLIN, B.J.: Isr. J. Bot. 31, 155-162 (1982). - CULLIMORE, J.V., LARA, M., LEA, P.J., MIFLIN, B.J.: Planta 157, 245-253 (1983).

DATKO, A.H., MUDD, S.H.: Plant Physiol. 69, 1070-1076 (1982). - DATKO, A.H., MUDD, S.H., GIOVANELLI, J.: J. Biol. Chem. 252, 3436-3445 (1977). - DIERKS-VENTLING, C., TONELLI, C.: Plant Physiol. 69, 130-134 (1982). - DÖRFFLING, K., TIETZ, D.: Progr. Bot. 44, 182-204 (1982). - DURZAN, D.J., STEWARD, F.C.: Nitrogen metabolism, 55- 261. In: Plant Physiology, ed. F.C. STEWARD, Vol. VIII. New York, London: Academic Press 1983. - DYVINIAK, S.D., DUNHAM, V.L. Biochem. Biophys. Res. Commun. 114, 604-611 (1983).

ELTHON, T.E., STEWART, C.R.: Plant Physiol. 67, 780-784 (1981); - ibid. 70, 567-572 (1982). - ENDRESS, P.: Plant Physiol. 68, 272-274 (1981).

FENTEM, P.A., LEA, P.J., STEWART, G.R.: Plant Physiol. 71, 496-501 (1983a); - ibid. 71, 502-506 (1983b). - FLORENCIO, F.J., VEGA, J.M.: Z. Naturforsch. 38c, 531-538 (1983). - FRANTZ, T.A., PETERSON, D.M., DURBIN, R.D.: Plant Physiol. 69, 345-348 (1982).

GAINES, C.G., BYNG, G.S., WHITAKER, R.J., JENSEN, R.A.: Planta 156, 233-240 (1982). - GIOVANELLI, J., MUDD, H.S., DATKO, A.H.: 453-505 in: The Biochemistry of Plants, Vol. 5. New York, London: Academic Press 1980a; - Natural Sulfur Compounds, eds. D. CAVALLINI, G.E. GAULL, V. ZAPPIA, p. 81-82. New York: Plenum Press 1980b; - Biochem. Biophys. Res. Commun. 100, 831-839 (1981). - GIOVANELLI, J., DATKO, A.H., MUDD, S.H., THOMPSON, G.A.: Plant Physiol. 71, 319-327 (1983). - GRAZIANA, A., BOUDET, A.M.: Plant Cell Physiol. 21, 793-802 (1980). - GRAZIANA, A., BOUDET, A., BOUDET, A.M.: Plant Cell Physiol. 21, 1163-1174 (1980). - GRAZIANA, A., RANJEVA, R., BOUDET, A.M.: FEBS Lett. 156, 325-328 (1983). - GROAT, R.G., SCHRADER, L.E.: Plant Physiol. 70, 1759-1761 (1982). - GURANOWSKI, A., PASZEWSKI, A.: Biochim. Biophys. Acta 717, 289-294 (1982). - GURANOWSKI, A.B., CHIANG, P.K., CANTONI, G.L.: Eur. J. Biochem. 114, 293-299 (1981).

HARTMANN, T., SCHOOFS, G., WINK, M.: FEBS Lett. 115, 35-38 (1980). - HIBBERD, K.A., GREEN, C.E.: Proc. Natl. Acad. Sci. USA 79, 559-563 (1982). - HIBBERD, K.A., GREEN, C.E., GENGENBACH, B.G.: Planta 148, 183-187 (1980). - HIREL, B., GADAL, P.: Physiol. Plant. 54, 69-74 (1982). - HIREL, B., PERROT-RECHENMANN, C., SUZUKI, A., VIDAL, J., GADAL, P.: Plant Physiol. 69, 983-987 (1982a). - HIREL, B., VIDAL, J., GADAL, P.: Planta 155, 17-23 (1982b). - HIREL, B., PERROT-RECHENMANN, C., MAUDINAS, B., GADAL, P.: Physiol. Plant. 55, 197-203 (1982c). - HIREL, B., LAYZELL, D.B., McCASHIN, B., McNALLY, S.F., CANVIN, D.T.: Can. J. Bot. 61, 2257-2259 (1983a). - HIREL, B., LAVERGNE, D., McNALLY, S.F., GADAL, P.: Plant Sci. Lett. 32, 169-175 (1983b). - HOLLÄNDER, H., AMRHEIN, N.: Plant Physiol. 66, 823-829 (1980). - HOLLÄNDER-CZYTKO, H., AMRHEIN, N.: Plant Sci. Lett. 29, 89-96 (1983). - HUBER, T.A., STREETER, J.G.: Plant Physiol. 74, 605-610 (1984).

IRELAND, R.J., JOY, K.W.: Plant Physiol. 72, 1127-1129 (1983a); - Arch. Biochem. Biophys. 223, 291-296 (1983b).

JOY, K.W., IRELAND, R.J., LEA, P.J.: Plant Physiol. 73, 165-168 (1983).

KEYS, A.J., BIRD, I.F., CORNELIUS, M.F., LEA, P.J., WALLSGROVE, R.M., MIFLIN, B.J.: Nature 275, 741-742 (1978). - KNIGHT, T.J., WEISSMAN, G.S.: Plant Physiol. 70, 1683-1688 (1982). - KOSHIBA, T.: Plant Cell Physiol. 20, 667-670 (1979).

LARA, M., CULLIMORE, J.V., LEA, P.J., MIFLIN, B.J., JOHNSTON, A.W.B., LAMB, J.W.: Planta 157, 254-258 (1983). - LAURENCO, E.J., NEVES, V.A.: Phytochemistry 23, 497-499 (1984). - LEA, P.J., MILLS, W.R., MIFLIN, B.J. FEBS Lett. 98, 165-168 (1979). - LEA, P.J., AWONAIKE, K.O., CULLIMORE, J.V., MIFLIN, B.J.: Isr. J. Bot. 31, 140-154 (1982). - LEASON, M., CUNLIFFE, D., PARKIN, D., LEA, P.J., MIFLIN, B.J.: Phytochemistry 21, 855-857 (1982).

MACNICOL, P.K., DATKO, A.H., GIOVANELLI, J., MUDD, S.H.: Plant Physiol. 68, 619-625 (1981). - MARTIN, F., WINSPEAR, M.J., MacFARLANE, J.D.,

OAKS, A.: Plant Physiol. 71, 177-181 (1983). - MATOH, T., TAKAHASHI, E.: Plant Cell Physiol. 22, 727-731 (1981); - Planta 154, 289-294 (1982). - MATTHEWS, B.F., WIDHOLM, J.M.: Can. J. Bot. 57, 299-304 (1979). - McCORMACK, D.K., FARNDEN, K., BOLAND, M.: Arch. Biochem. Biophys. 218, 561-571 (1982). - McNALLY, S.F., HIREL, B.: Physiol. Vég. 21, 761-774 (1983). - McNALLY, S.F., HIREL, B., STEWART, G.R.: New Phytol. 94, 47-56 (1983a). - McNALLY, S.F., OREBAMJO, T.O., HIREL, B., STEWART, G.R.: J. Exp. Bot. 34, 610-619 (1983b). - McNALLY, S.F., HIREL, B., GADAL, P., MANN, A.F., STEWART, G.R.: Plant Physiol. 72, 22-25 (1983c). - MIFLIN, B.J., LEA, P.J.: 5-64 in: Encyclopedia of Plant Physiology, N.S., Vol. 14A, Nucleic Acids Proteins Plants 1. Berlin, Heidelberg, New York: Springer 1982. - MILLS, W.R., LEA, P.J., MIFLIN, B.J.: Plant Physiol. 65, 1166-1172 (1980). - MOUSDALE, D.M., COGGINS, J.R.: Planta 160, 78-83 (1984).

NISHIMURA, M., BHUSAWANG, P., STRZALKA, K., AKAZAWA, T.: Plant Physiol. 70, 353-356 (1982).

PLATT, S.G., ANTHON, G.E.: Plant Physiol. 67, 509-513 (1981). - PLATT, S.G., RAND, L.: Plant Cell Physiol. 23, 917-921 (1982). - POWELL, G.K., DEKKER, E.E.: J. Biol. Chem. 258, 8677-8683 (1983). - PUSHKIN, A.V., DZHOKHARIDZE, T.Z., TSUPRUN, V.L., EVSTIGNEEVA, Z.G., KRETOVICH, V.L.: Dokl. Akad. Nauk. SSR. 271, 234-237 (1983).

RANJEVA, R., REFENO, G., BOUDET, A.M., MARMÉ, D.: Proc. Natl. Acad. Sci. USA 80, 5222-5224 (1983). - REFENO, G., RANJEVA, R., BOUDET, A.M.: Planta 154, 193-198 (1982a). - REFENO, G., RANJEVA, R., FONTAINE-DELVARE, S., BOUDET, A.M.: Plant Cell Physiol. 23, 1137-1144 (1982b). - ROGNES, S.E., LEA, P.J., MIFLIN, B.J.: Nature 287, 357-359 (1980). - ROGNES, S.E., BRIGHT, S.W.J., MIFLIN, B.J.: Planta 157, 32-38 (1983). - ROTHE, G.M., HENGST, G., MILDENBERGER, I., SCHARER, H., UTESCH, D.: Planta 157, 358-366 (1983). - RUBIN, J.L., JENSEN, R.A.: Plant Physiol. 64, 727-734 (1979). - RUBIN, J.L., GAINES, C.G., JENSEN, R.A.: Plant Physiol. 70, 833-839 (1982).

SAINIS, J.K., MAYNE, R.G., WALLSGROVE, R.M., LEA, P.J., MIFLIN, B.J.: Planta 152, 491-496 (1981). - SCHOOFS, G., TEICHMANN, S., HARTMANN, T., WINK, M.: Phytochemistry 22, 65-69 (1983). - SCHULTZ, G., BICKEL, H., BUCHHOLZ, B., SOLL, J.: Photosynthesis V, Chloroplast Development, 311-318, ed. G. AKOUYUNOGLOU. Philadelphia: Balaban Int. Sci. Serv. 1981. - SCHULZ, A., SOST, D., AMRHEIN, N.: Arch. Microbiol. 137, 121-123 (1984). - SCHWENN, J.D., SCHRIEK, U., KILTZ, H.H.: Planta 158, 540-549 (1983). - SHAPIRO, S.K., BARRETT, A.: Biochem. Biophys. Res. Commun. 102, 302-307 (1981). - SHAPIRO, S.K., SCHLENK, F.: Biochim. Biophys. Acta 633, 176-180 (1980). - SKOKUT, T.A., VARNER, J.E., SCHAEFER, J., STEJSKAL, E.O., McKAY, R.A.: Plant Physiol. 69, 308-313 (1982a); - ibid. 69, 314-316 (1982b). - SMITH, T.A.: 249-268 in: The Biochemistry of Plants, Vol. 7. New York, London: Academic Press 1981. - STEINRÜCKEN, H.C., AMRHEIN, N: Biochem. Biophys. Res. Commun. 94, 1207-1212 (1980). - STREIT, L., FELLER, U.: Physiol. Vég. 21, 103-108 (1983). - SUZUKI, A., GADAL, P.: Plant Physiol. 69, 848-852 (1982). - SUZUKI, A., VIDAL, J., GADAL, P.: Plant Physiol. 70, 827-832 (1982a). - SUZUKI, A., JACQUOT, J.P., MARTIN, F., GADAL, P.: Biochem. Biophys. Res. Commun. 106, 65-71 (1982b). - SUZUKI, A., NATO, A., GADAL, P.: Plant Sci. Lett. 33, 93-101 (1984).

TAKEBA, G.: Plant Cell Physiol. 24, 1477-1483 (1983). - THOMAS, M.D., LANGSTON-UNKEFER, P.J., UCHYTIL, T.F., DURBIN, R.D.: Plant Physiol. 71, 912-915 (1983). - THOMPSON, G.A., DATKO, A.H., MUDD, S.H.: Plant Physiol. 70, 1347-1352 (1982a). - THOMPSON, G.A., DATKO, A.H., MUDD, S.H., GIOVANELLI, J.: Plant Physiol. 69, 1077-1083 (1982b). - TISCHNER, R., SCHMIDT, A.: Plant Physiol. 70, 113-116 (1982).

WAHNBAECK-SPENCER, R., HENKE, R.R., MILLS, W.R., BURDGE, E.L., WILSON, K.G.: FEBS Lett. 104, 303-308 (1979). - WALLSGROVE, R.M., MAZELIS, M.: FEBS Lett. 116, 189-192 (1980); - Phytochemistry 20, 2651-2655 (1981). - WALLSGROVE, R.M., LEA, P.J. MIFLIN, B.J.: Planta 154, 473-476 (1982). - WALLSGROVE, R.M., KEYS, A.J., LEA, P.J., MIFLIN, B.J.: Plant Cell Environ. 6, 301-309 (1983a). - WALLSGROVE, R.M., LEA, P.J., MIFLIN, B.J.: Plant Physiol. 71, 780-784 (1983b). - WANG, S.Y., ADAMS, D.O., LIEBERMANN, M.: Plant Physiol. 70, 117-121 (1982). - WINK, M., HARTMANN, T.: Plant Physiol. 70, 74-77 (1982). - WINK, M., HARTMANN, T., WITTE, L.: Z. Naturforsch. 35c, 93-97 (1980). - WINTER, H.C., POWELL, G.K., DEKKER, E.E.: Plant Physiol. 68, 588-593 (1981); - ibid. 69, 41-47 (1982). - WINTER, H.C., SU, TI ZHI, DEKKER, E.E.: Biochem. Biophys. Res. Commun. 111, 484-489 (1983). - WOO, K.C., OSMOND, B.C.: Plant Physiol. 69, 591-596 (1982). - WOO, K.C., MOROT-GAUDRY, J.F., SUMMONS, R.E., OSMOND, C.B.: Plant Physiol. 70, 1514-1517 (1982).

YUNG, K.H., YANG, S.F., SCHLENK, F.: Biochem. Biophys. Res. Commun. 104, 771-777 (1982).

Professor Dr. THOMAS HARTMANN
Dr. ADELHEID EHMKE
Institut für Pharmazeutische
Biologie der Technischen
Universität Braunschweig
Mendelssohnstr. 1
D-3300 Braunschweig, FRG

VI. Secondary Plant Substances. Monoterpenes

By HORST-ROBERT SCHÜTTE

1. Introduction

Monoterpenes occur in a variety of acyclic and cyclic structural types and derivatives (cf. Fig. 1), and represent one of the largest and most diverse families of natural products. The so-called isoprene rule was the first unified concept for a common structural relationship among terpenoid natural products (RUZICKA 1953). The principles of the monoterpene biosynthesis have been summarized some years ago (cf. Progr. Bot. 38, 129 ff.). But the last decade has witnessed many well-documented studies on the monoterpenes utilizing both in vivo and in vitro experimental approaches, which have been well covered in some complimentary reviews (BUNTON and CORI 1978, CHARLWOOD and BANTHORPE 1978, BANTHORPE and CHARLWOOD 1980, CANE 1980, CORI et al. 1980, LOOMIS and CROTEAU 1980, CROTEAU 1981, POULTER and RILLING 1981, CORI et al. 1982, CORI 1983).

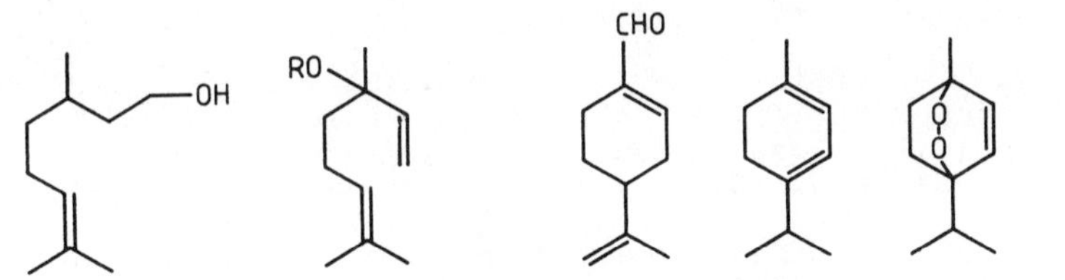

I Citronellol | II R = H Linalool, III R = PP Linaloyl pyrophosphate | IV Perillaldehyde | V α-Terpinene | VI Ascaridole

Fig. 1. Several simple monoterpenes

2. Substrates for Cyclization

Most monoterpenes exhibit the typical head-to-tail isoprenoid construction pattern found in geranyl pyrophosphate (VIII), whose formation of mevalonate is well established. The C-10 of geraniol (VII) is exclusively derived from C-2 of mevalonate (AKHILA and BANTHORPE 1980b). On the other hand, also leucine has been found to be incorporated into geraniol (VII) and citronellol (I) by *Pelargonium roseum* and into linalool (II) by *Cinnamomum camphora* preferentially in dimethylallyl pyrophosphate moiety in contrast to the predominant location of the activity on the isopentenyl pyrophosphate derived moiety from mevalonate (TANGE 1981, TANGE et al. 1981). Tracer studies on *Pelargonium graveolens* have shown

Progress in Botany, Vol. 46

that the activity of geranyl pyrophsophate synthetase passed through a sharp maximum in July, being then some 390-fold greater than the levels during winter. Under conditions of optimum incorporation, both moieties of geraniol were equivalently labeled from mevalonate. No evidence could be obtained to support the view that valine or leucine were precursors of the dimethylallyl pyrophosphate moiety (BANTHORPE et al. 1983b). A high specific geranyl pyrophosphate synthetase was isolated from the bacterium *Micrococcus lysodeikticus* as well as from orange flavedo (SAGAMI et al. 1978, CORI et al. 1980, PÉREZ et al. 1980a, 1983, DE LA FUENTE et al. 1981). Prenylphosphatases, exhibiting a certain degree of chain length specificity, have been partially purified from the flavedo of *Citrus sinensis* (PÉREZ et al. 1980b).

It is natural to suggest that geranyl pyrophosphate (VIII) as ubiquitous intermediate is a direct precursor of monoterpenes. However, the trans configuration of the Δ^2 double bound of geranyl pyrophosphate prevents the direct C-1 to C-6 cyclization to the cyclohexanoid types. Many investigators ascribed a central role to the cis isomer neryl pyrophosphate (XIII) as the appropiate acyclic progenitor (cf. BUNTON et al. 1979). Geraniol (VII) and nerol (XII) were specifically incorporated into pulegone (XIX) in *Mentha pulegium* such that C-1 of these precursor became the carbonyl C in the product. Tracer of C-10 of geraniol however became equally distributed between the gem-methyls of pulegone (AKHILA and BANTHORPE 1980a).

Three possible routes for the origin of neryl pyrophosphate (XIII) have been suggested (cf. CHARLWOOD and BANTHORPE 1978), involving either the direct condensation of isopentenyl pyrophosphate and dimethylallyl pyrophosphate to neryl pyrophosphate, a direct enzymatic isomerization of the trans double bond of geraniol or its phosphorylated derivatives to the cis configuration, or a redox interconversion from geraniol by way of the aldehyde.

Thus one hydrogen from C-5 of [5-^{3}H,2-^{14}C] mevalonate was lost in the conversion to iso-thujone (XLVII) in *Tanacetum vulgare* leaves (BANTHORPE et al. 1978a), and a similar loss of tritium from [1-^{3}H,^{14}C] geraniol (VII), but not [1-^{3}H,^{14}C] nerol (XII), was also demonstrated, implying redox conversion of geraniol to nerol before cyclization (BANTHORPE et al. 1978b). But no geraniol dehydrogenase has been found in this system, which would be necessary for a redox system for conversion of geraniol into nerol (BANTHORPE et al. 1976b). In other experiments cell-free extracts from leaves of *T. vulgare* synthesized geraniol and nerol separately from isopentenyl pyrophosphate and mevalonate respectively. The terpenols were predominantly labeled in the isopentenyl pyrophosphate derived moiety (BANTHORPE et al. 1976a). Cell-free extracts of *T. vulgare*, *Artemisia annua*, and *Santolina chamaecyparissus* contain seasonally dependent enzyme systems that convert geranyl pyrophosphate (VIII) and neryl pyrophosphate (XIII) into the corresponding epoxides (BANTHORPE et al. 1977a).

By double label experiments it could be shown that flowerheads of *Rosa damascena*, which accumulate geraniol (VII) and nerol (XII) as major monoterpenes (OHNO and TANAKA 1977), convert geraniol into nerol in a NADPH-dependent reaction with loss of the pro-(1 S) hydrogen, whereas the reverse isomerization involves loss of the pro-(1R) atom (BANTHORPE and POOTS 1979). However, there is evidence that an alternative nonredox route exists for the conversion of geraniol into nerol in higher plants, because [5-^{3}H,2-^{14}C] mevalonate was converted to cyclic monoterpenes such as limonene (XI), perillaldehyde (IV), and camphor (XL) in several other species without loss of tritium, thus ruling out the redox steps (SUGA et al. 1977, SUGA and SHISHIBORI 1979).

In addition to neryl pyrophosphate (XIII), linaloyl pyrophosphate (III) has been proposed as the immediate acyclic precursor of cyclic monoterpenes and tracer studies have been supportive of this role. Comparisons of the incorporations of [^{14}C]- and [^{3}H]-labeled nerol (XII), geraniol (VII), and linalool (II) and their pyrophosphates using a cell-free system prepared from growing tips of *Mentha spicata* (SUGA et al. 1980b, 1983b), a partially purified enzyme preparation from the flavedo of *Citrus limonum* (PORTILLA et al. 1982a,b), as well as from *Artemisia annua* shown a preferential incorporation of linalool and its pyrophosphate for the formation of cyclic monoterpenes.

But at least the latter was an artefact, as inhibition of endogenous phosphate-cleaving enzymes led to the preference of geranyl pyrophosphate. Consequently, previous conclusions drawn from measurements of the relative efficiencies of the esters as precursors for cyclic monoterpenes, indicating linaloyl pyrophosphate to be the preferred substrate, may be invalid (BANTHORPE et al. 1983a).

Reaction shemes for the cyclization of geranyl pyrophosphate (VIII) by way of bound intermediates, and without the involvement of any other prenyl pyrophosphate, can be readily envisioned. Several studies with partially purified monoterpene synthetases demonstrate that geranyl pyrophosphate is the preferred substrate for cyclization, in the absence of any competing phosphohydrolases and any detectable interconversions of geranyl pyrophosphate, neryl pyrophosphate (XIII), and linaloyl pyrophosphate (III), i.e., bornyl pyrophosphate synthetase (CROTEAU and KARP 1979a).

Soluble enzyme preparations from the epidermis of immature *Mentha piperita* leaves converted labeled geranyl pyrophosphate (VIII) to limonene (XI) as the major cyclic product (KJONAAS and CROTEAU 1983). No free intermediates were detected suggesting that limonene is the first cyclic intermediate to arise in the pathway. The epidermal oil glands are the presumed site of monoterpene biosynthesis (CROTEAU 1977). Earlier indications that neryl pyrophosphate (XIII) with preparations from whole *M. piperita* leaves was enzymatically cyclized to α-terpineol (XXXIII), which in turn, might be dehydrated, could not be substantiated with epidermis preparations.

Similar results are obtained with soluble enzyme preparations from *Salvia officinalis* which converted [1-$^{3}H_2$,G^{14}C] geranyl pyrophosphate (VIII) to cyclic monoterpenes of the pinane, isocamphane, p-menthane, and bornane type without loss of [^{3}H], and without significant conversion to other free acyclic intermediates. Similar results have been obtained with intact *S. officinalis* leaves. Also, these results support a scheme whereby the trans precursor is cyclized directly by way of a bound linaloyl intermediate (IX) (Fig. 2) (CROTEAU and FELTON 1981).

Other monoterpene cyclase preparations showed an apparent preference for neryl pyrophosphate (XIII) as substrate (CROTEAU and KARP 1977a), while others have been shown to utilize both geranyl pyrophosphate (VIII) and neryl pyrophosphate with roughly equal efficiency in the absence of interconversion of these substrates (CHAYET et al. 1977, POULOSE and CROTEAU 1978b). Soluble enzyme preparations from *Chenopodium ambrosioides* converted geranyl pyrophosphate (VIII) or neryl pyrophosphate (XIII) to α-terpinene (V) and limonene (XI) and α-terpinene to the unusual endoperoxide ascaridole (VI) (JOHNSON and CROTEAU 1980).

However, in none of these cases the removal of all competing phosphohydrolases was assured, and in several instances only crude preparations were utilized. Thus, it is uncertain whether the specificities

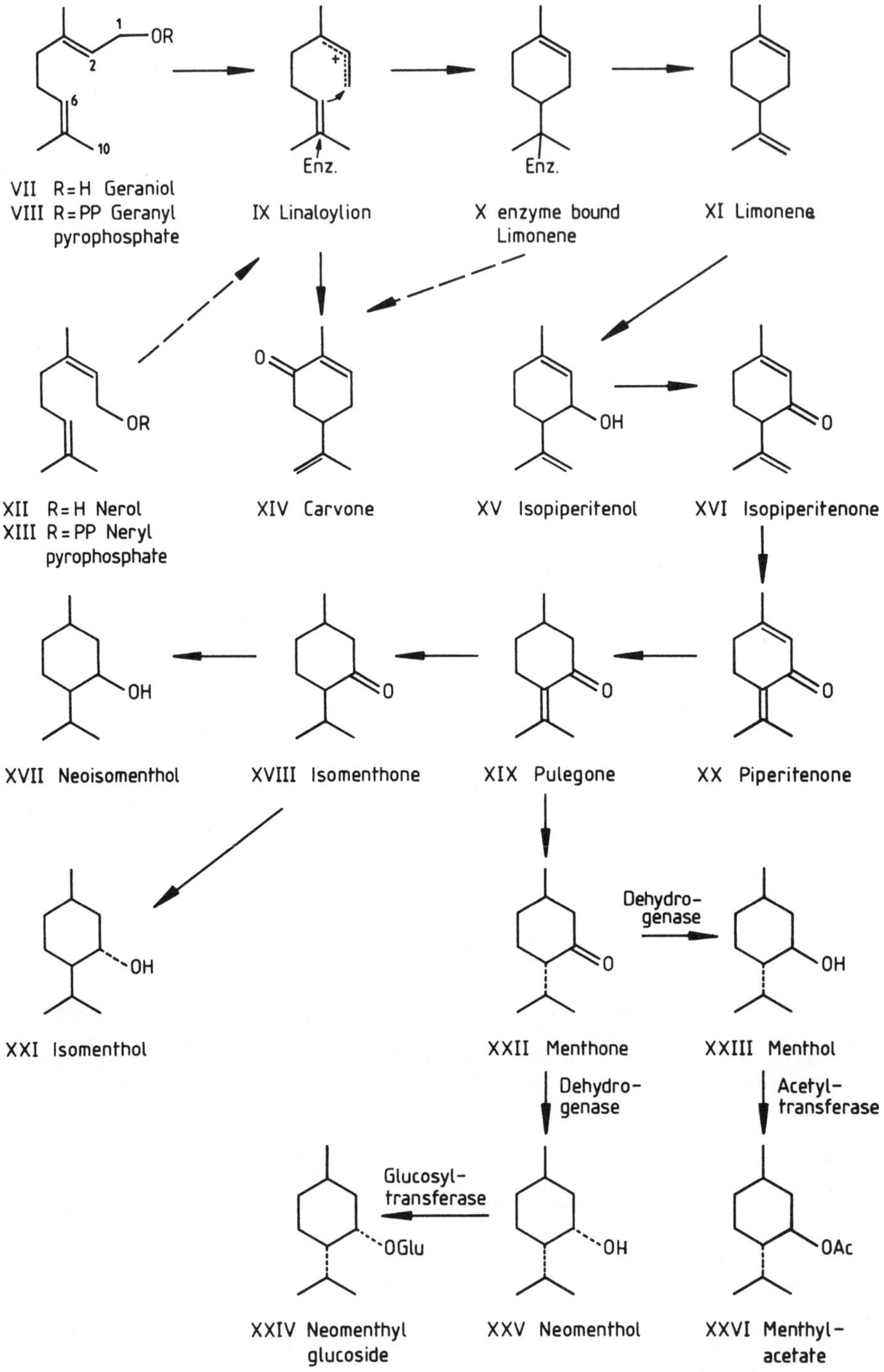

Fig. 2. General pathway for the menthane derivatives

observed are those of the monoterpene synthetases, or whether they reflect the presence of competing phosphohydrolases. Neryl- and geranyl pyrophosphate carbocyclase activities from *Citrus limonum* were partially purified and show participation of SH-groups in maintaining the activity of the enzyme (CORI et al. 1981, CHAYET et al. 1982).

3. Menthone-type Monoterpenes

The cyclic monoterpene dienes are considered to be the progenitors of a series of oxygenated compounds. Precursor product relationships were deduced from a variety of experiments, including time-course studies and direct feeding experiments with labeled monoterpenes discussed in the earlier review.

Piperitenone (XX) is considered to be an essential intermediate in the biosynthesis of menthone (XXII), menthol (XXIII), and other C-3-oxygenated p-menthane monoterpenes in *Mentha* species. The time course of incorporation of $|U-^{14}C|$ sucrose, an effective precursor of monoterpenes in leaf tissues (CROTEAU 1977) into the monoterpenes of *M. piperita* (peppermint) shoot tips was consistent with the initial formation of limonene as precursor for piperitenone (XX) presumably via isopiperitenol (XV) and isopiperitenone (XVI) and its subsequent conversion through pulegone (XIX) to menthone (XXII) and finally the menthols (KJONAAS and CROTEAU 1983) (Fig. 2). Furthermore limonene (XI) was readily incorporated into pulegone (XIX), menthone (XXII), and other oxygenated derivatives, whereas terpinolene was not appreciably incorporated into these compounds. Similarly, limonene (XI) was specifically incorporated into pulegone (XIX) in M. pulegium and into carvone (XIV) in *M. spicata*, confirming the role of this olefin as the essential precursor of oxygenated p-menthane monoterpenes.

All of the steps in the conversion of piperitenone (XX) to the menthol stereoisomers have been demonstrated in cell-free preparations from *M. piperita* leaves, and each of the NADPH-dependent reductases has been subjected to preliminary characterization. Thus cell-free extracts from *M. piperita* reduce the Δ^1 and Δ^4 carbon-carbon double bond of the 3-keto monoterpenes, piperitenone (XX), piperitone, and pulegone (XIX), but not carvone (XIV), characteristic of spearmint (BURBOTT and LOOMIS 1980) and soluble enzyme preparations from mature peppermint leaves catalyzed the reduction of 1-menthone (XXII) (KJONAAS et al. 1982).

Several in-vitro cultured *Mentha* cell lines are capable to transform menthone (XXII) into neomenthol (XXV) and pulegone (XIX) into isomenthone (XVIII), whereas none of these cell lines reduced isomenthone to the corresponding alcohol (AVIV and GALUN 1978, RODOV et al. 1983, AVIV et al. 1981). Some of these transformations have been carried out with immobilized cells (GALUN et al. 1983). Pulegone was reduced to menthone preferably in the blooming stage of several *Mentha* species (FUJITA et al. 1977). From analysis of crosses between *Mentha* species, the genetic basis for the production of most of these monoterpenes was determined (HEFENDEHL and MURRAY 1976, LINCOLN and MURRAY 1978, MURRAY et al. 1980). Thus the conversion of pulegone to menthone was due to a single gene.

Based on both short-term tracer studies and long-term periodic analysis, there is considerable evidence that monoterpenes are metabolically active and subject to rapid turnover in plants (BURBOTT and LOOMIS 1969). Thus menthone (XXII), the major monoterpene component of the essential oil of maturing peppermint leaves, is rapidly metabolized at the onset of flowering to menthol (XXIII) and menthyl acetate (XXVI) (CROTEAU and HOOPER 1978), which accumulates in the essential oil and to neomenthol (XXV), most of which is glucosylated and transported to the rhizome (Fig. 2) (CROTEAU and MARTINKUS 1979, KJONAAS et al. 1980). Monoterpenyl glycosides have been found in several essential oil-bearing plants, e.g., thymol- and carvacrolglucoside and -galatoside in *Thymus vulgaris* (SKOPP and HÖRSTER 1976) and in *Ocimum basilicum* (LANG and HÖRSTER 1977), menthyl-β-D-glucoside in *Mentha arvensis* (SAKATA and MITSUI 1975), and hydroxylinaloyl-β-D-glucoside (TSCHESCHE et al. 1977).

Such a reductive metabolism of monoterpene ketones appears to be a common feature of maturing essential oil plants (NAGASAWA et al. 1975, HENDRIKS and VAN OS 1976, FUJITA et al. 1977, SAKATA and KOSHIMIZU 1980). In peppermint, the metabolic disposition of the diastereomeric reduction products of menthone (XXII) is highly specific in that only menthyl acetate (XXVI) and neomenthyl-β-D-glucoside (XXIV) are formed. However, when either labeled menthol (XXIII) or neomenthol (XXV) was administered to peppermint, both the corresponding acetates and glucosides were formed. These and other experiments provided compelling evidence that compartmentation effects, and not the substrate specificity of the transferases, were responsible for the selectivity of pathways observed; this conclusion was confirmed by isolation and characterization of the acetyl CoA:monoterpenol acetyltransferase and the UDP-glucose: monoterpenol glucosyl transferase involved (CROTEAU and HOOPER 1978, MARTINKUS and CROTEAU 1981).

The transacetylase was located mainly in the 100,000 g supernatant fraction and had a molecular weight of about 37,000. The nature of the terpenyl moiety of the acetate and β-D-glucoside is, thus, determined at the menthone reduction step by compartmentation of these stereospecific reactions with the appropriate, relatively nonselective, transferase. On the other hand, two distinct and stereospecific dehydrogenases were shown to be responsible for the synthesis of menthol (XXIII) and neomenthol (XXV) from menthone (XXII) (KJONAAS et al. 1982). Thus, compartmentation of each stereospecific dehydrogenase with the corresponding nonspecific accessory enzyme is responsible for the metabolic selectivity observed. The neomenthol dehydrogenase and the glucosyl transferase resided specifically in the mesophyll layer, whereas the menthol dehydrogenase and substantial amounts of the acetyl transferase were located in the epidermis, presumably within the epidermal oil glands, suggesting that the compartmentation of the menthone metabolism in peppermint leaves is intercellular (CROTEAU and WINTERS 1982).

Incorporations of labeled geraniol and mevalonate by *M. spicata* (spearmint) indicated that oxidation of limonene (XI) or its biogenetic equivalent (e.g., X) to form carvone (XIV) involves a shift of the endocyclic double bond and that limonene and carvone are probably formed on divergent pathways from a common intermediate (AKHILA et al. 1980).

4. Thymol and Cineole

The major components of the volatile oil of thyme *(Thymus vulgaris)* are the aromatic monoterpenes, thymol (XXXII) and p-cymene (XXXI), and the cyclic diene, γ-terpinene (XXX). The time course of incorporation of [$^{14}CO_2$] into the volatile terpenoids of thyme cuttings as well as feeding experiments suggested a biosynthetic sequence by which γ-terpinene

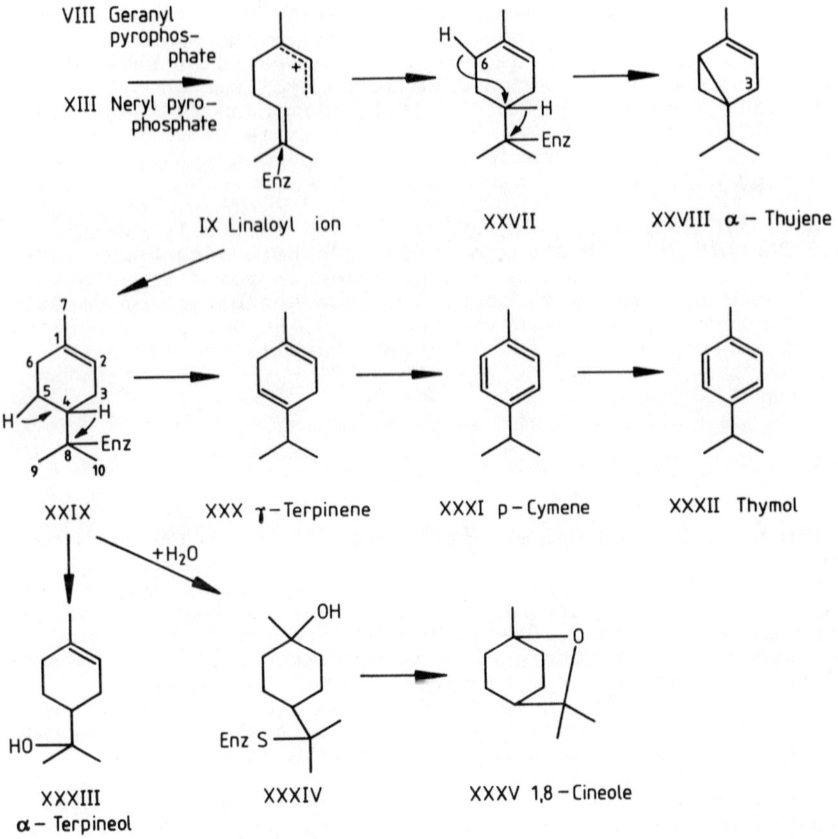

Fig. 3. Biosynthesis of α-thujene, thymol, α-terpineol, and cineole

gave rise to p-cymene, which in turn yielded thymol (Fig. 3) (POULOSE and CROTEAU 1978a). A soluble enzyme preparation obtained from young thyme leaves catalyzed the direct cyclization of both [1-^{3}H] neryl pyrophosphate (XIII) and [1-^{3}H] geranyl pyrophosphate (VIII) to γ-[3-^{3}H] terpinene (XXX) without interconversion of the acyclic precursors and without the involvement of any free intermediates (POULOSE and CROTEAU 1978b). In the fractionation steps, γ-terpinene synthetase has an apparent molecular weight of 96,000, shows a pH optimum at about 6.8, and requires Mg^{2+} for catalytic activity.

Based on the specific location of tritium in γ-terpinene (XXX) and α-thujene (XXVIII), a mechanism is proposed which involves the cyclization of the acyclic precursor after delocalization of charge with rotation about the C2-C3 bond of a linaloyl ion (IX) or its enzyme-stabilized equivalent leading to an enzyme-stabilized α-terpinyl ion (XXIX), loss of a proton from C-5 to form the Δ^4 double bond, and a 1,2-hydride shift from C-4 to C-8 to give γ-[3-^{3}H] terpinene (XXX) Fig. 3). The formation of α-terpineol can be readily envisioned as being due to the addition of a hydroxyl group to the cyclic cation. Similar hydride shift (in XXVII), but with loss of a proton from C-6 and cyclopropane ring formation would generate α-[3-^{3}H] thujene (XXVIII).

The heterocyclic monoterpene 1,8-cineole (XXXV) is one of the major component of the volatile oil produced by sage *(Salvia officinalis)*. A soluble enzyme system is obtained from immature leaves of this plant that catalyze the anaerobic, cation-dependent, cyclization of neryl pyrophosphate (XIII) to α-terpineol (XXXIII), limonene (XI) and 1,8-cineole (XXXV) (CROTEAU and KARP 1976a). These three p-menthane monoterpenes were derived independently from neryl pyrophosphate and were not intermediates of common reaction sequence.

The 1,8-cineole synthetase from *S. officinalis* was partially purified (CROTEAU and KARP 1977a). This enzyme catalyzes both a carbocyclization and a heterocyclization. The lack of evidence for a free intermediate in cineole biosynthesis, coupled to the inhibition of cineole synthetase activity by p-hydroxymercuribenzoate, suggests the possible involvement of an enzyme thiol-bound intermediate in the reaction (Fig. 3).

5. Bornane Type

The ketone camphor (XL) and the related alcohol borneol (XXXIX) are bicyclic monoterpenes commonly found together in the volatile oil produced by a large number of plant species, and both enantiomers of each occur in nature, i.e., (+)-borneol and (+)-camphor in sage *(Salvia officinalis)* and (-)-borneol and (-)-camphor in rosemary *(Rosmarinus officinalis)*. The camphor content of sage leaves increased as the leaves expanded, and the increase was roughly proportional to the number of filled peltate oil glands which appeared on the leaf surface during the expansion process, demonstrating that immature sage leaves synthesize and accumulate camphor most rapidly (CROTEAU et al. 1981).

A soluble enzyme preparation from sage leaves was shown to catalyze the cation-dependent formation of borneol (XXXIX) from an acyclic precursor, and the NAD-dependent oxidation of borneol to camphor (XL) (CROTEAU and KARP 1976b) in a manner consistent with the earlier hypothesis (RUZICKA 1953).

The 100-fold purified borneol dehydrogenase is a soluble enzyme with an apparent molecular weight of 91,000 and has a β-type stereospecificity in regard to the hydride transfer from borneol to NAD (CROTEAU et al. 1978). In addition to borneol oxidation, the enzyme readily catalyzed the oxidation of thujanol to thujone. Several types of evidence strongly suggested that the same enzyme catalyzes the oxidation step in the biosynthesis of both camphor (XL) and thujone, which are the major monoterpenes of sage.

Borneol (XXXIX) was synthesized in two discrete enzymatic steps, the cyclization of the acyclic precursor to bornyl pyrophosphate (XXXVIII) and its hydrolysis to borneol (CROTEAU and KARP 1977b). Such a pyrophosphorylated intermediate was only demonstrable after separation of the synthetase from a much greater quantity of bornyl pyrophosphate hydrolase activity. Detailed examination of the terpenyl pyrophosphate hydrolases of sage (CROTEAU and KARP 1979b) revealed that crude preparations also contained very high levels of an enzyme which preferentially hydrolyzed geranyl pyrophosphate, and which could, therefore, seriously interfere with studies on the monoterpene synthetase. Investigation of the bornyl pyrophosphate synthetase in crude preparations indicated that both neryl pyrophosphate (XIII) and geranyl pyrophosphate (VIII) could be cyclized to bornyl pyrophosphate (XXXVIII) (CROTEAU and KAPR 1979a), but on the removal of competing phosphatases,

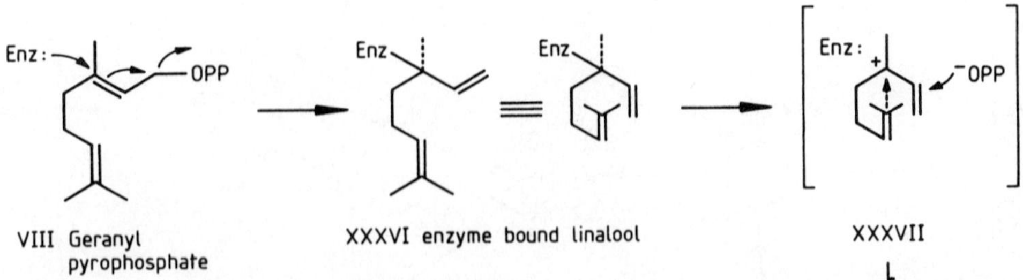

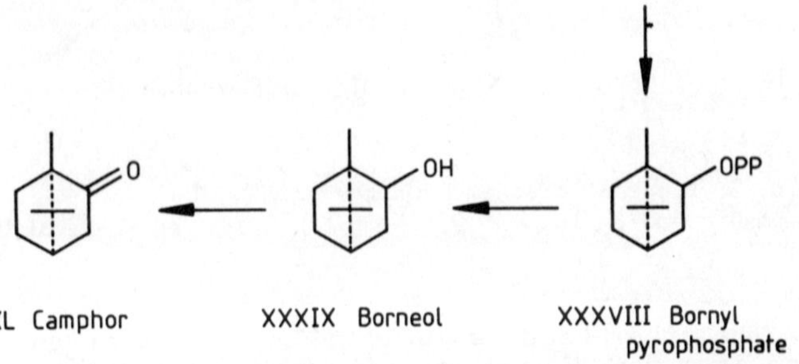

Fig. 4. A possible mechanism for the biosynthesis of borneol and camphor

the synthetase was shown to prefer geranyl pyrophosphate as substrate. The partially purified bornyl pyrophosphate synthetase had an apparent molecular weight of 95,000, and required Mg^{2+} for catalytic activity.

The results are consistent with a multistep mechanism whereby geranyl pyrophosphate is initially converted to an intermediate stereochemically suitable for cyclization, which could proceed by loss of the pyrophosphate with the assistance of a nucleophilic group of the enzyme, thus leading to an enzyme-stabilized linaloyl cation (XXXVI) as an intermediate (Fig. 4). Once the appropriate conformation is achieved, cyclization and release of the product could occur with insertion of the pyrophosphate and removal of the enzyme as indicated (according XXXVII).

6. Pinane Type (Fig. 5)

α-Pinene (XLIII) and β-pinene (XLII) are among the most widely distributed monoterpenes in the plant kingdom and are the major constituents of the various turpentines. The volatile oil of immature sage *(Salvia officinalis)* leaves was shown to contain (±)-α-pinene and (-)-β-pinene as major constituents.

In *Pinus pinaster* the terpene synthesis ability was closely associated with several membrane compartments and the terpene profile is genetically controlled. It is postulated that the formation of monoterpene hydrocarbons is photodependent and requires the integrity of the organs. There is therefore a cooperation between the light-activated sites and the monoterpene-synthesizing sites (CARDE and BARNARD-DAGAN 1982, BANTHORPE and NJAR 1984, BANTHORPE et al. 1984). On the basis of in-vivo time-course studies with *P. pinaster*, the acyclic trienes cis-ocimene and myrcene were suggested as precursors of α- and β-pinene, respectively (GLEIZES et al. 1982, WHITE 1983).

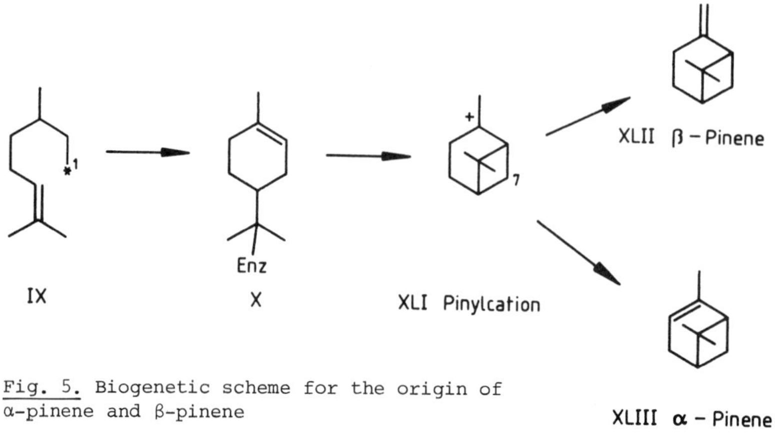

Fig. 5. Biogenetic scheme for the origin of α-pinene and β-pinene

On the other hand, soluble enzyme preparations from immature sage leaves and from the flavedo of *Citrus limonum* catalyzed the divalent cation-dependent conversion of [1-^{3}H] geranyl pyrophosphate (VIII) to α-pinene (XLIII) and β-pinene (XLII) together with limonene (XI) and camphene (CROTEAU and KARP 1976a, CHAYET et al. 1977, GAMBLIEL and CROTEAU 1980, 1982, CROTEAU and FELTON 1981). All the tritium was located on the methylene bridge carbon (C-7) of the pinane nucleus, as expected. The possibility of isomerization of β-pinene (XLII) to α-pinene (XLIII) was eliminated, as was the involvement of other free intermediates, demonstrating direct cyclization of geranyl pyrophosphate (VIII) probably via an appropiate cation such as (IX) and a monocyclic enzyme-bound α-terpinyl intermediate (X) which undergoes internal Markownikoff addition to a pinyl cation (XLI) that is subsequently stabilized by two variants of proton loss to the mixture of pinene isomers. The corresponding carbocyclase could be separated in two fractions, cyclase I (MW 95,000) and II (MW 57,000). The major biosynthetic product of cyclase I is α-pinene (XLIII), whereas cyclase II yields both α-pinene and β-pinene.

7. Thujane Type

Isothujone (XLVII) is one of the most common members of the thujone family of bicyclic monoterpenes. The manner of mevalonate incorporation into isothujone in *Tanacetum vulgare*, i.e., of C-2 label into C-3 of isothujone, specific incorporation of one hydrogen from C-2 of the precursor at C-4 of product, retention of the pro-(4R) hydrogen and loss of the pro-(4S)hydrogen and of one hydrogen from C-5, as well as the incorporation pattern of labeled α-terpineol (XXXIII), is consistent with a mechanism for the construction of isothujone by way of the α-terpinyl-derived ion (XLIV) which undergoes ring closure to (XLV). Two hydrogen migrations yield isothujanol (XLVI) which is dehydrogenated to isothujone (XLVII) (Fig. 6) (BANTHORPE et al. 1976a, 1977d, 1978a,b). A corresponding dehydrogenase has been isolated from *Salvia officinalis* (CROTEAU et al. 1978). This mechanism also readily accomodates the co-occurrence of sabinene (XLVIII) and isothujanol in *T. vulgare*. Sabinene has been obtained in a cell-free extract from *Citrus limonum* (CHAYET et al. 1977). Mevalonate was more effectively incorporated into isothujone during autumn or winter than in summer.

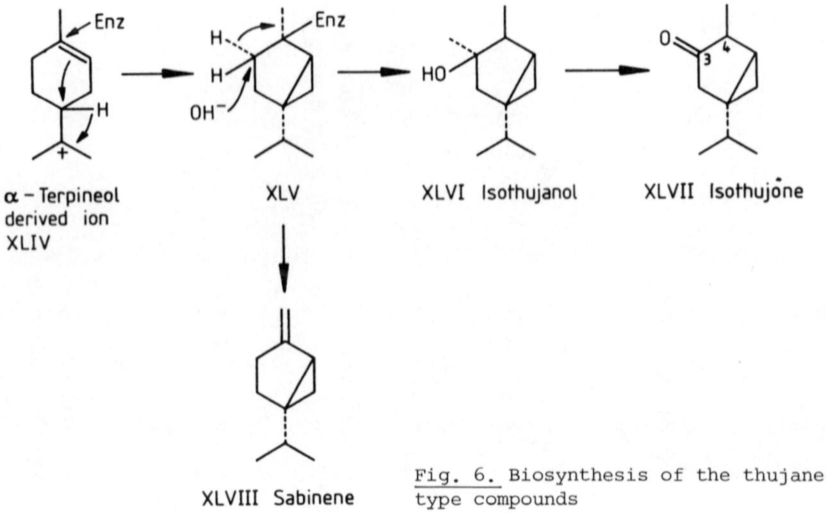

Fig. 6. Biosynthesis of the thujane-type compounds

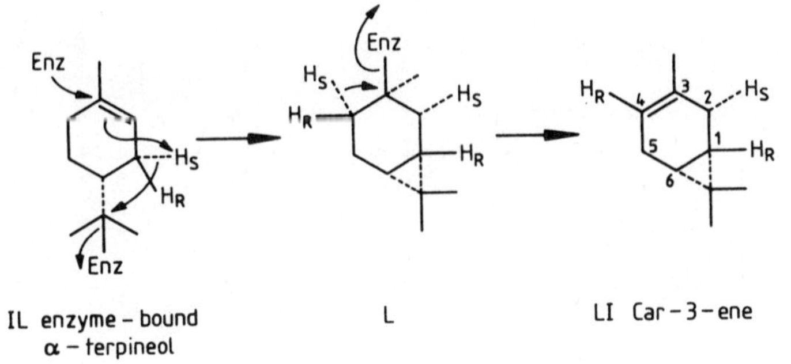

Fig. 7. Biosynthesis of carene

8. Carane Type

Car-3-ene (LI) is the only common occurring monoterpene with the carane skeleton. It is fairly widely distributed, mainly in the Coniferae. In car-3-ene biosynthesized from [2-^{14}C] mevalonate in *Pinus palustris* or *P. sylvestris* the label was found in C-4 and the pro-(2S) hydrogen of mevalonate is stereospecifically lost in the formation of the Δ^3-double bond. Further experiments are consistent with a scheme, after which the carene skeleton is constructed in a two-step mechanism from an enzyme-bound monocyclic intermediate (IL) involving migration of an olefinic bond and a 1,2-proton shift (L) to the site of the original double bond (Fig. 7).

9. Iridane Type

The progenitor of the iridane family of monoterpenoids involving iridoids and secoiridoids is considered to be loganin (LIV) or its deriva-

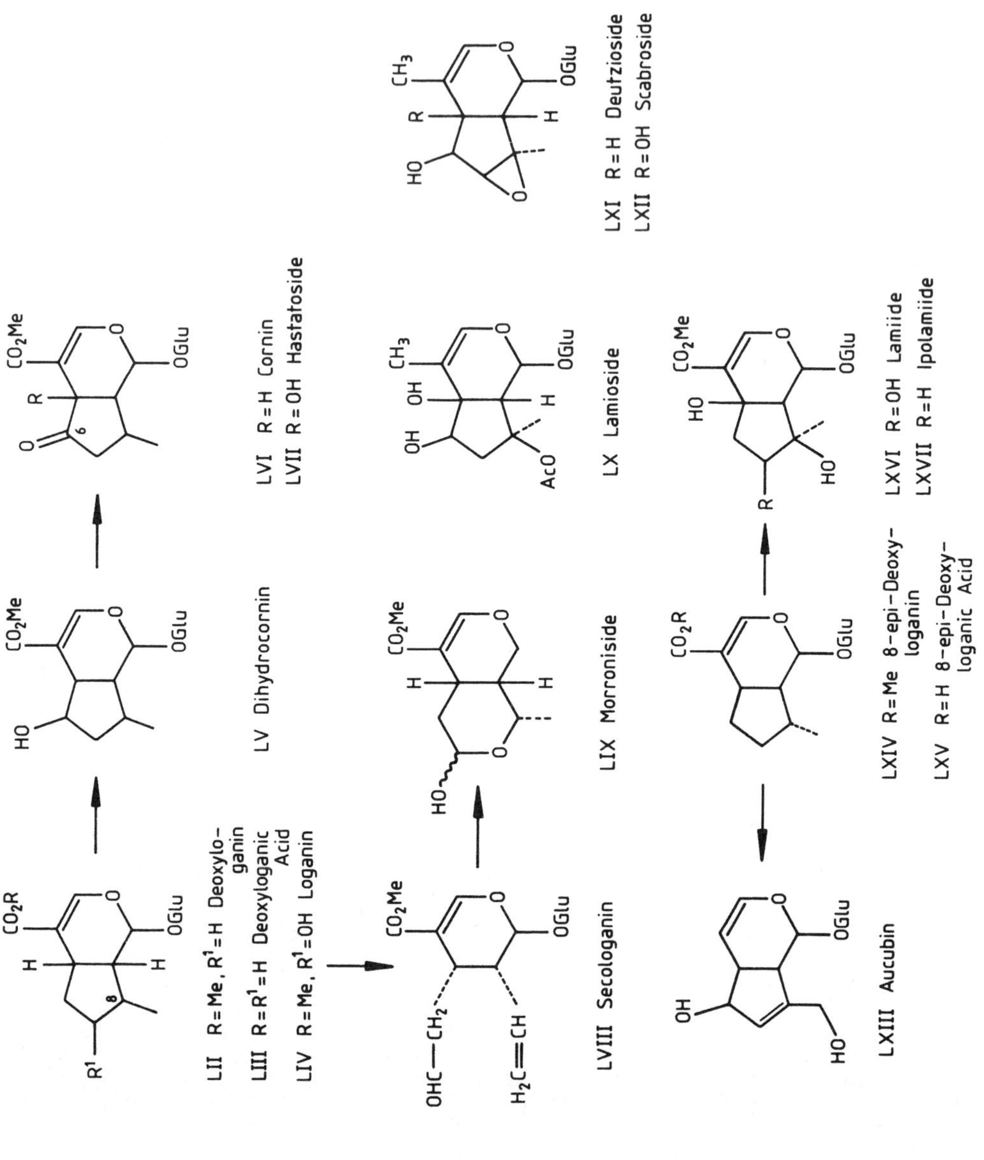

Fig. 8. Biosynthesis of some iridane-type monoterpenoids

tives (INOUYE et al. 1976, INOUYE 1978). Earlier investigations with *Verbena officinalis* have established that mevalonate, geraniol (VII), and deoxyloganic acid (LIII) were precursors of cornin (LVI) (HÜNI et al. 1966, HORODYSKI et al. 1969, INOUYE et al. 1972), and dihydrocornin (LV) has been demonstrated to be a biosynthetic intermediate (Fig. 8) (DAMTOFT et al. 1983a). The label at C-8 deoxyloganin (LII) retained during its conversion into cornin (LVI) and hastatoside (LVII) in *V. officinalis*, and *V. hastata*, suggesting that the 6 oxo group is produced by oxidation in an unactivated position (DAMTOFT et al. 1980, 1983b). In experiments with *Lonicera morrowii*, *Cornus officinalis* and *Gentiana thunbergii* the C-8 proton of deoxyloganic acid (LIII) was retained in loganin (LVI), secologanin (LVIII) and morroniside (LIX) (TAKEDA and INOUYE 1976) and deoxyloganic acid (LIII) is more efficiently incorporated into secologanin (LVIII) than loganin (LVI) (INOUYE et al. 1981).

In *Lamium amplexicaule*, *Deutzia crenata*, and *Galium spurium* ipolamiide (LXVII), lamiide (LXVI), lamioside (LX), deutzioside (LXI) and scabroside (LXII) are biosynthesized via 10-hydroxygeraniol and iridodial (LXXI) (INOUYE et al. 1977, 1978). A cytochrome P-450-dependent monoterpene hydroxylase isolated from 5-day old etiolated *Vinca rosea* seedlings as well as from rabbit liver microsomes was shown to catalyze the hydroxylation of geraniol and nerol exclusively to their corresponding 10-hydroxy-derivatives (MADYASTHA et al. 1976, LICHT and COSCIA 1978).

8-epi-Deoxyloganin (LXIV) rather than deoxyloganin (LII) is incorporated into lamiide (LXVI) and ipolamiide (LXVII) in *Hebenstreitia denata* (Scrophulariaceae) (DAMTOFT 1981) and 8-epi-deoxyloganic acid (LXV) rather than deoxyloganic acid (LIII) is a precursor for aucubin (LXIII) in *Plantago major* and *Scrophularia racemosa* and antirrinoside in *Antirrhinum majus* (Fig. 8) (DAMTOFT 1983).

Dolichodial (LXXII), a cyclopentanoid dialdehyde, is the main monoterpene component of *Teucrium marum*, a wild plant that growns in the Mediterranean area, and is accompanied by dolycholactone (LXXVII), a cyclopentanoid monoterpene lactone (PAGNONI et al. 1976). Feeding experiments are consistent with a biosynthetic pathway via nerol (XII), citronellol (I), its 10-hydroxy derivative (LXVIII) and probable the dialdehyde (LXIX) to dolichodial (Fig. 9) (BELLESIA et al. 1983). Iridodial (LXXI) is also incorporated. The C-2/C-3-saturated acyclic monoterpenes seem to be better precursors than the unsaturated ones. The same pathway is realized for the biosynthesis of nepetalactone (LXXVI) in *Nepeta cataria*, for which iridodial (LXXI) is found to be a key intermediate (BELLESIA et al. 1984). A hydride shift from C-1 to C-10 in iridodial through a Cannizzaro-type reaction yields (LXX) and the intermediate (LXXIII) is capable for lactonization to (LXXIV) which is a natural constituent of *N. cataria*.

Iridodial (LXXI) is also a very efficient precursor of dolicholactone (LXXVII) in *Teucrium marum* (GRANDI et al. 1983). In the biogenetic formation of the lactone again a hydride shift from C-1 to C-10 to (LXXV) is observed and as in the case of dolichodial citronellol and its 10-hydroxy derivative are preferred as precursors with respect to the C-2/C-3 unsaturated analogs.

10. Fenchane Type

The common herb fennel *(Foeniculum vulgare)* produces a volatile oil containing (1S)-fenchone (LXXXII) as a major monoterpene. A soluble enzyme

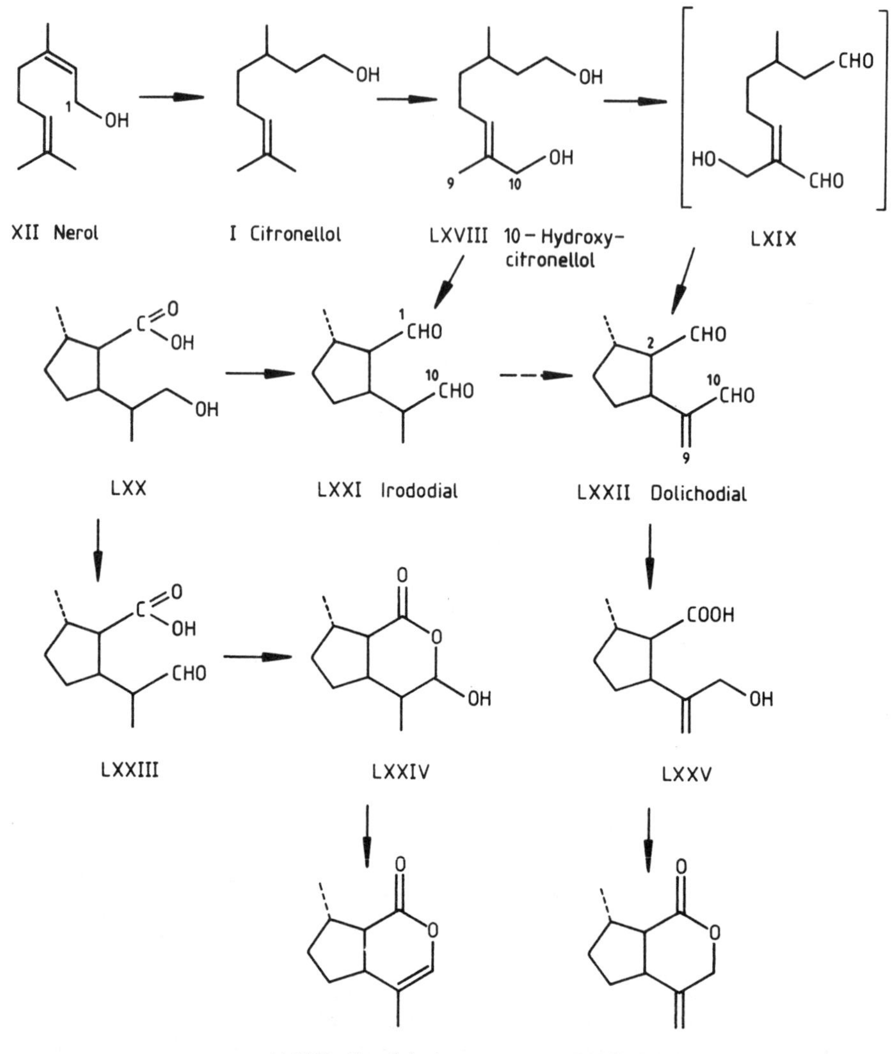

Fig. 9. Biosynthesis of nepetalactone, dolichodial, and dolicholactone

preparation obtained from homogenates of young fennel leaves converted geranyl pyrophosphate (VIII) as the preferred substrate to (1S)-endo-fenchol (LXXXI) in the presence of Mn^{2+}, and in the presence of pyridine nucleotide, dehydrogenated the fenchol formed to (1S)-fenchone (CROTEUA and FELTON 1980, CROTEAU et al. 1980a,b). The label of [1-^{3}H] geranyl pyrophosphate is located on the bridge carbon, demonstrating that the endo-fenchol is derived directly from the [1-^{3}H]-labeled acyclic precursor via a pinane-type intermediate (LXXX). α-Pinene (XLIII) is also a constituent of fennel oil, suggesting that pinanes and fenchanes may have a similar origin in this system. The endo-fenchol synthetase has an apparent molecular weight of 60,000 and shows a pH optimum near 7.0.

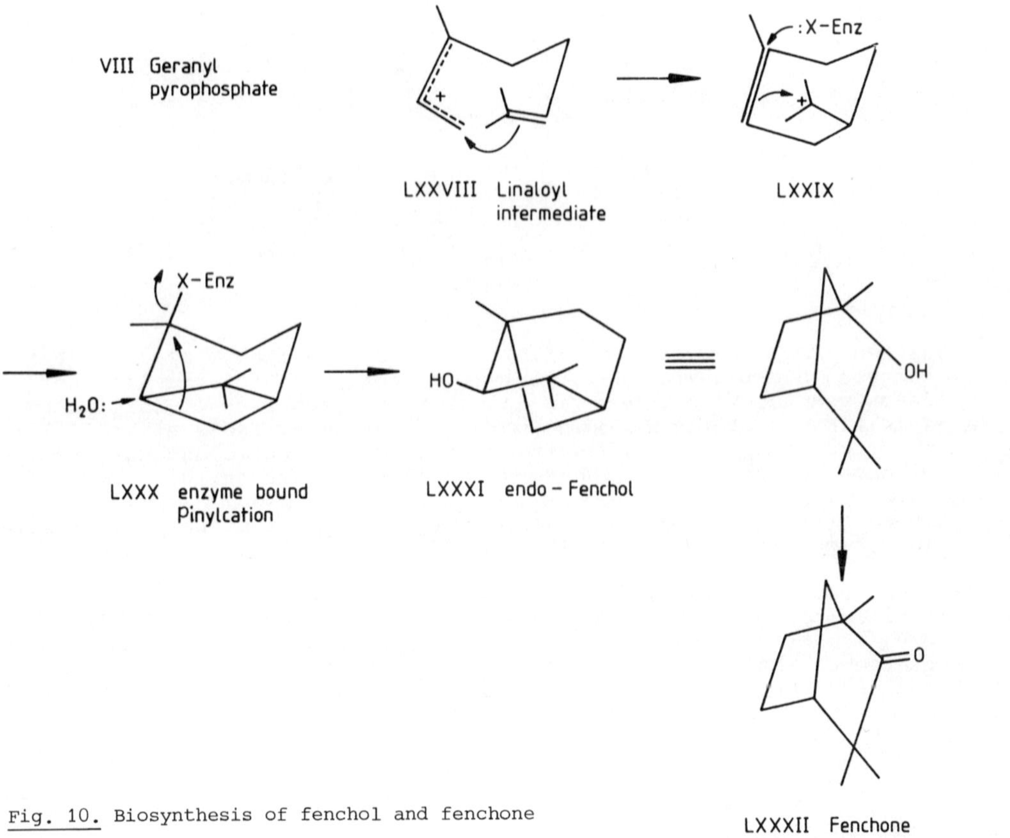

Fig. 10. Biosynthesis of fenchol and fenchone

The results are consistent with a scheme for fenchol synthesis (Fig. 10) after which geranyl pyrophosphate (VIII) has to be converted in a manner mentioned above via a bound linaloyl intermediate (LXXVIII) to the α-terpinyl structure (LXXIX). Subsequent attack of the tertiary ionic center of the α-terpinyl ion on the tri-substituted double bond results in a pinyl cation (LXXX). Binding of this to a nucleophilic enzyme group specifically promotes Wagner-Meerwein shift of the methylene group, by backside displacement of the binding nucleophile, to afford a fenchyl cation. If attack of water on the incipient cation is concomitant with the rearrangement, then stereoelectronic factors dictate the release of endofenchol (LXXXI) from the enzyme surface.

11. Irregular Type (Fig. 11)

Artemisia ketone (LXXXIII) is probably the most widely distributed monoterpene of the irregular type that does not obey the classical isoprene rule. Mevalonate, isopentenyl pyrophosphate, and dimethylallyl pyrophosphate were incorporated into artemisia ketone by *Artemisia annua* in a specific manner so that the C-5 moiety not containing the carbonyl group was preferentially labeled (ALLEN et al. 1977, BANTHORPE et al. 1977c). Geraniol (VII) is incorporated only after extensive de-

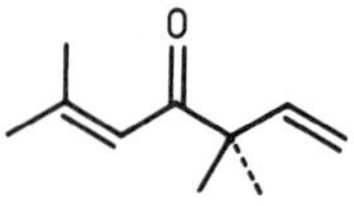

LXXXIII
Artemisia ketone

LXXXIV
Chrysanthemyl
alcohol

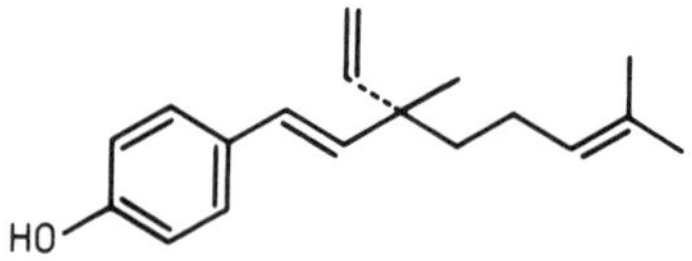

LXXXV Bakuchiol

Fig. 11. Irregular monoterpenoids

gradation (BANTHORPE et al. 1977b). Trans-chrysanthemyl alcohol (LXXXIV) and chrysanthemates were negligibly incorporated into artemisia keton. Trans-chrysanthemyl alcohol was an effective precursor of chrysanthemic acid in flowers of *Chrysanthemum cinerariaefolium* (PATTENDEN et al. 1975). Thus for construction of these types many biosynthetic aspects are still open to speculation. Whereas undifferentiated callus tissue of *Tanacetum cinerariifolium* showed no synthesis of pyrethrins shoots which had been differentiated from callus revealed the presence of six pyrethrin esters (CASHYAP et al. 1978).

Many natural products are of mixed origin and are derived by condensation of isoprenoid units with non-terpenoidal moieties. Bakuchiol (LXXXV) isolated from *Psoralea corylifolia* (Leguminosae) belongs to such meroterpenes (MEHTA et al. 1973). Using phenylalanine and mevalonate it has been demonstrated that bakuchiol is derived from one phenylpropane with loss of one carbon atom and two isoprenoid units (BANERJI and CHINTALWAR 1983).

12. Some Biological Aspects of Monoterpenes

The plastids of all *Pinus* species investigated are regarded to be the site of terpene synthesis (VASSILIEV and CARDE 1976, BOSABALIDIS and TSEKOS 1982). In *Pinus maritima* needles, a preferential site of monoterpene synthesis was the epithelial cells of the resin ducts, with the basal part of the needles having a higher content of monoterpenes than the upper parts (BERNARD-DAGAN et al. 1979, 1982). In *Poncirus trifoliata*, a Rutaceae, there are probably two different types of glands in the endocarp and the exocarp, respectively, producing essential oils differing in their contents of mono- and sesquiterpenes. Electron microscopy gives some evidence for plastidial synthesis of terpenes in both types of glands (HEINRICH 1970, HEINRICH et al. 1980, 1983). In *Monarda fistulosa*, one of the Lamiaceae, the essential oil predominantly occurs in the peltate trichomes of leaves. A leucoplasts preparation isolated from the exocarpium of young *Citrofortunella* fruits was able to synthesize monoterpene hydrocarbons when incubated with isopentenyl pyrophosphate (GLEIZES et al. 1983).

The variability in essential oil composition is determined both by genetic and environmental factors, e.g., in *Achillea collina* (CERNAJ et al. 1983). A diurnal variation was found for the volatile terpenoids of *Juniperus scopularum* (ADAMS and HAGERMAN 1977). *Pseudomonas putida*, which is able to grow with p-cymene (XXXI), transforms this monoterpene by oxidation in three steps to give p-cumic acid and further by ring cleavage of 2,3-dihydroxy-p-cumate (DE FRANK and RIBBONS 1977a,b).

Several investigations have been done on the formation of monoterpenes in cultured cells. Cell suspension cultures of "Muscat de Frontignan" grapes *Vitis vinifera* are able to convert citral, a mixture of neral and geranial, into the corresponding monoterpenic alcohols, nerol and geraniol (AMBID et al. 1982, FALLOT et al. 1982). The characteristic aroma of Muscat grapes has been shown to be attributed to some monoterpenic compounds such as linalool (II), nerol (XII), geraniol (VII), and α-terpineol (XXXIII) (HARDY 1970, RIBEREAU-GAYON et al. 1975). Tissue cultures of the Damask rose contained glucoside-bound terpenic alcohols similar to those found in intact plants (KIREEVA et al. 1977). Callus from the stem of *Rosa damascena* did not accumulate monoterpenes. However, cell-free extracts of the cultures converted isopentenyl pyrophosphate into geraniol (VII) and nerol (XII) some 300-fold more efficiently than did the optimum extracts from the parent plants, showing that the crucial enzymes for monoterpene synthesis were present in the cultures (BANTHORPE and BARROW 1983). Callus tissue and suspension cells of *Perilla frutescens* produce the same monoterpenes as the original plant leaves (NABETA et al. 1983, SUGISAWA and OHNISHI 1976).

In-vitro treatment of *Mentha piperita* tissue with colchicine increased essential oil production due to an increase in the number of neoformed secretory glands (BRICOUT and GARCIA-RODRIGUEZ 1978). In *M. piperita* and *M. rotundifolia*, the products synthesized in vitro were more oxidized than the natural forms (BRICOUT et al. 1978). Pulegone analogs are biotransformed by *Mentha* cell suspensions only in certain cases (AVIV et al. 1983).

The ability of cultured cells of plants to metabolize foreign substrates is of considerable interest because of the specificity of the transformations which may be effected by such cells.

Cultured cells of *Nicotiana tabacum* were able to transform foreign substrates such as monoterpenoid alcohols by regio- and steroselective hydroxylation at the C=C bond as well as its allylic positions, and to catalyze other enantioselective bioconversions (SUGA et al. 1976, 1980a, 1982, 1983c, HIRATA et al. 1981, 1982a,b, SUGA and HIRATA 1983). Thus cultured suspension cells of *N. tabacum* transform preferentially one of each enantiomeric pair of borneol (XXXIX), isoborneol, and isopinocampheol to the corresponding carbonyl compounds (SUGA et al. 1983a).

References

ADAMS, R.P., HAGERMAN, A.: Am. J. Bot. 64, 278-285 (1977). - AKHILA, A., BANTHORPE, D.V.: Z. Pflanzenphysiol. 99, 277-282 (1980a); - Phytochemistry 19, 1429-1431 (1980b); - ibid. 19, 1691-1694 (1980c). - AKHILA, A., BANTHORPE, D.V., ROWAN, M.G.: Phytochemistry 19, 1433-1437 (1980). - ALLEN, K.G., BANTHORPE, D.V., CHARLWOOD, B.V., VOLLER, C.M.: Phytochemistry 16, 79-83 (1977). - AMBID, C., MOISSEEFF, M., FALLOT, J.: Plant Cell Rep. 1, 91-93 (1982). - AVIV, D., GALUN, E.: Planta Med. 33, 70-77 (1978). - AVIV, D., KROCHMAL, E., DANTES, A., GALUN, E.: Planta Med. 42, 236-243 (1981). - AVIV, D., DANTES, A., KROCHMAL, E., GALUN, E.: Planta Med. 47, 7-10 (1983).

BANERJI, A., CHINTALWAR, G.J.: Phytochemistry 22, 1945-1947 (1983). - BANTHORPE, D.V., BARROW, S.E.: Phytochemistry 22, 2727-2728 (1983). - BANTHORPE, D.V., CHARLWOOD, B.V.: pp. 185-220 in: Encyclopedia of Plant Physiology, N.S. Vol. 8, eds. E.A. BELL, B.V. CHARLWOOD. Berlin, Heidelberg, New York: Springer 1980. - BANTHORPE, D.V., EKUNDAYO, O.:

Phytochemistry 15, 109-112 (1976). - BANTHORPE, D.V., NJAR, V.C.O.: Phytochemistry 23, 295-299 (1984). - BANTHORPE, D.V., POOTS, I.: Phytochemistry 18, 1297-1299 (1979). - BANTHORPE, D.V., BUCKNALL, G.A., DOONAN, H.J., DOONAN, S., ROWAN, M.G.: Phytochemistry 15, 91-100 (1976a). - BANTHORPE, D.V., CARDEMIL, E., CONTRERAS, M.C.: Phytochemistry 15, 391-394 (1976b). - BANTHORPE, D.V., BUCKNALL, G.A., GUTOWSKI, J.A., ROWAN, M.G.: Phytochemistry 16, 355-358 (1977a). - BANTHORPE, D.V., CHARLWOOD, B.V., GREAVES, G.M., VOLLER, C.M.: Phytochemistry 16, 1387-1392 (1977b). - BANTHORPE, D.V., DOONAN, S., GUTOWSKI, J.A.: Phytochemistry 16, 85-92 (1977c). - BANTHORPE, D.V., MANN, J., POOTS, I.: Phytochemistry 16, 547-550 (1977d). - BANTHORPE, D.V., EKUNDAYO, O., ROWAN, M.G.: Phytochemistry 17, 1111-1114 (1978a). - BANTHORPE, D.V., MODAWI, B.M., POOTS, I., ROWAN, M.G.: Phytochemistry 17, 1115-1118 (1978b). - BANTHORPE, D.V., CHRISTOU, P.N., PINK, C.R.: WATSON, D.G.: Phytochemistry 22, 2465-2468 (1983a). - BANTHORPE, D.V., LONG, D.R.S., PINK, C.R.: Phytochemistry 22, 2459-2463 (1983b). - BANTHORPE, D.V., EKUNDAYO, O., NJAR, V.C.O.: Phytochemistry 23, 291-294 (1984). - BELLESIA, F., GRANDI, R., PAGNONI, U.M., PINETTI, A., TRAVE, R.: Phytochemistry 23, 83-87 (1984). - BELLESIA, F., PAGNONI, U.M., PINETTI, A., TRAVE, R.: Phytochemistry 22, 2197-2202 (1983). - BERNARD-DAGAN, C., CARDE, J.P., GLEIZES, M.: Can. J. Bot. 57, 255-263 (1979). - BERNARD-DAGAN, C., PAULY, G., MARPEAU, A., GLEIZES, M., CARDE, J.P.: Physiol. Veg. 20, 775-795 (1982). - BOSABALIDIS, A., TSEKOS, I.: Protoplasma 112, 63-70 (1982). - BRICOUT, J., GARCIA-RODRIGUEZ, M.J.: C.R. Acad. Sci. Ser. D. 286, 1585-1588 (1978). - BRICOUT, J., GARCIA-RODRIGUEZ, M.T., PAUPARDIN, C., SAUSSAY, R.: C.R. Acad. Sci. Ser. D. 287, 611-613 (1978). - BUNTON, C.A., CORI, O.: Interciencia 3, 291-297 (1978). - BUNTON, C.A., CORI, O., HACHEY, D., LERESCHE, J.P.: J. Org. Chem. 44, 3238-3244 (1979). - BURBOTT, A.J., LOOMIS, W.D.: Plant Physiol. 44, 173-179 (1969); - ibid. 65,(Suppl.) 96 Abstr. 522 (1980).

CANE, D.E.: Tetrahedron 36, 1109-1159 (1980). - CARDE, J.P., BERNARD-DAGAN, C.: Actual. Bot. 1982, 52-70, C.A. 98, 176202 (1983). - CASHYAP, M.M., KUEH, J.S.H., MACKENZIE, I.A., PATTENDEN, G.: Phytochemistry 17, 544-545 (1978). - CERNAJ, P., REPCAK, M., HONCARIU, R., TESARIK, K.: Biologia (Bratislava) 38, 865-872 (1983), C.A. 100 3548 (1984). - CHARLWOOD, B.V., BANTHORPE, D.V.: Progr. Phytochem. 5, 65-125 (1978). - CHAYET, L., ROJAS, C., CARDEMIL, E., JABALQUINTO, A.M., VINCUNA, R., CORI, O.: Arch. Biochem. Biophys. 180, 318-327 (1977). - CHAYET, L., HOLUIGUE, L, CORI, O.: Bol. Soc. Chil. Quim. 27, 207-208 (1982), C.A. 35316 (1982). - CORI, O.: Phytochemistry 22, 331-341 (1983). - CORI, O., CHAYET, L., DE LA FUENTE, M., FERNANDEZ, L.A., HASHAGEN, U., PÉREZ, L.M., PROTILLA, G., ROJAS, C., SANCHEZ, G., VIAL, M.V.: Mol. Bio. Biochem. Biophys. 32, 97-110 (1980). - CORI, O., CHAYET, L., PÉREZ, L.M., ROJAS, M.C., PORTILLA, G., HOLUIGUE, L., FERNANDEZ, L.A.: Arch. Biol. Med. Exp. 14, 129-137 (1981). - CORI, O., PORTILLA, G., CHAYET, L.: Arch. Biol. Med. Exp. 15, 357-364 (1982). - CROTEAU, R.: Plant Phyisol. 59, 519-520 (1977). - CROTEAU, R.: pp. 225-282 in: Biosynthesis of Isoprenoid Compounds, eds. J.W. PORTER, S.L. SPURGEON, Vol.I. New York: Wiley 1981. - CROTEAU, R., FELTON, N.M.: Phytochemistry 19, 1343-1347 (1980); - Arch. Biochem. Biophys. 207, 460-464 (1981). - CROTEAU, R., HOOPER, C.L.: Plant Physiol. 61, 737-742 (1978). - CROTEAU, R., KARP, F.: Arch. Biochem. Biophys. 176, 734-746 (1976a); - Biochem. Biophys. Res. Commun. 72, 440-447 (1976b); - Arch. Biochem. Biophys. 179, 257-265 (1977a); - ibid. 184, 77-86 (1977b); - ibid. 198, 512-522 (1979a); - ibid. 198, 523-532 (1979b). - CROTEAU, R., MARTINKUS, C.: Plant Physiol. 64, 169-175 (1979). - CROTEAU, R., WINTERS, J.N.: Plant Physiol. 69, 975-977 (1982). - CROTEAU, R., HOOPER, C.L., FELTON, M.: Arch. Biochem. Biophys. 188, 182-193 (1978). - CROTEAU, R., FELTON, M., RONALD, R.C.: Arch. Biochem. Biophys. 200, 534-546 (1980a); - ibid. 200, 524-533 (1980b). - CROTEAU, R., FELTON, M., KARP, F., KJONAAS, R.: Plant Physiol. 67, 820-824 (1981).

DAMTOFF, S.: Chem. Commun. 1981, 228-229; - Phytochemistry 22, 1929-1930 (1983). - DAMTOFF, S., JENSEN, S.R., NIELSEN, B.J.: Chem. Commun. 42-43 (1980). - DAMTOFF, S., JARS, M.U., JENSEN, S.R., KIRKO, O., NIELSEN, B.J.: Phytochemistry 22, 695-697 (1983a). - DAMTOFF, S., JENSEN, S.R., NIELSEN, B.J.: J. Chem. Soc. Perkin Trans. I, 1943-1948 (1983b). - DE FRANK, J.J., RIBBONS, D.W.: J. Bacteriol. 129, 1356-1364 (1977a); - ibid. 129, 1365-1374 (1977b). - DE LA FUENTE, M., PÉREZ, L.M., HASHAGEN, U., CHAYET, L., ROJAS, C., PORTILLA, G., CORI, O.: Phytochemistry 20, 1551-1557 (1981).

FALLOT, J., MOISSEEFF, M., AMBID, C.: Plant Tissue Cult., Proc. Int. Congr. Plant Tissue Cell Cult., 5th, 389-390 (1982). - FUJITA, S., NAKANO, T., FUJITA, Y.: J. Agric. Chem. Soc. Jn. 51, 405-408 (1977).

GALUN, E., AVIV, D., DANTES, A., FREEMAN, A.: Planta Med. 49, 9-13 (1983). - GAMBLIEL, H., CROTEAU, R.: Plant Physiol. 65 (Suppl.) 98, Abstr. 525 (1980); - J. Biol. Chem. 257, 2335-2342 (1982). - GLEIZES, M.: C.R. Hebd. Seances Acad. Sci., Ser. D 283, 97-100 (1976). - GLEIZES, M., MARPEAU, A., PAULY, G., BERNARD-DAGAN, C.: Phytochemistry 21, 2641-2644 (1982). - GLEIZES, M., PAULY, G., CARDE, J.P., MARPEAU, A., BERNARD-DAGAN, C.: Planta 159, 373-381 (1983). - GRANDI, R., PAGNONI, U.M., PINETTI, A., TRAVE, R.: Phytochemistry 22, 2723-2726 (1983).

HARDY, P.J.: Phytochemistry 9, 709-715 (1970). - HEFENDEHL, F.W., MURRAY, M.J.: Lloydia 39, 39-52 (1976). - HEINRICH, G.: Protoplasma 69, 15-36 (1970). - HEINRICH, G., SCHULTZE, W., WEGENER, R.: Protoplasma 103, 115-129 (1980). - HEINRICH, G., SCHULTZE, W., PFAB, I., BÖTTGER, M.: Physiol. Veg. 21, 257-268 (1983). - HENDRIKS, H., OS, F.H.L. van: Phytochemistry 15, 1127-1130 (1976). - HIRATA, T., AOKI, T., HIRANO, Y., ITO, T., SUGA, T.: Bull. Chem. Soc. Jn 54, 3527-3529 (1981). - HIRATA, T., HAMADA, H., AOKI, T., SUGA, T.: Phytochemistry 21, 2209-2212 (1982a). - HIRATA, T., LEE, Y.S., SUGA, T.: Chem. Lett. 671-674 (1982b). - HORODYSKY, A.G., WALLER, G.R., EISENBRAUN, E.J.: J. Biol. Chem. 244, 3110-3116 (1969). - HÜNI, J.E.S., HILTEBRAND, H., SCHMID, H., GRÖGER, D., JOHNE, S., MOTHES, K.: Experientia 22, 656 (1966).

INOUYE, H.: Planta Med. 33, 193-216 (1978). - INOUYE, H., UEDA, S., AOKI, Y., TAKEDA, Y.: Chem. Pharm. Bull. 20, 1287-1296 (1972). - INOUYE, H., UEDA, S., TAKEDA, Y.: Heterocycles 4, 527-565 (1976). - INOUYE, H., UEDA, S., UESATO, S.: Phytochemistry 16, 1669-1675 (1977). - INOUYE, H. UEDA, S., UESATO, S., KOBAYASHI, K.: Chem. Pharm. Bull. 26, 3384-3394 (1978). - INOUYE, K., TAKEDA, Y., TANAHASHI, T., INOUYE, H.: Chem. Pharm. Bull. 29, 981-990 (1981).

JOHNSON, M., CROTEAU, R.: Plant Physiol. 65 (Suppl.), 96 Abstr. 523 (1980).

KIREEVA, S.A., BUGORSKII, P.S., REZNIKOWA, S.A.: Fiziol. Rast. 24, 824-831 (1977). - KJONAAS, R., CROTEAU, R.: Arch. Biochem. Biophys. 220, 79-89 (1983). - KJONAAS, R., MARTINKUS-TAYLOR, C., CROTEAU, R.: Plant Physiol. 65 (Suppl.), 96 Abstr. 524 (1980). - KJONAAS, R., MARTINKUS-TAYLOR, C., CROTEAU, R.: Plant Physiol. 69, 1013-1017 (1982).

LANG, E., HÖRSTER, H.: Planta Med. 31, 112-118 (1977). - LICHT, H.J., COSCIA, C.J.: Biochemistry 17, 5638-5646 (1978). - LINCOLN, D.E., MURRAY, M.J.: Phytochemistry 17, 1727-1730 (1978). - LOOMIS, W.D., CROTEAU, R.: pp. 363-418 in: The Biochemistry of Plants, ed. P.K. STUMPF, Vol. 4. New York: Academic Press 1980.

MADYASTHA, K.M., MEEHAN, T.D., COSCIA, C.J.: Biochemstry 15, 1097-1102 (1976). - MARTINKUS, C., CROTEAU, R.: Plant Physiol. 68, 99-106 (1981). - MEHTA, G., NAYAK, U.R., DEV, S.: Tetrahedron 29, 1119-1125 (1973). - MURRAY, M.J., LINCOLN, D.E., HEFENDEHL, F.W.: Phytochemistry 19, 2103-2110 (1980).

NABETA, K., OHNISHI, Y., HIROSE, T., SUGISAWA, H.: Phytochemistry 22, 423-425 (1983). - NAGASAWA, T., UMEMOTO, K., TSUNEYA, T., SHIGA, M.: J. Agric. Chem. Soc. Jpn 49, 491-493 (1975).

OHNO, Y., TANAKA, S.: Agric. Biol. Chem. 41, 399-401 (1977).

PAGNONI, U.M., PINETTI, A., TRAVE, R., GARANTI, L.: Aust. J. Chem. 29, 1375-1381 (1976). - PATTENDEN, G., POPPLESTONE, C.R., STORER, R.: Chem. Commun. 290-291 (1975). - PÉREZ, L.M., CORI, O., CHAYET, L., ROJAS, C., PORTILLA, G.: Plant Physiol. 65 (Suppl.) 98 Abstr. 537 (1980a). - PÉREZ, L.M., TAUCHER, G., CORI, O.: Phytochemistry 19, 183-187 (1980b). - PÉREZ, L.M., LOZADA, R., CORI, O.: Phytochemistry 22, 431-433 (1983). - PORTILLA, G., ROJAS, M.C., CHAYET, L., CORI, O.: Arch. Biochem. Biophys. 218, 614-618 (1982a). PORTILLA, G., ROJAS, M.C., CORI, O.: Bol. Soc. Chil. Quim. 27, 191-193 (1982b), C.A. 97, 36157 (1982). - POULOSE, A.J., CROTEAU, R.: Arch. Biochem. Biophys. 187, 307-314 (1978a); - Arch. Biochem. Biophys. 191, 400-411 (1978b). POULTER, C.D., RILLING, H.C.: pp. 161-224 in: Biosynthesis of Isoprenoid Compounds, eds. J.W. PORTER, S.L. SPURGEON, Vol. 1. New York: Wiley 1981.

RIBEREAU-GAYON, P., BIODRON, J.N., TERRIER, A.: J. Agric. Food Chem. 23, 1042-1047 (1975). - RODOV, V.S., REZNIKOVA, S.A., MELNIKOV, V.N.: Rastit. Resur. 19, 92-97 (1983), C.A. 98, 104 395 (1983). - RUZICKA, L.: Experientia 9, 357-367 (1953).

SAGAMI, H., OGURA, K., SETO, S., KUROKAWA, T.: Biochem. Biophys. Res. Commun. 85, 572-578 (1978). - SAKATA, I., KOSHIMIZU, K.: J. Agric. Chem. Soc. Jpn 54, 1037-1043 (1980). - SAKATA, I. MITSUI, T.: Agric. Biol. Chem. 39, 1329-1330 (1975). - SKOPP, K., HÖRSTER, H.: Planta Med. 29, 208-215 (1976). - SUGA, T., AOKI, T., HIRATA, T., LEE, Y.S., NISHIMURA, O., UTSUMI, M.: Chem. Lett. 229-230 (1980a). - SUGA, T., HIRATA, T.: Nippon Kagaku Kaishi 1345-1352 (1983), C.A. 99, 172 950 (1983). - SUGA, T., SHISHIBORI, T.: Chem. Lett. 1157-1158 (1979). - SUGA, T., HIRATA, T., HIRANO, Y., ITO, T.: Chem. Lett. 1245-1248 (1976). - SUGA, T., SHISHIBORI, T., HIRATA, T.: Chem. Lett. 937-938 (1977). - SUGA, T., SHISHIBORI, T., MORINAKA, H.: Chem. Commun. 167-168 (1980b). - SUGA, T., HIRATA, T., LEE, Y.S.: Chem. Lett. 1595-1598 (1982). - SUGA, T.; HIRATA, T., HAMADA, H., FUTATSUGI, M.: Plant Cell Rep. 2, 186-188 (1983a). - SUGA, T., HIRATA, T., OKITA, H., TSUJI, K.: Chem. Lett. 1491-1494 (1983b). - SUGA, T., LEE, Y.S., HIRATA, T.: Bul. Chem. Soc. Jpn 56, 784-787 (1983c). - SUGISAWA, H., OHNISHI, Y.: Agric. Biol. Chem. 40, 231-232 (1976).

TAKEDA, Y., INOUYE, H.: Chem. Pharm. Bull. 24, 79-84 (1976). - TANGE, K.: Bull. Chem. Soc. Jpn 54, 2763-2769 (1981). - TANGE, K., OKITA, H., NAKAO, Y., HIRATA, T., SUGA, T.: Chem. Lett. 777-780 (1981). - TSCHESCHE, R., CIPER, F., BREITMAIER, E.: Chem. Ber. 110, 3111-3117 (1977).

VASSILIEV, A.E., CARDE, J.P.: Protoplasma 89, 41-48 (1976).

WHITE, E.E.: Phytochemistry 22, 1399-1405 (1983).

Professor Dr. HORST-ROBERT SCHÜTTE
Akademie der Wissenschaften der DDR
Institut für Biochemie der
Pflanzen Halle
DDR-4010 Halle (Saale)

VII. Developmental Physiology

By Bertold Hock

From the time an egg cell is fertilized onward, an organism undergoes a series of developmental changes which continue until maturity and senescence. The following review is addressed to physiological changes during development in higher plants. Except for general statements, the article is mainly based upon progress reports published during the last year.

1. Dormancy

Plants have to cope with adverse conditions, e.g., drought or cold. This is achieved by dormancy, a period of developmental arrest which is characterized by dehydration and termination of growth. The release from dormancy often requires a complicated interaction between exogenous and endogenous factors. The best-studied example is provided by seed dormancy.

The issue has been reviewed by TAYLORSON and HENDRICKS (1977), and more recently by BEWLEY and BLACK (1982) and KHAN (1982).

a) Seed Dormancy

During seed dormancy, the germination of a viable seed is prevented. Rehydration does not suffice for germination, as in quiescent seeds. Rather, additional effects have to act on the seed for the transition to the next developmental stage. Dormancy is not only a means for adaptation to unfavorable growth conditions, it is also a device for the extension of germination in time which enhances the spread and the survival of a particular species.

There are three basic types of dormancy. During the *primary (true) dormancy*, when the seed is shed from the mother plant, even the excised embryo is incapable of growing. The *whole seed dormancy* comprises coat-, testa-, or endosperm-imposed cases.

The nondormant embryo starts growing when it is decoated. The *secondary (induced) dormancy* is induced naturally or by artificial means in a mature, inbibed seed following the breaking of primary dormancy. It develops in the absence of the factors necessary for germination (e.g., light, oxygen, or a suitable temperature). It then does not allow germination under conditions which previously led to germination. The mechanisms for the entry into dormancy, its maintenance and its breakage are therefore quite different for these types. The physiological basis has been studied in greater detail in cocklebur *(Xanthium pennsylvanicum)* seeds which provide a classical example of seed dormancy since

Progress in Botany, Vol. 46

the early work of CROCKER (1906). Cocklebur seeds are characterized by a polymorphism: The dispersal unit contains a smaller seed produced by the upper floret with a long and deep dormancy, and a larger, less dormant seed of the lower floret. When freshly harvested, both types of seeds exhibit primary dormancy. Only after a certain period of dry storage become their embryos nondormant (WAREING and FONDA 1957). Succeeding events diverge, as it is explained by ESASHI et al. (1983a). The lower seeds germinate under favorable conditions, they can also enter secondary dormancy when they are exposed to high temperature in hypoxia. The upper seeds, however, display coat-imposed dormancy, which gradually proceeds to secondary dormancy if they do not germinate under ordinary conditions.

The responses depend upon the different stages of dormancy. Whereas after-ripened upper seeds with coat-imposed dormancy are capable of germinating in response to chilling, anoxia, KCN or NaN_3, seeds with primary, and especially secondary, dormancy hardly respond. Thiourea, benzyladenin, or ethylene, however, releave primary, but not secondary dormancy. The most effective means for breaking secondary dormancy is the combined application of ethylene either with enriched O_2 or CO_2, or subsequent to a KCN treatment (ESASHI et al. 1983b). A further possibility is a thermoperiodic treatment by alternating cool and warm phases (ESASHI et al. 1983c).

A physiological explanation emerges only very slowly. The coat-imposed dormancy is related to the presence of an inhibitor within the embryo (WAREING and FONDA 1957). The concentration probably can be reduced by the interaction of several treatments. Primary and secondary dormancy have been related so far to an impairment of growth potential coinciding with the ability of the embryo to form chlorophyll (ESASHI et al. 1983b). Interestingly, the establishment of secondary dormancy is accomplished by a decrease in respiration and ethylene production, which seems to be necessary for germination.

An involvement of the alternative respiration in the germination-promoting system of secondary dormant cockleburs (ESASHI et al. 1982) resembles another situation, where NaN_3 stimulates germination of dormant wild oat *(Avena fatua)*, as described by UPADHYAYA et al. (1983). This effect is correlated to an increase of oxygen uptake by as much as 200% - 400%. It is completely abolished by salicyl-hydroxamic acid, an inhibitor of the alternative respiration (cf. HOCK 1984). ESASHI et al. (1983) give evidence for an increase of the adenylate pool and the energy charge during thermoperiodic treatment of secondarily dormant cocklebur seeds.

A great deal of work has been carried out on the control of dormancy by the interaction of promoting and inhibiting hormones. The most frequently cited examples deal with abscisic acid (ABA) as a dormancy-inducing hormone, and cytokinins and gibberellins as terminators of dormancy and stimulators of seed germination. However, the situation appears rather bewildering if different species are compared with respect to the hormones mentioned above.

The role of endogenous ABA in the induction of seed dormancy has been demonstrated in elegant studies by KARSSEN et al. (1983) for *Arabidopsis thaliana*. They used ABA-deficient mutants which are characterized by symptoms of withering and increased transpiration and a precocious germination of their seeds. In wild-type seed development, there is an induction of dormancy correlating with the presence of high ABA levels in the seeds. Reciprocal crosses of wild-type and ABA-deficient mutants permitted to identify the regulatory site of the dormancy induction and the dual origin of endogenous ABA in developing seeds. The genotype of the mother sporophyte regulates a sharp rise in ABA content during the early seed development (maternal ABA). The genotype of the embryo and endosperm is responsible for a second ABA fraction (embryonic ABA) The onset of dormancy correlated well with the presence of embryonic ABA fraction, but not with the maternal ABA. Dormancy developed independently from the absence or presence of maternal ABA in the seeds. In this respect, it resembled exogenously applied ABA, which did not induce dormancy in ABA-deficient seeds. The role of this ABA type is seen in a stimulation of the formation of a mucilage layer around the testa during the inhibition of the mature seeds.

From the experiments with *Arabidopsis* it can be deduced that after the onset of dormancy endogenous ABA is not required for its maintenance. This interpretation corresponds with other findings which question a central role of ABA in the regulation of dormancy breakage and of germination. TILLBERG (1983) showed in her experiments with *Rosa rugosa* that there is no difference in the decrease of ABA levels during early stages of stratification between samples kept at 40°C and the controls at 17°C. Although the latter declined to even lower values, they practically did not germinate, in contrast to the cold-stratified achenes with a very high yield. The author suggests that the ABA decrease is a prerequisite for activation of growth promotors that initiate dormancy breakage. TEGELMAN (1983) detected in the same plant material early changes in cytokinin-like substances during cold stratification, which are related to the breakage of dormancy.

A general picture which explains the role of endogenous factors being involved in the breakage of dormancy probably does not exist.

METZGER (1983) identified GA_1 in seeds (caryopses) of wild oat *(Avena fatua)* as the major biologically active gibberellin. Both unimbibed dormant and nondormant seeds contained similar amounts of GA_1, and the level declined in both dormant and nondormant seeds. The latter, when imbibed in the presence of CCC, an inhibitor of GA biosynthesis, germinated equally well as untreated seeds. In this case, GA clearly is no initiator of processes leading to germination.

Regarding the enormous progress of analytical techniques, it seems to be possible to assay simultaneously all three groups of endogenous hormones involved in the induction and breakage of dormancy. This strategy should yield a clearer picture of the general rules governing the participation of phytohormones and their interference with other factors in the control of dormancy.

The molecular basis for the entry into dormancy and its maintenance is still under debate. Already BONNER et al. (1963) suggested changes in the structure of chromatin as a template for RNA polymerase. This concept is supported by MORENO et al. (1983) who used differences in chromatin susceptibility to nuclease digestion as an indicator. Digestion proved to be more rapid in proliferating meristems from adventitious roots of *Allium cepa* bulbs, compared with dormant ones being dissected from the unsprouted bulb. The greater accessibility of chromatin in the nuclei of proliferating meristems is related to the lesser amount of dense chromatin and may be linked to their increased replication and transcription processes. EICHHOLTZ et al. (1983) observed a distinct change of polypeptide profiles in the embryonic axis of apple seeds *(Malus domestica)* during stratification at 50°C, which allowed a release from dormancy.

b) Bud Dormancy

Much less work has been carried out with respect to the dormancy of buds, probably because of greater difficulties being connected with the plant material and proper experimental conditions. The recovery of active growth in dormant buds of ash *(Fraxinus excelsior)* is highly correlated with a rise in water content above a critical treshold (COTTIGNIES 1983a). The bud break is preceded by an increase of the mitotic index in the shoot apex. Earlier, the dormant cells were arrested in the noncycling state of G_o (COTTIGNIES 1983b). It would be a promising goal to find the targets of phytohormones which interfere with the cell cycle during the regulation of dormancy.

In pea *(Pisum sativum)* buds, GA_3 strongly increases RNA synthesis. This effect was related to an enhancement of the template activity of chromatin (TOMI et al. 1983a) and to an enhancement of RNA polymerase activities (1983b). Loosely bound proteins, which can be extracted from chromatin, probably are responsible for GA_3-enhanced template activity due to their relaxing the interaction of core histones and DNA.

2. Seed Germination

The early development of higher plants is interrupted by a period of rest. During this stage, the embryonic plant is protected within its coverings and provided with food reserves. The entire unit, the seed, is characterized by a low water content and extremely slow metabolic activities, which enable the young organism to resist the vicissitudes of the environment. If a quiescent dry seed is exposed to water, it will germinate and establish a seedling. The transition from the dehydrated to the fully functional state requires a set of changes which operate at all levels of organization. Recent research has concentrated on a better understanding of the molecular events taking place during germination. Reviews were given by MAYER and POLJAKOFF-MAYBER (1975), BEWLEY and BLACK (1978), and KHAN (1982).

a) Nucleic Acid and Protein Synthesis

Protein synthesis does not occur in dry seeds. It is, however, a prerequisite for the radicle emergence, the first morphological criterium for seed germination. Therefore, it starts immediately when cells are sufficiently hydrated, and is most closely linked to early germination (DATTA et al. 1983). There are still uncertainties whether this protein synthesis is dependent upon prior RNA synthesis.

The presence of preexisting RNA, which can be used during the initial phase of germination, is supported by several pieces of evidence. DATTA et al. (1983) checked in wheat *(Triticum aestivum)* the effect of specific inhibitors, the efficacy of which was established. Although protein synthesis is strongly required for the early growth of embryonic axes, an inhibition of DNA, rRNA, and mRNA synthesis has no, or only little, effect on this stage of embryo growth. These observations suggest that preexistent mRNA can maintain the onset of growth. The occurrence in dry wheat embryos of poly (A)-rich RNA, which can be translated in a cell-free system, has been demonstrated by HANLEY-BOWDOIN and LANE (1983). A striking proportion of this message is channeled into the production of a single, agglutinin-like protein, which is synthesized shortly after the onset of water imbibition of dry embryos. The presence of stored mRNA in embryonic axes of *Vigna unguiculata* seeds has been shown by SUZUKI and MINAMIKAWA (1983). Poly (A)-RNA is newly synthesized to some extent there as early as 4 h after the onset of imbibition. In contrast to cotyledons, axes at early stages of germination may degrade the reserve proteins albumin and globulin, and utilize them for the synthesis of new proteins (MINAMIKAWA et al. 1983).

The presence of preformed mRNA does not exclude an early de novo mRNA synthesis. DOMMES and VAN DE WALLE (1983) followed the time course of polysome formation in the embryonic axis of germinating maize *(Zea mays)*. Analysis of polysomes prepared 0.5 and 1 h after the start of imbibition suggests that translation of newly synthesized mRNA probably occurs within the first hour of inhibition, i.e., well before the completion of initial water uptake. Similarly, in the radicle of germinating lentil embryos *(Vicia lens)* newly synthesized RNA of all major types is detected 3 h after the start of imbibition, again before completion of water uptake. The authors stress the simultaneous synthesis of all RNA species which, however, might differ from patterns observed in some other systems than maize embryos.

Transcription and translation are controlled by several exogenous and endogenous factors. Gibberellic acid (GA_3) stimulates polysome formation during the first 3 days of castor bean *(Ricinus communis)* germination (MARTIN and NORTHCOTE 1983). The main effect seems to be due to an increase in the amount of cellular mRNA. Again, during the period 0 - 24 h, polysome formation is independent of mRNA levels. In pea *(Pisum sativum)* seedlings, GA_3 enhances both DNA-dependent RNA polymerase I and II activities in apical buds (TOMI et al. 1983). In this system, the GA_3-induced increase in RNA synthesis is related to an increase of RNA polymerase activities, and to an enhancement of the template activity of chromatin.

The action of GA at the transcriptional level, which also extends to the rRNA, is in contrast to the increase of polymerase formation by kinetin in cucumber *(Cucumis sativus)* cotyledons (GWÓŹDŹ and WOŹNY 1983). This effect most likely acts at the translational level by increasing the rate of initiation through an enhancement of the affinity of ribosomes for the mRNA. It is not clear, at present, how such an interference relates to other cytokinin-controlled physiological processes, e.g., promotion of radicle growth, nutrient mobilization, or cotyledon expansion, which have been studied intensively during the past (for review see VAN STADEN 1982 and VAN STADEN et al. 1983).

The inhibitory effect of ABA on RNA synthesis of bean *(Phaseolus vulgaris)* embryo axes, as measured by $[^3H]$-uridine incorporation into total RNA, has been related to the degradation of RNA species (SEN et al. 1983). Moreover, light has been found to influence gene expression. This subject is treated in more detail later.

b) Control of Enzyme Activity: Biosynthesis and Secretion of α-Amylase

Since the classical experiments of HABERLANDT (1890), the formation and secretion of α-amylase by *aleurone cells* of germinating cereals has been intensively analyzed (for review see AKAZAWA and MIYATA 1982, AMRHEIN 1983). As gibberellic acid can substitute for the embryo in the induction of α-amylase (YOMO 1960, PALEG 1960), the following hypothesis has become generally accepted: During seed germination GAs produced by the embryo or the scutellum diffuse to the aleurone layer where they induce the formation of hydrolytic enzymes, among them α-amylase. This concept has recently been disproved by ATZORN and WEILER (1983).

By means of sensitive and selective immunoassays, they demonstrated in germinating barley caryopses *(Hordeum vulgare)* that the enzyme induction is due to a production of GA_4 in situ, i.e., by the aleurone tissue itself. The presence of the embryo is required only from about 24 - 36 h after imbibition. It does not provide, however, GAs or their precursors, although exogenous GAs (as GA_1) can induce GA_4 release and α-amylase production by isolated aleurone layers. GA_3 is only present in traces during the early period of germination. The nature of the embryo-derived stimulus is not yet clear. The authors favor a water-soluble factor, released by the embryo, since isolated embryos, incubated together with caryopses from which the embryos and scutella were excised, are still effective in producing the enzyme.

Since the production and the target sites of GA seem to be identical in the germinating barley caryopsis, this issue does not fit any more into the classical hormone concept. It does not disprove, however, the concept of phytohormones in general. They only resemble the cell and tissue hormones more closely than the secretory hormones both of which have been described in animal systems.

The increase of α-amylase mRNA as an intermediate step in the hormone-induced de novo synthesis of α-amylase (HIGGINS et al. 1983) has been compared with changes of other species of poly $(A)^+$RNA by means of cloned cDNA probes (BAULCOMBE and BUFFARD 1983). The time course of expression of α-amylase mRNA differs from other GA-regulated mRNA species by its longer persistence. These analyses revealed three regulatory classes of mRNA. This emphasizes the role of phytohormones as triggers which do not, however, change the general developmental pattern which is predetermined by the genome.

The effect of ABA, which strongly inhibits GA-induced α-amylase formation by aleurone layers (CHRISPEELS and VARNER 1967), is due to (1) a reduction of the level of GA_4 in the aleurone layer and (2) to an additional interference in α-amylase formation at a later stage (ATZORN and WEILER 1983), probably by an interference with the translation of α-amylase mRNA. Based upon inhibitor studies, VARTY et al. (1983) relate the ABA effect in isolated wheat *(Triticum aestivum)* aleurone layers to an inhibition of both the transcription and translation of α-amylase mRNA. Only the latter site of action is susceptible to reversal by ethylene.

The release of α-amylase from aleurone cells is regulated by cations. JONES and JACOBSEN (1983) demonstrated in isolated barley aleurone layers that Ca^{2+} selectively controls the secretion of two out of four α-amylase forms. It does not seem, however, to influence the isoenzyme pattern and activities in the aleurone cells themselves. The other α-amylase isoenzymes are secreted independently from Ca^{2+}. Furthermore, the secretion of other acid hydrolases is also strongly influenced by Ca^{2+}.

The intracellular route of amylase after biosynthesis has been studied extensively by the group of AKAZAWA in the scutellar epithelium of rice and other cereals. This tissue is the major site of amylase production during early stages of germination (for review cf. AKAZAWA and MIYATA 1982). Only in later stages, aleurone layers contribute to the production of this enzyme. α-amylase as a secretory protein is synthesized at the rER as higher molecular weight precursor and is cotranslationally inserted into the lumen of the ER. Cell-free experiments with isolated microsomes (MIYATA and AKAZAWA 1983) indicate that proteolytic processing and subsequent glycosylation occur at the luminal site of the ER. This mode of vectorial transport contrasts with the posttranslational sequestration of proteins by cell organelles, as shown for example for glyoxysomal and mitochondrial malate dehydrogenase (GIETL and HOCK 1982).

3. Photomorphogenesis

Photomorphogenesis is defined as the strategy of plant development in light, whereas etiolation ("skotomorphogenesis") is the strategy of choice in the absence of light (cf., MOHR and SHROPSHIRE Jr. 1983). Phytochrome is the most important and best-studied photoreceptor for many light-mediated developmental responses. The subject matter has been reviewed in a comprehensive way in the two volumes of *Photomorphogenesis* edited by SHROPSHIRE and MOHR (1983).

a) Properties of Phytochrome

The chromoprotein phytochrome consists of a linear tetrapyrrole chromophore covalently linked to a polypeptide. It acts as the primary photoreceptor for red light-mediated responses. The physiologically inactive (red absorbing) P_r form is transformed by red light into the physiologically active (far-red-adsorbing) P_{fr} form, which in turn is reconverted by far-red light into the P_r form. Since phytochrome does not possess any enzyme activity, photoreversibility has been for a long

I

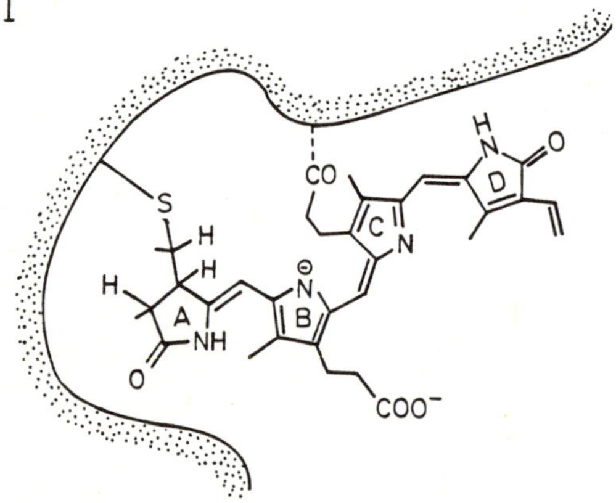

II

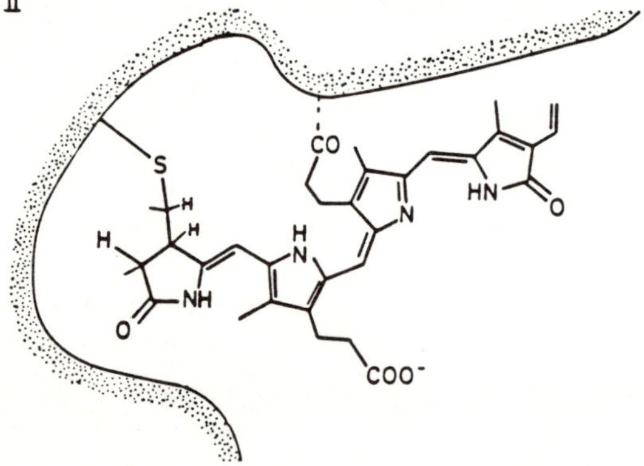

Fig. 1. Chemical structure of the P_{fr} *(I)* and the P_r *(II)* chromatophore. Adapted from RÜDIGER (1983) and RÜDIGER et al. (1983)

time the only criterium that could be used for a spectrophotometric assay. Only recently have very sensitive immunochemical methods been added. The purification and properties of phytochrome are treated by PRATT (1982). The molecular mechanism(s) of P_{fr} action is unknown.

Major breakthroughs in phytochrome research have been achieved lately in the chemistry of photoconversion and in the characterization of native phytochrome.

RÜDIGER et al. (1983) have identified the chemical basis for the photoconversion. It is due to a cis, trans (Z,E) isomerization at C15 of the chromophore (Fig. 1). The analogy to the visual pigment rhodopsin is apparent, in which cis, trans isomerization of the chromophore is considered as the primary photoprocess. RÜDIGER (1983) discusses the basis for a different dichroic orientation of P_r and P_{fr}, which has been described for several organisms (HAUPT 1973). It might be due to a different orientation of the whole phytochrome molecule within the membrane, or of the chromophore within the apoprotein. These changes are expected to be a part of the signal chain leading from P_{fr} to the biological responses. The localization of active phytochrome in the plasma-membrane or ectoplasm has been proved by WADA et al. (1983) for protonemata of the fern *Adiantum capillus-veneris*, where organelles and oil droplets can be removed from the apical region by centrifugation.

Before 1982, phytochrome preparations yielded the chromophore in a proteolytically degraded form ("120 kD" phytochrome) which differed from in vivo phytochrome in several respects, e.g., the phototransformation difference spectrum. Starting from P_{fr} and working in the presence of a (serine) proteinase inhibitor, the undegraded phytochrome with a monomeric molecular weight of 124,000 (instead of 120,000 or less) could be obtained (KERSCHER and NOWITZKI 1982, VIERSTRA and QUAIL 1982a,b). With respect to the spectral properties (Fig. 2), it is indistinguishable from in-vivo phytochrome if care is taken to avoid denaturation during the purification procedure (VIERSTRA and

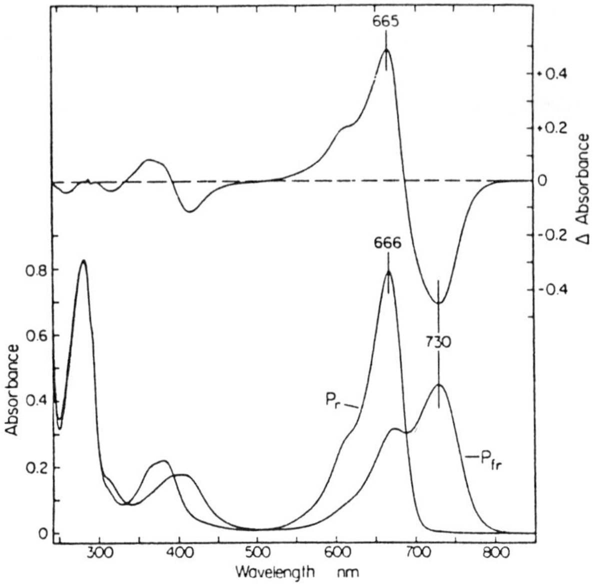

Fig. 2. Absorbance and difference spectra of native (124,000 MW) phytochrome. Absorbance spectra were measured at 3°C after saturating red (P_{fr}) and far-red (P_r) irradiation. The difference spectrum was determined by substracting the spectrum of P_{fr} from that of P_r. From VIERSTRA and QUAIL (1983)

QUAIL 1983). These technical improvements showed that P_r is more susceptible to proteolysis than P_{fr} and that the removal of a polypeptide segment by partial proteolysis changes the spectral properties. Because of the differential accessibility of the proteolytically sensitive site in the two forms, it can be concluded that this domain is involved in a change in molecular topology upon phototransformation. Data from BARON and EPEL (1983) on the differential sensitivity of the native P_{fr} and P_r to permanganate oxidation suggest that the chromophore of *both* forms is in a relatively protected environment. Only following partial proteolytic degradation of the phytochrome does the chromophore of P_{fr} become relatively more exposed. This also explains why it has not been possible, yet, to raise specific antibodies which discriminate between P_r and P_{fr}.

Such results require a reexamination of earlier data based upon experiments with 120,000 MW phytochrome, especially with respect to in vitro binding studies which yielded significant differences for P_r and P_{fr} (e.g., KIM et al. 1983, SUNDQUIST and WIDELL 1983).

The model of SONG (e.g., HAHN and SONG 1981) postulates an intramolecular proton transfer as primary photoreaction. The consequence is a reorientation of the P_r chromophore, which is located on the surface of the apoprotein, away from the surface, thereby exposing a hydrophobic area of the protein. This is then the active site for the binding of phytochrome to receptors in the plant cell. Since the model is based upon experiments with 120,000 MW phytochrome, it remains to be seen whether it is compatible with present views on native phytochrome. SONG (1983) points out that even the undergraded protein exposes the additional hydrophobic surface upon the $P_r \rightarrow P_{fr}$ phototransformation, similarly to the case of degraded phytochrome. However, in the former case, the P_{fr} chromophore might be shielded by a 6000 MW peptide segment. Membrane-binding studies of phytochrome will prove most valuable in the future. Some of the fastest P_{fr} responses can be related to changes of membrane properties which could provide the key for an identification of primary P_{fr} reactions.

The availability of antibodies, especially of the monoclonal type (CORDONNIER et al. 1983, NAGATANI et al. 1983), directed against phytochrome not only allows a very sensitive quantitation of the chromoproteins (SHIMAZAKI et al. 1983), it also provides the basis for a immunocytochemical localization in the tissue. SAUNDERS et al. (1983) examined epicotyls of dark grown pea *(Pisum sativum)* seedlings. Phytochrome is observed in both cortical cells and stomatal guard cells where it exhibits a homogeneous cytoplasmic distribution. After red light irradiation, phytochrome becomes transiently inaccessible to antibodies. This is thought to be due to an association of phytochrome with a subcellular component that interferes with antibody binding. If the phytochrome is maintained in the P_{fr} form for 10 min before fixation, at least a portion of the phytochrome pool becomes accessible to antibodies and assumes a "sequestered" distribution. Both red light effects are prevented or reversed by subsequent far-red light irradiation.

Although in many cases a perfect correlation between P_{fr} concentration and the physiological response has been verified, SMITH (1983) emphasizes the occurrence of other examples which lack such a correlation. He points out that dark- and light-grown plants in principle differ in their responses: dark-grown plants show high sensitivity to small changes in P_{fr} concentration at low P_{fr} levels, whereas light-grown plants show high sensitivity at high P_{fr} levels. SMITH concludes that P_{fr} is not the only active form of phytochrome. TANADA (1983a,b) on the basis of the nyctinastic closing of *Albizzia julibrissin* pinnulus and its delay by light suggests that in addition to phytochrome an unknown photoreceptor *(heliochrome)* with far-red, green photoreversibility is involved.

It is important to realize that in some examples the competence to respond to P_{fr} might be more essential than the absolute P_{fr} concentration. SCHAER et al. (1983) showed for coleoptiles of etiolated oat *(Avena sativa)* seedlings that the sites of photoperception for red-light induced cell elongation do not match the areas where phytochrome is more heavily concentrated. The situation calls to mind the discussion whether plant development is regulated by changes in the concentration of growth substances or by changes in the sensitivity to growth substances (CLELAND vs. TREWAVAS 1983b).

b) High-Irradiance Response (HIR)

The classical HIR, formerly called high energy reaction, is exhibited by etiolated tissues after their subjection over long periods to high irradiances. This treatment appears to be more effective in generating photomorphogenic responses than the use of inductive conditions (brief red pulses), as mentioned above. The action maxima of the HIR lie in the far-red and in the blue spectral regions, although the latter might involve an additional photoreceptor (cryptochrome). The HIR is characterized by a lack of photoreversibility and the failure to obey the reciprocity law. Although this continuous light response distinctly differs in its characteristics from the induction response (brief irradiation with red light), both types can be interpreted on the basis of the phytochrome system. The issue has been recently reviewed by FUKSHANSKY and SCHÄFER (1983).

The main difficulty in the interpretation of HIR results from the prolonged period of continuous irradiation during which further parameters dependent upon development are expected to change. WALL and JOHNSON (1983) performed a detailed analysis of phytochrome action in the HIR of mustard *(Sinapsis alba)* seedlings with respect to hypocotyl extension and anthocyanin synthesis. The P_{fr} level, determined as a function of irradiation time (0 to 24 h) and wavelength (in the range of 660 to 760 nm) is correlated with the physiological response. For a given irradiation period, the wavelength which produces the highest P_{fr} level yields the highest response.

At the onset of irradiation, red light (660 nm) produces most P_{fr}, and far-red light (760 nm) the least. Between 6 and 24 h after the beginning of irradiation, the maximum P_{fr} level is maintained by wavelengths between 710 and 720 nm, which delimit the region where the far-red maximum is seen in the action spectrum.

Although the absolute P_{fr} levels decrease with increasing irradiation, the physiological response increases. The most obvious influence is exerted by the fluence rate expressed by μMol quanta/$m^2 \cdot s$. If, for example, conditions are chosen to produce the same P_{fr} concentrations throughout a 24-h period by red or far-red light irradiation, respectively, much higher fluence rates (by the order of three magnitudes) are required in the far-red irradiation (HOLMES and SCHÄFER 1981). WALL and JOHNSON (1983) relate these differences to a difference in cycling rate, i.e., the rate at which the photoconversions between P_r and P_{fr} occur. Any treatment that results in a greater phytochrome cycling rate yields greater response, too. A critical consideration of cycling rates and their rigorous definition has been given, however, by FUKSHANSKY and SCHÄFER (1983). For the future, it seems to be important to consider the physiological response as a function of P_{fr}, wavelength, fluence rate, and time as parameters. It would be easier then to identify the postulated components which in addition to P_{fr} are needed to explain the observed response.

c) Cryptochrome

Blue light effects a great variety of physiological and biochemical responses which are mediated by a blue/UV-A light photoreceptor. The term cryptochrome reminds one of cryptogams, where it is particularly important, although this photoreceptor is found in the entire plant kingdom. According to MOHR and SHROPSHIRE (1983), the relative significant of this photoreceptor has decreased in the course of evolution, whereas the significance of phytochrome has correspondingly increased. The action spectrum shows a peak near 370 nm, and three peaks or shoulders between 400 and 500 nm. In higher plants, responses, such as phototropic reactions, opening of stomata, protoplasmic streaming, and sometimes hypocotyl growth are mediated by cryptochrome. In most cases, however, complex interactions with phytochrome and HIR are observed (cf. BEGGS et al. 1983, SCHÄFER and HAUPT 1983). In some cases, a blue-light effect can be ascribed solely to phytochrome (LERCARI et al. 1982).

Therefore, the study of cryptochrome seems to be easier in system where an interference with phytochrome is not observed, e.g., in fungi. Responses, such as induction of carotinoid synthesis, suppression and phase shifting of circadian conidiation, induction of protoperithecia formation, phototropism of sporangiophores and fruiting bodies are best studied. Although the precise nature of cryptochrome is not known, the most likely candidate is a flavin or flavoprotein photoreceptor. Reviews have been published by SCHMIDT (1980), SENGER (1980), SENGER and BRIGGS (1981), and SENGER (1982).

Current efforts in the characterization and identification make use of different approaches. In *Neurospora crassa*, several mutants with an altered sensitivity to light are available. HARDING and MELLES (1983) studied carotinoid biosynthesis and the phototropism of perithecial beaks. In some mutants both photoinduction processes are blocked. Therefore, the defects are either due to a defective photoreceptor or a modified component of the sensory transduction pathway. A study of such mutants allows the identification of the light receptor and the signal chain leading to the response. Earlier genetic approaches yielded correlations between a reduced light sensitivity and cytochrome b (BRAIN et al. 1977) or riboflavin (PAIETTA and SARGENT 1981) deficiency, respectively. These findings suggest a role for a cytochrome b-flavin complex in photoreception.

Care has to be taken, however, in the interpretation of data obtained with riboflavin auxotrophs which require more riboflavin for blue-light induced responses than for growth. It is necessary to differentiate between the need for elevated flavin levels for photoreception, or for subsequent dark reactions in the sequence leading to the response. HORWITZ and GRESSEL (1982) proved for *Trichoderma* that the extra supplement, required for maximum photoresponse, is not needed prior to or during exposure to light, but can be supplied later during the subsequent dark period. Therefore, it does not participate in photoinduction itself.

Further support for flavin involvement comes from the demonstration that riboflavin analogs can act as photoreceptors in a riboflavin requiring auxotroph of *N. crassa* (PAIETTA and SARGENT 1983). Two analogs with a red-shifted absorption (1-deazariboflavin and roseoflavin) replace the natural photoreceptor for two blue-light responses with respect to conidiation. Since they are ineffective for the induction of carotinoid synthesis, a different photoreceptor is postulated for this response. Although the data are consistent with an uptake of the flavin analog and its incorporation into a flavoprotein photoreceptor, it cannot be ruled out that the added flavins act like photosensitizing dyes. A well-known example is methylene blue, which photoreduces cyt b and acts as artificial photoreceptor for carotenogenesis (LANG-FEULNER and RAU 1975).

Nitrate reductase, a potentially photoactive flavoprotein, does not qualify as photoreceptor in *Neurospora*, since a number of photoresponses were found to be normal in nitrate reductase mutants (BELOZERSKAYA et al. 1982, PAIETTA and SARGENT 1983). On the other hand, nitrate reductase of spinach *(Spinacia oleracea)*, which contains FAD, cytochrome b and molybdenum as prosthetic groups, is influenced by blue light: After inactivation in vitro by acetylene, a reactivation by irradiation with blue light is observed, which is due to an excitation of FAD (MAURIÑO et al. 1983).

An analysis of *flavin* in *Phycomyces blakesleeanus* has been performed by DOHRMANN (1983). This fungus contains as endogenous soluble flavins only riboflavin, riboflavin-5'-phosphate and FAD, with highest (and sufficient) concentrations in the apical part of this sporangiophore to which blue-light sensitivity is restricted. In addition, membrane-derived heat-labile proteins which covalently bind riboflavin were found exclusively in this zone. In vitro binding studies with soluble flavins yielded specific, reversible binding to membrane particles in sporangiophores and mycelia. Riboflavin showed the highest binding affinity. It is not clear, yet, whether soluble flavins such as riboflavin or covalently bound flavins, which both are membrane associated, are involved in blue-light protoreception. Theoretically, both types qualify as candidates.

In the slime mold *Physarum polycephalum* one of the blue-light responses, which were studied in greater detail, is the contractile behavior of plasmodial strands (KOROHADA et al. 1983). Irradiation rapidly induces a prolongation of the period of the contraction-relaxation cycle – an effect that is related to mitochondrial respiration. Evidence is given for the involvement of pyruvate and α-glutarate-dehydrogenase complexes, two flavin linked enzymes, in the primary blue-light reaction.

Fern gametophytes have been known for a long time for their morphogenetic responses to blue light. Upon irradiation, the apical cell of the filament starts lateral swelling, which marks the transition from filamentous to two-dimensional growth. COOKE et al. (1983) presented evidence for the involvement of a plasma-bound flavin-cytochrome complex and ATP-driven proton pump in the initial events of this growth response to blue light, based upon specific inhibitors of these components.

d) Molecular Analysis of the Photomorphogenic Responses

Advanced techniques in molecular biology allow a new approach to the question, how light influences gene expression. Especially the analysis of phytochrome-directed responses has contributed to a better understanding of the general problem, how gene expression is regulated.

MÖSINGER and SCHOPFER (1983) investigated the effect of HIR on cellular polysome levels in mustard *(Sinapsis alba)* cotyledons. They concluded that in these organs photomorphogenesis is related to a massive increase of newly synthesized mRNA leading to a correspondingly increased recruitment of ribosomal subunits into polysomes. Therefore, photomorphogenesis would seem to involve mainly quantitative changes in the production of many proteins rather than qualitative changes in a few key enzymes.

The availability of complementary DNA (cDNA) clones representing individual mRNA sequences has greatly facilitated the analysis of how particular genes respond to light. By these molecular probes, changes in the rare mRNA sequences can be detected. DE VRIES et al. (1983) showed for pea *(Pisum sativum)* seedlings that the number of average-sized polysomal mRNA's, which should represent the number of expressed genes, is similar in light-grown and etiolated seedlings both in roots and in shoots. It was estimated at around 20,000. No evidence was obtained for massive regulation of rare mRNA sequences during photomorphogenesis. Only regulation of a limited number of abundant mRNA's was observed. This corresponds to a limited number of genes coding for more abundant mRNA's, which are expressed in shoots after illumination. In plant systems, the extent for regulation observed in the complex class of mRNA's, if occurring, is therefore smaller than that observed during animal development. THOMPSON et al. (1983) examined phytochrome effects on the abundance of transcripts from several nuclear and chloroplast genes in the buds of dark-grown pea seedlings where phytochrome strongly stimulates leaf development, and primary leaves of dark-grown mung bean *(Vigna radiata)* seedlings, which are less responsive to phytochrome. Apart from several patterns of induction which could be identified, the most general observation, concerning the differences between the responses of the two species, is related to the more extensive development of mung bean seedlings in the dark. Almost all of the prominent mRNA's that are assayed in these seedlings are present in 7-day dark-grown leaves. In contrast, most nuclear-coded RNA's in pea buds are present at very low levels initially and only increased upon illumination. Most chloroplast RNA's also increase significantly upon illumination in pea, but not in mung bean tissue. The hypothesis follows that these differences are attributable to differences in the stage at which development is arrested in the dark.

SASAKI et al. (1983) demonstrated the induction by phytochrome of ribulose-1,5-bisphosphate carboxylase/oxygenase (RuBPCase) in etiolated pea seedlings on the mRNA levels. This chloroplast enzyme is composed of eight large and eight small subunits. The large subunit is encoded by chloroplast DNA and synthesized on chloroplast ribosomes. The small subunit is encoded in nuclear DNA, synthesized as a precursor on cytosolic ribosomes, and transported into chloroplasts. P_{fr} increases both mRNA levels. The effect is reversed by far-red light. The lag time of induction, however, is more than 24 h. A more rapid increase of mRNA after a red light pulse (15 s) has been reported by GOLLMER and APEL (1983) for the light harvesting chlorophyll a/b protein (LHPC) of barley *(Hordeum vulgare)*. Already after 2.5 h, the increase of steady-state concentration of the mRNA sequence could be detected. Far-red light (5 min) reversed substantially the effect of red light. These results suggest that phytochrome rapidly affects the transcription of specific genes. Experiments with isolated nuclei can further substantiate this hypothesis. The distribution and expression of mRNA for RuBCase and LHPC within the developmental gradient in barley leaves has been studied in greater detail by VIRO and KLOPPSTECH (1983). STIEKEMA et al. (1983) demonstrated a rapid effect of phytochrome on the expression of genes for the small subunit of RuBCase and LHPC in *Lemna gibba* by the aid of hybridization probes. The increase of the messengers could be detected 30 min after the start of red illumination. Phytochrome is expected to change either the transcription rate or the rate of degradation of these mRNA's.

Blue-light effects which are mediated by cryptochrome could also recently be related to a regulation of transcriptional processes (PUTZER et al. 1983). Blue light, which inhibits the plasmodium-spherule transition in the slime mold *Physarum polycephalum*, suppresses also the synthesis of spherule-specific mRNA together with the synthesis of spherule proteins. Only three protein components are induced by blue light, indicating that illumination does not induce a novel differentiated plasmodial state.

A most interesting question concerns the control of mRNA for the photoreceptor apoprotein by light. Major advances have been made with respect to phytochrome mRNA levels in *Avena sativa* seedlings by COLBERT et al. (1983) and GOTTMANN and SCHÄFER (1983). The former group proved that phytochrome exerts autoregulatory control over its own translatable mRNA level. Irradiation of etiolated seedlings with 5-s red light induces a decline in this mRNA that is detectable within 15 - 30 min. It results in a >95% reduction within 2 h. This effect is reversible by a far-red impulse. The observed change in translatable phytochrome mRNA level is one of the most rapid phytochrome-induced alterations in any cellular mRNA yet recorded. According to these results, it is necessary to revise existing concepts of how phytochrome concentrations are modulated in vivo: Conversion to P_{fr} not only enhances degradation of the chromoprotein but also decreases the rate of synthesis due to a concomitant reduction in the level of translatable phytochrome mRNA. Of course, cDNA probes will have to be used to confirm these results for steady-state levels of the phytochrome messenger. GOTTMANN and SCHÄFER (1983) found substantially reduced rates of phytochrome synthesis in white light, as compared to dark-grown oat seedlings, by measuring the activity of mRNA coding for the phytochrome apoprotein. Under continuous white light, also very slow rates of phytochrome destruction are inferred. Phytochrome levels were simultaneously established by radioimmunoassays and spectrophotometrical methods. Again, it was concluded that the rate constants of the phytochrome dark reactions are different in the dark and during white light irradiation. In this system, phytochrome was inferred as the photoreceptor which controls phytochrome synthesis.

The alteration of gene expression by phytochrome emphasizes the necessity of identifying the primary reaction(s) after light-signal perception. This subject matter deserves special attention, since phytohormones are also known to influence gene expression. During the past, complex interactions between phytochrome and phytohormones have been postulated. This opinion is not supported by recent results.

In maize *(Zea mays)* coleoptiles, red light inhibits the production of free IAA at the coleoptile tip; at the same time it stimulates the elongation in the apical region (IINO 1982). Since the endogenous IAA concentrations is not in the inhibitory range, stimulation of coleoptile elongation cannot be the result of changes in free IAA levels. TONG et al. (1983) carefully analyzed the effect of exogenously applied zeatin and photomorphogenetically active light on various responses of mustard seedlings. The authors conclude that the action of phytochrome on developmental processes is not related to cytokinin levels. Similarly, in a three-factor analysis of the rate of chlorophyll a accumulation in excised mustard cotyledons, it has been concluded that P_{fr}, kinetin, and an excision factor operate independently through a common intermediate (presumably δ-aminolevulinate) in the biosynthetic chain leading to chlorophyll (KASEMIR and MOHR 1982). FREDERICQ et al. (1983) studied the synergism between GA_3 and P_{fr} in the light-requiring seeds of *Kalanchoe blossfeldiana*. GA_3 alone does not induce germination, but it reduces dramatically the threshold level of P_{fr} required for germination. Besides an amplification mechanism exerted by GA_3, the possibility is discussed that the phytohormone increases the number or efficiency of P_{fr}-receptor sites.

For the time being, it cannot be deduced that the phytohormones mentioned above are involved in the early sequence of events initiated by P_{fr}. These results rather suggest an independent coaction than an interaction of P_{fr} phytohormones.

4. Elongation Growth

Growth of the plant is based on cell division and cell elongation. Major efforts have been directed at a better understanding of the molecular events taking place during elongation. The majority of studies in this field have concentrated on shoot axes in developmental stages when cell divisions are completed.

The enlargement of plant cells is governed by two factors: Water uptake (regulated by hydraulic conductivity and the osmotic difference between the cell contents and the water source) and the extensibility of the cell wall which surrounds the protoplast to accomodate the inflowing water. Most studies have used the auxin-stimulated promotion of cell elongation to analyze details of this growth reaction. COSGROVE and CLELAND (1983) used the pressure probe technique to demonstrate that water transport does not limit cell expansion. Auxin does not alter the hydraulic conductance, either at the cellular or the whole tissue level, as measured in pea *(Pisum sativum)* internode tissue. Similar conclusions have been reached by COSGROVE (1983) on the basis of growth inhibition by blue light in seedlings. Blue irradiation suppresses growth by acting on the yielding characteristics of the cell wall. The hydraulic conductivity does not impede the rate of cell enlargement.

The auxin-stimulated wall extension is the fastest response observed so far. Since acidic solutions promote short-term elongation as effectively as IAA (BONNER 1934, RAYLE and CLELAND 1970), it has been proposed that the primary effect of auxin is stimulation of an outwardly directed proton pump in the plasma membrane of target cells (HAGER et al. 1971, CLELAND 1971). As predicted by this acid growth hypothesis, auxin stimulates cell wall acidification (CLELAND 1973, RAYLE 1973). A loosening of the cell wall by direct breakage of acid-labile cell wall bonds or by activation of wall-loosening enzymes located in the cell wall would be the immediate consequence. After this rapid but short-lived response other processes of greater duration take place which require RNA and protein snythesis and have been already noticed by KEY (1969). A differential tissue response to auxin must be taken into account (PEARCE and PENNY 1983).

a) Fast Membrane Reactions

The view of primary processes during IAA induced elongation growth has experienced a considerable shift, mainly by the noting of an involvement of Ca^{2+} ions. BRUMMER and PARISH (1983) suggest that the primary effect of auxins is to lower the cytosolic pH. This could be achieved by a binding of auxins to their receptors on the plasma membrane and the ER, which induces an increase of the cytosolic levels of free Ca^{2+}. This Ca^{2+} is taken up by mitochondria, which simultaneously secrete protons into the cytosol. The externally directed proton pump would be stimulated by the lowered cytosolic pH rather than directly by auxin. The next consequences are: (1) a change in transmembrane $\Delta\Psi$ leading to changes in membrane structure, which effect permeability and enzyme activity, and (2) a change in transmembrane ion gradient influencing as a second messenger processes such as secretion, protein synthesis and gene transcription. HERTEL (1983) constructed a model which stresses the similarity between the auxin response system and the auxin transport, both of which involve the mediation of Ca^{2+}. The primary auxin action is expected to take place at an inner compartment, e.g., the tonoplast and perhaps the ER, which contain high concentrations of Ca^{2+}. The membrane would harbor a Ca^{2+}-

gate at an auxin exit carrier into the vacuole. Passage of cytosolic IAA opens a transient "leakage" gate for Ca^{2+} which is released into the cytosol and might stimulate a vesicle flow from Golgi to the wall, resulting in an enhanced elongation rate. The increased Ca^{2+} concentration would immediately increase H^+ extrusion, too, which however is not rate controlling and is not in the main causal chain from auxin to elongation. The model additionally contains an electrogenic symport cotranslocating $2H^+$ with 1 IAA back into the cytosol. Interestingly, polar auxin transport operating in the plasma membrane can be described by a similar but not identical mechanism. Auxin transport requires a symport and an exit carrier. The accumulation of auxin into the cytosol is driven by the strong promotive force, due to the pH gradient across the plasmalemma, and mediated by the symport carrier. The efflux of the auxin anions takes place through polarly distributed exit carriers in the plasmalemma. It is coupled to a Ca^{2+} counterflow. The transport mechanism at the plasmalemma is a mirror image to the primary mechanism of auxin action at the tonoplast (or ER). The carrier-mediated uptake of IAA into closed and sealed vesicles derived from plasma membranes of zucchini *(Cucurbita pepo)* hypocotyls has been demonstrated by HERTEL et al. (1983); it is driven by a pH gradient. The carrier-mediated efflux of IAA is inhibited by 1-N-naphthylphthalamic acid and by triiodobenzoic acid. Evidence has been obtained for two different binding sites on the same carrier complex (DEPTA et al. 1983). Presumptive auxin transport carriers could be localized by means of immunocytochemistry (monoclonal antibodies to auxin anion carrier) in the plasma membranes at the basal ends of parenchyma cells sheating the vascular bundles of pea stem tissue (JACOBS and GILBERT 1983). Detailed binding studies of auxin (1-naphthylacetic-acid) to membrane vesicles from tobacco *(Nicotiana tabacum)* pith callus yielded the binding of four hormone molecules to one receptor molecule (MAAN et al. 1983). It is evident that the presence of hormone receptors is an absolute requirement for the specificity of the hormone response, and it might be conceivable that receptor concentration could be influenced by the hormone itself.

b) Changes in RNA Synthesis

Continued elongation growth requires RNA and protein snythesis. The longer-term effects of auxin are clearly linked to changes in gene expression. Earlier investigations by the group of KEY used specific inhibitors of protein and RNA synthesis to demonstrate this relation. Recently, a more direct approach has been chosen relying on advanced techniques of molecular biology. WALKER and KEY (1982) isolated auxin-responsive cDNA clones from a cDNA library prepared from elongating soybean *(Glycine max)* hypocotyl poly $(A)^+$ RNA. By the aid of cloned cDNA blot analyses, an auxin-mediated selective increase in mRNA concentration could be demonstrated in tissue undergoing enhanced rates of cell elongation. The hormone selectively and dramatically alters the concentration of a few poly (A^+)RNA sequences in growing tissue. It will be most interesting to see how auxin alters the expression of the corresponding genes. mRNA for ribosomal proteins could be identified among the gene products, which are greatly enhanced after auxin induction (GANTT and KEY 1983). The dramatic increase in nucleolar volume after auxin treatment of quiescent mature hypocotyl tissue of soybean could be related to a large increase in ribosomal proteins and rRNA, which accumulate and assemble in the nucleoli in the form of preribosomal particles (CHEN et al. 1983). The coordinate expression of different genes in response to the hormone signal deserves special attention.
A rapid alteration of the population of translatable mRNA by IAA has been demonstrated by ZURFLUH and GUILFOYLE (1982) in elongating maize *(Zea mays)* coleoptile sections. Within 10 min of hormone treatment, the

amounts of two in-vitro translation products are increased, suggesting a comparable increase of the corresponding messengers within the latent period which precedes the onset of cell elongation. It is not clear, yet, whether the observed translation products are involved in the auxin-induced cell extension.

c) Changes in Cell Wall Structure

In spite of advances in the knowledge of plant cell wall composition, there is still a considerable lack of information as far as the changes are concerned that enable wall extensibility during elongation. Wall extensibility depends upon the rate of biochemical wall-loosening events and upon the amount of extension per unit stress that occurs as a consequence of each wall-loosening event (CLELAND 1971). Wall loosening can be induced by an acidic wall pH, although in some systems growth can be promoted in the absence of detectable proton excretion by IAA (see e.g., POPE 1983). In the common case where acid-induced wall loosening is involved, the capacity for acid-induced wall loosening (CAWL) might be an important factor in the control of growth rate. According to CLELAND (1983a), CAWL might be a function of the cell wall proteins rather than the polysaccharides. The decrease of CAWL in the presence of the protein synthesis inhibitor cycloheximide should be noticed in this context, and it is used as an argument for a direct response of a change in CAWL to auxin rather than to elongation or to a auxin-induced H^+-pumping.

d) Growth Response to Hormones Other than IAA

Elongation of shoot axes is not only controlled by auxin but also by other phytohormones. Gibberellin has been recognized for a long time as another principal regulator of stem elongation. However, no clear picture of the kind of interaction between the two types of phytohormones has emerged. During the past, synergistic, additive, subadditive, and antagonistic relationships have been described. ADAMS and ROSS (1983) demonstrated the promotion of growth in stem segments of oat *(Avena sativa)* by GA_3 and IAA, when given separately. But when applied together, IAA strongly inhibits GA_3-induced growth. In addition, IAA modifies GA_3-induced growth by shortening the response time to GA_3. The data show that IAA and GA_3 individually induce divergent and unnatural growth patterns, but together they interact to give an intermediate growth rate and cell pattern more typical of normal growth. The hypothesis is put forward that in the presence of endogenous IAA a factor, required for GA_3-induced growth, is more rapidly produced in the tissue.

Ethylene exerts an entirely different influence on growth of most plants: Primary elongation is changed to a more isodiametric expansion. In excised internode segments of etiolated pea *(Pisum sativum)* seedlings, the alteration of growth pattern has an initial lag phase of ca. 1 h, but the reorganization appears to be completed by 3 h (EISINGER et al. 1983). The initial inhibition of elongation is supposed to result from a combination of longitudinally deposited cellulose microfibrils and decreased xyloglucan turnover. TAIZ et al. (1983) investigated the role of wall acidification and acid-induced wall extension in the presence of ethylene. The effect of ethylene is not associated with any alteration of auxin-stimulated proton extrusion. Lateral expansion in response to ethylene requires an acidified wall, since it is prevented by strong neutral buffers and by the ATPase inhibitor orthovanadate. Obviously, ethylene dramatically reduces the capacity of pea stems to elongate in response to acid pH. Simultaneously, the sensitivity of the outer cortical parenchyma cells to acid in transverse direction is increased. The reduction in transverse microfibrils coupled with the increased turgor pressure results in lateral cell expansion. A detailed review on the pea system, discussed so far, has been given by EISINGER (1983).

An entirely different response to ethylene has been reported for some aquatic plants where elongation growth is stimulated by ethylene. MÉTRAUX and KENDE (1983) identified the internode elongation in rice *(Oryza sativa)* after submergence as an ethylene effect. When ethylene synthesis in partially submerged plants was blocked with amino-oxy-acetic acid and aminoethoxyvinylglycine, internode elongation was inhibited. This growth inhibition was reversed when ethylene biosynthesis was restored with the ethylene precursor ACC. The increase of ethylene synthesis after submergence is due to an enhanced ACC formation. Elongation of coleoptile segments from rice is promoted by exogenous ethylene as well as by IAA (ISHIZAWA and ESASHI 1983). The effects are additive. The data suggest that the elongation of rice coleoptiles is regulated cooperatively by endogenous ethylene and auxin. In the aquatic plant *Nymphoides peltata*, ethylene only promotes growth of petioles in the presence of auxin; IAA alone induces a lesser promotion (MALONE and RIDGE 1983). Both growth promotors stimulate medium acidification, which is taken as support for the acid growth hypothesis. The data favor the same mechanism of wall loosening during promotion of growth by auxin and ethylene, although different pathways must lead to this step. The ecological advantage of ethylene as growth promotor in aquatic plants is related to the ability of these plants to accomodate to change in water depths.

5. Flowering

The transition from vegetative to reproductive growth is a radical cut in the life of a seed plant. It involves distinct changes in the apical meristem, e.g., a rise in DNA synthesis and mitotic activity. During the broadening and elongation, floral primordia develop. Those cases are most suitable for analysis in which a change of environmental factors, such as light or temperature, trigger the floral response. In some plants, however, floral morphogenesis is determined entirely by endogenous factors.

a) Photoperiodism

Photoperiodism is the response to the length of the day. In *short-day* plants, flowering is triggered or promoted when days are shorter than a critical duration, in *long-day* plants, however, when days are longer than a critical duration in 24-h cycles. The perception of day-length is largely confined to younger leaves. Phytochrome is the photoreceptor responsible for *induction* processes in the leaves. They generate a (still unknown) transmissible stimulus ("florigen") which is conducted to the shoot apical or lateral meristems where the *evocation* takes place. The subject matter has been reviewed by VINCE-PRUCE (1983a).

BÜNNING (1936) recognized that time measurement depends upon a comparison of the external light 24-h periodicity *(circadian rhythm)*. Special attention has been given by VINCE-PRUCE (1983b) to the perception of light-dark transitions which serve as *Zeitgeber* to synchronize circadian rhythms, and which are required for the timing of critical night-length. Evidence has been provided that changes in irradiance rather than changes in light quality (P_r/P_{fr} ratio) are used for the detection of light-dark transitions. These transitions could be sensed by a decrease in P_{fr} levels in darkness, and dark-light by a rapid formation of P_{fr} in the light.

The first part of this hypothesis, however, does not correspond then to the relative stability of P_{fr} in light-grown plants. The author discusses the existence of kinetically separate pools of phytochromes with different physiological functions: (1) an unstable pool of P_{fr} coupled to time measurement; a decrease in irradiance below a threshold level then could constitute the light-off signal, (2) a stable pool of P_{fr} required for the induction of flowering. This property of phytochrome, which is used for a discrimination between light and dark, clearly differs from other functions: the detection of light quality in response to shade (see e.g., HOLMES 1983) or to detect light direction (see e.g., HAUPT 1983).

A great number of experiments on the flower induction of short-day plants substantiates the concept that night-length is measured rather than the length of day. An investigation of the floral inhibition in the duckweed *Lemna paucicostata* by OOTA (1983) comes to a different conclusion. This short-day plant has a critical day-length of ca. 14 h. Experiments on the daily night interruption by a light pulse were performed with different periods of day-length (12-, 9-, 6-, 3-, or 1-h days). Flowering was prohibited only when night interruption was linked to either the preceding or the subsequent main photoperiod to form a *skeleton photoperiod* whose length was equal to, or longer than, 14 h. A skeleton photoperiod which begins and ends with a short light pulse is usually physiologically equivalent to a complete photoperiod of the same length. Therefore, the author assume that a light pulse falling on the inhibition zone prohibits flowering, not because it curtails the dark period but because it is associated with the adjacent main photoperiod. A day-length measurement on the basis of light-on oscillation (light-on clock) is postulated.

b) Involvement of Growth Regulators

Control of flowering by the exogenous application of hormones or hormone-like factors is of considerable practical importance. The promotive effect of gibberellins under noninductive conditions is well known for plants which need long-day and/or low temperature treatment for their flowering, although these phytohormones certainly are not identical with florigen. In several conifers enhanced flowering was observed after gibberellin treatment. In Sitka spruce *(Picea sitchensis)*, $GA_{4/7}$ was applied by stem injections (PHILIPSON 1983). Only in conjunction with cultural stress treatments, as heat and drought, did $GA_{4/7}$ treatment produce a distinct stimulation of the production of male and female cones. The role of heat and drought is seen as root stress factors. A reduction in root activity could be involved in the induction of flowering. It is not known, however, how the environmental stresses facilitate the response to $GA_{4/7}$. Interestingly, in systems such as *Pharbitis nil* or *Lemna paucicostata* , a strong correlation has been observed between flowering and a suppression of root elongation under a wide variety of environmental factors. The metabolism of GA_4 in Norway spruce *(Picea abies)* was studied by DUNBERG et al. (1983). It is converted into GA_2 and GA_3 as the major metabolite. Cultural treatments known to enhance flowering slow down the metabolization of exogenous GA_4. A promotion of flowering through the build-up of this relatively apolar phytohormone by interfering with its conversion to more polar GA might well reflect the situation in vivo. – Recent interest in the promotion of flowering in conifers is related to the observation that an early symptom of the prevailing forest diseases is a premature flowering.

When GA_3 application to shoot apex induces flower formation, its site of action is expected to be in the axial region where the rate of cell division and DNA synthesis are significantly increased. In the long-day plant *Silene armeria* grown under noninductive conditions, treatment with GA_3 caused stem elongation, but no flowering (BESNARD-WIBAUT et al. 1983). The hormone strongly stimulated mitotic activity and increase in the proportion of nuclei at the 4C level (S + G_2 phase of the cell cycle) in the rib meristem, and to a lesser extent in the lateral zone. The failure of the axial zone to respond explained the ineffectiveness of GA_3 in inducing flowering.

The promotive effect on flowering of salicyclic acid and its analogs, such as benzoic acid, in some plant systems is known through the investigation of CLELAND and AJAMI (1974). Endogenous levels of benzoic acid in *Lemna* species have been determined by FUJIOKA et al. (1983). No direct correlation could be found between the endogenous levels and flowering. In two strains of *L. paucicostata*, which are very sensitive to benzoic acid and flower readily on application of this compound, a strong synergistic effect of cytokinins, which alone induce no flowering, has been observed (FUJIOKA et al. 1983). Gibberellins, IAA, and abscisic acid were inhibitory. The authors assume that the balance between endogenous levels of benzoic acid and phytohormones contributes to the regulation of flowering in these systems. Similar findings were reported for *Lemna minor*, where structure-activity relationships for the flower-inducing activity of substituted benzoic acid derivatives in the presence of cytokinin were elucidated (WATANABE et al. 1983).

An interesting hypothesis, how growth substances could influence flowering, points to a redirection of the flow of assimilates. This *nutrient diversion hypothesis*, which was proposed by SACHS and HACKETT (1977), suggests that assimilate supply limits reproductive development, and that promotion of development should result from treatments that increase assimilate supply to the inflorescence, and from treatments that decrease assimilate supply to competing sinks. Evidence for this concept is claimed by LEONARD et al. (1983) from their studies on the inflorescence development in tomato plants *(Lycopersicon esculentum)*. The promotion by an exogenous application of benzyladenine + $GA_{4/7}$ is preceded by an altered pattern of assimilate distribution: Assimilate supply to the treated inflorescence increases concomitantly with a decrease in the assimilate import into the apical shoot, reflecting a competition between these two plant parts. It must be realized, however, that by these data rather the development of young inflorescences than the evocation is concerned.

c) Changes in the Apical Meristem During Transition to Flowering and Flower Differentiation

During the period of evocation, one or more cycles of synchronous cell division are observed (BERNIER 1971). With the aid of water-soluble fluorescent probes, GOODWIN and LYNDON (1983) demonstrated in *Silene coeli-rosa* that a strong synchronization of cell division in the apex is due to an increased cell isolation, leading to the de-repression of cell division. The alternative, the formation of a syncytium by a major increase in cell/cell permeability, could be ruled out. It can be speculated that the restricted movement of a division inhibitor is the signal triggering synchrony.

Another interpretation, which does not necessarily contradict the experimental observations mentioned above, relates the evocation of the apical meristem to an aging process (cf., BERNIER 1979). This would explain flowering in such species as finally become reproductive regardless of the environmental factors required for flowering in earlier stage of development.

LYNDON et al. (1983) detected in shoot meristems of *Synapsis alba* changes in protein composition during floral evocation. This effect is observed 50 to 52 h after the beginning of the inductive photoperiod in the meristem, distal to the primordia. It occurs *before* morphological changes in the meristem, and before initiation of flower bud primordia. The view is supported that in evocation there is not simply a general activation of protein synthesis, but an expression of genes specific to flowering.

The development of unisexual flowers in the tassel and ear of maize *(Zea mays)* involves selective developmental abortions in bisexual initials. CHENG et al. (1983) showed that male flowers arise in the tassel through selective vacuolation and abortion of cells of the early gynoecium. The single female flower in each ear spikelet arises through vacuolation and abortion of stamens in the upper flower, and the repression of growth and the eventual regression of the lower flower in each spikelett. Apart from the practical implications with respect to a production of male-sterile plants, it seems to be most important to identify those factors which trigger the selective organ abortions.

An important role in the future could be played by in vitro studies allowing analysis of the induction of flower differentiation. RAJASEKARAN et al. (1983) used callus from hypocotyl explants of cucumber *(Cucumis sativus)* grown in a medium with benzyladenine and 2,4-D. After transfer to a medium without hormones, somatic embryos and adventitious buds are formed. After a second transfer, normal plantlets develop which flower within 2 weeks. The flowers are either staminate or pistillate. COUSSON and TRAN TRANH VAN (1983) studied the nature of organs neoformed from tobacco *(Nicotiana tabacum)* thin cell layers excised from the basal segment of floral ramifications at the terminal green fruit stage of the inflorescence. During a critical initiation phase (from day 6 to day 11 of culture) light and carbon energy sources are needed for flower initiation. Light per se is not sufficient to induce flower differentiation. A combination of glucose and sucrose almost substitutes for the light requirement. The effect of light is seen in an action on the energy-dependent sugar uptake and metabolism, and on the increase of reducing potential of chloroplasts.

In conclusion, the events following the induction of flowering cannot be ascribed to the action of a single substance, but require the interaction of different hormones and nutrients as well, which can become limiting to flowering in different plants under different conditions.

6. Senescence and Aging

The life span of multicellular organisms is restricted within more or less narrow limits by genetically fixed controls. This is evident for annual or biennial plants. But, in principle, it also holds true for perennials, although there might be different mechanisms restricting lifetime. The continuous turnover of organisms is necessary if natural selection and consequently evolution should work. Without a limited lifetime of individuals there would be no chance for the adaption of populations to changing environments.

The deteriorative processes, which are initiated at full maturity and ultimately lead to death, are summarized by the term *senescence*. Consequently, senescence represents the final phase of development. Although the terms aging and senescence are frequently used synonymously, *aging* is referred here more aptly to the wider array of processes of accruing maturity with the passage of time (MEDAWAR 1957). LEOPOLD (1975) considers aging as the natural consequence of the absence of the common sources of biological turnover (i.e., predation, stress, or programmed senescence). Clearly, programmed senescence is more accessible for experimental research than aging. Higher plants provide a wealth of examples for the former type. Programmed senescence not only refers to entire plants, as in annuals and biennals, but also to organs which are shed or replaced during the life cycle. The latter aspect is dealt here with respect to cotyledons, mature leafs, flowers and fruits. More general reviews were recently given by LEOPOLD (1975), WOOLHOUSE (1980), and TANNER (1980). The abscission is treated by ADDICOTT (1982) and SEXTON (1982), leaf senescence by THOMAS and STODDART (1980). Advances in the senescence of fungi, which are beyond the scope of this article, are summarized by ESSER and TUDZYNSKI (1980) and MUNKRES (1983).

a) Seeds

The survival of dormant or quiescent seeds is limited by their *viability*. The time period after which a gradual loss can be observed, varies for different species. It also depends upon the physiological stage during the long-term storage. ROBERTS (1973) recognized two categories of seeds with respect to their survival in storage: *Orthodox seeds* can be dried to low moisture contents without damage. The period of viability increases approximately exponentially with decrease in temperature and moisture content. Probably seeds of all available crops and the majority of herbaceous species are in this category. *Recalcitrant seeds* are killed by desiccation. They last longest when fully hydrated. Even then, the storage lasts only for a few weeks or months. Several woody species belong to this group.

For lettuce achenes *(Lactuca sativa)*, IBRAHIM and ROBERTS (1983) demonstrated a critical moisture content of ca. 15%. Below this level, the rules for the orthodox type apply, and there is a detrimental effect of oxygen. Above this level oxygen becomes beneficial to survival, and the decrease in longevity with increase in moisture content becomes less marked. By a careful control of all major factors which influence viability IBRAHIM et al. (1983) obtained data which support the concept of *Villiers* (1983). Repair and turnover mechanisms are absent from dry seeds but are activated on hydration. The critical level in lettuce would be ca. 15% moisture content. Below this level there would be only subcellular damages which accumulate at a rate dependent on temperature, moisture content, and oxygen partial pressure. Above this level the rate of subcellular damage would increase but, simultaneously, repair mechanisms would begin to operate which depend on aerobic conditions. This explains the considerable longevity of fully hydrated seeds.

These considerations correspond to the high viabilities of lettuce achenes (up to 10 months) when maintained in the state of secondary dormancy, i.e., under fully imbibed conditions at 25°C in the dark as compared to samples stored in the dry state (POWEL et al. 1983). During secondary dormancy the achenes do not respond to red light or gibberellic acid, but to a combination of benzyladenine and red light. During the dormant state the metabolism is reduced to very low levels.

These data are strongly supported by experiments on the storage of *Datura ferox* seeds (SANCHEZ and DE MIGUEL 1983). Again, the viability in hydrated seeds is related to metabolism-dependent repair mechanisms.

The loss of viability in the orthodox seeds of Norway maple *(Acer platanoides)*, stored after imbibition at higher temperatures, is correlated to an electrolyte leakage (PUKACKA 1983). Under these conditions, the proportions of different phospholipid species changed: The decrease of phosphatidyl choline, the main membrane component, is probably responsible for the loss of integrity.

The loss of viability is due to a deterioration which is manifested in a variety of ways, e.g., chromosome aberrations, depletion of nutrients, reduced capacity for protein synthesis, low ATP content, fatty acid peroxidation, irreversible membrane leakages (BEWLEY and BLACK 1982), or an enrichment of inhibitors, e.g., phenolic acids (SREERAMULU 1983). Any one of these events might be sufficient to impede germination.

b) Seedlings

Cotyledons have been used extensively in physiological investigations of senescence, probably because of the ease and the short time needed to obtain suitable plant material. Most studies concentrate on systems where senescence can be reversed by experimental manipulation (e.g., wounding, light, or hormone treatment). In principle, the senescence of cotyledons from epigeal seedlings follows the same pattern as in foliage leaves. Chlorophyll loss is accompanied by a burst in CO_2 output, an increase in ethylene production, and an accumulation in H_2O_2. Most of the present data support the concept that senescence is controlled by genetically programmed mechanisms.

Evidence is piling up that the generation of toxic oxygen radicals, which deteriorate the structure and function of plant tissue, is an important step in the chain of events leading to the senescence of cotyledons (FERGUSON et al. 1983, PERCIVAL and DODGE 1983). It is not clear, however, whether such a process is due to a loss of formerly present protective mechanisms or to an increased production of oxygen radicals.

Senescence is not a mere weakening of cellular functions. It rather involves the appearance of new compounds. SKADSEN and CHERRY (1983) demonstrated in light-grown soybean seedlings *(Glycine max)* that the loss of chlorophyll, RNA, protein, and polyribosomes during senescence of cotyledons is accompanied by quantitative shifts in individual species of mRNA population or their translation rates. Therefore, senescence does not involve a complete gene repression or depression. Even after 90% of nucleic acids and proteins have been lost, the senescence can be reversed by epicotyl removal. As shown by in-vivo and in-vitro studies, rejuvenation also causes a reversal in the level of various proteins, i.e., an increase of roughly half of the products and a decrease of ca. one-third.

Similar shifts in polypeptide patterns have been described for aging pea *(Pisum sativum)* epicotyls (SCHUSTER and DAVIS 1983a). This process is reversed by wounding and by auxin treatment. In general, the products which increased after aging were lower molecular weight and those that decreased were higher molecular weight. Interestingly, the rejuvenation effect of wounding and of auxin treatment is mediated by different mechanisms: Wounding acts primarily at the level of translation, whereas auxin has a greater effect on transcription (SCHUSTER and DAVIS 1983b). Consequently, the authors emphasize the importance of using intact plants for senescence studies. This demand is observed in an ideal way if light can be used for the retardation of senescence. In mustard cotyledons *(Sinapsis alba)*, P_{fr} retards the differential loss of chlorophyll a, b, and carotinoids, and it preserves the fine structure of chloroplasts (BISWAL et al. 1983).

c) Foliage Leaves

Recent studies of leave senescence have concentrated on cereals, as wheat and oat. The time course involves a loss of photosynthetic activity as an early event in senescence. In wheat leaves *(Triticum aestivum)* this could be related to a loss of whole chloroplasts rather than to a general degradation of the entire chloroplast population (CAMP et al. 1983). Relations between respiration and senescence have been elucidated by SATLER and THIMANN (1983) in excised oat leaves *(Avena sativa)* which undergo a rapid deterioration in the dark. The loss of chlorophyll is closely paralleled by a respiratory rise. All factors which delay senescence (e.g., light or cytokinins) correspondingly delay the respiratory rise. It resembles the climacteric rise in respiration during fruit maturation or in aging tuber slices, although the mechanisms might be different. The alternative respiration pathway, which already functions in the initial states of the leaves and accounts for the bulk of the respiration, is overlaid during the respiratory rise by an entry of the cytochrome pathway and consequently by a large increase in ATP content (over 200%).

In detached flag leaves of wheat, where the export of assimilates is prevented, senescence is paralleled by a differential accumulation of soluble carbohydrates (LAZAN et al. 1983). Such a closed system, of course, also prevents the import of cytokinins, and it remains to be seen which factors are directly related to senescence. This system, at least, is inconsistent with a mere nutrient depletion hypothesis of senescence. It favors an induction of senescence by actively controlled processes.

The mechanisms involved in the individual steps of senescence can be analyzed by studying the effects of different exogenous factors which influence senescence. The group of Galston (SHIH et al. 1982, FUHRER et al. 1982) used 1,3-diaminopropane (Dap), a natural oxidation product of the polyamines spermidine and spermine, which occur in cereals. This compound inhibits senescence of excised oat leaves in the dark. The effect seems to require an initial attachment of Dap to a membrane site shared with Ca^{2+} which competitively reduces the effect of Dap. This potent inhibitor prevents the sharp rise of protease activity, which precedes chlorophyll loss, and the increase of *ethylene* formation by promoted ACC synthetase activity. It could be ruled out, however, that the inhibition of ethylene production is the cause for the prevention of chlorophyll loss. The authors favor a model where senescence and ethylene synthesis as membrane-localized processes are independently affected by compounds which interfere with membrane properties. They assume that the basic step in senescence is the formation of oxygen radicals (e.g., OH·) which promote simultaneously lipid peroxidation and oxidation of ACC, the precursor of ethylene. Both processes depend upon specific physical properties of membranes.

In this context, the interaction of light and ethylene production, which was examined by BASSI and SPENCER (1983) in three different dicot species, deserves attention. When cut leaf segments are incubated in the presence of ACC in closed flasks, light strongly inhibits ethylene production. With intact plants, however, the effect of light on ethylene production is only marginal. Therefore, the light effect in the former case must be attributed to the techniques used for these measurements and to the effect of wounding. Unfortunately, no data for the speed of senescence have been reported.

In detached rize *(Oryza sativa)*, however, ethylene seems to be involved in the regulation of senescence (KAO and YANG 1983). A wound or stress effect is probably ruled out in this system. The situation is similar to senescing flowers and ripening fruits where an increase in both ACC content and ethylene production are observed. Although ABA promotes senescence, it decreases ethylene production. Benzyladenine retards senescence but promotes ethylene production. A simple explanation which accounts for these observations is that ABA increases tissue sensitivity to ethylene, whereas benzyladenine treatment decreases it.

The role of photooxidation in senescence is stressed by SATLER and THIMANN (1983). They showed that under conditions favoring the formation of toxic oxygen species, senescence is promoted in oat. This effect can be demonstrated in light, which normally prevents senescence, only if CO_2 levels are reduced. It is prevented under 100% N_2. Senescence in light minus CO_2 occurs as though in dark, although the stomata are open. Interestingly, the action of ABA as senescence promoter and kinetin as senescence inhibitor is unaffected by the conditions mentioned above, suggesting a different level of control.

The inhibition of senescence by light in air and its promotion by light minus CO_2 are due to different effects: In the first case it is due to phytochrome, and in the second, to the transfer of electrons from the excited chlorophyll to O_2 because of the inhibited electron transfer.

The hypothesis of a destruction by toxic oxygen species of the photosynthetic apparatus is strongly supported by the work on a catalase-deficient mutant of barley *(Hordeum vulgare)*, isolated by KENDALL et al. (1983). The mutant grows normally in air enriched to 0.2% CO_2, but poorly in natural air where leaves rapidly deteriorate. The death of leaves is explained by the generation of peroxide inside the peroxisome during the conversion of glycolate into glyoxylate and its subsequent release to the rest of the cell. Lipid peroxidation and deterioration of membranes would be the next step. The severe damage of chloroplasts in these mutants kept in air has been demonstrated by PARKER and LEA (1983) at the ultrastructural level. The effect can be mimicked by certain herbicides (e.g., Difunon, Norflurazon), which by inhibition of carotinoid synthesis cause self-photooxidation of chlorophyll and chlorophyll sensitized photooxidative destruction of plastid constituents in light (REIß et al. 1983).

The dependence of foliar senescence upon fruit development has been investigated by NEUMANN et al. (1983) in soybean *(Glycine max)*. Excised leaf discs remain green even after the corresponding intact leaves have turned yellow on fruiting plants. Both systems were kept under short-day conditions. The acceleration of senescence by maturing pods can even be demonstrated with podded explants, which therefore appear to be good analogs of the corresponding parts of the intact plants: Leaf yellowing and abscission are delayed by removal of pods or seeds. Minerals and especially zeatin can retard senescence. The loss of these compounds in senescent leaves of the intact system seems to be an indirect result of pod development.

A thorough analysis of the senescence syndrome in leaves is the key for a better understanding of the events leading to disease and death of trees, which is recently threatening large areas of the forests in Europe. A premature senescence seems to be the common denominator for all physiological changes observed so far.

d) Flowers

Senescence of flower petals precedes and facilitates the subsequent developmental step, the formation of seeds and fruits. The entire process of aging has recently been analyzed in more detail in carnation petals *(Dianthus caryophyllus)*, where three phases can be distinguished (MOR et al. 1983): The first starts at the beginning of anthesis and is characterized by the enlargement of the inner petals and growth of the styles. There is a low, steady rate of ethylene production and a decrease of respiration. The second phase starts several days after anthesis. Senescence processes begin, showing typical inrolling of petals concomitantly with a burst of ethylene production and climacteric respiration. The third phase is characterized by advanced senescence: petals wilt, ethylene production and respiration decline. The events during the second state are correlated with a sharp decrease in apparent sucrose uptake (ADAM et al. 1983). The authors suppose as primary event a decline in membrane fluidity leading to a corresponding reduction in membrane-bound ATPase activity and sucrose uptake.

Changes in membrane properties during senescence also provide a link in the understanding of ethylene production which seems to require membrane-coupled processes. The onset of the second, climacteric phase can be accelerated by exposure of carnation flowers to ethylene or by application of ACC (SACALIS et al. 1983). Pretreatment of detached petals by cytokinins delays senescence and blocks the conversion of externally supplied ACC (MOR et al. 1983). Cytokinins as antisenescence agents are only effective if they are applied at an early preclimacteric stage. The situation resembles fruit tissus in their response to cytokinin. The authors propose that these phytohormones influence ethylene biosynthesis at two sites, namely the activity or synthesis of ACC synthetase and the conversion of ACC into ethylene.

The typical events occurring during ACC or ethylene-induced senescence of carnation petals can also be provoked by UV irradiation (BOROCHOV and FARAGHER 1983). The authors conclude that the irradiation effects on senescence are not mediated by stimulation of ethylene production. Rather, both ethylene and UV promote a sequence of events similar to those of natural senescence. The primary sites of action, however, were proved to be different, although the targets are most likeley membrane components. A promising model for the occurrence of ethylene synthesis at plasma membranes which requires a transmembrane, electrogenic proton flux, has been proposed by JOHN (1983). Interactions with P_{fr} and auxin are interpreted by changes in the membrane polarization.

e) Fruits

Fruit ripening is another example for senescence in plants. On the basis of their respiratory behavior during ripening, fruits can be separated into two classes: (1) *Climacteric fruits* (e.g., apples) which undergo a large increase in respiration (climacteric rise) accompanied by marked changes in composition and texture. Ethylene, which is consistently produced at or before the onset of the climacteric, is dramatically increased at the time of the climacteric. Premature senescence

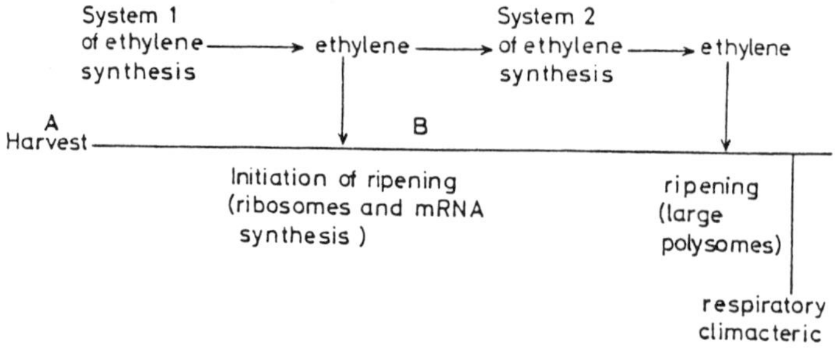

Fig. 3A,B. Sequence of events which occur before the climacteric peak of Passe-Crassane pear and Golden Delicious apple. *A* Early harvest (100 d from bloom); *B* Commercial harvest (150 d from bloom). From DROUET et al. (1983)

can be induced by exposing fruits to ethylene. Removing of ethylene from fruits delays ripening. (2) *Nonclimacteric fruits* (e.g., cherries) show no changes in respiration associated with distinct changes in the composition of the fruit.

DROUET et al. (1983) analyzed the ripening of apple *(Pyrus malus)* and cherry *(Prunus avium)*. Ethylene biogenesis during ripening of apples is regulated by two systems: The first during preclimacteric is inactive at 15°C, while the second one, which is related to a burst in mRNA and ribosome production in pulp tissues, is active in a wider range of temperatures and is followed by massive ethylene synthesis and by the rise of respiration. As a result, the commercial harvest of apples must take place between the two periods of ethylene synthesis (Fig. 3). When apples are harvested 100 days after bloom, they do not ripen. A storage at 4°C enables a later ripening at 15°C.

For cherry fruit, growth and ripening are concurrent. Obviously, the system 2 of ethylene synthesis is lacking. Ethylene production remains low and respiration rate decreases progressively. The amount of ribosomes and mRNA continuously decrease with time. In this case, maturation can hardly be a part of senescence, which only begins after physiological maturity.

LURIE and BEN-ARIE (1983) followed changes in membrane properties during different stages of ripening in apples. The progressive loss in membrane function is manifested by a sharp increase in membrane viscosity and in permeability. These changes take place between the mid-climacteric and climacteric peaks. They are correlated with changes in the lipid composition of the membranes, especially in the sterol to phospholipid ratio. This is mainly due to an increase in sterol content, in contrast to flowers and seeds where a loss phospholipids has been observed. Since there are no large changes at earlier stages of ripening, these membrane changes are a consequence, and not a cause, of the ripening process.

In conclusion, ripening of climacteric fruits has many processes in common with senescence of other plant organs. The main target for the initiating signals seems to be membranes. The precise role of ethylene in these processes has to be elucidated.

References

ADAM, Z., BOROCHOV, A., MAYAK, S., HALEVY, A.H.: Physiol. Plant 58, 257-262 (1983). - ADAMS, P.A., ROSS, M.A.: Plant Physiol. 73, 566-568 (1983). - ADDICOTT, F.T.: Abscission. 369 pp. Berkeley, Los Angeles, London: University of California Press 1982. - AKAZAWA, T., MIYATA, S.: Essays in Biochemistry, Vol. 18, eds. P.N. CAMPBELL, R.D. MARSHALL. New York: Academic Press 1982. - AMRHEIN, N.: Progr. Botany 45, 136-165 (1983). - ATZORN, R., WEILER, E.W.: Planta 159, 289-299 (1983).

BARON, O., EPEL, B.L.: Plant Physiol. 73, 471-474 (1983). - BASSI, P.K., SPENCER, M.S.: Plant Physiol. 73, 758-760 (1983). - BAULCOMBE, D.C., BUFFARD, D.: Planta 157, 493-501 (1983). - BEGGS, C.J., BARTH, J., SCHÄFER, E.: Physiol. Plant 57, 114-118 (1983). - BELOZERSKAYA, T.A., BURIKHUNOV, S.S., CHERNYSHOVA, E.K., KRITSKY, M.S., LVOV, N.P.: Neurospora Newsl. 29, 14-15 (1982). - BERNIER, G.: Can. J. Bot. 49, 803-819 (1971); - In: La physiologie de la floraison. Colloques Internationaux du C.N.R.S. No. 285 CNRS. Paris: 1979. - BESNARD-WIBAUT, C., NOIN, M., ZEEVAART, J.A.D.: Plant Cell Physiol. 24, 1269-1279 (1983). - BEWLEY, J.D., BLACK, M. (eds.): Physiology and Biochemistry of Seeds in Relation to Germination. Vol. 1, 306 pp. Berlin, Heidelberg, New York: Springer 1978; - Physiology and Biochemistry of Seeds in Relation to Germination. Vol. 2, 375 pp. Berlin, Heidelberg, New York: Springer 1982. - BISWAL, U.C., BERGFELD, R., KASEMIR, H.: Planta 157, 85-90 (1983). - BONNER, J.: Protoplasma 21, 406-423 (1934). - BONNER, J., HUANG, R.C., GILDEN, R.: Proc. Natl. Acad. Sci. USA 50, 893-900 (1963). - BOROCHOV, A., FARAGHER, J.: Plant Physiol. 71, 536-540 (1983). - BRAIN, R.D., WOODWARD, D.O., BRIGGS, W.R.: Carnegie Inst. Wash. Publ. 76, 295-299 (1977). - BRUMMER, B., PARISH, R.W.: FEBS Lett. 161, 9-13 (1983). - BÜNNING, E.: Ber. Dtsch. Bot. Ges. 54, 590-607 (1936).

CAMP, P.J., HUBER, S.C., BURKE, J.J., MORELAND, D.E.: Plant Physiol. 70, 1641-1646 (1983). - CHENG, P.C., GREYSON, R.I., WALDEN, D.B.: Am. J. Bot. 70, 450-462 (1983). - CHEN, Y.-M., HUANG, D.-H., LIN, S.-F., LIN, C.-Y., KEY, J.L.: Plant Physiol. 73, 746-753 (1983). - CHRISPEELS, M.J., VARNER, J.E.: Plant Physiol. 42, 1008-1016 (1967). - CLELAND, C.F., AJAMI, A.: Plant Physiol. 54, 904-906 (1974). - CLELAND, R.E.: Annu. Rev. Plant Physiol. 22, 197-222 (1971); - Proc. Natl. Acad. Sci. USA 70, 3092-3093 (1973); - J. Exp. Bot. 34, 676-680 (1983a); - TIBS 8, 354-357 (1983b); - COLBERT, J.T., HERSHEY, H.P., QUAIL, P.H.: Proc. Natl. Acad. Sci. USA 80, 2248-2252 (1983). - COOKE, T.J., RACUSEN, R.H., BRIGGS, W.R.: Planta 159, 300-307 (1983). - CORDONNIER, M.-M., SMITH, C., GREPPIN, H., PRATT, L.H.: Planta 158, 369-376 (1983). - COSGROVE, D.J.: Phil. Trans. R. Soc. Lond. B 303, 453-465 (1983). - COSGROVE, D.J., CLELAND, R.E.: Plant Physiol. 72, 332-338 (1983). - COTTIGNIES, A.: C.R. Acad. Sci. Paris, Ser. III, t. 296, 829-832 (1983a); - Z. Pflanzenphysiol. 111, 133-139 (1983b). - COUSSON, A., TRAN THANH VAN, K.: Plant Physiol. 72, 33-36 (1983). - CROCKER, W.: Bot. Gaz. 42, 265-291 (1906).

DATTA, K., MARSH, L., MARCUS, A.: Plant Physiol. 72, 394-397 (1983). - DEPTA, H., EISELE, K.-H., HERTEL, R.: Plant Sci. Lett. 31, 181-192 (1983). - DE VRIES, S.C., SPRINGER, J., WESSELS, J.G.H.: Planta 158, 42-50 (1983). - DOMMES, J., VAN DE WALLE, C.: Plant Physiol. 73, 484-487 (1983). - DOHRMANN, U.: Planta 159, 357-365 (1983). - DROUET, A., NIVET, C., HARTMANN, C.: Plant Physiol. 73, 754-757 (1983). - DUNBERG, A., MALMBERG, G., SASSA, T., PHARIS, R.P.: Plant Physiol. 71, 257-262 (1983).

EICHHOLTZ, D.A., ROBITAILE, H.A., HERRMANN, K.M.: Plant Physiol. 72, 750-753 (1983). - EISINGER, W.: Annu. Rev. Plant Physiol. 34, 225-240

(1983). - EISINGER, W., CRONER, L.J., TAIZ, L.: Plant Physiol. 73, 407-412 (1983). - ESASHI, Y., LEOPOLD, A.C.: Plant Physiol. 43, 871-876 (1968). - ESASHI, Y., KOMATSU, H., USHIZAWA, R., SAKAI, Y.: Aust. J. Plant Physiol. 9, 97-111 (1982). - ESASHI, Y., KURAISHI, R., TANAKA, N., SATOH, S.: Plant Cell Environ. 6, 493-499 (1983a). - ESASHI, Y., KURAISHI, R., TANAKA, N., NAGAYAMA, T.: Plant Cell Physiol. 24, 267-271 (1983b). - ESASHI, Y., SATOH, M., SAIJOH, K., SATOH, S.: Plant Cell Physiol. 24, 17-26 (1983c). - ESSER, K., TUDZYNSKI, P.: Senescence in fungi, pp. 67-83. In: Senescence in Plants, ed. K.V. THIMANN. Boca Raton, FL: CRC Press 1980.

FERGUSON, I.B., WATKINS, C.B., HARMAN, J.E.: Plant Physiol. 71, 182-186 (1983). - FREDERICQ, H., RETHY, R., ONCKELEN, VAN, H., DE GREEF, J.A.: Physiol. Plant. 57, 402-406 (1983). - FUHRER, J., KAUR-SAWHNEY, R., SHIH, L.-M., GALSTON, A.W.: Plant Physiol. 70, 1597-1600 (1982). - FUJIOKA, S., YAMAGUCHI, I., MUROFUSHI, N., TAKAHASHI, N., KAIHARA, S., TAKIMOTO, A.: Plant Cell Physiol. 24, 235-239 (1983). - FUKSHANSKY, L., SCHÄFER, E.: Models in photomorphogenesis, pp. 69-95. In: Photomorphogenesis. Encyclopedia of Plant Physiology, N.S., Vol. 16A, eds. W. SHROPSHIRE Jr., H. MOHR. Berlin, Heidelberg, New York, Tokyo: Springer 1983.

GANTT, J.S., KEY, J.L.: Biochemistry 22, 4131-4139 (1983). - GIETL, C., HOCK, B.: Plant Physiol. 70, 483-487 (1982). - GOLLMER, I., APEL, K.: Eur. J. Biochem. 133, 309-313 (1983). - GOODWIN, P.B., LYNDON, R.F.: Protoplasma 116, 219-222 (1983). - GOTTMAN, K., SCHÄFER, E.: Planta 157, 392-400 (1983). - GRESSEL, J., HORWITZ, B.A.: Plant Physiol. 71, 200-204 (1983). - GWÓŹDŹ, E.A., WOŹNY, A.: Physiol. Plant 59, 103-110 (1983).

HABERLANDT, G.: Ber. Dtsch. Bot. Ges. 8, 40-48 (1890). - HAGER, A., MENZEL, H., KRAUSS, A.: Planta 100, 47-75 (1971). - HAHN, T.-R., SONG, P.-S.: Biochemistry 20, 2602-2609 (1981). - HANLEY-BOWDOIN, L., LANE, B.G.: Eur. J. Biochem. 135, 9-15 (1983). - HARDING, R.W., MELLES, S.: Plant Physiol. 72, 996-1000 (1983). - HAUPT, W.: Localization of phytochrome within the cell. pp. 553-569. In: Phytochrome, eds. K. MITRAKOS, W. SHROPSHIRE Jr., London, New York: Academic Press 1973. - HAUPT, W.: Philos. Trans. R. Soc. Lond. B 303, 467-478 (1983). - HERTEL, R.: Z. Pflanzenphysiol. 112, 53-67 (1983). - HERTEL, R., LOMAX, T.L., BRIGGS, W.R.: Planta 157, 193-201 (1983). - HIGGINS, T.J.V., ZWAR, J.A., JACOBSEN, J.V.: Nature 260, 166-169 (1983). - HOCK, B.: Antibiotica und ähnliche Verbindingen, pp. 189-223. In: Pflanzentoxikologie, eds. B. HOCK, E. ELSTNER. Mannheim, Wien, Zürich: Bibliographisches Institut 1984. - HOLMES, M.G.: Philos. Trans. R. Soc. Lond. B 303, 503-521 (1983). - HOLMES, M.G., SCHÄFER, E.: Planta 153, 267-272 (1983). - HORWITZ, B.A., GRESSEL, J.: Plant Phyiol. 71, 200-204 (1982).

IBRAHIM, A.E., ROBERTS, E.H.: J. Exp. Bot. 34, 620-630 (1983). - IBRAHIM, A.E., ROBERTS, E.H., MURDOCH, A.J.: J. Exp. Bot. 34, 631-640 (1983). - IINO, M.: Planta 156, 21-32 (1982). - ISHIZAWA, K., ESASHI, Y.: J. Exp. Bot. 34, 74-82 (1983).

JACOBS, M., GILBERT, S.F.: Science 220, 1297-1300 (1983). - JOHN, P.: FEBS Lett. 152, 141-143 (1983). - JONES, R.L., JACOBSEN, J.V.: Planta 158, 1-9 (1983).

KAO, C.H., YANG, S.F.: Plant Physiol. 73, 881-885 (1983). - KARSSEN, C.M., BRINKHORST-van der SWAN, D.L.C., BREEKLAND, A.E., KOORNNEEF, M.: Planta 157, 158-165 (1983). - KASEMIR, H., MOHR, H.: Planta 156, 282-

288 (1982). - KENDALL, A.C., KEYS, A.J., TURNER, J.C., LEA, P.J., MIFLIN, B.J.: Planta 159, 505-511 (1983). - KERSCHER, L., NOWITZKI, S.: FEBS Lett. 146, 173-176 (1982). - KEY, J.L.: Annu. Rev. Plant Physiol. 20, 449-474 (1969). - KHAN, A.A. (ed.): Physiology and biochemistry of seed development, dormancy, and germination, 547 pp. Amsterdam: Elsevier Biomedical Press 1982. - KIM, I.-S., KIM, E.S., SONG, P.-S.: Biochem. Biophys. Acta 747, 55-64 (1983). - KOROHODA, W., SHRAIDEH, Z., BARANOWSKI, Z., WOHLFARTH-BOTTERMANN, K.E.: Planta 158, 54-62 (1983).

LANG-FEULNER, J., RAU, W.: Photochem. Photobiol. 21, 179-183 (1975). - LAZAN, H.B., BARLOW, E.W.R., BRADY, C.J.: J. Exp. Bot. 34, 726-736 (1983). - LEONARD, M., KINET, J.-M., BODSON, M., BERNIER, G.: Physiol. Plant. 57, 85-89 (1983). - LEOPOLD, A.C.: Bioscience 25, 659-662 (1975). - LERCARI, B., SODI, F., FASTAMI, C.: Planta 156, 546-552 (1982). - LURIE, S., BEN-ARIE, R.: Plant Physiol. 73, 636-638 (1983). - LYNDON, R.F., BERNIER, G.: Physiol. Plant. 59, 476-480 (1983).

MAAN, A.C., VREUGDENHIL, D., BOGERS, R.J., LIBBENGA, K.R.: Planta 158, 10-15 (1983). - MALONE, M., RIDGE, I.: Planta 157, 71-73 (1983). - MARTIN, C., NORTHCOTE, D.H.: Planta 158, 16-26 (1983). - MAURINO, S.G., VARGAS, M.A., APARICIO, P.J., MALDONADO, J.M.: Physiol. Plant. 57, 411-416 (1983). - MAYER, A.M., POLJAKOFF-MAYBER, A.: The Germination of Seeds. 2nd ed., 192 pp. Oxford, New York, Toronto, Sydney, Paris, Braunschweig: Pergamon Press 1975. - MEDAWAR, P.B.: The Uniqueness of the Individual, 191 pp. London: Methuen 1957. - MÉTRAUX, J.-P., KENDE, H.: Plant Physiol. 72, 441-446 (1983). - METZGER, J.D.: Plant Physiol. 73, 791-795 (1983). - MINAMIKAWA, T., SUZUKI, Y., KOSHIBA, T.: Plant Cell Physiol. 24, 1361 1369 (1983). - MIYATA, S., AKAZAWA, T.: J. Cell Biol. 96, 802-806 (1983). - MOHR, H., SHROPSHIRE Jr., W.: An introduction to photomorphogenesis for the general reader, pp. 24-38. In: Photomorphogenesis. Encyclopedia of Plant Physiology, N.S. Vol. 16A, eds. W. SHROPSHIRE Jr., H. MOHR. Berlin, Heidelberg, New York, Tokyo: Springer 1983. - MÖSINGER, E., SCHOPFER, P.: Planta 158, 501-511 (1983). - MOR, Y., SPIEGELSTEIN, H., HALEVY, A.H.: Plant Physiol. 71, 541-546 (1983). - MORENO, M.L., PAGÈS, M., PALAU, J., DE LA TORRA, C.: Ann. Bot. 51, 347-352 (1983). - MUNKRES, K.D.: Rev. Biol. Res. Aging 1, 25-28 (1983).

NAGATANI, A., YAMAMOTO, K.T., FURUYA, M., FUKUMOTO, T., YAMASHITA, A.: Plant Cell Physiol. 24, 1143-1149 (1983). - NEUMANN, P.M., TUCKER, A.T., NOODÉN, L.D.: Plant Physiol. 72, 182-185 (1983).

OOTA, Y.: Plant Cell Physiol. 24, 327-332 (1983).

PAIETTA, J., SARGENT, M.L.: Proc. Natl. Acad. Sci. USA 78, 5573-5577 (1981); - Photochem. Photobiol. 35, 853-855 (1982); - Plant Physiol. 72, 764-766 (1983). - PALEG, L.: Plant Physiol. 35, 902-906 (1960). - PARKER, M.L., LEA, P.J.: Planta 159, 512-517 (1983). - PEARCE, D., PENNY, D.: Plant Sci. Lett. 30, 347-353 (1983). - PERCIVAL, M.P., DODGE, A.D.: J. Exp. Bot. 34, 47-58 (1983). - PHILIPSON, J.J.: J. Exp. Bot. 34, 291-302 (1983). - POPE, D.G.: Physiol. Plant. 58, 269-274 (1983). - POWELL, A.D., LEUNG, D.W., BEWLEY, J.D.: Planta 159, 182-188 (1983). - PRATT, L.H.: Annu. Rev. Plant Physiol. 33, 557-582 (1982). - PUKACKA, S.: Z. Pflanzenphysiol. 112, 199-205 (1983). - PUTZER, H., WERENSKIOLD, K., VERFUERTH, C., SCHRECKENBACH, Th.: EMBO J. 2, 261-267 (1983).

RAJASEKARAN, K., MULLINS, M.G., NAIR, Y.: Ann. Bot. 52, 417-420 (1983). - RAYLE, D.L., CLELAND, R.: Plant Physiol. 46, 250-253 (1970); - Planta 114, 63-73 (1973). - REIß, T., BERGFELD, R., LINK, G., THIEN, W., MOHR,

H.: Planta 159, 518-528 (1983). - ROBERTS, E.H.: Seed Sci. Technol. 1, 515-527 (1973). - RÜDIGER, W.: Philos. Trans. R. Soc. Lond. B 303, 377-386 (1983). - RÜDIGER, W., THÜMMLER, F., CMIEL, E., SCHNEIDER, S.: Proc. Natl. Acad. Sci. USA 80, 6244-6248 (1983).

SACALIS, J., WULSTER, G., JANES, H.: Z. Pflanzenphysiol. 112, 7-14 (1983). - SACHS, M.R., HACKETT, W.P.: Acta Hortic. 68, 29-49 (1977). - SAKURAI, N., INOUHE, M., MASUDA, Y., KURAISHI, S.: Plant Cell Physiol. 24, 317-325 (1983). - SANCHEZ, R.A., DE MIGUEL, L.C.: Z. Pflanzenphysiol. 110, 319-329 (1983). - SASAKI, Y., SAKIHAMA, T., KAMIKUBO, T., SHINOZAKI, K.: Eur. J. Biochem. 133, 617-620 (1983). - SATLER, S.O., THIMANN, K.V.: Plant Physiol. 71, 67-70 (1983); - ibid. 72, 540-546 (1983). - SAUNDERS, M.J., CORDONNIER, M.-M., PALEVITZ, B.A., PRATT, L.H.: Planta 159, 545-553 (1983). - SCHÄFER, E., HAUPT, W.: Blue-light effects in phytochrome-mediated responses, pp. 723-744. In: Photomorphogenesis. Encyclopedia of Plant Physiology. N.S. Vol. 16 B, eds. W. SHROPSHIRE Jr., H. MOHR. Berlin, Heidelberg, New York, Tokyo: Springer 1983. - SCHAER, J.A., MANDOLI, D.F., BRIGGS, W.R.: Plant Physiol. 72, 706-712 (1983). - SCHMIDT, W.: Struct. Bonding 41, 1-44 (1980). - SCHUSTER, A.M., DAVIS, E.: Plant Physiol. 73, 809-916 (1983a); - ibid. 73, 822-827 (1983b). - SEN, S., RAHMANI, M.A., AIKENS, I.: Z. Pflanzenphysiol. 110, 1-6 (1983). - SENGER, H. (ed.): The Blue Light Syndrome. Berlin, Heidelberg, New York: Springer 1980. - SENGER, H.: Photochem. Photobiol. 35, 911-920 (1982). - SENGER, H., BRIGGS, W.R.: The blue light receptor(s): Primary reactions and subsequent metabolic changes, pp. 1-38. In: ed. K.C. SMITH, Photochemical and Photobiological Review, Vol. 6. New York: Plenum 1981. - SEXTON, R.: Annu. Rev. Plant Physiol. 33, 133-162 (1982). - SHIH, L.-M., KAUR-SAWHNEY, R., FUHRER, J., SAMAANTA, S., GALSTON, A.W.: Plant Physiol. 70, 1592-1596 (1982). - SHIMAZAKI, Y., CORDONNIER, M.-M., PRATT, L.H.: Planta 159, 534-544 (1983). - SHINOZAKI, M., TAKIMOTO, A.: Plant Cell Physiol. 24, 433-439 (1983). - SHROPSHIRE Jr., W., MOHR, H.: Photomorphogenesis. Encyclopedia of Plant Physiology, eds. A. PIRSON, M.H. ZIMMERMANN, Vol. 16A. Berlin, Heidelberg, New York, Tokyo: Springer 1983. SKADSEN, R.W., CHERRY, J.H.: Plant Physiol. 71, 861-868 (1983). - SMITH, H.: Philos. Trans. R. Soc. Lond. B 303, 443-452 (1983). - SONG, P.-S.: Annu. Rev. Biophys. Bioeng. 12, 35-68 (1983). - SREERAMULU, N.: Ann. Bot. 51, 209-216 (1983). - STIEKEMA, W., WIMPEE, C.F., SILVERTHORNE, J., TOBIN, E.M.: Plant Physiol. 72, 717-724 (1983). - SUNDQUIST, C., WIDELL, S.: Physiol. Plant. 59, 35-41 (1983). - SUZUKI, Y., MINAMIKAWA, T.: Plant Cell Physiol. 24, 1371-1377 (1983).

TAIZ, L., RAYLE, D.L., EISINGER, W.: Plant Physiol. 73, 413-417 (1983). - TANADA, T.: Physiol. Plant. 57, 42-46 (1983a); - ibid. 58, 475-478 (1983b). - TANNER, W.: Biol. unserer Zeit 10, 45-51 (1980). - TAYLORSON, R.B., HENDRICKS, S.B.: Annu. Rev. Plant Physiol. 28, 331-354 (1977). - TEGELMAN, A.J.: Z. Pflanzenphysiol. 111, 379-388 (1983). - THOMAS, H., STODDART, J.L.: Annu. Rev. Plant Physiol. 31, 83-111 (1980). - THOMPSON, W.F., EVERETT, M., POLANS, N.O., JORGENSEN, R.A., PALMER, J.D.: Planta 158, 487-500 (1983). - TILLBERG, E.: Physiol. Plant. 58, 243-248 (1983). - TOMI, H., SASAKI, Y., KAMIKUBO, T.: Plant Cell Physiol. 24, 587-592 (1983); - Plant Sci. Lett. 30, 155-164 (1983a); - Plant Cell Physiol. 24, 587-592 (1983b); - ibid. 24, 1087-1092 (1983c). - TONG, Z., KASEMIR, H., MOHR, R.: Planta 159, 136-142 (1983). - TREWAVAS, A.J.: TIBS 8, 354-357 (1983). - TUAN, D., BONNER, J.: Plant Physiol. 39, 768-772 (1964).

UPADHYAYA, M.K., NAYLOR, I.M., SIMPSON, G.M.: Physiol. Plant 58, 119-123 (1983).

VAN DE WALLE, C., DELTOUR, R., FORGEUR, G.: Physiol. Plant 57, 181-188 (1983). - VAN STADEN, J.: Physiol. Plant. 58, 340-346 (1983). - VAN

STADEN, J., DAVEY, J.E., BROWN, N.A.C.: Cytokinins in seed development and germination. In: The Physiology and Biochemistry of Seed Development, Dormancy, and Germination, ed. A.A. KHAN. Amsterdam, New York: Elsevier Biomedical Press 1982. - VARTY, K., ARREGUÍN, L., GOMÉZ, T., LOPEZ, T.P.J., GOMÉZ, L.M.A.: Plant Physiol. 73, 692-697 (1983). - VIERSTRA, R.D., QUAIL, P.H.: Biochemistry 22, 2498-2505 (1983); - Planta 156, 158-165 (1982a); - Proc. Natl. Acad. Sci. USA 79, 5272-5276 (1982b). - VILLIERS, T.A.: Ageing and longevity of seeds in field conditions, pp. 265-288. In: Seed ecology, ed. W. HEYDECKER. London: Butterworths 1973. - VINCE-PRUCE, D.: Photomorphogenesis and flowering, pp. 457-490. In: Photomorphogenesis. Encyclopedia of Plant Physiology, N.S. Vol. 16B, eds. W. SHROPSHIRE Jr., H. MOHR. Berlin, Heidelberg, New York, Tokyo: Springer 1983a; - Philos. Trans. Roy. Soc. Lond. B 303, 523-536 (1983b). - VIRO, M., KLOPPSTECH, K.: Planta 157, 202-208 (1983).

WADA, M., KADOTA, A., FURUYA, M.: Plant Cell Physiol. 24, 1441-1447 (1983). - WALKER, J.C., KEY, J.L.: Proc. Natl. Acad. Sci. USA 79, 7185-7189 (1982). - WALL, J.K., JOHNSON, C.B.: Planta 159, 387-397 (1983). - WAREING, P.F., FONDA, H.A.: Physiol. Plant. 10, 266-280 (1957). - WATANABE, K., TAKIMOTO, A., IWAMURA, H., FUJITA, T.: Plant Cell Physiol. 24, 889-897 (1983). - WOOLHOUSE, H.W.: Bio Science 28, 25-31 (1980).

YAMAOKA, T., CHIBA, N.: Plant Cell Physiol. 24, 1281-1290 (1983). - YOMO, H.: Hakko Kyokaishi 18, 600-602 (1960).

ZURFLUH, L.L., GUILFOYLE, T.J.: Planta 156, 525-527 (1982).

Professor Dr. BERTOLD HOCK
Lehrstuhl für Botanik
Technische Universität München
D-8050 Freising-Weihenstephan, FRG

VIII. Physiology of Movement

By WOLFGANG HAUPT and KONRAD SEITZ

Intracellular Movement

Usually two facets of intracellular movement can be distinguished: Continuous movement of cytoplasm with or without chloroplasts, which does not change the overall pattern of organelle distribution, and redistribution of chloroplasts to patterns which are controlled by the light direction or, more generally, by patterns of intracellular light absorption. In both types of movement substantial progress has been made in elucidating the mechanism of movement as well as the regulatory processes.

1. Cytoplasmic Streaming

Several reviews have appeared during the recent years on cytoplasmic streaming, which is known from green and nongreen plant cells as well as from slime molds (cf. KAMIYA 1981, NAGAI 1979, ROBARDS 1983, SEITZ 1979). For the study of cytoplasmic streaming new videomicroscopic techniques are available (ALLEN and ALLEN 1981).

This report will consider only publications on cytoplasmic streaming in green cells. One can distinguish here cells of characean algae (e.g., *Chara* and *Nitella*), in which chloroplasts normally do not participate in cytoplasmic streaming, and cells of other algae (e.g., *Acetabularia* and *Vaucheria*) or higher plants (e.g., *Vallisneria*), in which chloroplasts move together with the cytoplasm. In both cases the motive force of cytoplasmic streaming is generated at the interface between the moving endoplasm and the stationary ectoplasm by an interaction of an endoplasmic myosin with actin microfilaments anchored to the ectoplasm (KAMIYA 1981, SEITZ 1979, WILLIAMSON 1980). By electron microscopy, membrane-bound endoplasmic structures can be seen in *Chara*, linked to actin microfilament bundles by myosin-containing protuberances (NAGAI and HAYAMA 1979).

In cells of *Nitella* local heating stops cytoplasmic streaming apparently by an effect on the endoplasm, since the light microscopic structural appearance of ectoplasmic filament bundles remains unmodified (KAMITSUBO 1981). Actin and myosin filaments can also be observed extending as a three-dimensional network throughout the endoplasm (ALLEN 1980). Actin and myosin have been isolated not only from characean cells (cf. KAMIYA 1981), but recently also from tomato cells (VAHEY et al. 1982). As an alternative to the classical theory of actin-myosin interaction, a screw-mechanical model has been proposed by FOISSNER and JAROSCH (1981) for the motive mechanism.

The large cells of *Chara* and *Nitella*, allowing perfusion techniques (cf. Fortschr. Bot. 29, 164), are especially well suited for an experimental

Progress in Botany, Vol. 46

analysis (cf. KAMIYA 1981). They can even be used, after proper manipulation, as membrane-free cell models (see below).

As reported earlier (Fortschr. Bot. 29, 164; 35, 175), streaming in characean cells depends on the osmotic value of the cell sap. This has been investigated in more detail with different methods. In intact cells of *Chara*, increase of cell-sap concentration by plasmolysis causes a transient inhibition of cytoplasmic streaming presumably by evoking an action potential; interestingly, photosynthesis is inhibited in parallel (LUCAS and ALEXANDER 1981). Similarly, if transcellular osmosis in *Nitella* is performed, local depolarization and local cessation of streaming is induced. This effect is found even if the external medium is free of calcium; thus, Ca^{2+} is not an essential link connecting membrane depolarization and cessation of streaming (HAYAMA and TAZAWA 1978), although calcium can inhibit streaming (see below). Moreover, electrogenic H^+ extrusion seems not to be closely linked to streaming: In *Nitellopsis*, vanadate, which is an inhibitor of H^+ ATPase, hardly affects cytoplasmic streaming, even if it is introduced into the vacuole (SHIMMEN and TAZAWA 1982c).

Great progress in the analysis of cytoplasmic streaming was possible by the model systems of membrane-free characean cells (cf. KAMIYA 1981). Tonoplast-free cells are obtained by perfusing the cell with EGTA medium. Again, an increase in the osmolarity of the cell sap by perfusion leads to a decrease in streaming rate, and streaming is reversibly stopped by Ca^{2+} (TOMINAGA and TAZAWA 1981a,b). Experimental change of the internal pH reveals a rate optimum for streaming at pH 7 (TAZAWA and SHIMMEN 1982).

Endoplasm has been squeezed out of a tonoplast-free *Chara* cell and introduced in a tonoplast-free *Nitella* cell, in which the endoplasmic myosin had been inactivated by N-ethylmaleimide. The *Chara* endoplasm started to stream along the ectoplasm of the host cell and became attached to it (SHIMMEN and TAZAWA 1982b). In tonoplast-free cells, furthermore, the motile endoplasmic organelles can be stripped off from the ectoplasm by perfusion with ATP. In such a preparation it was possible to demonstrate that cytochalasin B stabilizes the ectoplasmic actin filament bundles against a solution of low ionic strength (WILLIAMSON 1978).

Plasmalemma-free models have been prepared in *Nitella*. By removal of Ca^{2+} and plasmolysis at low temperature, the cell membrane becomes permeabilized, which is equivalent to its complete absence. In these models, streaming can be stopped by bathing in a medium without ATP. Streaming is reactivated by Mg^{2+}-ATP and can be reversibly stopped again by depletion of ATP (SHIMMEN and TAZAWA 1982a, 1983). In such a cell streaming occurs in absence of Ca^{2+} and can be reversibly inhibited by Ca^{2+}. These results indicate that an endoplasmic Mg^{2+}-ATPase interacting with ectoplasmic actin microfilaments is involved in the motive mechanism of cytoplasmic streaming, and that streaming is controlled by ATP, Mg^{2+}, and Ca^{2+} (TOMINAGA et al. 1983). In this respect, changes in the intracellular concentration of adenylates have been found to be correlated with changes in the rate of cytoplasmic streaming (REID and WALKER 1983).

Some of the experiments in characean cells have been extended to green cells of other organisms and important results have been confirmed, but there are also a few controversies. In *Acetabularia*, endoplasm with chloroplasts has been separated from the ectoplasm by centrifugation. After centrifugation filaments remain at the ectoplasm, and endoplasm resumes streaming along the stationary filaments. By a combination of centrifugation and double-chamber techniques it was possible to treat the endo- and ectoplasm separately with inhibitors. The results show that the actin inhibitor cytochalasin B inhibits streaming by affecting the ectoplasm, while the myosin inhibitor N-ethylmaleimide is acting on the endoplasm (NAGAI and FUKUI 1981).

In addition to actin microfilaments also microtubuli somehow seem to be involved in the motive mechanism of cytoplasmic streaming in *Acetabularia*, since streaming is partly inhibited by colchicine (KOOP and KIERMAYER 1980a,b, KOOP 1981).

Cytochalasin B reversibly inhibits rotational streaming also in *Vallisneria*, apparently by modifying the structural integrity of actin filaments, since streaming direction was found to be altered after recovery (ISHIGAMI and NAGAI 1980). Similarly as in *Nitellopsis*, vanadate and other membrane-ATPase inhibitors do not affect the movement responses of chloroplasts in *Vallisneria* (SEITZ and MAURER 1984).

Rotational streaming of chloroplasts in *Vallisneria gigantea* can be controlled by Ca^{2+}: An increase in Ca^{2+} concentration inhibits streaming as in characean cells (see above), and this effect can be promoted by far-red light; on the other hand, lowering of Ca^{2+} by EGTA in the medium induces streaming (TAKAGI and NAGAI 1983). In contrast, no significant effect of EGTA and of Ca^{2+} has been found in studies on movement responses of chloroplasts in *Vallisneria spiralis* (SEITZ, unpublished).

Also in staminal hairs of *Tradescantia* cytoplasmic streaming is not affected by EGTA and Ca^{2+}, yet the Ca^{2+}-ionophore A-23187 was found to inhibit streaming here, even in absence of external Ca^{2+}, which has been explained by a release of membrane-bound calcium into the cytosol (DOREE and PICARD 1980).

In higher plant cells cytoplasmic streaming can be observed as rotational streaming of chloroplasts along the anticlinal walls. This movement depends upon the motive force of cytoplasmic streaming and upon the adhesional resistance of chloroplasts at the ectoplasm, both factors being determined by the molecular properties of the actomyosin system of the motive mechanism of cytoplasmic streaming (SEITZ 1978, 1979). Actin filaments have been studied in relation to rotational streaming of chloroplasts by electron microscopy in *Vallisneria*. Filament bundles are found to be located parallel to streaming direction at those cell sides where streaming occurs (YAMAGUCHI and NAGAI 1981). The micrographs also show that more cytoplasm and chloroplasts are located at the anticlinal cell wall when streaming has been induced. These pictures thus nicely prove the occurrence of an orientation movement of chloroplasts to the anticlinal cell side as a precondition of rotational streaming, as was assumed earlier (SEITZ 1979).

Rotational streaming of chloroplasts in *Vallisneria* and in other plants is controlled by the external factor light. Irradiation with high intensity of blue light induces streaming (SEITZ 1979, 1982). This has also been found for cytoplasmic streaming in *Acetabularia* (DAZY et al. 1981). Blue light of low intensity, locally applied to *Vaucheria*, causes a stop of movement and aggregation of chloroplasts. This effect is mediated by a change in the structural arrangement of microfilament bundles (BLATT and BRIGGS 1980), as will be described in greater detail in the next section.

2. Chloroplast Redistribution as Controlled by Light

Several reviews on this topic have appeared recently (BRITZ 1979, HAUPT 1982, 1983, HAUPT and WAGNER 1984). We therefore can restrict this report to four systems in which the most interesting progress has been made, with single observations on other systems mentioned only briefly in the proper connection.

a) *Funaria* and *Lemna*

Although moss and duckweed are quite unrelated organisms, the pattern of chloroplast orientation is very similar in both, and they are usually investigated as the most typical examples of chloroplast orientation.

According to an analysis of GABRYŚ-MIZERA (1976), in *Funaria* and *Lemna* the chloroplast distribution coincides nicely with the computed intracellular distribution of light intensity, with the brightest regions occupied by chloroplasts in low-intensity orientation and emptied in high-intensity orientation. From this coincidence it is concluded that the absorption gradient is generated mainly by light refraction at the interfaces surrounding medium/cell wall and cell wall/cytoplasm, but light attenuation in the chloroplasts and dichroic orientation of photoreceptor molecules are modifying factors. The observations are consistent with the assumption of the photoreceptor molecules being localized in the outermost layer of the cytoplasm, probably associated with the cytoplasmic membrane.

In *Lemna*, the effect of short light pulses has been investigated. Usually it is reported that the proper light condition must prevail during the whole time of response, if this is to come to its completion, with the only exception of *Mougeotia* (see below). Incomplete orientation movements, however, can be induced by short light pulses, as observed by SEITZ (1967) already in *Vallisneria*. Such induction effects have been analyzed more thoroughly in *Lemna* (GABRYŚ et al. 1981, ZURZYCKI et al. 1983). The actual response pattern depends on the fluence rate and on the time of the light pulse. If its product, i.e., the fluence, is sufficiently high, the chloroplasts start with a high-intensity movement, which is followed, then, by a tendency to low-intensity distribution. Thus, the cell "remembers" first the "high-intensity signal", and as it fades away its after-effect is perceived as a low-intensity signal. A model calculation is proposed, assuming sequences of events after light perception. The results from kinetic studies were fitted to the model, thus allowing to estimate its parameters, e.g., lifetimes of the intermediate steps. It is a challenge for further investigations to relate actual biophysical or biochemical processes to the formal steps of the model.

b) Vaucheria

In "low-intensity response", chloroplasts of *Vaucheria sessilis* accumulate at those regions where cell wall and cortical cytoplasm are nearly perpendicular to the light direction, i.e., where light absorption is highest. To facilitate quantitative measurements, in experiments usually this response is replaced by chloroplast accumulation in a microbeam which irradiates part of the coenocyte. With fiber optics for irradiation and for measurements of absorption changes due to chloroplast aggregation, BLATT and BRIGGS (1983) could automatize and seriously improve the quantitation of the results.

An action spectrum of this response differs from the older one of FISCHER-ARNOLD (1963), with the main peak around 470 nm instead of 450 nm. From these data, BLATT (1983) proposes a kinetic model with steady-state photoreceptor cycling during continuous irradiation. But since the new action spectrum is not consistent with the absorption characteristics of flavins, the problem of the photoreceptor pigment is open again. As to its location, however, its association with the cell membrane is a reasonable assumption, especially after ZURZYCKI and LEŁATKO (1969) have found an action dichroism in *V. sessilis*.

Important observations and experiments of BLATT contribute to our understanding of the mechanisms of chloroplast aggregation. We already knew that the autonomous streaming of cytoplasm with chloroplasts is locally stopped where the microbeam is placed (Fortschr. Bot. 26, 274). As BLATT and BRIGGS (1980) have shown, this cessation of movement is preceded by structural changes, as observed in the interference-contrast microscope: cables of fibers, along which cell organelles are moving, desintegrate in the illuminated area and form a network. This light-induced reticulation appears to have the same spectral dependence as the chloroplast aggregation. Moreover, phalloidin, which inhibits the reticulation, also inhibits the aggregation. Cytochalasin B, on the other hand, causes reticulation in light and dark regions as well, thus making any movement impossible, and hence also aggregation in the "light trap". And the cytochalasin effect is abolished again by simultaneously given phalloidin. The inhibiting effect of cytochalasin B on movement and the antagonizing effect of phalloidin suggest actin-myosin interaction as underlying the mechanism of movement. And indeed, the classical arrowhead decoration with heavy meromyosin (HMM) and its reversal by ATP were successful, thus proving actin as the main component of the fibers (BLATT et al. 1980).

What are the connecting steps between light absorption and local inactivation of actin microfilaments? An early light effect, which precedes both chloroplast aggregation and fiber reticulation, is an efflux of protons at the lighted area and their reentering at some distance in the dark area (BLATT et al. 1981). It is not yet clear whether the local hyperpolarization or the outward current is causally connected with the actin inactivation. But it is tempting to assume that calcium is involved in these processes, as the composition of the surrounding medium with a proper calcium concentration is important for the effects.

There is one apparent paradox in this system: In contrast to the localized light effect as reported above, no significant inhibition of movement is observed upon total irradiation of the coenocyte. But a speculative interpretation could assume that no substantial proton efflux happens if no reentering of protons is possible because of lack of dark regions. Experiments along this line are still needed.

c) *Mougeotia*

A transduction chain from light perception to response has been proposed by HAUPT (1980) and by WAGNER and KLEIN (1981) for the low-intensity orientation, based on various approaches, mainly by the groups of SCHÖNBOHM, of WAGNER, and of WEISENSEEL.

In the low-intensity response of *Mougeotia*, the directional light signal is perceived by phytochrome and transduced into an intracellular gradient of P_{fr}. As an early effect of P_{fr}, transport properties of the cell membrane are altered, as indicated by a red-light-enhanced Ca^{2+} uptake (DREYER and WEISENSEEL 1979). This does not necessarily mean that Ca^{2+} uptake indeed is important, but additional observations favor the view of Ca^{2+} redistribution in the cells being a link in the transduction chain: starvation of Ca^{2+} reduces the ability of the cells to perform the response, concomitantly with the disappearance of Ca^{2+} from a water-insoluble cell fraction (WAGNER and KLEIN 1978; see however the different view below); Ca^{2+} is mainly contained in specific microbodies, usually called tannin vesicles, which are most abundant close to the chloroplast's edge in the peripheral cytoplasm (WAGNER and ROSSBACHER 1980). P_{fr} is assumed to control the sequestering-release cycle of Ca^{2+} in and out of these microbodies, thus also controlling the cytoplasmic level of Ca^{2+} (WAGNER et al. 1984).

One step further, calcium is assumed to interact with calmodulin, which has been found in *Mougeotia*, too, and accordingly calmodulin inhibitors can stop the chloroplast movement (WAGNER et al. 1984, GROLIG et al. 1983). Other calmodulin inhibitors have been used by SERLIN and ROUX (1984) with the same effect; thus, unspecific effects of the inhibitors are unlikely. Calcium and calmodulin in turn could control actin microfilaments and their interaction with myosin.

Some evidence for actin microfilaments as underlying the mechanism of movement had already been presented in this series (SCHÖNBOHM 1978), and it has been seriously strengthened since: electron microscopic preparations show structures in the cytoplasm adjacent to the chloroplast's edge, which correspond in size to actin microfilaments (WAGNER and KLEIN 1978); decoration of microfilaments with heavy meromyosin (HMM) is possible (KLEIN et al. 1980); staining with fluorescent phallotoxin, which binds specifically to actin, corresponds to the other findings (WAGNER: in HAUPT and WAGNER 1984).

A direct approach to the calcium involvement in chloroplast reorientation has been undertaken by SERLIN and ROUX (1984). If a calcium ionophore is locally applied to the *Mougeotia* cell, close to the chloroplast's edge, chloroplast reorientation is observed with the edge avoiding the sites of calcium influx. Thus, local formation of P_{fr} can be replaced by local calcium influx. The result is only obtained if the surrounding medium contains calcium, and the Ca^{2+} ionophore cannot be replaced by the K^+ ionophore valinomycin. Thus, the specificity of the calcium effect is ensured.

In spite of these obvious supports for the P_{fr}-calcium-calmodulin-actin transduction chain, there are some open questions, not only concerning details, but also concerning the basic concept. In recent investigations SCHÖNBOHM and SCHÖNBOHM (1984a) could not find a strong effect, if the cells were depleted of Ca^{2+}. In fact, the low-intensity response was completely uninfluenced, and for the high-intensity movement it was possible to obtain full response even under calcium starvation by slightly increasing the intensity and/or the duration of the light signal. This led the authors to another approach.

The above-mentioned microbodies contain, besides Ca^{2+}, high amounts of phenolic compounds. These compounds are leached out, together with Ca^{2+}, very fast, if the cells are surrounded by distilled water or by sorbitol solution. Moreover, the efflux of the phenolic compounds appears to be controlled by calcium. Finally, and most importantly, some of these compounds strongly inhibit the high-intensity movement of the chloroplast, but leave the low-intensity movement uninfluenced. It would be very interesting to investigate the influence of red and far-red light on the one hand, and of high-intensity blue light on the other hand, on the efflux of phenolic compounds.

Taken these observations together, there is no need to abandon completely the calcium hypothesis. But very probably the function of calcium in the transduction chain is more complicated than the simple model suggests. Phenolic compounds may well be an additional link in the transduction chain, comparable, e.g., to their involvement in seismo- or thigmonastic trugor movements (cf. the discussion in SCHÖNBOHM and SCHÖNBOHM 1984a). A consistent new concept, however, cannot be presented yet. It seems therefore desirable to analyze in detail the interaction of calcium and phenolic compounds in the transduction chain.

There is another, apparently controversial point, which calls for improving the hypothesis. According to WAGNER and KLEIN (1981), P_{fr} is assumed to loosen the anchoring of actin microfilaments. The centrifugation experiments by SCHÖNBOHM and SCHÖNBOHM (1984b; see also references to earlier papers), however, clearly show fastening of the chloroplast's edge to the peripheral cytoplasm as an effect of P_{fr} formation, and this process depends on ATP in the same way as the low-intensity movement. Further interpretations should have in mind a dual function of actin-myosin interaction: Linking together the stationary and the moving compartment as a prerequisite for a shearing force to become effective, and generating the shearing force.

Detailed experiments have been performed also on the perception level, inducing the cells with single or repeated flashes in the millisecond range. Although P_{fr} certainly is the effector in low-intensity movement, accumulation of intermediates of the $P_r \rightarrow P_{fr}$ phototransformation can seriously influence the temporal pattern of P_{fr} formation and thus indirectly be important for the response (KRAML and SCHÄFER 1983; see also SCHÖNBOHM 1978, in this series). Moreover, with flashes of polarized light the kinetics of the dichroic reorientation of phytochrome molecules has been investigated (KRAML et al. 1984). At least one early intermediate, detectable until 5 ms after the flash, has still the surface-parallel orientation of P_{fr}, but after 30 ms all phytochrome molecules have attained the surface-normal orientation of P_{fr}. These results may be more interesting for phytochrome research in general than for elucidating chloroplast movement.

A more specific effect concerns the effects of blue light in *Mougeotia*. Besides the high-intensity response, which has an absolute requirement for blue light, also the low-intensity response can be induced by blue light, in this case replacing the usual red-light effect. It was an unsolved problem whether this blue-light effect is mediated by phytochrome, too, or by a separate and additional photoreceptor pigment (this series, vol. 40, 1978). A new approach by GABRYŠ et al. (1984) and WALCZAK et al. (1984) seems to give an unambiguous answer: The low-intensity response is brought about by 30-to-60 min continuous unilateral blue light, with or without a background of strong undirected far-red (FR); no difference can be found between the two treatments. As a control, a corresponding effect of red light (R) is completely blocked by FR, even if R is given high in saturation. Thus, the blue-light effect in low-intensity movement is certainly mediated by a separate photoreceptor pigment rather than by phytochrome. A very preliminary action spectrum points to a maximum near 450 nm (WALCZAK et al. 1984).

It should be noted, however, that these experiments have been performed with long-lasting irradiations, acting during the whole time of response. An extrapolation to inductive irradiations, with the movement proceeding in darkness, seems reasonable, but its validity still has to be tested.

Finally, the energetics of chloroplast movement in *Mougeotia* have been reinvestigated and earlier conclusions completed (SCHÖNBOHM and SCHÖNBOHM 1983). Accordingly, ATP as the general energy source is derived from oxidative phosphorylation in the usual low-intensity movement, when the response proceeds in darkness after a short pulse; it is derived from oxidative as well as from cyclic phosphorylation, if low-intensity movement is initiated by continuous light; and cyclic phosphorylation as the only ATP-providing process is responsible for the high-intensity movement. In all cases external ATP can substitute for the respiratory and/or assimilatory ATP.

d) Brown Algae

Recent data, obtained by the group of NULTSCH, mainly for *Dictyota dichotoma* and the European species of *Fucus* (NULTSCH et al. 1979, PFAU 1977, PFAU et al. 1979, PFAU et al. 1974, RÜFFER et al. 1978, RÜFFER et al. 1981), are a sound basis for future physiological analysis.

Presently, the most interesting results in brown algae deal with the ecological significance of the chloroplast movements (NULTSCH and PFAU 1979, NULTSCH et al. 1981). The usual interpretation of the low-intensity arrangement as a tool to optimize photosynthesis is not really questioned. In contrast, the high-intensity arrangement seems to have not always the function to protect the photosynthetic structures against damaging irradiations. There are several observations along this line (NULTSCH and PFAU 1979, NULTSCH et al. 1981). One important argument is derived from kinetic observations of chloroplast arrangement and photosynthesis under different light conditions. Upon transfer to high-intensity light, decrease in photosynthesis (with measurements taken in uniform light conditions) develops much faster than rearrangement of chloroplasts, the factor being at least 5:1. Similarly, recovery of photosynthetic capacity upon transfer back to low-intensity light again is much faster than the arrangement of chloroplasts to the low-intensity pattern. Moreover, a decrease in photosynthetic capacity develops in strong red light also, which cannot cause any chloroplast redistribution. As a further support of independence of the two light effects, the brown alga *Alaria* cannot exhibit chloroplast rearrangements at all; yet, strong light causes photosynthetic capacity to decline.

Although this is not of primary interest for the physiology of movement, it is a challenge to find another ecological meaning of the apparent "luxurious" response to high-intensity light, which nevertheless has not been contraselected in a long evolution.

There is, however, one observation which is closely related to the causality of chloroplast orientation in brown algae and which calls for further elucidation: The orienting effect of intracellular light gradients can be seriously modified by neighboring cells (RÜFFER et al. 1978, 1981). We do not know yet the nature of those endogenous factors; but it is reasonable to attribute the effects to chemical gradients, as has been proposed already by SENN (1908), who reported similar observations.

3. General Remarks

It is not yet possible to interprete all examples of intracellular movement by a unifying hypothesis. As far as light is the signal, controlling cytoplasmic streaming and orienting chloroplast arrangement, at least two completely different photoreceptor pigments are involved, viz. phytochrome and a blue-light absorbing pigment (usually assumed to be a flavin); but it is still open to discussion whether all blue-light effects are mediated by the same pigment (cf. the cycling hypothesis referred to for *Vaucheria*).

Transduction of light signals to the visible response probably involves changes in ion-transport properties of membranes; but we are far from establishing a general hypothesis, and additional messengers may be important (cf. the phenolic compounds in *Mougeotia*).

Finally, the mechanism of movement seems to be uniform throughout, viz. actin-myosin interaction. However, this assumption must not be generalized beyond the movements which are controlled directly by light. The circadian chloroplast movement of some green marine algae, on which light acts only indirectly via synchronizing the physiological clock, may behave differently. There is good evidence, especially in *Ulva*, that microtubules are more important in these systems than actin-myosin interaction (BRITZ 1979; cf. also references in HAUPT 1983), and a possible contribution of microtubuli in addition to actin as the main effector has been reported, in this review, for *Acetabularia*. Detailed experiments in additional systems are badly needed.

References

ALLEN, N.S.: Can. J. Bot. 58, 786-796 (1980). - ALLEN, R.D., ALLEN, N.S.: Protoplasma 109, 209-216 (1981).

BLATT, M.R.: Planta 159, 267-276 (1983). - BLATT, M.R., BRIGGS, W.R.: Planta 147, 355-362 (1980); - Photochem. Photobiol. 38, 347-353 (1983). - BLATT, M.R., WESSELS, N.K., BRIGGS, W.R.: Planta 147, 363-375 (1980). - BLATT, M.R., WEISENSEEL, M.H., HAUPT, W.: Planta 152, 513-526 (1981). - BRITZ, S.J.: Chloroplast and nuclear migration, pp. 170-205. In: Encyclopedia of Plant Physiology, N.S., Vol. VII, eds. W. HAUPT, M.E. FEINLEIB. Berlin, Heidelberg, New York: Springer 1979.

DAZY, A.C., PUSIEUX-DAO, S., DURAND, M., SANTA MARIA, A.: Protoplasma 105, 354 (1981). - DOREE, M., PICARD, A.: Experientia 36, 1291-1292 (1980). - DREYER, E.M., WEISENSEEL, M.H.: Planta 146, 31-39 (1979).

FISCHER-ARNOLD, G.: Protoplasma 56, 495-520 (1963). - FOISSNER, J., JAROSCH, R.: Cell Motility 1, 371-385 (1981).

GABRYŚ-MIZERA, H.: Photochem. Photobiol. 24, 453-461 (1976). - GABRYŚ, H., WALCZAK, T., ZURZYCKI, J.: Planta 152, 553-556 (1981). - GABRYŚ, H., WALCZAK, T., HAUPT, W.: Planta 160, 21-24 (1984). - GROLIG, F., VALENTIN, P., WAGNER, G.: Eur. J. Cell Biology (Suppl.) 2, 12 (1983)

HAUPT, W.: Sensory transduction and photobehaviors: Final considerations and emerging themes, pp. 397-404. In: Photoreception and Sensory Transduction in Aneural Organisms, eds. F. LENCI, G. COLOMBETTI. New York, London: Plenum Press 1980; - Annu. Rev. Plant Physiol. 33, 205-23 (1982). - Progr. Phycol. Res. 2, 227-281 (1983). - HAUPT, W., WAGNER, G.: Chloroplast movement, pp. 331-375. In: Membranes and Sensory Transduction, eds. G. COLOMBETTI, F. LENCI. New York, London: Plenum Press 1984 - HAYAMA, T., TAZAWA, M.: Cell Struct. Funct. 3, 47-60 (1978).

ISHIGAMI, M., NAGAI, R.: Cell Struct. Funct. 5, 13-20 (1980).

KAMITSUBO, E.: Protoplasma 109, 3-12 (1981). - KAMIYA, N.: Annu. Rev. Plant Physiol. 32, 205-236 (1981). - KLEIN, K., WAGNER, G., BLATT, M.R.: Planta 150, 354-356 (1980). - KOOP, H.U.: Protoplasma 109, 143-157 (1981). - KOOP, H.U., KIERMAYER, O.: Protoplasma 102, 147-166 (1980a); - ibid. 102, 295-306 (1980b). - KRAML, M., SCHÄFER, E.: Photochem. Photobiol. 38, 461-467 (1983). - KRAML, M., ENDERS, M., BÜRKEL, N.: Planta 161, 216-222 (1984).

LUCAS, W.J., ALEXANDER, J.M.: Plant Physiol. 68, 553-559 (1981).

NAGAI, R.: What's New in Plant Physiol. 10, 45-48 (1979). - NAGAI, R., FUKUI, S.: Protoplasma 109, 79-89 (1981). - NAGAI, R., HAYAMA, T.: J. Cell Sci. 36, 121-136 (1979). - NULTSCH, W., PFAU, J.: Marine Biol. 51, 77-82 (1979). - NULTSCH, W., RÜFFER, U., PFAU, J.: Helgoländer Wiss. Meeresunters. 32, 228-238 (1979). - NULTSCH, W., PFAU, J., RÜFFER, U.: Marine Biol. 62, 111-117 (1981).

PFAU, J.: Z. Pflanzenphysiol. 81, 17-25 (1977). - PFAU, J., THROM, G., NULTSCH, W.: Z. Pflanzenphysiol. 71, 242-260 (1974). - PFAU, J., RÜFFER, U., NULTSCH, W.: Ber. Dtsch. Bot. Ges. 92, 695-715 (1979).

REID, R.J., WALKER, N.H.: Aust. J. Plant Physiol. 10, 373-384 (1983). - ROBARDS, A.W.: Progr. Bot. 45, 1-17 (1983). - RÜFFER, U., NULTSCH, W., PFAU, J.: Helgoländer Wiss. Meeresunters. 31, 333-346 (1978). - RÜFFER, U., PFAU, J., NULTSCH, W.: Z. Pflanzenphysiol. 101, 283-293 (1981).

SCHÖNBOHM, E.: Progr. Bot. 40, 185-196 (1978). - SCHÖNBOHM, E., SCHÖNBOHM, E.: Biochem. Physiol. Pflanzen 178, 157-176 (1983); - ibid. 179, 489-505 (1984a); - ibid. 179, 473-487 (1984b). - SEITZ, K.: Z. Pflanzenphysiol. 57, 96-104 (1967); - Ber. Dtsch. Bot. Ges. 91, 455-458 (1978); - Cytoplasmic streaming and cyclosis of chloroplasts, pp. 150-169. In: Encyclopedia of Plant Physiology, N.S., Vol. VII, eds. W. HAUPT, M.E. FEINLEIB. Berlin, Heidelberg, New York: Springer 1979. - SEITZ, K.: Chloroplast motion in response to light in aquatic vascular plants, pp. 89-101. In: Studies on Aquatic Plants, eds. J.J. SYMOENS, S.S. HOOPER, P. COMPÈRE. Brüssel: Roy. Soc. Belg. 1982. - SEITZ, K., MAURER, B.: Influence of ATPase inhibitors on light dependent movements of chloroplasts in *Vallisneria*?, pp. 460-462. In: Blue Light Effects in Biological Systems, ed. H. SENGER. Berlin, Heidelberg, New York: Springer 1984. - SENN, G.: Die Gestalts- und Lageveränderung der Pflanzen-Chromatophoren. 397 S. Stuttgart: Engelmann 1908. - SERLIN, B.S., ROUX, S.J.: Proceed. Nat. Acad. Sci. USA, in press (1984). - SHIMMEN, T., TAZAWA, M.: Protoplasma 112, 101-106 (1982a); - ibid. 113, 127-131 (1982b); - Plant Cell Physiol. 23, 669-677 (1982c); - Protoplasma 115, 18-24 (1983).

TAKAGI, S., NAGAI, R.: J. Cell Sci. 62, 385-405 (1983). - TAZAWA, M., SHIMMEN, T.: Bot. Mag. Tokyo 95, 147-154 (1982). - TOMINAGA, Y., TAZAWA, M.: Protoplasma 109, 103-111 (1981a); - ibid. 109, 113-125 (1981b). - TOMINAGA, Y., SHIMMEN, T., TAZAWA, M.: Protoplasma 116, 75-77 (1983).

VAHEY, M., TITUS, M., TRAUTWEIN, R., SCORDILIS, S.: Cell Motility 2, 131-147 (1982).

WAGNER, G., KLEIN, K.: Photochem. Photobiol. 27, 137-140 (1978); - Protoplasma 109, 169-185 (1981). - WAGNER, G., ROSSBACHER, R.: Planta 149, 298-305 (1980). - WAGNER, G., VALENTIN, P., DIETER, P., MARMÉ, D.: Planta 162, 62-67 (1984). - WALCZAK, T., GABRYŚ, H., HAUPT, W.: Flavin-mediated weak-light chloroplast movement in *Mougeotia*, pp. 454-459. In:

Blue Light Effects in Biological Systems, ed. H. SENGER. Berlin, Heidelberg, New York: Springer 1984. - WILLIAMSON, R.E.: Cytobiologie 18, 107-113 (1978). - Can. J. Bot. 58, 766-772 (1980).

YAMAGUCHI, Y., NAGAI, R.: J. Cell Sci. 48, 193-205 (1981).

ZURZYCKI, J., LEŁATKO, Z.: Acta Soc. Bot. Pol. 38, 493-506 (1969). - ZURZYCKI, J., WALCZAK, T., GABRYŚ, H., KAJFOSZ, J.: Planta 157, 502-510 (1983).

Professor Dr. WOLFGANG HAUPT
Professor Dr. KONRAD SEITZ
Institut für Botanik und
Pharmazeutische Biologie der
Universität Erlangen-Nürnberg
Schloßgarten 4
D-8520 Erlangen, FRG

C. Genetics

I. Replication

By RUDOLF EICHENLAUB

1. Introduction

In continuation to my previous review (EICHENLAUB 1982) I would like to present recently published data on special subjects, restricting myself mainly on regulation and termination of DNA replication and partitioning of genomes in prokaryotes.

For further reading on this field of research the *1982 Supplement to DNA Replication* by KORNBERG (1982) and review articles by NOSSAL (1983), NORDSTRÖM (1983) and SCOTT (1984) are recommended.

2. Cloning of Replication Proteins

A central problem in the biochemical and physical characterization of the numerous proteins involved in the initiation and elongation stages of DNA replication is the preparation of large quantities of purified proteins. Based on the specific activity of purified proteins and the activity found in cells, the calculated number of molecules per cell of many of the DNA replication proteins is only between 20 and 200 (see KORNBERG 1980), resulting in low yields, preventing many desirable studies. One way to overcome this problem is to transfer the respective gene of a protein on a phage or plasmid molecule. Overproduction by temperature induction of lysis-defective lambda phages has been achieved for DNA polymerase I (Kelley et al. 1977), DNA ligase (PANASENKO et al. 1977), *dnaZ*-protein, the gamma subunit of DNA polymerase III, primase (*dnaG*-protein), and *dnaC*-protein from *E. coli* (ROWEN et al. 1982).

High copy number vectors like pBR322, or plasmid R1 derivatives, with temperature-inducible runaway replication (UHLIN et al. 1979) resulting in a strong gene dosage effect, were employed for the cloning of *dnaB* (ARAI et al. 1981), *dnaA*, *dnaZ*, and ssb-proteins (YASUDA and TAKAGI 1983), and *dnaE* (alpha subunit of DNA polymerase III) (WELCH and MCHENRY, 1982).

Gene technology has derived great benefit from this work, and more detailed information on the function and primary structure of such proteins has now been obtained (*dnaC:* KOBORI and KORNBERG 1982a, 1982b, 1982c, *dnaA* and *dnaN:* HANSEN et al. 1982; *dnaB:* GÜNTHER et al. 1984).

Now that stringently expressed replication proteins have become available in large quantities, it has to be expected that more information on the properties will emerge and that enrichment of cell extracts with these proteins will allow the establishment of more in vitro replication systems.

Progress in Botany, Vol. 46

Progress of this kind has recently been reported for the in vitro replication of oriC plasmids (FULLER et al. 1981, KAGUNI et al. 1982), bacteriophage lambda (WOLD et al. 1982) lamda-dv plasmids (TSURIMOTO and MATSUBARA 1982, ANDERL and KLEIN 1982) and some bacterial plasmids (DIAZ and STAUDENBAUER 1982, DIAZ et al. 1981, SCHERZINGER et al. 1984).

An in vitro replication system has also been established for the *Bacillus subtilis* phage phi29 (PENALVA and SALAS 1982, WATABE et al. 1982) and plasmids of Gram-negative bacteria will certainly follow.

3. Regulation of DNA Replication

During the past years a large body of information on this subject has been accumulated (see NORDSTRÖM 1983, SCOTT 1984), showing that there are some common mechanisms in the regulation of the initiation of DNA replication, although the total picture is still puzzling.

On the regulation of DNA replication of the *E. coli* chromosome little is yet known. Several years ago plasmids replicating by the chromosomal origin, oriC, were constructed (YASUDA and HIROTA 1977, MESSER et al. 1978, ZYSKIND et al. 1979) but further investigations have not revealed anything about replication control. It seems that oriC plasmids may not at all provide any insights into this process. The chromosome has a strict replication pattern in which replication is initiated at a critical cell mass (DONACHIE 1968), and underreplication at fast growth rates is prevented by a premature simultaneous initiation at the two origins before a round of replication is terminated. In contrast, oriC plasmids show a broad variation in copy number, replication throughout the cell cycle (LEONARD et al. 1982), and they do not express incompatibility with the host chromosome (for review on incompatability see TIMMIS 1979). All these characteristics suggest that oriC plasmids do not carry any regulatory elements. If such elements exist in *E. coli*, they seem to map some distance from the origin region.

In this respect plasmids have been more yielding objects of investigation. Some systems studied in detail revealed a negative control of initiation; an example of this kind, ColE1, has been reviewed previously (EICHENLAUB 1982). Negative control, which has been originally proposed by PRITCHARD (1969) has now been demonstrated in different plasmids of the incFII group represented by plasmids R1, R100, and R6-5 (DANBARA et al. 1981, STOUGAARD et al. 1981a,b, LIU et al. 1983). The common principles observed are: (1) the inhibitory elements are either small, basic polypeptides of about 80-100 amino acids or small RNA molecules of 80-110 nucleotides, (2) the inhibitors act in trans, (3) their target site on the DNA is not the origin itself. It is a striking observation that the small, high copy number plasmid ColE1 and the large, low copy number R factor R1 have very similar control mechanisms involving an inhibitory RNA molecule and a small inhibitor protein. However, the consequences of inhibition in both systems exerted by the negative control elements are different.

In plasmid R1 adjacent to the origin of replication three strong promotor sites are located (LURZ et al. 1981, EASTON et al. 1981).

Site 1 and site 2 promote overlapping transcription toward the origin producing RNA-C and RNA-II, respectively (EASTON et al. 1981, ROSEN et al. 1981). RNA-C encodes the genes *copB* and *repA* (STOUGAARD et al.

1981a,b, DANBARA et al. 1981), while RNA-II only encodes the *repA* protein.

The third promoter is located between the gene loci *repA* and *copB*, giving rise to RNA-I, which is transcribed to the opposite direction, thus consisting of a complementary nucleotide sequence of RNA-II and RNA-C (LIGHT and MOLIN 1981, 1982, 1983). The *repA* protein is required for initiation of replication in plasmid R1, as demonstrated by in vitro replication (DIAZ et al. 1981, DIAZ and STAUDENBAUER 1982, MASAI et al. 1983) and its expression is negatively regulated by protein *copB* and RNA-I (LIGHT and MOLIN 1981, 1982, 1983, STOUGAARD et al. 1981a,b).

The inhibitor RNA-I exerts a post-transcriptional regulation of *repA* expression probably by base pairing with the complementary sequence of RNA-C outside the coding region of the *repA* protein (DANBARA et al. 1981, LIGHT and MOLIN 1983). This interaction may change the secondary structure of RNA-C in a way to prevent translation or processing of the RNA molecule.

The *copB* repressor proten (9700 MW): (ROSEN et al. 1980, RYDER et al. 1981, STOUGAARD et al. 1981a,b, LIU et al. 1983) on the other hand prevents formation of RNA-II by binding to the promoter of RNA-II (LIGHT and MOLIN 1981, 1982), thus also inhibiting the production of the essential *repA* protein.

The interaction of the components of this dual control mechanism, which in plasmid R1 regulates the expression of the *repA* protein and in ColE1 (*rop*-protein and RNA-1) regulates RNA-primer formation (ITOH and TOMIZAWA 1980, CESARENI et al. 1982), is up to now not clearly understood.

In contrast, in some plasmids a positive control of replication has been observed (F: MUROTSU et al. 1981, TOLUN and HELINSKI 1981, EICHENLAUB 1984; R6K: SHAFFERMAN et al. 1982; phage P1: AUSTIN et al. 1982, FROEHLICH et al. 1983).

The minimal replicon region of the F plasmid is confined to a region of about 2250bp (F-coordinates 44.1-46.35F) (MUROTSU et al. 1981). This region, which has been sequenced, contains a functional origin of replication, encodes a polypeptide of 29000 MW, protein E, and has two region flanking the protein E gene containing four and five repeated nucleotide sequences of 19 base pairs, respectively (MUROTSU et al. 1981, TOLUN and HELINSKI 1981). Recent studies have elucidated possible functions of the E protein and the repeats. Protein E is strictly required for F replication as was demonstrated in trans-complementation experiments (EBBERS and EICHENLAUB 1981, TOLUN and HELINSKI 1982, WATSON et al. 1982) and by the isolation of amber mutations within the protein E gene (EICHENLAUB 1984, MAKI et al. 1983). Furthermore, there are also mutations in protein E which affect copy number (EICHENLAUB 1984, KLINE and TRAWICK 1983). In three such Cop-mutants sequence analysis disclosed the amino acid exchanges within the E protein. This suggests that the E protein is also responsible for copy number control, a surprising fact, since this was believed to be the domain for negatively acting control elements.

Cloning of the 19 base pair repeats on high copy number plasmid vectors resulted in an incompatibility reaction against another F plasmid (TOLUN and HELINSKI 1981, TSUTSUI et al. 1983). The intensity of the incompatibility reaction increases with the copy number of the vector and number of repeats cloned. The interpretation is that the exclusion of another F plasmid in the cell (incompatibility) by the repeats is caused by binding of the E protein to the repeats, thus dropping the concentration of the E protein below the critical level required for the initiation of DNA replication. Furthermore, binding of the E protein to the repeats adjacent to the promoter of the E gene seems to repress its expression (SOGAARD-ANDERSEN et al. 1984). This indicates that the E protein has also an autoregulatory function, in accordance with the autorepressor model (SOMPAYRAC and MAALOE, 1973). However, it is not clear how binding to the repeats downstream from the E gene stimulates initiation of replication at the origin. Although some data suggest that negative control may also exist in F (KLINE and SEELKE 1982), tentative repressors have not yet been identified. It may be the case that the small polypeptides described recently mapping close to the origin region, have such function (BEX et al. 1983).

A regulatory system for the initiation of DNA synthesis similar to F is found in plasmid R6K, an antibiotic resistance factor having a size of 38 kb and an intermediate copy number of about 15 (KONTOMICHALOU et al. 1970).

Replication of R6K is depending on a protein, termed π, encoded by the gene *pir*, mapping close to the gamma origin of replication, as shown by transcomplementation experiments (KOLTER et al. 1978). The π-protein is a 34.000 MW basic protein (GERMINO and BASTIA 1982, STALKER et al. 1982) and thus may be suited for binding to DNA, which presumably is neccessary for its function in the initiation of R6K replication. The π-protein is unstable, but has recently been purified by the construction of a gene fusion, linking the region of the alpha fragment of the β galactosidase to the *pir* gene (GERMINO and BASTIA 1983). The fusion protein was shown to bind specifically to seven direct repeats of 22 base pairs in the region of origin gamma, which are required for replication (GERMINO and BASTIA 1983, KOLTER and HELINSKI 1982). The π-protein furthermore binds to a lone repeat at the promoter of the *pir* gene (GERMINO and BASTIA 1983) supporting the idea that this single repeat may function as an operator for autoregulation of *pir* gene expression (STALKER et al. 1981).

Recent experiments indicated that the π-protein does not regulate the frequency of initiation of replication (SHAFFERMAN et al. 1982), since a variation of the π-protein concentration up to almost 100-fold did not affect the copy number of R6K.

Surprisingly, however, the isolation of copy number mutants which map within the π-protein coding sequence was recently reported (STALKER et al. 1983). This can either mean that the π-protein of R6K has a dual function in initiation of replication and regulation of plasmid copy number, or that the DNA sequence encoding the π-protein may in addition promote the formation of a negative regulatory RNA. As is apparent, the components involved in the regulation of replication of plasmid R6K resemble very much those described for plasmid F.

4. Termination

It is well documented that the initiation of DNA replication requires a unique nucleotide sequence, the origin, providing the special features for interaction with a variety of proteins. Sequence analysis of origin regions has revealed common features among origins of various sources (for references see EICHENLAUB 1982).

Not as clear is the status of the antipode of the origin, the site of termination of DNA replication, the terminus. Two systems in which termination occurs at specific sites have been investigated in some detail.

In plasmid R6K the terminus has been cloned and subsequently its nucleotide sequence has been determined (BASTIA et al. 1981a,b). Disappointigly, neither study so far gave any clues on the possible mechanism by which the replication forks are delayed within the terminus region.

In *Escherichia coli* the terminus *(terC)* has been mapped in a region between map positions 30 to 35 min close to the locus *rac* (for references see EICHENLAUB 1982). Unfortunately, this region is located within a cotransductional gap, which is the largest silent region of the *Escherichia coli* chromosome (BACHMANN and LOW 1980). Recent efforts brought some light at least into the physical organization of this region and mapped the terminus within a 100 kb region at map position 31.2 min (BOUCHE 1982, BOUCHE et al. 1982). However, deletions created by use of transposon Tn10 extending into the *terC* region were not accompanied by any phenotypic alteration in the mutants (HENSON et al. 1984). All *terC* deletion mutants exhibited a normal growth rate, viability, and cell morphology. Therefore, HENSON et al. (1984) discuss the possibility that other sites close to the *terC* region in the *E. coli* chromosome may also function as termini when the primary terminus is deleted.

So far it is not clear whether a terminus of DNA replication is required in *E. coli* at all. The *terC* region may slow down fork movement just by a peculiar nucleotide sequence.

On the other hand, if one remembers the remarkable instability of plasmids replicating by the oriC of *E. coli*, which seems to be due to the lack of a partitioning function, one is tempted to speculate that the *terC* site may be involved in the partitioning event.

5. Partitioning

The observation that deletion of regions within plasmid genomes not affecting replication but rendering the plasmids inheritably unstable initiated closer examination. It became apparent that those regions encoded partitioning functions *(par)* (NORDSTRÖM 1980a,b, MIKI et al. 1980, IKE et al. 1981, PHUA et al. 1982, HOGAN et al. 1982).

At present, the partitioning mechanism of the F plasmid has been disclosed in some detail. OriC plasmids, as mentioned before, using the replication origin of the *E. coli* chromosome are not stably maintained during cell growth (YASUDA and HIROTA 1977, MIKI et al. 1978, VON MEYENBURG et al. 1978, MESSER et al. 1978). This instability is possibly due to the lack of the partitioning mechanism of the chromosome. Construction of an oriC: miniF hybrid plasmid in which the F origin was deleted resulted in stable inheritance of the composite plasmid. This suggestet that (1) oriC plasmids are stabilized by partitioning functions of F and (2) that the partitioning mechanism of F acts independently from F replication (OGURA et al. 1980).

Insertions of transposon Tn3 and deletion of restriction fragments from the miniF genome have helped to identify a region of about 3 kb (F-coordinates 46.35-49.3F) to be required for stable inheritance of the F plasmid (BERGQUIST et al. 1981, EBBERS and EICHENLAUB 1981), which is outside of the region essential for replication (MUROTSU et al. 1981).

The *par* region encodes two polypeptides of 44.000 MW and 36.000 MW (WEHLMANN and EICHENLAUB 1980, KOMAI et al. 1982) and a gene locus termed *incD* within a 360 bp region, responsible for incompatibility with plasmids of the incFI group (GARDNER et al. 1982, PHUA et al. 1982).

All three components of the *par* region seem to be required for equipartitioning (OGURA and HIRAGA 1983, AUSTIN and WIERZBICKI 1983). In transcomplementation experiments it was found that the 44.000 MW and 36.000 MW polypeptides (also termed *sopA* and *sopB*, respectively) specify trans-acting functions and that the *incD* region (termed *sopC)* is acting only in cis. Based on recent experiments, a model for the partioning of the F plasmid was proposed (OGURA and HIRAGA 1983).

The 360 op *sopC* region is designed to act as the specific site where partitioning actually occurs, derived from its exclusive activity in cis. The *sopB* protein specifically binds to the *sopC* site. This site however, seems to interact not only with the *sopB* protein but possibly with two more proteins of 75.000 MW and 33.000 MW specified by the *E. coli* host, to constitute the partitioning apparatus which is believed to reside in or at the cellular membrane. The function of the *sopA* protein is still unclear. It is possible that this polypeptide either also interacts with the *sopC* region or that it regulates the expression of the *sopB* gene.

In contrast, partitioning of plasmid pSC101 does not involve any plasmid-encoded trans-acting function, but only requires a cis-acting small DNA region of about 270 bp (MEACOCK and COHEN 1980). In case protein factors are involved in pSC101 partitioning as well, which seems to be indispensable, these have to be supplied by the bacterial host. In this context, it is worthwhile to mention that pSC101 is one of the few plasmide requiring the initiation function *dnaA* of the *E. coli* host (HASUNUMA and SEKIGUCHI 1977) and thus may also depend completely on host function with respect to partitioning.

These encouraging results are currently initiating experiments in other genetic systems, such as phage P1 and various plasmids, in order to determine a common feature for partitioning of DNA molecules in *Escherichia coli*. The outcome may be that for partitioning the mechanisms developed are as numerous and complicated as for the initiation and regulation of DNA replication.

References

ANDERL, A., KLEIN, A.: Nucl. Acids Res 10, 1733-1740 (1982). - ARAI, K., YASUDA, S., KORNBERG, A.: J. Biol. Chem 256, 5247-5252 (1981). - AUSTIN, S., WIERZBICKI, A.: Plasmid 10, 73-81 (1983). - AUSTIN, S., HART, F., ABELES, A., STERNBERG, N.: J. Bacteriol. 152, 63-71 (1982).

BACHMANN, B., LOW, K.B.: Microbiol. Rev. 44, 1 56 (1980). - BASTIA, D., GERMINO, J., CROSA, J.H., HALE, P.: Gene 14, 81-89 (1981a). - BASTIA, D., GERMINO, J., CROSA, J.H., RAM, J.: Proc. Natl. Acad. Sci. USA 78, 2095-2099 (1981b). - BERGQUIST, P.L., DOWNARD, R.A., CAUGHEY, P.A., GARDNER, R.C., LANE, H.E.D.: J. Bacteriol. 147, 888-899 (1981). - BEX, F., KAROUI, H., ROKEACH, L., DREZE, P., GARCIA, L., COUTURIER, M.: EMBO J. 2, 1853-1861 (1983). - BOUCHE, J.P.: J. Mol. Biol. 154, 1-20 (1982). - BOUCHE, J.P., GELUGNE, J.P., LOUARN, J., LOUARN, J.M.: J. Mol. Biol. 154, 21-32 (1982).

CESARENI, G., MÜSING, M.A., POLISKY, B.: Proc. Natl. Acad. Sci. USA 79, 6313-6317 (1982).

DANBARA, H., BRADY, G., TIMMIS, J.K., TIMMIS, K.N.: Proc. Natl. Acad. Sci. USA 78, 4699-4703 (1981). - DIAZ, R., STAUDENBAUER, W.L.: Nucl. Acids Res. 10, 4687-4702 (1982). - DIAZ, R., NORDSTRÖM, K., STAUDENBAUER, W.L.: Nature 289, 326-328 (1981). - DONACHIE, W.D.: Nature 219, 1077-1079 (1968).

EASTON, A.M., SAMPATHKUMAN, P., ROWND, R.H.: Incompatibility of IncFII R plasmid NR1, 125-141. In: Initiation of DNA Replication, ed. D.S. RAY. New York: Academic Press 1981. - EBBERS, J., EICHENLAUB, R.: J. Bacteriol. 147, 736-743 (1981). - EICHENLAUB, R.: Progr. Bot. 44, 231-243 (1982); - Development of plasmids and cloning procedures. In: Developments in Molecular Virology, ed. Y. BECKER. The Hague: Nijhoff 1984 (in press).

FROEHLICH, B.J., TATTI, K., SCOTT, J.R.: J. Bacteriol. 156, 205-211 (1983). - FULLER, R.S., KAGUNI, J.M., KORNBERG, A.: Proc. Natl. Acad. Sci. USA 78, 7370-7374 (1981).

GARDNER, R.C., MALCOLM, L., BERGQUIST, P.L., LANE, H.E.D.: Mol. Gen. Genet. 188, 345-352 (1982). - GERMINO, J., BASTIA, D.: Proc. Natl. Acad. Sci. USA 79, 5475-5479 (1982); - Cell 32, 131-140 (1983). - GÜNTHER, E., BAGDASARIAN, M., SCHUSTER, H.: Mol. Gen. Genet. 193, 225-230 (1984).

HANSEN, B.E., HANSEN, F.G., VON MEYENBURG, K.: Nucl. Acids Res. 10, 7373-7385 (1982). - HASUNUMA, K., SEKIGUCHI, M.: Mol. Gen. Genet. 154, 225-230 (1977). - HENSON, J.M., KOPP, B., KUEMPEL, P.L.: Mol. Gen. Genet. 193, 263-268 (1984). - HOGAN, J.E., KLINE, B.C., LEVY, S.B.: Plasmid 8, 36-44 (1982).

IKE, Y., HASHIMOTO, H., MITSUHASHI, S.: J. Bacteriol. 147, 578-588 (1981). - ITOH, T., TOMIZAWA, J.: Proc. Natl. Acad. Sci. USA 77, 2450-2454 (1980).

KAGUNI, J.M., FULLER, R.S., KORNBERG, A.: Nature 296, 623-627 (1982). - KELLEY, W.S., CHALMERS, K., MURRAY, N.E.: Proc. Natl. Acad. Sci. USA 74, 5632-5636 (1977). - KLINE, B.C., SEELKE, R.W.: Mol. Gen. Genet. 187, 218-224 (1982). - KLINE, B.C., TRAWICK, J.: Mol. Gen. Genet. 192, 408-415 (1983). - KOBORI, J.A., KORNBERG, A.: J. Biol. Chem. 257, 13757-12762 (1982a); - ibid. 257, 13763-13769 (1982b); - ibid. 257, 13770-13775 (1982c). - KOLTER, R., HELINSKI, D.R.: J. Mol. Biol. 161, 45-56 (1982). - KOLTER, R., INUZUKA, M., HELINSKI, D.R.: Cell 15, 1199-1208 (1978). - KOMAI, N., NISHIZAWA, T., HAYAKAWA, Y., MUROTSU, T., MATSUBARA, K.: Mol. Gen. Genet. 186, 193-203 (1982). - KONTOMICHALOU, P., MITANI, M., CLOWES, R.C.: J. Bacteriol. 104, 34-44 (1970). - KORNBERG, A.: DNA Replication. San Francisco: Freeman 1980; - 1982 Supplement to DNA Replication. San Francisco: Freeman 1982.

LEONARD, A.C., HUCUL, J.A., HELMSTETTER, C.E.: J. Bacteriol. 149, 499-507 (1982). - LIGHT, J., MOLIN, S.: Mol. Gen. Genet. 184, 56-61 (1981); - ibid. 187, 486-493 (1982); - EMBO J. 2, 93-98 (1983). - LIU, C., CHURCHWARD, G., CARO, L.: Plasmid 10, 148-155 (1983). - LURZ, R., DANBARA, H., RÜCKER, B., TIMMIS, K.N.: Mol. Gen. Genet. 183, 490-496 (1981).

MAKI, S., KURIBAYASHI, M., MIKI, T., HORIUCHI, T.: Mol. Gen. Genet. 191, 231-237 (1983). - MASAI, N., KAZIRO, Y., ARAI, K.-J.: Proc. Natl. Acad. Sci. USA 80, 6814-6818 (1983). - MEACOCK, P.A., COHEN, S.N.: Cell 20, 529-542 (1980). - MESSER, W., BERGMANS, H.E.N., MEIJER, M., WOMACK, J.E., HANSEN, J.E.: Mol. Gen. Genet. 162, 269-275 (1978). - MIKI, T., HIRAGA, S., NAGATA, T., YURA, T.: Proc. Natl. Acad. Sci. USA

75, 5099-5103 (1978). - MIKI, T., EASTON, A.M., ROWND, R.H.: J. Bacteriol. 141, 87-99 (1980). - MUROTSU, T., MATSUBARA, K., SUGISAKI, H., TAKANAMI, M.: Gene 15, 257-271 (1981).

NORDSTRÖM, K., MOLIN, S., AAGAARD-HANSEN, H.: Plasmid 4, 332-349 (1980a); - ibid. 4, 215-227 (1980b). - NORDSTRÖM, K.: Plasmid 9, 1-7 (1983). - NOSSAL, N.G.: Annu. Rev. Biochem. 53, 581-615 (1983).

OGURA, T., HIRAGA, S.: Cell 32, 351-360 (1983). - OGURA, T., MIKI, T., HIRAGA, S.: Proc. Natl. Acad. Sci. USA 77, 3993-3997 (1980).

PANASENKO, S., CAMERON, J., DAVIS, R.W., LEHMAN, J.R.: Science 196, 188-189 (1977). - PENALVA, M.A., SALAS, M.: Proc. Natl. Acad. Sci. USA 79, 5522-5526 (1982). - PHUA, S.H. BERGQUIST, P.L., LANE, H.E.D.: Mol. Gen. Genet. 188, 353-355 (1982). - PRITCHARD, R.H., BARTH, P.T., COLLINS, J.: Symp. Soc. Gen. Microbiol. 19, 263-297 (1969).

ROSEN, J., RYDER, T., INOKUCHI, H., OHTSUBO, H., OHTSUBO, E.: Mol. Gen. Genet. 179, 527-537 (1980). - ROSEN, J., RYDER, T., OHTSUBO, H., OHTSUBO, E.: Nature 290, 794-799 (1981). - ROWEN, L., KOBORI, J.A., SCHERER, S.: Mol. Gen. Genet. 187, 501-509 (1982). - RYDER, T., ROSEN, J., ARMSTRONG, K., DAVIDSON, D., OHTSUBO, E., OHTSUBO, H.: Dissection of the replication region controlling incompatibility, copy number, and initiation of DNA synthesis in the resistance plasmids R100 and R1, 91-111. In: Initiation of DNA Replication, ed. D.S. RAY. New York: Academic Press 1981.

SCHERZINGER, E., BAGDASARIAN, M.M., SCHOLZ, P., LURZ, R., RÜCKERT, B., BAGDASARIAN, M.: Proc. Natl. Acad. Sci. USA 81, 654-658 (1984). - SCOTT, J.R.: Microbiol. Rev. 48, 1-23 (1984). - SHAFFERMAN, A., KOLTER, R., STALKER, D., HELINSKI, D.R.: J. Mol. Biol. 161, 57-76 (1982). - SOGAARD-ANDERSEN, L., ROKEACH, L.A., MOLIN, S.: EMBO J. 3, 257-262 (1984). - SOMPAYRAC, L., MAALLOE, O.: Nature 241, 133-135 (1973). - STALKER, D.M., THOMAS, C., HELINSKI, D.R.: Mol. Gen. Genet. 181, 8-12 (1981). - STALKER, D.M., KOLTER, R., HELINSKI, D.R.: J. Mol. Biol. 161, 33-43 (1982). - STALKER, D.M., FILUTOWICZ, M., HELINSKI, D.R.: Proc. Natl. Acad. Sci. USA 80, 5500-5504 (1983). - STOUGAARD, P., MOLIN, S., NORDSTRÖM, K.: Proc. Natl. Acad. Sci. USA 78, 6008-6012 (1981a). - STOUGAARD, P., MOLIN, S., NORDSTRÖM, K., HANSEN, G.: Mol. Gen. Genet. 181, 116-122 (1981b).

TIMMINS, K.N.: Mechanisms of plasmid incompatibility, 13-22. In: Plasmids of Medical, Environmental, and Commercial Importance, eds. K.N. TIMMIS, A. PÜHLER. New York: Elsevier/North-Holland Biomedical Press 1979. - TOLUN, A., HELINSKI, D.R.: Cell 24, 687-694 (1981); - Mol. Gen. Genet. 186, 372-377 (1982). - TSURIMOTO, T., MATSUBARA, K.: Proc. Natl. Acad. Sci. USA 79, 7639-7643 (1982). - TSUTSUI, H., FUJIYAMA, A., MUROTSU, T., MATSUBARA, K.: J. Bacteriol. 155, 337-344 (1983).

UHLIN, B.E., MOLIN, S., GUSTAFSSON, P., NORDSTRÖM, K.: Gene 6, 91-106 (1979).

VON MEYENBURG, K., HANSEN, F.G., NIELSEN, L.D., RIISE, E.: Mol. Gen. Genet. 160, 287-295 (1978).

WATABE, K., SHIH, M.-F., SUGINO, A., ITO, J.: Proc. Natl. Acad. Sci. USA 79, 5245-5248 (1982). - WATSON, L.A., PHUA, S.H., BERGQUIST, P.L., LANE, H.E.D.: Gene 19, 173-178 (1982). - WEHLMANN, H., EICHENLAUB, R.: Mol. Gen. Genet. 180, 205-211 (1980). - WELCH, M.M., MCHENRY, C.S.: J. Bacteriol. 152, 351-356 (1982). - WOLD, M.S., MALLORY, J.B., ROBERTS, J.D., LEBOWITZ, J.H., MCMACKEN, R.: Proc. Natl. Acad. Sci. USA 79, 6179-6180 (1982).

YASUDA, S., HIROTA, Y.: Proc. Natl. Acad. Sci. USA 74, 5458-5462 (19 (1977). - YASUSDA, S.- TAKAGI, T.: J. Bacteriol. 154, 1153-1161 (1983).

ZYSKIND, J.W., DEEN, L.T., SMITH, D.W.: Proc. Natl. Acad. Sci. USA 76, 3097-3101 (1979).

Professor Dr. RUDOLF EICHENLAUB
Universität Hamburg
Institut für Allgemeine Botanik
Arbeitsbereich Genetik
Ohnhorststraße 18
D-2000 Hamburg 52, FRG

II. Recombination

a) DNA Transformation in Cyanobacteria, Algae and Fungi: Trends and Perspectives

By ULRICH KÜCK and BERNARD J. MULLIGAN

1. Introduction

During the last few years several reviews in this series have described genetic recombination by DNA transfer in eukaryotes, especially in the yeast *Saccharomyces cerevisiae* and other fungi. Because of the importance of transformation as a tool for basic and applied research, in blue-green and green algae, we will extend this chapter to review recent transformation experiments in both cyanobacteria[1] and algae.

In cyanobacteria, especially, numerous vectors have been constructed to serve as vehicles for transferring genetic material into the host organism. In addition, we shall discuss current trends in transformation in *Saccharomyces cerevisiae* and other yeasts and, further, review the transformation systems recently developed in various filamentous fungi. The latter organisms are of great importance from a commercial point of view in providing materials for use in the production of various drugs and foodstuffs.

Previous reviewers (HOWELL 1982, WENZEL 1983) have remarked on the difficulties associated with different eukaryotic transformation systems. Compared with bacterial systems, the transformation frequency and stability of eukaryotic transformants seem to be low, but there several successful attempts have been made to optimize the replication efficiency of recombinant vector molecules and the expression of genes carried on those vectors. For example, various host-specific sequences, either promoting autonomous replication or increasing the rate of transcription and translation, have been used in the construction of versatile vectors. The difficulties of achieving recombination by DNA transfer in higher plants are similar to those mentioned above.

Plant systems have been reviewed by HOWELL (1982) and by WENZEL (1983) in recent issues of this series. The transfer of DNA by transformation has been reviewed by STRUHL (1983).

2. Principles of Transformations in Cyanobacteria

The cyanobacteria are a group of photosynthetic prokaryotes of great diversity (STANIER and COHEN-BAZIRE 1977). Their oxygenic mode of

[1] Formerly cyanobacteria has been named as "prokaryotic algae" or "blue-green algae"

Progress in Botany, Vol. 46

photosynthesis, very similar to that of higher plants, and the ability of some species to fix molecular nitrogen make these organisms important subjects for research. An understanding of cyanobacterial genetics would be of immense value in probing the processes of nitrogen fixation and photosynthesis at the molecular level. The genetics of cyanobacteria is the subject of a recent, comprehensive review by HERDMAN (1982). In this article we will concentrate on transformation methods in cyanobacteria since this branch of cyanobacterial genetics is currently developing at a great rate.

SHESTAKOV and KHYEN (1970) were the first to provide a convincing demonstration of transformation of cyanobacteria by purified DNA. Using the unicellular organism *Anacystis nidulans* strain 602, they were able to show stable transfer of *ery*, *str* and *fil* (filamentous growth character) markers to wild-type cells following incubation with chromosomal DNA from three mutant strains of the organism.

Transformation by purified chromosomal DNA has also been achieved in other unicellular cyanobacteria including *Anacystic nidulans* UTEX 625 (HERDMANN 1973), *Aphanocapsa* 6714 (ASTIER and ESPARDELLIER 1976), *Gloeocapsa alpicola* (DEVILLY and HOUGHTON 1977) *Agmenellum quadruplicatum* (STEVENS and PORTER 1980) and *Synechocystis* 6803 (GRIGORIEVA and SHESTAKOV 1982). In these papers, transformation efficiency varied over a range of approximately 1×10^{-6} to 4×10^{-3} depending on the organsism and the marker under study.

Transformation by purified DNA has not yet been described in filamentous cyanobacteria. However, gene transfer may occur in mixed cultures of some filamentous strains (see HERDMANN 1982). In *Nostoc muscorum*, for example, marker transfer between wild type and a mutant, *het- eth^R val-* (nonheterocyst-producing, ethionine-resistant, valine auxotroph) occurred with frequencies between 10^{-5} - 10^{-3}.

The transfer was bidirectional, and, apparently, DNase-sensitive and occurred even when the donor cell was killed by heat treatment (TREHAN and SINHA 1981). Presumably, this result reflected transformation by donor DNA which had accumulated in the growth medium. This phenomenon had previously been amply documented by HERDMANN (1973, 1982) for cultures of *Anacystis nidulans* UTEX 625.

The widespread occurrence of restriction endonucleases in filamentous cyanobacteria (e.g., DUYVESTEYN et al. 1983) goes some way to explaining the lack of a well-described transformation system for these organisms. Recently, WOLK et al. (1984) have described a method of conjugative transfer to introduce plasmid vectors into *Anabaena* strains. This "indirect transformation" system may point the way to a general method for the introduction of DNA into these organisms; it is described at the end of this section.

In all cyanobacteria successfully transformed to date, the uptake of exogenous DNA is a "natural" process, the recipient cells possessing an inherent competence for transformation. The availability of a highly transformable strain, *Anacystis nidulans R-2* (GRIGORIEVA and SHESTAKOV 1976) has facilitated recent studies. This organism houses two plasmids (VAN DEN HONDEL et al. 1980) which, like all plasmids of cyanobacteria, are genetically cryptic (HERDMAN 1982). One of these, pUH24 (7.9 Kb), has formed the basis for the development of plasmid shuttle vectors capable of replication in both *Anacystis nidulans* and *E. coli* (GOLDEN and SHERMAN 1983, KUHLEMEIER et al. 1981, SHERMAN and VAN DE PUTTE 1982, VAN DEN HONDEL et al. 1980) and also cosmid shuttle vectors (TANDEAU DE MARSAC et al. 1982).

Various antibiotic resistance markers (ampicillin, chloramphenicol, streptomycin), derived from *E. coli* plasmids and a β-lactamase transposon (Tn 901), have proved valuable as genetic markers in *Anacystis nidulans R-2* (CHAUVAT et al. 1983, GOLDEN and SHERMAN 1983, KUHLEMEIER et al. 1981, SHERMAN and VAN DE PUTTE 1982, TANDEAU DE MARSAC et al. 1982, VAN DEN HONDEL et al. 1980); but these markers may confer much lower levels of antibiotic resistance on the cyanobacterium than on *E. coli*.

For example, the Ap^R gene of *Tn901* confers a resistance level to ampicillin of 0.5-2 µg/ml and 50 µg/ml in *A. nidulans* and *E. coli*, respectively. GENDEL et al. (1983) have taken vector construction a stage further by ligating a polylinker-containing plasmid into the unique *BamHI* site of pUH24. The resulting hybrid plasmid pLAN82 can be stably maintained in *Anacystic nidulans R-2* and *E. coli* and the polylinker provides a variety of cloning sites. A smaller, deletion derivative of pLAN82 was also constructed.

CHAUVAT et al. 1983 have used genetically marked derivatives of pUH24 to determine the optimal conditions for transformation in *Anacystis nidulans R-2*. The two most important factors are the physiological state of the organism and the methods employed to select for the transformed genetic marker.

The first is optimum when the cells are at the transition between the first and second exponential phases of growth (a characteristic feature of the growth curve) (CHAUVAT et al. 1983, ORKWISZEWSKI and KANEY 1976). The second requires newly transformed cells to grow for at least 16 h on solid, nonselective medium prior to the addition of the selective antibiotic, which is then layered under the agar. This procedure subjects the transformed cells to a progressively increasing drug concentration and maximizes expression of the transformed resistance gene. Another method of selection involves growing newly transformed cells on a nitrocellulose filter on the surface of nonselective agar. After a suitable grow-out period, the filter is simply replaced on antibiotic-containing agar to select for transformants (GENDEL et al. 1983, WILLIAMS and SZALAY 1983).

Whether or not light is required for maximum efficiency of transformation in *Anacystis nidulans R-2* remains an open question (CHAUVAT et al. 1983, WILLIAMS and Szalay 1983). Similarly, with transformation in other cyanobacteria the role of light has not been described - in *Agmenellum quadruplicatum*, however, it is not required (STEVENS and PORTER (1980).

$CaCl_2$ treatment is not necessary for transformation in *Anacystis nidulans R-2* but is essential for DNA uptake in *Gloeocapsa alpicola* and improves transformation in *Agmenellum quadruplicatum* (DEVILLY and HOUGHTON 1977, STEVENS and PORTER 1980).

CHAUVAT et al. (1983) have suggested that maximum transformation efficiency in *Anacystis nidulans R-2* is only achieved when pUH24 is present in the recipient cells. Cultures cured of this plasmid were transformed about 30 times less efficiently than the parent strain. Using pUC3 as donor DNA, a pUH24-derived plasmid carrying a Sm^R marker, wild-type cultures were transformed with an efficieny of 20 x 10^{-5}, contrasting with a frequency of 6.2 x 10^{-6} in a cured strain. What pUH24-derived functions are important in DNA uptake has not yet been determined.

Recently, KUHLEMEIER et al. (1983) constructed a new vector, pUC303 (CmR SmR), again derived from pUH24, which transformed with high efficiency both *Anacystis R-2* and *R2-SPc*, a cured strain which lacked pUH24. pUC303 was stably maintained in strain *R2-SPc*. In strain *R-2*, however, the unselected marker was frequently lost, presumably following a recombination event between pUH24 and the pUC303 vector. This suggests that the stable maintenance of cloned DNA sequences in cyanobacteria, for example during experiments to complement mutations in the cyanobacterial genome, may be facilitated by the use of cured hosts. This advantage may outweigh any slight loss of transformation efficiency with cured strains, as observed by CHAUVAT et al. (1983).

Transformation has opened the way for the application of other techniques of bacterial genetics to cyanobacteria. VAN DEN HONDEL et al. (1980) introduced the β-lactamase transposon Tn901 of *E. coli* into pUH24 and showed that the transposition functions were expressed in *Anacystis nidulans R-2*. The pUH24/Tn901 hybrid was subsequently employed to generate transposition-induced mutants of *Anacystis nidulans* with, in this case, an impaired ability for methionine biosynthesis (TANDEAU DE MARSAC et al. 1982). The presence of the transposon within the cyanobacterial chromosome allowed the identification by hybridization of the mutant *met* DNA and its subsequent isolation.

This DNA sequence was then used as a hybridization probe to recover the wild-type *met* DNA from a cosmid library of the genome of wild-type *Anacystis nidulans R-2*. Interestingly, transformation of *met*- cells with a recombinant cosmid containing the wild-type *met* DNA resulted in restoration of the *met*+ phenotype. The cosmid vector itself, a shuttle vector capable of replication in *Anacystis nidulans*, was eliminated. This result implied that the region of chromosomal DNA containing the transposon was replaced by wild-type DNA by homologous recombination.

WILLIAMS and SZALAY (1983) have studied in detail the fate of DNA transformed into *Anacystis nidulans R-2*. Several donor molecules were constructed containing chromosomal DNA sequences of *Anacystis nidulans* inserted into unique *BAMHI* site of pBR322. The cyanobacterial DNA was either interrupted by or linked to an *E. coli* plasmid DNA fragment conferring chloramphenicol resistance. Transformation of *Anacystis nidulans* by these plasmids resulted in insertion of donor cyanobacterial DNA into the chromosome at the site of homology.

The yield of transformants increased with DNA concentration up to 0.3 μg/ml. Transformation efficiency showed 20-fold variation between different experiments but the best achieved was one transformant per 200 cells. Donor molecules in which the *Anacystis nidulans* DNA was interrupted by the Cm^R gene gave rise to a high proportion of Cm^R colonies (type I), few Ap^R (type II), and even fewer colonies resistant to both antibiotics (type III). The ration of transformants I : II : III was 107 : 8 : 1. Type I transformants were completely stable - types II and III were much less so. After growth on nonselective medium, 14% of type II transformants lost the Ap^R marker, while 25% type III cells lost both Ap^R and Cm^R markers.

Restriction and hybridization analysis showed that type I transformants arose following nonreciprocal exchange between plasmid and chromosome, resulting in the replacement of chromosomal DNA by the homologous interrupted sequence and consequent elimination of the vector DNA. Transformants of types II and III were produced by addition of total donor plasmid DNA either with (type III) or without (type II) the Cm^R gene. This addition gave rise to flanking copies of homologous DNA within the cyanobacterial chromosome. Such transformants would be expected to be unstable, frequent structural rearrangements occurring via homologous recombination.

Both linear and circular DNA were efficient at transformation. Foreign DNA was introduced into the chromosome as long as it was linked at both ends to cyanobacteria DNA, regardless of whether linear or circular molecules were presented to the recipient cells. The authors suggest this may indicate similarities between the transformation process in *Anacystis nidulans R-2* and *Bacillus subtilis*, since, in the latter, circular DNA is cleaved concomitant with uptake.

Subsequent work by KOLOWSKY et al. (1984) showed that the yield of type I and III transformants was strictly dependent on the length of the interrupting DNA within the cyanobacterial DNA portion of the donor molecule. Transformation efficiency was reduced 1000-fold as the interrupting DNA increased in length from 1.3 to 20 Kbp. An increase in length of 2-3 Kbp halved the transformation efficiency.

It is also of interest to note in the work of WILLIAMS and SZALAY (1983) that one donor plasmid, pKW1065, produced mutant colonies in type I transformation. The mutant phenotype was manifested as very small colonies which died on subculturing. Presumably, this reflected an insertion of the Cm^R gene into an essential region of the cyanobacterial chromosome. KOLOWSKY et al. (1984) circumvented this problem by ensuring that the foreign, interrupting DNA was located in a nonessential region of the donor cyanobacterial DNA. Mutant colonies were, therefore, not produced by type I transformation in their experiments.

Interestingly, these type I transformants were very stable after prolonged growth on nonselective medium - regardless of the size of the foreign DNA insert. As expected however, types II and III were unstable. The present of 20 Kbp of foreign DNA resulted in 97%-99% loss of ApR and CmR markers from type III cells after growth on nonselective medium. In type II transformation, the Cm^R gene plus any associated interrupting DNA are eliminated (WILLIAMS and SZALAY 1983, KOLOWSKY et al. 1984). Thus, type II stability is not influenced by the length of interrupting DNA - marker loss occurred with about the same frequency (13% or 19%), as reported by WILLIAMS and SZALAY (1983). It should be stressed here that the donor molecules described by WILLIAMS and SZALAY (1983) and KOLOWSKY et al. (1984) do not contain a replication origin functional in *Anacystis R-2*. Transformed DNA is stably maintained as a result of insertion into the host chromosome.

As mentioned earlier, filamentous cyanobacteria have not yet been successfully transformed by chemically purified DNA. A transformation system for these organisms is essential to allow a genetic study of heterocyst differentiation and nitrogen fixation. WOLK et al. (1984) have capitalized on the ability of the broad host range plasmid RP-4 to mediate, in the presence of a suitable helper plasmid, the transfer of pBR322 between *E. coli* and other Gram-negative bacteria. A series of pBR322-derived shuttle vectors were constructed, named pRL vectors.

They contain the replication origin of pBR322, the *bom* region of pBR322 (needed for RP-4-mediated transfer) and the plasmid, pDU1, a plasmid from the filamentous cyanobacterium *Nostoc* PCC 7524. (This replicon was shown to function in Anabaena.) The restriction endonucleases AvaI and AvaII are endogenous to *Anabaena;* recognition sites for these enzymes were thus deleted, where possible, from the vector molecule. Vector pRL1 (CmR) contained sites for neither AvaI nor AvaII; pRL5 (Cm^R Sm^R), pRL6 (Cm^R Km^R) and pRL8 (Cm^R Em^R) contained a single site for one or other enzyme.

The pRL vectors were then transformed into *E. coli* HB101, which also contained a helper plasmid (either pGJ28 or pDS4101) with the necessary functions to promote conjugative transfer. Triparental matings were then carried out between (1) strains of *Anabaena*, (2) *E. coli* HB101 containing the helper plasmid plus a pRL vector and (3) *E. coli* J58 which contains plasmid RP-4. Exconjugant cyanobacteria were selected on antibiotic-containing media, then freed of contaminating *E. coli* by streaking on simple cyanobacterial medium. Plasmid DNA isolated from exconjugant cyanobacteria showed the same restriction pattern as the starting pRL vector. The frequency of DNA transfer from *E. coli* to the cyanobacterium was about 1×10^{-3}. Also of interest is the observation that a plasmid vector containing several recognition sites for AvaI and AvaII was not efficiently transferred to *Anabaena*. This indicates that endogenous restriction endonucleases may indeed reduce transformation in cyanobacteria, though whether this prohibits extensive cloning/transformation experiments in the filamentous strains remains to be determined.

3. Cloning of Recombinant DNA in Algae

Transformation in green algae has not been much developed since these systems have been reviewed in this series (KÜCK 1982). The first transformation of a green alga was performed by ROCHAIX and VAN DILLEWIJN 1982, who transformed an arginine auxotrophic strain of *Chlamydomonas rheinhardii* with an *E. coli*- yeast shuttle vector, carrying the *arg4* gene of yeast. This vector transforms with a frequency of about one transformant per µg of DNA.

Several attempts have been undertaken to increase this low transformation efficiency by constructing special vectors, carrying autonomous replicating sequences from *Chlamydomonas*. Furthermore, several recent reports describe examples where the yeast transformation system was used to select for autonomous replication sequences (ars) from algae and higher plants; these sequences were then used for construction of efficiently replicating vectors.

Ars originating from nuclear as well as from chloroplast DNA have been selected from the green algae *Chlamydomonas rheinhardii* (LOPPES and DENIS 1983, VALLET et al. 1984) and *Chlorella* (YAMADA et al. 1983) and from wheat *(Triticum aestivum)* (ANDRE et al. 1983) and tobacco *(Nicotiana tabacum)* (UCHIMIYA et al. 1983).

Recent studies with *Chlamydomonas reinhardii* have provided evidence that the chloroplast genome contains more than one sequence promoting self-replication in yeast (LOPPES and DENIS 1983, VALLET et al. 1984).

Thus, this situation is similar to experiments with yeast and filamentous fungi where multiple ars have been identified on mitochondrial genomes (BLANC and DUJON 1981, HYMAN et al. 1982, LAZDINS and CUMMINGS 1982, TUDZYNSKI and ESSER 1983).

Sequence analysis of ars from chloroplast DNA of *Chlamydomonas rheinhardii*, VALLET et al. (1984) have shown that all contain at least one consensus sequence typical for ars functioning in yeast (STINCHOMB et al. 1981, BROACH et al. 1982). It is noteworthy, however, that the efficiencies of transformation and multiplication of recombinant plasmids carrying chloroplast ars are both rather variable; transformation efficiency is 8 to 1920 yeast transformants per µg DNA; plasmid doubling time under selective growth conditions is between 3 and 8.5 h.

Very recently ROCHAIX et al. (1984) have constructed plasmids able to replicate both in *Chlamydomonas rheinhardii* and *Escherichia coli* by "shotgun" cloning.

Restriction fragments from total DNA of *Chlamydomonas rheinhardii* were integrated randomly into the yeast-*E. coli* vector containing the *arg4* gene.

When this plasmid pool was used for transforming *Chlamydomonas rheinhardii*, four different hybrid plasmids were isolated from individual transformants. Surprisingly, all contained sequences with homology to the chloroplast DNA. By DNA sequencing, consensus sequences (12 bp or 19 bp in length) could be determined in all autonomous replicating sequences.

Mapping and sequencing analysis revealed that only one of the ars functioning in the yeast system promotes self-replication in the algal host. When transformation frequencies (10^{-6} to 10^{-7} transformants per treated cell) were compared with data from experiments where the yeast-*E. coli* shuttle vector had been used, no increase in efficiency could be detected.

4. Trends in Yeast Systems

An important trend in yeast transformation systems has been the attempt to develop general vectors which are not restricted in use to genetically constructed laboratory strains.

One example is the development of a nonselective transformation system for *Saccharomyces cerevisiae* by HOLLENBERG and co-workers (REIPEN et al. 1982). Using the bacterial β-lactamase gene, they were able to overproduce this enzyme in yeast transformants at a sufficient level for detection on indicator plates.

A fusion of the ADH promoter (ADH= alcohol dehydrogenase gene) with the *E. coli* β-lactamase gene was cloned on a yeast-*E. coli* shuttle vector. This hybrid plasmid provided a 100-fold expression of the bacterial gene in yeast transformants, compared with vectors, where the eukaryotic promotor had been omitted (BREUNIG et al. 1982, ROGGENKAMP et al. 1981).

Other systems have been developed which do not require complementation of auxotrophic markers as a selection method for transformants. These systems are based on the application/use of resistance markers, originating from bacteria or from yeast itself. For example the bacterial chloramphenicol, kanamycin (G418) and hygromycin genes resistance have been overproduced in yeast, leading to resistance of the eukaryotic host cell.

Chloramphenicol resistance gene: COHEN and co-workers (1980, 1983) have transformed yeast cells with a yeast-*E. coli* vector carrying the *Tn9*-derived chloramphenicol resistance gene. Several plasmid-containing yeast mutants were isolated; these overproduced the plasmid-coded chloramphenicol acetyltransferase. The overproducing plasmid was analyzed in detail to check for sequence alternations. Surprisingly, the plasmid studied showed a deletion of 127 bp just in front of the ATG start of the chloramphenicol resistance gene. This result implies that the post-transcriptional expression of the gene was affected.

Kanamycin (G418) resistance gene: The enzyme aminoglycoside 3'-phosphotransferease is coded by the kanamycin (G418) gene of Tn903. Construction of several yeast-*E. coli* shuttle vectors containing this gene made it possible to transform and select G418-resistant cells of *Saccharomyces cerevisiae*. For expression of the resistant phenotype, both the composition of the growth medium and the time point at which the selective pressure is administered are of importance. Using this vector, the transformation frequencies ranged from 100 to 700 transformants per μg DNA. In some transformants, integration of the drug resistance gene was observed. This event resulted in a high stability of the resistance phenotype, even under nonselective conditions (WEBSTER and DICKSON 1983).

Hygromycin resistance gene: This gene codes for the hygromycin B phosphotransferase. The enzyme modifies covalently the fungicidal drug hygromycin B (GONZALEZ et al. 1978, RAO et al. 1983). After fusion of the promoters and amino terminal coding regions from various eukaryotic genes with the hygromycin gene, the plasmid-encoded gene product reached a high expressional level in the appropriate yeast transformants. Hence, this system allows the selection of transformants directly, without the use of yeast mutants as the recipient strain (KASTER et al. 1984).

For their plasmid constructions, KASTER et al. (1984) used activator sequences from three yeast genes: (1) the heat shock gene; (2) a heat shock cognate gene; (3) the phosphoglycerate kinase gene.

Last but not least, there has also been a successful attempt to isolate and express a yeast gene conferring resistance against copper (FOGEL and WELCH 1982, FOGEL et al. 1983, WELCH et al. 1983).

The *cup* gene of yeast codes for a low-molecular-weight copper binding protein. The gene was isolated by transforming a copper-sensitive recipient strain with a pool of hybrid plasmids representing yeast genomic DNA and selecting for copper-resistant colonies.

One impressive aspect of yeast transformation is the considerable success which has been achieved in expression of heterologous genes. There are numerous examples where genes from pro- and other eukaryotes have been cloned and expressed in yeast.

Specific examples are as follows: the modification methylase gene from *Bacillus spaericeus R* (FEHER et al. 1983); human leukocyte interferon gene (HITZEMAN et al. 1981); human interferon alpha gene (DOBSON et al. 1983); synthetic human proinsulin gene (STEPIEN et al. 1983); hepatitis B surface antigen (VALENZUELA et al. 1982, MIYANOHARA et al. 1983).

In this section we will survey the problems involved in the expression of heterologus mosaic genes in the yeast *Saccharomyces cerevisiae*.

The results of sequencing studies of different yeast introns imply that splicing of mRNA in yeast is not very different from the splicing process in other eukaryotes. However, preliminary reports suggest that the expression of foreign mosaic genes in yeast might be hampered in certain cases; such genes from higher eukaryotes are not properly spliced in yeast (BEGGS et al. 1980).

Introns from the following genes have been sequenced in recent years: actin gene (GALLWITZ and SURES 1980; NG and ABELSON 1980), and genes coding for various ribosomal proteins (TEEM and ROSBASH 1983, LEER et al. 1982, KÄUFER et al. 1983).

Extensive studies by GALLWITZ and co-workers have shown that the details of the splicing process in yeast is different compared to that of other eukaryotes (GALLWITZ 1982, LANGFORD and GALLWITZ 1983, LANGFORD et al. 1983).

Integrating a foreign intron from duck and amoeba into the mosaic actin gene from yeast, GALLWITZ and co-workers demonstrated that transformed yeast strains do not remove the foreign introns, although the yeast intron is properly excised from the same transcript (LANGFORD et al. 1983).

The discovery by LANGFORD and GALLWITZ (1983) of a sequence near the 3'end of introns, critical for precise splicing of nuclear genes in yeast makes it obvious that introns of higher eukaryotes must be modified before precise processing in yeast can occur.

With these observations in mind, it may well be that transformation in other yeast species or in filamentous fungi might provide a more versatile way to express mosaic genes from vertebrates. These transformation hosts posses a more developed splicing machinery than *S. cerevisiae* which may be able to splice correctly the mRNA transcribed from cloned foreign mosaic genes.

The first successful transformation in the yeast *Schizosaccharomyces pombe* was accomplished by BEACH and NURSE (1981), who showed that a shuttle vector functioning in *Saccharomyces cerevisiae* and in *Escherichia coli* is capable of expression and self-replication in *Schizosaccharomyces pombe*. Surprisingly, this vector did not contain any DNA sequence from *Schizosaccharomyces pombe* at all. Subsequent work has led to several improvements in the transformation system for this yeast.

After complementation of an auxotrophic mutant of *Escherichia coli*, SAKAGUCHI and YAMAMOTO (1982) isolated the ura1 structural gene from *Schizosaccharomyces pombe*. After integration into a bacterial plasmid, this gene was used for high-frequency transformation (up to 44,000 transformants per µg DNA). Unfortunately, the mitotic instability of the transformants was very high (after 12 generations more than 50% of the cells have the *ura*-phenotype). Furthermore in contrast to the situation in *Saccharomyces cerevisiae*, these hybrid plasmids were found to multiply in *Schizosaccharomyces pombe* in a polymeric form.

This group of workers has since developed a high-frequency cotransformation system to introduce nonreplicating circular DNA into the fission yeast. The system is based on the fact that the vector DNA described above will recombine with the nonreplicating DNA within the fission yeast when both are cotransferred. Thus, every foreign DNA can be introduced into *Schizosaccharomyces pombe* and tested for expression (SAKAI et al. 1984).

Another method for high-frequency transformation in fission yeast involves taking an extrachromosomal 3µm plasmid from *Saccharomyces pombe* and fusing this to the *ura3* gene from *Saccharomyces cerevisiae* (FOURNIER et al. 1982). The hybrid molecule is highly efficient at transformation.

Besides the examples described above, transformation has been successfully accomplished in the yeast *Kluyveromyces lactis* (DAS and HOLLENBERG 1982), in the Basidiomycete *Ustilago maydis* (BANKS 1983) and the myxomycete *Dictyostelium discoideum* (BARCLAY and MELLER 1983). The resistance gene for aminoglycosides, described above, was used as the genetic marker in these experiments.

As in *Saccharomyces cerevisiae*, the kan resistance gene can be used for selecting transformants in *Kluyveromyces lactis* (about four transformants per µg plasmid DNA). The gene product is able to inactivate the aminoglycoside G418 (DAS and HOLLENBERG 1982). Using a similar approach, BANKS (1983) has transferred the yeast-*E. coli* vector pMP81 into spheroplasts of *Ustilago maydis*. Expression of the kan resistance gene permitted the selection of 10 to 15 neomycin resistant cells per µg DNA. When transformants were analyzed after vegetative propagation, no plasmid DNA rearrangement was observed. However, the resistant phenotype was mitotically unstable, when the drug was omitted from the media.

The same situation occurs with transformation in *Dictyostelium discoideum*. Vector pAG60 has been successfully used for the transformation of mammalian cells. In this vector the kan resistance gene has been fused with the promotor region of herpes simplex virus type 1. Integration of DNA fragments from *Dictyostelium discoideum* resulted in pAG60 derivates, which allowed expression of the bacterial gene conferring G418 resistance in the myxomycete (BARCLAY and MELLER 1983).

Recently, METZ et al. (1983) have discovered an endogenous plasmid in a cobalt-resistant strain of *Dictyostelium discoideum*. This circular plasmid might be a versatile tool in improving transformation in this myxomycete.

5. New Transformation Vectors in Filamentous Fungi

In the last 2 years, several transformation systems in filamentous fungi have been developed and others improved. The transformation systems developed for each particular organism will be discussed seperately.

Aspergillus nidulans: Because of its capability for sexual propagation, much information is available about the genetics of this fungus. In addition, a large number of physiological and morphological mutants are available, so satisfying one of the necessary prerequisites for the development of a successful transformation system. Last but not least, this fungus might be used as a model system representing species of *Aspergillus* and *Penicillium* in commercial use. The *trpC* gene of Aspergillus was isolated by complementation in *Escherichia coli*. of the corresponding mutant (YELTON et al. 1983) and has been used subsequently for transformation of *Aspergillus nidulans* (YELTON et al. 1984). Using a similar approach, these authors have also used the *arg B* gene from *Aspergillus nidulans* (BERSE 1981) for transformation of this filamentous fungus.

After integration of the *trpC* gene from *Aspergillus nidulans* into the bacterial vector pBR329, TIMBERLAKE and co-workers succeeded in transforming an *Aspergillus trpC*- strain to *trpC*+. A yield of less than 20 integrative stable transformants per μg DNA was obtained (YELTON et al. 1984). As in the other cases described below, total DNA from *Aspergillus* transformants did not contain any freely replicating plasmids. To overcome the difficulties in analyzing integrated plasmid DNA, chromosomal DNA from *Aspergillus* transformants was digested by a restriction enzyme, then treated with ligase. This pool of circular DNA molecules, recombined in vitro, was used to transform *Escherichia coli*. In some experiments, the original plasmid used for transformation of the fungus was recovered. This procedure permitted the molecular analysis of plasmid DNA after passage through the fungal host.

Other genes which have been used for successful transformation of *Aspergillus nidulans* are the *amdS* gene (acetamidase gene) from *Aspergillus nidulans* (TILBURN et al. 1984) and the *pyrG* gene (orotidine-5'- phosphate decarboxylase gene) from *Neurospora crassa* (BALLANCE et al. 1983).

In this experiments transformation yield was less than one transformant per μg DNA.

In all cases of successful transformation in *Aspergillus nidulans*, the transforming DNA has been found to be integrated to the host genome. One explanation for this fact will be discussed in the following chapter.

Neurospora crassa: The first successful transformation using recombinant DNA in filamentous fungi was reported by GILES, CASE and co-workers (CASE et al. 1979). In their initial experiments they used the qa+2 gene from *Neurospora crassa* (coding for catabolic dehydrogenase), integrated in a bacterial vector, to transform a qa-2-mutant of *Neurospora crassa.* About 5-30 transformants per μg DNA were obtained in these experiments. Integrative transformants were produced in all cases.

More recently, other *Neurospora* genes, such as the *am* gene (coding for glutamate dehydrogenase), have been successfully used for high-frequency transformation in *Neurospora crassa* (KINSEY and RAMBOSEK 1984).

The *am* gene integrated in a lambda vector was isolated by the use of a synthetic oligonuleotide probe (KINNAIRD et al. 1982). After transformation of point and deletion mutants of *Neurospora crassa* with the recombinant lambda DNA or with plasmid subclones, more than 10,000 transformants per μg DNA were obtained. In all the cases tested, the recombinant DNA had integrated into the genome. Tetrad analysis revealed that over half of the transformants had integrated the *am* gene and adjacent lambda DNA on chromosomes other than linkage group V, the normal resident of the *am* gene.

To date there is but one report conclusively demonstrating the use of newly constructed, self-replicating plasmids in high-frequency transformation in *Neurospora crassa.* For their construction, LAMBOWITZ and co-workers used a mitochondrial plasmid originating from *Neurospora* strain P405-Labelle, the *Neurospora* qa-2+ gene and the bacterial vector pBR325 (STOHL and LAMBOWITZ 1983).

Even a deletion from the hybrid molecule of much of the sequence originating from the mitochondrial plasmid did not abolish the ability for self-replication of the hybrid plasmid. However, the replication efficiency was reduced when compared with the original plasmid (STOHL and LAMBOWITZ 1983).

Another report about a hybrid plasmid capable of self-replication in both *Escherichia coli* and *Neurospora crassa* by HUGHES et al. (1983a) was later discounted by the same investigators. Repetition of the original protocols of their experiments suggested that, in fact, the *Neurospora* transformants were contamined by *Escherichia coli* plasmids (HUGHES et al. 1983b).

Recent technical improvements for the transformation of *Neurospora crassa* have been described by DHAWALE et al. (1984). High-frequency transformation could be demonstrated after treating germinating conidia with lithium acetate.

In yeast, treatment of cells with lithium salts facilitates the uptake of recombinant DNA without prior removal of the cell wall (ITO et al. 1983). This technique, therefore, supersedes those procedures in which protoplasts had been presumed to be a necessary requirement in transforming unicellular microorganisms.

Podospora anserina: The use of a self-replicating recombinant plasmid in transformation of a fungus was first described for *Podospora anserina.*

In constructing the self-replicating molecule, STAHL et al. (1982) have integrated a mitochondrial plasmid as well as homologous restriction fragments from the chondriome into the bacterial vector pBR322.

Since this first report, other sequences promoting self-replication in *Podospora anserina* have been isolated. For example an ars from the mtDNA of *Cephalosporium acremonium* functions in yeast, as well as in *Podospora anserina* and *Cephalosporium acremonium* (TUDZYNSKI and ESSER 1983, TUDZYNSKI, pers. comm.). Thus far, this is the only example where an ars acting in yeast has been shown to support self-replication in another fungal system. The efficiency of transformation in *Podospora* is rather low (less than ten transformants per μg DNA) compared with others fungal systems. However, an advantage of the system is that it provides an opportunity for isolating free replicating plasmids from the fungal host for further detailed investigations.

Thus mitochondrial sequences can be used for the constructions of eukaryotic vectors. This new concept in genetic engineering has been reviewed recently by ESSER and co-workers (1983). Another interesting attempt to transform *Podospora anserina* has been undertaken by DEBUCHY and BRYGOO (1984). Using a suppressor gene in the construction of a hybrid vector, they succeeded in transforming the leu1-1 mutant of *Podospora anserina*, and thus complemented the nonsense mutation which was the cause of the auxotrophy.

A genomic bank from suppressor strains of *Podospora anserina* was used to complement strains of *Schizosaccharomyces pombe* containing opal mutations in the *ade6* and *glu1* genes. DEBUCHY and BRYGOO (1984) were able to select prototrophic transformants of *Schizosaccharomyces pombe*, carrying hybrid plasmids with the *Podospora* suppressor gene. In subsequent experiments, they were able to transform the *Podospora* strain *leu1-1*. In all transformants studied the suppressor had integrated into the chromosomal DNA.

6. Conclusions

In reviewing the current status of transformation systems in cyanobacteria it is clear that, at least in the case of several unicellular strains, it is now possible to achieve transformation efficiencies very like those obtained in bacteria other than *E. coli*. In addition, work with *Anacystis nidulans* A2 has demonstrated that foreign DNA may be inserted and stably maintained in selected regions of the chromosome or, alternatively, be maintained as part of a stably replicating plasmid within the cyanobacterial cell. Perhaps the main difficulty to be overcome, a problem of special relevance to the study of nitrogen fixation and heterodevelopment, is the presence of restriction endonucleases (perhaps several) in many filamentous cyanobacteria, as these reduce the stability of foreign DNA within the host organism. Recent work with *Anabaena* has suggested that methods for circumventing or alleviating this problem already exist, and may be refined in the not too distant future.

In complete contrast is the situation with transformation in eukaryotic algae. In green algae, the only gene yet shown to be a useful marker for DNA transfer is the *arg 4* gene of yeast. Moreover, the only plasmids available for transformation studies have a very low copy number in the algal host. Obviously, major efforts are required, developing new methods for selection of transformed cell, in exploring new genetic markers and in constructing new vectors. Without doubt, the most sophisticated transformation system available in eukaryotes is that developed for the yeast *Saccharomyces cerevisiae*. Two areas of improvement in the system can be expected. In particular, new expression vectors, similar to those currently available for the expression of foreign genes in prokaryotes, would be of value. In addition, attention is now being given to the construction of recombinant mini-chromosomes. The features required in an "optimal mini-chromosome" in yeast have been summarized recently in a brief review (ZAKIAN 1983). Perhaps the main drawback to the *Saccharomyces cerevisiae* system is the problem, described above, of obtaining expression of cloned mosaic genes of higher eukaryotes, owing to incorrect mRNA splicing.

In addition, it is still an open question whether recombinant vectors propagated in yeast are unaffected by intracellular recombination events during mitosis. One way out of such difficulties may be in the use of other yeast or yeast-like systems for the expression of heterologous eukaryotic DNA. However, it still remains to be seen whether these other species are in fact capable of carrying out the processing steps necessary to produce an accurate gene product from a foreign mosaic gene.

Last but not least, considerable advances in the transformation of filamentous fungi have been achieved in the last 2 years. However, while the use of auxotrophic strains as recipients has given satisfactory transformation frequencies in *Aspergillus nidulans* and *Neurospora crassa*, the recovery of a freely replicating, transforming plasmid has not been possible in most cases. To date this has only been achieved in *Neurospora crassa* and *Podospora anserina*, where recombinant molecules incorporating mitochondrial plasmids as eukaryotic replication sequences have been successfully used as self-replicating vectors. This success may reflect the fact that mitochondrial DNA sequences are not able to integrate into the chromosomal DNA of the host strains. In contrast, marker genes of nuclear origin are highly recombigenic with homologous sequences of the recipient chromosomal DNA. It might be expected, therefore, that bacterial antibiotic restistance genes might be useful as selectable markers with a reduced possibility of intracellular recombination.

References

ANDRE, D., JACQUEMIN, J.M., MASSON, P.: Plant Cell Rep. 2, 175-178 (1983). - ASTIER, C., ESPARDELLIER, F.: C.R. Acad. Sci. Paris. 282, 795-797 (1976).

BALLANCE, D.J., BUXTON, F.P., TURNER, G.: Biochem. Biophys Res. Commun. 112, 284-289 (1983). - BANKS, G.R.: Curr. Genet. 7, 73-77 (1983). - BARCLAY, S.L., MELLER, E.: Mol. Cell. Biol. 3, 2117-2130 (1983). - BEACH, D., NURSE, P.: Nature 290, 140-142 (1981). - BEGGS, J.D., VAN DEN BERG, J., VAN OOYEN, A., WEISSMANN, C.: Nature 283, 835-840 (1980). - BERSE, B.: Aspergillus Newsl. 15, 24-27 (1981). - BLANC, H., DUJON, B.: 279-294 in: Mitochondrial Genes, eds. P. SLONIMSKI et al., New York: Cold Spring Harbor Laboratory Press 1981. - BREUNIG, K.D., MACKEDONSKI, V., HOLLENBERG, C.P.: Gene 20, 1-10 (1982). - BROACH, J.R., LI, Y.Y., FELDMAN, J., JAYARAM, M., ABRAHAM, J., NASMYTH, K.A., HICKS, J.B.: Cold Spring Harbor Symp. Quant. Biol. 47, 1165-1173 (1982).

CASE, M.E., SCHWEIZER, M., KUSHNER, S.R., GILES, N.H.: Proc. Natl. Acad. Sci. USA 76, 5259-5263 (1979). - CHAUVAT, F., ASTIER, C., VEDEL, F., JOSET-ESPARDELLIER, F.: Mol. Gen. Genet. 191, 39-45 (1983). - COHEN, J.D., ECCLESHALL, T.R., NEEDLEMAN, R.B., FEDEROFF, H., BUCHFERER, B.A., MARMUR, J.: Proc. Natl. Acad. Sci. USA 77, 1078-1082 (1980). - COHEN, J.D., ABRAMS, E., ECCLESHALL, T.R., BUCHFERER, B., MARMUR, J.: Mol. Gen. Genet. 191, 451-459 (1983).

DHAWALE, S.S., PAIETTA, J.V., MARZLUF, G.A.: 8, 77-79 (1984). - DEVILLY, C.I., HOUGHTON, J.A.: J. Gen. Microbiol. 98, 277-280 (1977). - DAS, S., HOLLENBERG, P.C.: Curr. Genet. 6, 123-128 (1982). - DEBUCHY, R., BRYGOO, Y.: Abstr. EMBO workshop: Gene Expression in Filamentous Fungi, P-17 (1984). - DOBSON, M.J., TUITE, M.F., MELLOR, J., ROBERTS, N.A., KING, R.M., BURKE, D.C., KINGSMAN, A.J., KINGSMAN, S.M.: Nucl. Acids Res. 11, 2287 (1983). - DUYVESTEYN, M.G.C., KORSUIZE, J., DE WAARD, A., VONSHAK, A., WOLK, C.P.: Arch. Microbiol. 134, 276-281 (1983).

ESSER, K., KÜCK, U., STAHL, U., TUDZYNSKI, P.: Curr. Genet. 7, 239-243 (1983).

FEHER, Zs., KISS, A., VENETIANER, P.: Nature 302, 266-268 (1983). - FOGEL, S., WELCH, J.W.: Proc. Natl. Acad. Sci. USA 79, 5342-5346 (1982). - FOGEL, S., WELCH, J.W., CATHALA, G., KARIN, M.: Curr. Genet. 7, 347-355 (1983). - FOURNIER, P., GAILLARDIN, C., LOUVENCOURT, DE, L., HESLOT, H., LANG, B.F., KAUDEWITZ, F.: Curr. Genet. 6, 31-38 (1982).

GALLWITZ, D.: Proc. Natl. Acad. Sci. USA 79, 3493-3497 (1982). - GALLWITZ, D., SURES, I.: Proc. Natl. Acad. Sci. USA 77, 2546-2550 (1980). - GENDEL, S., STRAUS, N., PULLEYBLANK, D., WILLIAMS, J.: J. Bacteriol. 156, 148-154 (1983). - GOLDEN, S.S., SHERMAN, L.A.: J. Bacteriol. 155, 966-972 (1983). - GONZALES, A., JIMENEZ, A., VAZQUESZ, D., DAVIES, J.E., SCHINDLER, D.: Biochim. Biophys. Acta. 521, 459-469 (1978). - GRIGORIEVA, G.A., SHESTAKOV, S.V.: 369-370 in: Proceedings of the 2nd International Symposium on Photosynthetic Prokaryotes, eds. G.A. CODD, W.D.P. STEWART (1976). - FEMS Microbiol. Lett. 13, 367-370 (1982).

HERDMAN, M.: Mol. Gen. Genet. 120, 369-378 (1973); - 263-305 in: The Biology of Cyanobacteria, eds. N.G. CARR, B.A. WHITTON. Oxford: Blackwell (1982). - HITZEMAN, R.A., HAGIE, F.E., LEVINE, H.L., GOEDDEL, D. V., AMMERER, G., HALL, B.D.: Nature 293, 717-722 (1981). - HOWE, CH.J., BOWMAN, C.M., DYER, R.A., GRAY, J.C.: Mol. Gen. Genet. 190, 51-55 (1983). - HOWELL, S.H.: Annu. Rev. Plant Physiol. 33, 609-650 (1982). - HUGHES, K., CASE, M.E., GEEVER, R., VAPNEK, D., GILES, N.H.: Proc. Natl. Acad. Sci. USA 80, 1053-1057 (1983a); - Proc. Natl. Acad. Sci. USA 80, 7678 (1983b). - HYMAN, B.C., CRAMER, J.H., ROWND, R.H.: Proc. Natl. Acad. Sci. USA 79, 1578-1582 (1982).

ITO, H., FUKUDA, Y., MURATA, K., KIMURA, A.: J. Bacteriol. 153, 163-168 (1983).

KASTER, K.R., BURGETT, S.G., INGOLIA, T.D.: Curr. Genet. (in press) (1984). - KÄUFER, N.F., FRIEOL, H.M., SCHWINDINGER, W.F., JASIN, M., WARNER, J.R.: Nucl. Acids Res. 11, 3123-3135 (1983). - KINNAIRD, J.H., KEIGHREN, M.A., KINSEY, J.A., EATON, M., FINCHAM, J.R.S.: Gene 20, 387-396 (1982). - KINSEY, J.A., RAMBOSEK, J.A.: Mol. Cell. Biol. 4, 117-122 (1984). - KOLOWSKY, K.S., WILLIAMS, J.G.K., SZALAY, A.A.: Gene 27, 289-299 (1984). - KÜCK, U.: Prog. Bot. 44, 244-251 (1982). - KUHLEMEIER, C.J., BORRIAS, W.E., VAN DEN HONDEL, C.A.M.J.J., VAN ARKEL, G.A.: Mol. Gen. Genet. 184, 249-254 (1981). - KUHLEMEIER, C.J., THOMAS, A.A.M., VAN DER ENDE, A., VAN LEEN, R.W., BORRIAS, W.E., VAN DEN HONDEL, C.A.M.J.J., VAN ARKEL, G.A.: Plasmid 10, 156-163 (1983).

LANGFORD, C.J., GALLWITZ, D.: Cell 35, 519-527 (1983). - LANGFORD, C., NELLEN, W., NIESSING, J., GALLWITZ, D.: Proc. Natl. Acad. Sci. USA 80, 1496-1500 (1983). - LAZDINS, I.B., CUMMINGS, D.J.: Curr. Genet. 6, 173-178 (1982). - LEER, R.J., VAN RAAMSDONK-DUIN, M.M.C., MOLENAAR,

C.M.T., COHEN, L.H., MAGER, W.H., PLANTA, R.J.: Nucl. Acids Res. 10, 5869-5878 (1982). - LOPPES, R., DENIS, C.: Curr. Genet. 7, 473-480 (1983).

METZ, B.A., WARD, T.E., WELKER, D.L., WILLIAMS, K.L.: EMBO J. 2, 515-519 (1983). - MIYANOHARA, A., TOH-E, A., NOZAKI, C., HAMADA, F., OHTOMO, N., MATSUBARA, D.: Proc. Natl. Acad. Sci. USA 80, 1-5 (1983).

NG, R., ABELSON, J.: Proc. Natl. Acad. Sci. USA 77, 3912 (1980)

ORKWISZEWSKI, V.G., KANEY, A.R.: Arch. Microbiol. 98, 31-37 (1976).

RAO, R.N., ALLEN, N.E., HOBBS, J.N., ALBORN, W.E., KIRST, H.A., PASCHAL, J.W.: Antimicrob. Agents Chemother. 24, 689-695 (1983). - REIPEN, G., ERHARDT, E., BREUNIG, K.D., HOLLENBERG, C.P.: Curr. Genet. 6, 189-193 (1982). - ROCHAIX, J.D., VAN DILLEWIJN, J.: Nature 296, 70-72 (1982). - ROCHAIX, J.D., VAN DILLEWIJN, J., RAHIRE, M.: Cell 35, 925-931 (1984). - ROGGENKAMP, R., KUSTERMANN-KUHN, B., HOLLENBERG, C.P.: Proc. Natl. Acad. Sci. USA 78, 4466-4470 (1981).

SAKAGUCHI, J., YAMAMOTO, M.: Proc. Natl. Acad. Sci. USA 79, 7819-7823 (1982). - SAKAI, K., SAKAGUCHI, J., YAMAMOTO, M.: Mol. Cell. Biol. 4, 651-656 (1984). - SHERMANN, L.A., VAN DE PUTTE, J.W.: J. Bacteriol. 150, 410-413 (1982). - SHESTAKOV, S.V., KHYEN, N.T.: Mol. Gen. Genet. 107, 372-375 (1970). - STAHL, U., TUDZYNSKI, P., KÜCK, U., ESSER, K.: Proc. Natl. Acad. Sci. USA 79, 3641-3645 (1982). - STANIER, R.Y., COHEN-BAZIRE, G.: Annu. Rev. Microbiol. 31, 225-274 (1977). - STEPIEN, P.P., BROUSSEAU, R., WU, R., NARANG, S., THOMAS, D.V.: Gene 24, 289-297 (1983). - STEVENS, S.E., Jr., PORTER, R.D.: Proc. Natl. Acad. Sci. USA 77, 6052-6056 (1980). - STINCHCOMB, D., MAN, C., SELKER, E., DAVIS, R. 473-488 in: ICN-UCLA Symp. Molecular and Cellular Biology, vol. 22, eds. D.S. RAY, C.C. FOX (1981). - STOHL, L.L., LAMBOWITZ, A.M.: Proc. Natl. Acad. Sci. USA 80, 1058-1062 (1983). - STRUHL, K.: Nature 305, 391-397 (1983).

TANDEAU DE MARSAC, N., BORRIAS, W.E., KUHLEMEIER, C.J., CASTERS, A.M., VAN ARKEL, G.A., VAN DEN HONDEL, C.A.M.J.J.: Gene 20, 111-119 (1982). - TEEM, J.L., ROSBASH, M.: Proc. Natl. Acad. Sci. USA 80, 4403-4407 (1983). - TILBURN, J., SCAZZOCCHIO, C., TAYLOR, G.T., ZABICKY-ZISSMAN, J.H., LOCKING, R.A., DAVIES, R.W.: Gene 26, 205-221 (1984). - TREHAN, K., SINHA, U.: J. Gen. Microbiol. 124, 349-352 (1981). - TUDZYNSKI, P., ESSER, K.: Curr. Genet. 7, 165-166 (1983).

UCHIMIYA, H., OHTANI, T., OHGAWARA, T., HARADA, H., SUGITA, M., SUGIURA, M.: Mol. Gen. Genet. 192, 1-4 (1983).

VALENZUELA, P., MEDINA, A., RUTTER, W.J., AMMERER, G., HALL, B.D. (1982). Nature 298, 347-350 (1982). - VALLET, J.M., RAHIRE, M., ROCHAIX, J.D.: EMBO J. 3, 415-421 (1984). - VAN DEN HONDEL, C.A.M.J. J., VERBEEK, S., VAN DER ENDE, A., WISBEEK, P.J., BORRIAS, W.E., VAN ARKEL, G.A.: Proc. Natl. Acad. Sci. USA 77, 1570-1574 (1980).

WEBSTER, T.D., DICKSON, R.C.: Gene 26, 243-252 (1983). - WELCH, J.W., FOGEL, S., CATHALA, G., KARIN, M.: Mol. Cell. Biol. 3, 1353-1361 (1983). - WENZEL, G.: Progr. Bot. 45, 174-183 (1983). - WILLIAMS, J. G.K., SZALAY, A.A.: Gene 24, 53-59 (1983). - WOLK, C.P., VONSHAK, A., KEHOE, P., ELHAI, J.: Proc. Natl. Acad. Sci. USA 81, 1561-1565 (1984).

YAMADA, T., FUKUDA, Y., SAKAGUCHI, K.: 6th International Protoplast Symposium (Brüssel) 134 pp (1983). - YELTON, M.M., HAMER, J.E., DE SOUZA, E.J., MULLANEY, E.J., TIMBERLAKE, W.E.: Proc. Natl. Acad. Sci.

USA 80, 7576-7580 (1983). - YELTON, M.M., HAMER, J.E., TIMBERLAKE, W. E.: Proc. Natl. Acad. Sci USA 81, 1470-1474 (1984).

ZAKIAN, V.A.: Nature 305, 275 (1983).

Dr. ULRICH KÜCK
Ruhr-Universität Bochum
Lehrstuhl für Allgemeine Botanik
D-4630 Bochum 1, FRG

Dr. BERNARD J. MULLIGAN
University of Nottingham
Department of Botany
GB Nottingham NG7 2RD

III. Mutagenesis

By Christiane Gatz and Wolfgang Hillen

The purpose of this article is to give an overview of the methods which are now widely used to achieve specific mutagenesis. No attempt has been made to completely cover the literature regarding this field. Instead, our intention was to make reference to articles describing the respective method in detail, and also to some examples of its use. More detailed information may be taken from some excellent recent reviews (ZOLLER and SMITH 1983, SHORTLE et al. 1981, WEISSMANN et al. 1979, SHORTLE and BOTSTEIN 1983).

Whereas general mutagenesis has been used in microbiology as a well-known genetic method, the techniques of recombining DNA in vitro along with the use of mutagenic chemicals offer the possibility of limiting mutagenesis to a particular DNA sequence of interest. The main improvement is to generate DNA, expose a part as small as possible to the mutagenic activity, then to transform or transfect the DNA to an organism where it is replicated, to study the effect of the created mutations, and finally to analyze it on the nucleotide level.

1. Segment-specific Mutagenesis

This chapter describes methods to expose a segment of DNA as small as possible to a mutagenic activity. The majority of methods lead to a small segment of DNA in the single stranded form, which then reacts with either chemicals or enzymes to introduce the mutations. Then the entire replicon is restored and introduced into suitable *E. coli* strains.

a) Nick-induced Gap Misrepair

The principle of this method is to introduce a single nick at the site to be mutagenized, to convert this nick into a short gap by exonucleolytic removal of a limited number of nucleotides, and to misrepair this gap afterwards (SHORTLE and NATHANS 1978, SHORTLE et al. 1980, SHORTLE and BOTSTEIN 1982a) (Fig. 1). Single nicks can be introduced by either of two ways.

When incubating duplex DNA with type II restriction endonucleases in the presence of the intercalating agent ethidium bromide (SHORTLE and NATHANS 1978, PARKER et al. 1977, RAWLINS and MUZYCKA 1980) the cleavage reaction is inhibited after only one nick has been induced. Using purified supercoiled DNA and ethidium bromide at an optimal concentration, it is sometimes possible to nick 50%-90% of the input DNA. The reaction product is an open circular molecule with a single nick at the restriction enzyme's normal cleavage site.

Progress in Botany, Vol. 46

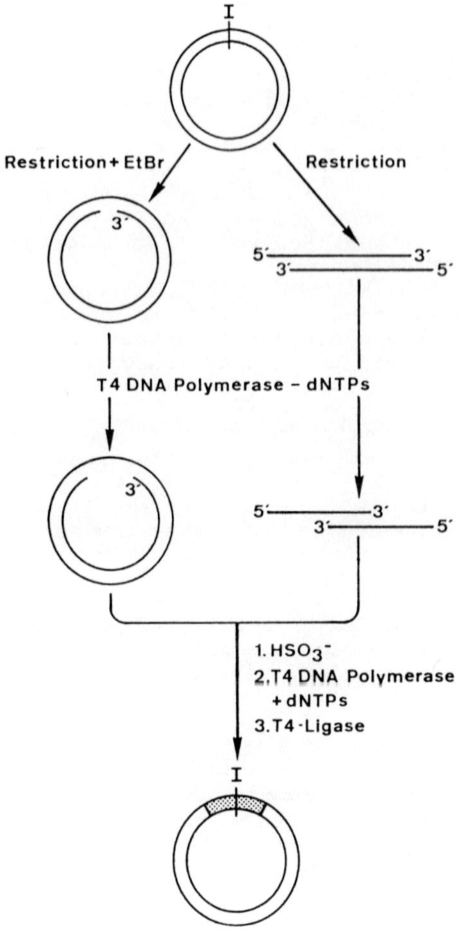

Fig. 1

Another possibility is to linearize a plasmid near the site to be mutagenized with a restriction enzyme and to create single-stranded regions at the two ends with the 3'→ 5' exonucleolytic activity of DNA polymerase I (CIAMPI et al. 1982). The extent of digestion can be limited by addition of only one of the deoxynucleoside triphosphates. These single-stranded ends can then be exposed to sodium bisulfite. The gapped DNA is converted to a complete double-stranded duplex by incubation with Klenow polymerase I in the presence of all four deoxynucleoside triphosphates.

If an easily scorable phenotype is available to detect mutants of interest, another simple approach to generate gaps is to nick the DNA with DNase I in the presence of ethidium bromide (GREENFIELD et al. 1975).

After open circular DNA with a single, site-specific nick has been prepared, the next step is to convert the nick into a short gap by exonucleolytic removal of a small number of nucleotides (Fig. 1). *Micrococcus luteus* DNA polymerase I and T4 DNA polymerase initiate hydrolysis of one strand of DNA starting from a nick, and the extent of hydrolysis can be controlled in the experiment.

The *M. luteus* enzyme has a weak 3'→5' and 5'→3' exonuclease activity (MILLER and WELLS 1972). An average of five to six nucleotides can be removed under defined conditions within 1 h. The exonuclease activity of T4 DNA polymerase can be used to remove nucleotides in the 3'→5' direction from a nick (RAWLINS and MUZYCKA 1980). This gapping reaction has the advantage that the extent of hydrolysis can be limited by addition of one or more deoxyribonucleoside triphosphates to the reaction mixture (ENGLUND et al. 1974).

The resulting gapped duplex DNA can be mutagenized with sodium bisulfite efficiently, inducing C to T transitions. Sodium bisulfite catalyzes the deamination of cytosine to uracil under mild conditions of temperature and pH in single-stranded DNA (HAYATSU 1976), whereas cytosine residues embedded within the Watson-Crick helix are essentially unreactive. DNA molecules that have undergone cytosine deamination within a single-stranded region contain the nonphysiological base deoxyuridine.

Before such molecules can give rise to pure mutant clones, the uracil base must function as a template for incorporation of a complementary A residue during synthesis of the opposite strand, either in vitro or in vivo after DNA transformation. It is advisable to repair the mutagenized gaps before transformation, using DNA polymerase I and T4 DNA ligase (GIZA et al. 1981), and to transform afterwards an *E. coli* ung$^-$ strain which has a deficient repair system for uracil excision (LINDAHL 1979).

Besides mutagenizing the short single-stranded gaps with sodium bisulfite, they can be used for mutagenesis by nucleotide misincorporation (SHORTLE et al. 1982b). This can be stimulated by omission of one of the four deoxynucleoside triphosphates.

Under these conditions only noncomplementary nucleotides are available for incorporation at those positions where the omitted nucleotide would otherwise be correctly incorporated. Infrequently, DNA synthesis proceeds past such a point by mis-incorporation. If the remainder of the single-stranded gap is "filled in" and the strand is immediately circularized by DNA ligase, the misincorporated nucleotide will be trapped in a state, where it cannot be excised by the 3'→5' "editing" exonuclease activity of DNA polymerase I.

An alternative misrepair scheme, described in the same paper, involves the misincorporation of an excision-resistant α-thiophosphate modified nucleotide, followed by gap "filling in" in the presence of all four natural nucleotides.

With both methods, the product of a completed reaction is a covalently closed circular DNA with one or more mismatched base pairs at the site of the original single stranded gap. When applied to short gaps constructed at one of several unique restriction sites on pBR322, both gap misrepair methods yielded mutations within the targeted restriction site at a frequency of up to 42%.

b) Mutagenesis at Displacement Loops

The number of sites accessible to mutagenesis via restriction enzyme catalyzed nicking (see sect 1a) is limited by the relative paucity of restriction targets. Mutagenesis at displacement loops is a method that efficiently introduces nicks into any segment of a small circular DNA molecule, given only that the target DNA can be isolated as a short single-stranded fragment (SHORTLE et al. 1980). These nicks introduced by this method can then be misrepaired as described above (Sect. 1a).

The procedure of this segment-specific nicking runs as follows (Fig. 2). Covalently closed circular DNA molecules and a homologous single-stranded fragment are incubated with the rec A protein of *E. coli*. This enzyme catalyzes the pairing of the single stranded fragment to the complementary sequence on the supercoiled DNA (SHIBITA et al. 1979, McENTEE et al. 1979). One of the two strands of the covalently closed circular molecule is thereby displaced, forming an arc of single-stranded DNA called a "displacement loop" or "D-loop." Because one negative superhelical turn is removed for every 10 bp displaced, the D-loop structure is stable under the conditions of reaction.

The next step involved incubation with the single-strand specific endonuclease S1. The primary substrate for this enzyme is the displaced strand of the D-loop.

The extent of the reaction with S1 is limited because the superhelicity stabilizing the D-loop is relaxed after the first S1 cleavage. After the loss of the superhelical stress, the D-loop decays in a fraction of a second (RADDING et al. 1977) by displacement of the single-stranded fragment, yielding a nicked circular DNA that is not a substrate for the S1 enzyme anymore. The relaxed circular molecules produced by this procedure have a nick somewhere in the region defined by the single-stranded fragment.

One problem of this approach is to obtain a moderate quantity (0.5-2μg) of the single-stranded restriction fragment. A purified duplex DNA can be converted to a heterogeneous mixture of single-stranded subfragments by limited digestion with exonuclease III, which is a double-strand specific exonuclease that degrades each strand in the 5'→3' direction.

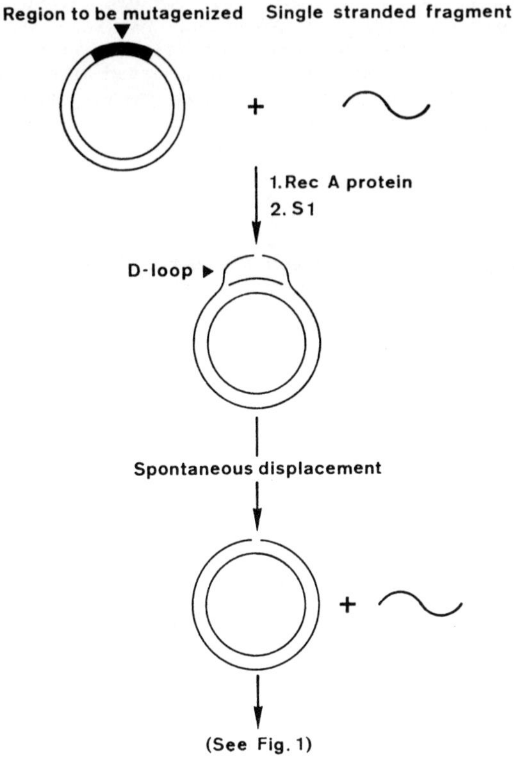

Fig. 2

Alternatively, one specific strand can be completely degraded with exonuclease III by first incorporating an α-thiophosphate nucleotide (which is resistant to hydrolysis by this enzyme) into the other end of the fragment in a reaction with DNA polymerase I (PUTNEY et al. 1981). Also, with some restriction fragments, the two DNA strands can be seperated electrophoretically and recovered from the gel matrix (MAXAM and GILBERT 1980). Using this procedure, coupled with local mutagenesis using the bisulfite method, point mutations were generated in two predetermined regions of the β-lactamase gene of pBR322 (SHORTLE et al. 1980).

c) Mutagenesis at Insertion Loops

Five different methods have been developed to generate point mutations within deletion loops of DNA heteroduplexes (PEDEN and NATHANS 1982, KALDERON et al. 1982, EVERETT and CHAMBON 1982, GIZA et al. 1981, HIROSE et al. 1982). They all have in common that a deletion mutant from the part of the gene to be mutagenized is available. Wild-type and deletion mutant DNA are cloned in a bacterial plasmid. Heteroduplexes are formed between the two DNAs to produce circular molecules containing a single-stranded deletion loop, which can be exposed to sodium bisulfite to deaminate cytosine residues, as described in Sect. 1a. Transformation of the mutagenized plasmid DNA into the uracil repair-deficient *E. coli* strain yielded CG to TA transitions at high frequency in the segment of DNA corresponding to the deletion loop.

A special method described by PEDEN and NATHANS 1982 is designed to increase the mutation frequency (Fig. 3). It requires that two single restriction sites should be removed from the wild-type plasmid and be present on the deletion derivative.

After mutagenesis of the heteroduplex it is transformed to *E. coli* ung$^-$. Assuming that the deletion loop is not excised prior to replication, each cell contains the two different plasmids. To enrich for the full-length plasmid with the introduced transition, DNA prepared from transformants is digested with the two restriction enzymes whose target sites had been removed from the wild-type plasmid. Because these sites are eliminated in the full-length molecule, this digestion selectively cleaves the deletion mutant DNA. In a second transformation only the wild-type DNA, now containing segment-specific transitions, is taken up by *E. coli*.

Alternatively, as described by KALDERON et al. 1982 (Fig. 4), the plasmids containing the wild-type DNA and the deletion mutant DNA are each linearized with different restriction enzymes. Heteroduplexes are formed. Because of the stagger in the restriction enzyme cleavage sites these heteroduplexes are circular, whereas the homoduplexes that also form simultaneously are linear.

After sodium bisulfite treatment and transformation half of the recovered plasmids are of wild-type length and the great majority of these contain one or more A:T base pairs in place of G:C base pairs in the target region. This procedure has been used to mutagenize eight out of nine sites in a 21 bp region that forms part of the coding region of the SV40 large antigen.

A similar method to generate heteroduplexes with single-stranded regions is described by GIZA et al. 1981 (Fig. 5).

A recombinant plasmid is linearized at a restriction site outsite the region to be mutagenized. The region to be mutagenized is cut out by flanking restriction sites in a different reaction. The linear molecules are then strand-separated by preparative agarose gel electrophoresis. The fast strand of the full-length plasmid and the slow strand of the deleted plasmid, or vice versa, are

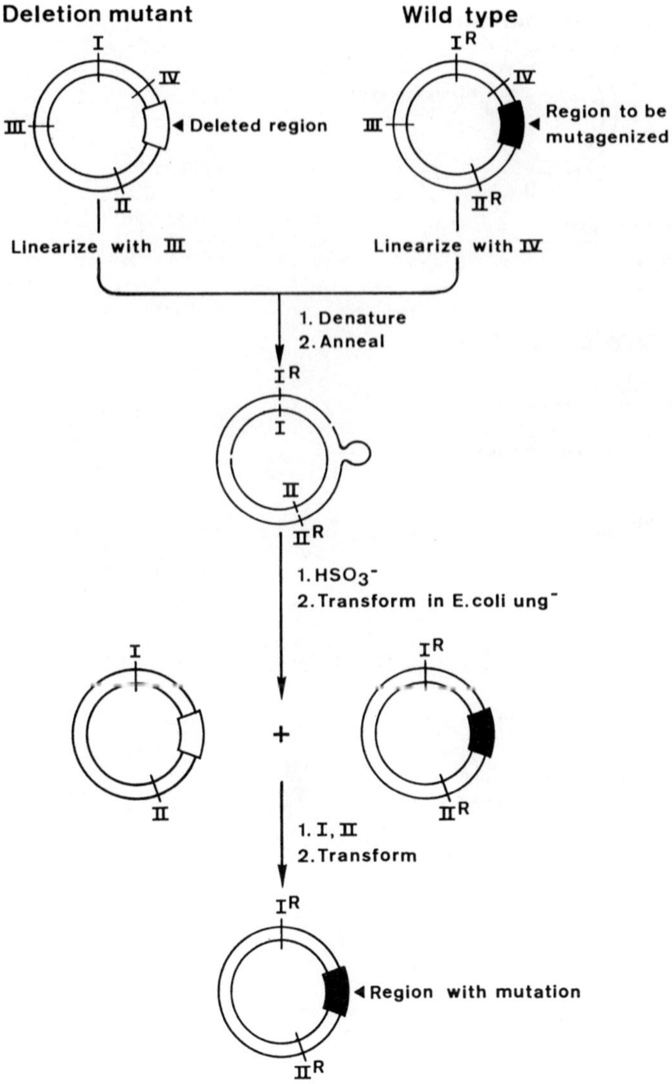

Fig. 3

annealed. The resulting gapped plasmids are mutagenized with sodium bisulfite and "filled in" with DNA polymerase I.

The fourth method (EVERETT and CHAMBON 1982) (Fig. 6) using insertion loops for segment-directed mutagenesis exploits the advantages of the *E. coli* phage M13, which can be isolated as a single-stranded circular DNA molecule as well as in the double-stranded replicative form. The DNA fragment to be investigated is cloned into M13, the region to be mutagenized is deleted and a linker containing a single restriction site is inserted at this position. The single stranded viral DNA of the wild-type M13 phage recombinant is now hybridized with the double-stranded replicative form DNA of the M13 phage containing the deletion, which is linearized at the restriction site introduced by the linker.

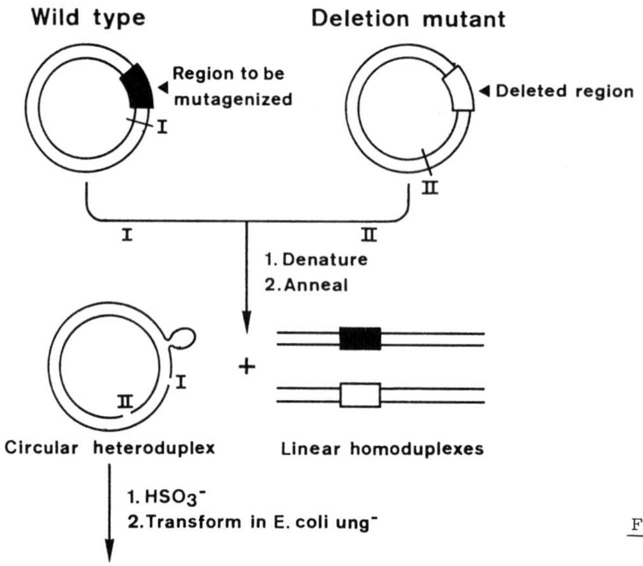
Wild type
Deletion mutant
Region to be mutagenized
I
Deleted region
II
I
II
1. Denature
2. Anneal
I
II
+
Circular heteroduplex
Linear homoduplexes
1. HSO_3^-
2. Transform in E. coli ung⁻

Fig. 4

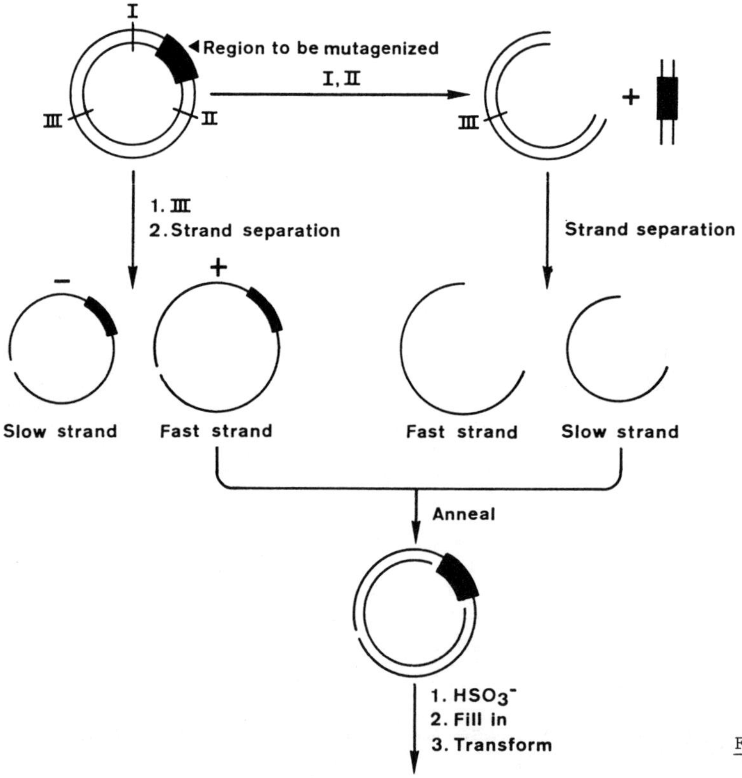
I
Region to be mutagenized
I, II
III
II
III
+
1. III
2. Strand separation
Strand separation
−
+
Slow strand
Fast strand
Fast strand
Slow strand
Anneal
1. HSO_3^-
2. Fill in
3. Transform

Fig. 5

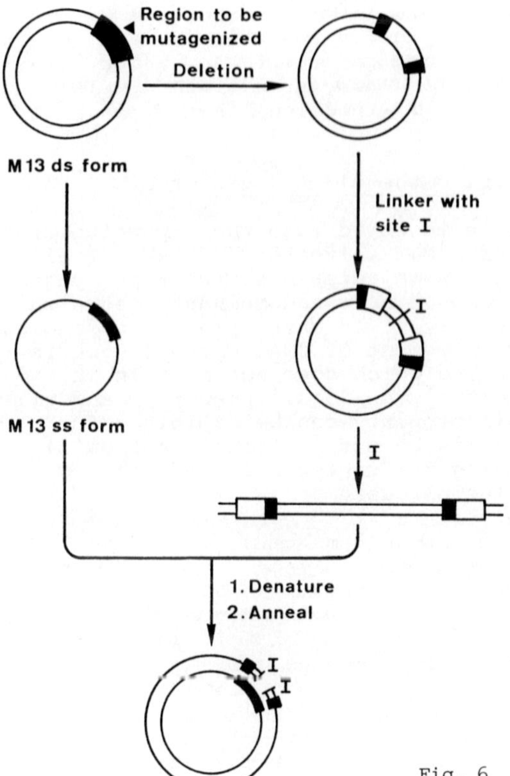

Fig. 6

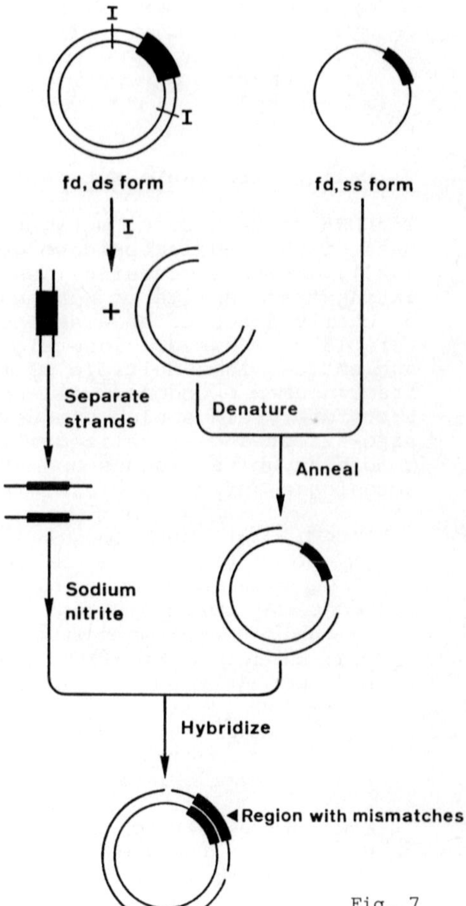

Fig. 7

This heteroduplex contains an exposed single-stranded region and a single nick on the opposite strand.

It is mutagenized with sodium bisulfite and the single-stranded region is filled in with DNA polymerase I, which incorporates adenine opposite the mutated uracils. Another possibility to mutate specific bases is to use nucleotide analogs during the fill-in reaction. After transfection into the M13 host *E. coli* JM103, viral DNA from the resultant plaques is used for the rapid dideoxy DNA-sequencing procedure. All of the plaques studied in this work contained point mutations within the desired area. One important advantage of this method is that the mutants can be rapidly sequenced and that their phenotypes are determined later. Therefore, potentially silent mutations are not overlooked.

HIROSE et al. (1982) produced a gapped duplex by making use of the phage vector f1 (Fig. 7), which can exist as a double strand in its replicative form and also as a single strand. The replicative form was digested so that the fragment of interest was cut out. The digested DNA was mixed with the single-stranded complementary f1 clone, heat denatured and quickly chilled to allow the formation of gapped duplexes.

The restriction fragment to be mutagenized was isolated, strand-separated by polyacrylamide gel electrophoresis and subjected to sodium nitrite treatment inducing C to T and A to G transitions. The mutagen-treated DNA strand was hybridized with the complementary gapped circular DNA. After treatment with DNA polymerase I and T4 DNA ligase the DNA could be transfered into *E. coli*.

d) Easily Detectable Mutagenesis Using the M13 *Lac* Vector

The DNA segment to be mutagenized is inserted into the α-peptide fragment of the β-galactosidase gene of a M13 *lac* vector (MESSING et al. 1981), generally causing loss of the β-galactosidase function by generating frame shifts or nonsense codons. These recombinant phages can be easily detected because they usually yield white plaques on indicator plates. However, insertion of a segment of DNA, whose length in nucleotides is a multiple of three and which does not contain in frame nonsense codons, will not destroy the β-gal$^+$ phenotype and will therefore yield a blue plaque. This property can be exploited for site-directed or localized mutagenesis by mutagenizing DNA from recombinant white plaques and screening for restauration of the β-galactosidase activity, visible as blue plaques.

TRABONI et al. (1982) have applied this method to mutagenize an eukaryotic tRNA gene promotor. They have used several mutagenizing agents, e.g., ICR191 (MILLER 1972), which causes insertion of nucleotides within runs of GC residues. To obtain a more random collection of insertion and deletion mutants, purified double-stranded DNA was treated with DNase I in the presence of Mn^{2+}. DNase I introduces multiple nicks, which, under carefully controlled conditions, will result in linearization of the plasmid DNA with small regions of protruding single-stranded DNA. These can be converted into blunt ends following incubation with DNA polymerase I holoenzyme in the presence of all four nucleoside triphosphates. This procedure yielded one blue out of 10^3 white plaques, whereas the ICR191 treatment yielded two out of 10^4 plaques.

A nonsense codon within a segment of DNA in frame with the α-fragment will cause the formation of a white plaque; such a codon can itself be the target of site-directed mutagenesis, because mutations converting the nonsense into a sense triplet will cause the formation of blue plaques. In addition, the sequence surrounding the nonsense triplet influences the efficiency of suppression. By using base pair substituting mutagenizing agents, such as aminopurine or nitrosoguanidine, and a strain with a sup E suppressor, nucleotides of the stop codon and neighbored nucleotides can be altered specifically. Another possible way is to grow the phages in mutator strains (LACATENA and CESARENI 1981).

e) Segment-specific Mutagenesis Using the Single-stranded Phage Vector fd

The procedure developed by WEIHER and SCHALLER (1982) involves in the first step the transfer of the fragment of interest to a single-stranded phage vector, such as fd 109 (HERRMANN et al. 1980). Cloning in either orientation is desirable because it provides both complementary single strands (Fig. 8). Base changes are introduced into the single-stranded DNA of the recombinant phage by chemical mutagenesis, e.g., with sodium bisulfite. A stable DNA copy is established from the altered DNA sequence of the fragment of interest by fragment-primed repair synthesis in vitro. The mutagenized DNA double strand is cleaved from the mutated vector, recloned in an unmutated vector and analyzed by sequencing and functional tests.

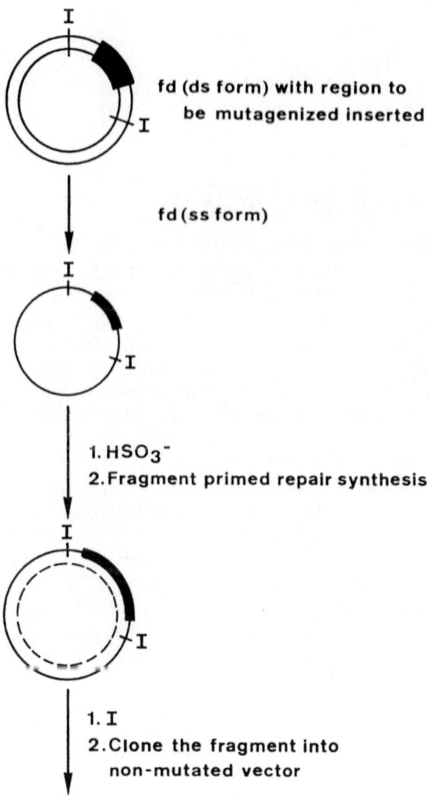

Fig. 8

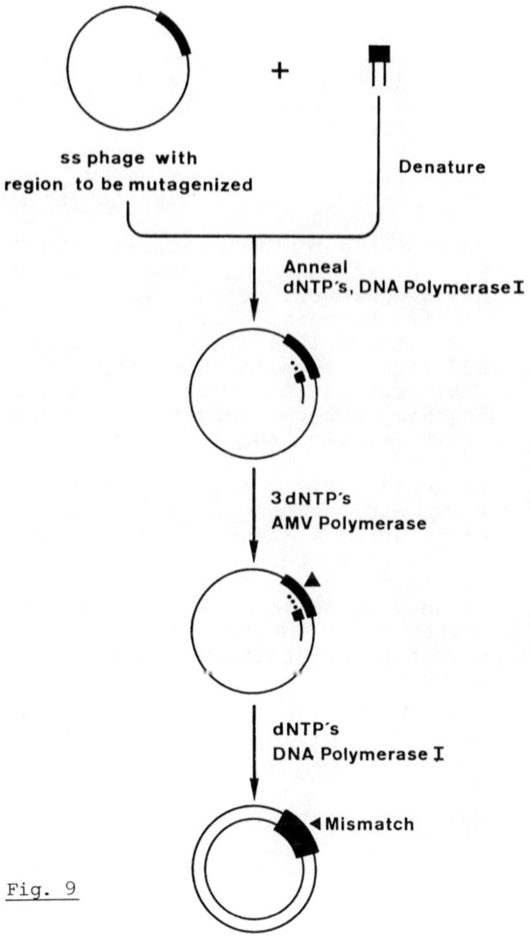

Fig. 9

Using this method, the nucleotide sequence of a 109-base restriction fragment containing the *lac* promotor/operator from *E. coli* was extensively modified. More than 90% of the 235 isolates obtained showed a change in phenotype; all of the 22 candidates sequenced contained multiple C to T point mutations in up to 60% of the possible target positions.

f) Site-specific Mutagenesis by Error-Directed DNA Synthesis

Zakour and Loeb (1981) approached the problem of specific nucleotide changes at defined sites, using the mistakes made by DNA polymerase enzymes in copying a sequence (Fig. 9).

DNA synthesis is initiated from an isolated restriction fragment, which is hybridized to its complementary sequence of a single-stranded phage or a plasmid made single-stranded by digestion of one strand with exonuclease III. The fragment is elongated to a designated position by sequentially adding complementary nucleotides and using *E. coli* DNA polymerase I, containing the proofreading 3'→5' exonuclease activity that enhances its accuracy in catalysis. Then the unincorporated nucleotides and DNA polymerase are removed and noncomplementary nucleotides as well as error-prone DNA polymerase from AMV RNA tumor virus

(BATTULA and LOEB 1974, KUNKEL and LOEB 1981) are added. After incubation of the reaction mixture in the presence of the specific nucleotides, all four deoxynucleoside triphosphates are added and faithful synthesis is allowed to proceed. It is demonstrated by DNA sequence analysis that both transitions and transversions can be produced at a designated position on the DNA template, using this method.

g) Mutagenesis by UV Irradiation of Isolated DNA Fragments

Mutagenesis of UV-irradiated DNA fragments (LIVNEH 1983) has been developed for analysis of mutagen specificity. DNA fragments to be mutagenized are cut out of the plasmid vector, separated by agarose gel electrophoresis, eluted from the matrix, dephosphorylated and irridiated with UV light. The damaged fragment is religated into the vector and transformed to SOS-induced *E. coli* (Fig. 10). Phenotypically scorable mutants are subjected to sequence analysis. The data revealed in this work the induction of transitions, a transversion, a frameshift mutation and deletions.

Segment-specific mutagenesis, described in Sects. 1a-1g has the advantages and disadvantages of highly localized mutagenesis without complete specification of the mutant sequence in advance. Therefore, this approach may be particularly useful for genetic investigations, whereas construction of predetermined variant sequences may be best accomplished by using chemical synthesis (see below). With the methods described here, the frequency of mutants is high enough to encourage screening of individual isolates one by one, even by biochemical methods.

2. Segment-specific Deletions

Specific deletions described here involve an initial directed incision of a circular DNA duplex, which usually is a recombinant bacterial plasmid, enzymatic removal of nucleotides from the ends of the incised mo-

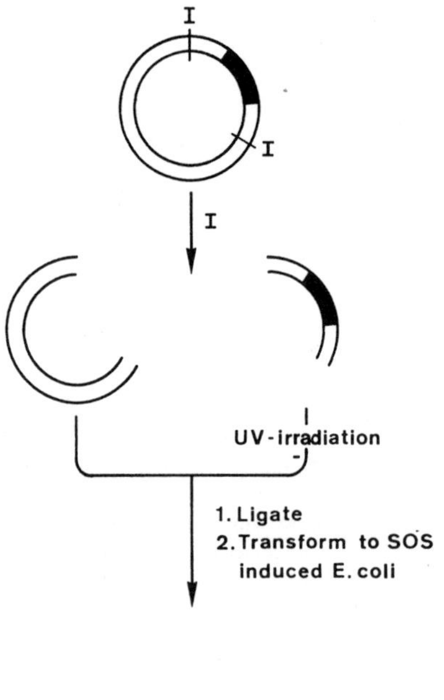

Fig. 10

lecules followed by ligation and cloning of the mutant vector. Deletions can as well be generated by using synthetic oligonucleotides as described below (WALLACE et al. 1980).

a) Deletions Created by Endonuclease S1

An obvious tool for target-directed incisions are restriction endonucleases recognizing between four and six base pairs long nucleotide sequences. After specific cleavage many of them leave various nucleotides single-stranded, which may be the substrate for the single-strand specific endonuclease S1 (BACKMAN et al. 1976).

Deletions smaller than the entire single-stranded region are achieved by a combined fill in in the presence of DNA polymerase I and a selected part of the necessary nucleoside triphosphates and "chew back" by endonuclease S1 (BACKMAN et al. 1976, WARTELL and REZNIKOFF 1980). Protruding ends can also be removed by *E. coli* DNA polymerase I (BENOIST and CHAMBON 1981).

Single-stranded gaps, introduced by one of the ways described in Sect. 1a, are opened by S1 endonuclease (PIPAS et al. 1980) and can then be used as specific sites for the construction of small deletions.

GREEN and TIBBETS (1980) have reported a method for directed deletion that eliminates the need for a conveniently placed restriction site. They found that endonuclease S1 is capable of linearizing a closed circular DNA at branch points of a D-loop (see Sect. 1b) (Fig. 2). Afterwards, die S1 "nibbling effect" removes a limited number of nucleotides at the ends of the linearized molecules (SHENK et al. 1975). Following cleavage by S1, electrophoretically purified linear DNA is circularized with T4 DNA ligase and used to transform bacterial cells.

In experiments using purified fragments from the pBR322 tetracycline resistance gene, an average of about 10% of the transformants contained deletions causing tetracycline sensitivity, and all of them mapped to the segment of the gene covered by the fragment originally used for strand uptake. Restriction analysis of these mutants indicated the presence of deletions of less than 15 base pairs in the targeted region.

Once a deletion mutant is obtained it is often desirable to extend the original deletion systematically. SHENK (1977) described a general procedure for accomplishing this by forming a circular mutant/wild-type DNA heteroduplex, opening the circle, and removing nucleotides at the deletion loop with S1 nuclease.

In this instance a viable deletion mutation lacking about 35 base pairs has been increased in size to produce a series of new mutants lacking from 45 to 90 bp.

b) Deletions Created by Nuclease Bal31 and Exonuclease III

Bal31 nuclease is a highly specific single-strand endonuclease, as well as an exonuclease that simultaneously degrades both the 3' and 5' termini of duplex DNA without internal single-stranded scissions (Legerski et al. 1978). Exonuclease III catalyzes the stepwise release of 5' mononucleotides from the 3' hydroxyl end of duplex DNA (HEFFRON et al. 1977).

Recently BANNER et al. (1983) used Bal31 to construct a series of deletion mutations that extend into a complex promotor region from the downstream and upstream directions. HILLEN et al. (1983) deleted specifically one of two neighbored operators of the Tn*10* encoded tetracycline resistance gene regulatory region. To create a series of progressively longer deletions both upstream and downstream from the *E. coli* galactose operon regulatory region, the combined action of Bal31, S1, exonuclease III, filling in of restriction targets and insertion of synthetic linkers was applied by BUSBY et al. (1983).

3. Site-specific Mutagenesis Directed by Oligonucleotides

The ultimate goal to be achieved by constructing and analyzing mutants is to assess the role of a single base pair in a regulatory region or a single amino acid in a protein. A recent method for the specific alteration of a preselected base pair in the DNA sequence is the oligonucleotide-directed mutagenesis (ZOLLER and SMITH 1983). This method is applicable regardless of the presence of restriction sites or other features of the DNA sequence (SMITH 1980). Prerequisite for this experiment is the knowledge of the DNA sequence surrounding the base pair of interest and the insertion of this DNA into a plasmid or phage.

An oligonucleotide about 20 bp in length is synthesized by chemical methods which is complementary to the sequence surrounding the target base pair. A single mismatch is introduced into the oligonucleotide, which is hybridized to the single-stranded wild-type DNA sequence. After synthesis of the remaining complementary strand the circular double-stranded plasmid or phage containing the mismatch is transformed into *E. coli*. Semiconservative replication leads then to circular DNA containing either the wild-type or the mutant sequence. Clones with the mutation are identified by colony hybridization, dot hybridization or by a genetic screen.

a) Mutagenesis After Cloning in Single-stranded *E. coli* Phages

Whereas in general oligonucleotide-directed mutagenesis is always achieved by hybridization of an oligonucleotide containing a mismatch to the single-stranded form of the target sequence, specific methods differ in the approach to obtain the sequence of interest in the single-stranded form. A convenient way is to use the *E. coli* phage M13 and its derivatives as a cloning vehicle (Fig. 11).

This phage has been optimized by introducing various polylinker sequences so that almost any DNA fragment can be inserted in either orientation (HEIDECKER et al. 1980, HU and MESSING 1982, MESSING 1982). Furthermore, genetic screens are available to detect recombinant M13 phages easily (MESSING et al. 1981). After cloning the DNA fragment of interest into the respective M13 derivative, the single-stranded form of the recombinant phage is isolated and hybridized to the oligonucleotide. The DNA polymerase I is used to extend the oligonucleotide sequence to the complete second strand of the phage.

This procedure often leads to degradation of the oligonucleotide, thus decreasing the mutation frequency drastically. To reduce this unfavorable event, a second universal primer DNA may be hybridized to the single-stranded phage which protects the oligonucleotide (NORRIS et al. 1983). In extending this approach consequently, the entire complementary strand of the M13 vector may be hybridized to the single-stranded form of the recombinant phage.

This procedure leaves only a small gap containing the target sequence. Now only a few nucleotides are left for the fill-in reaction employing DNA polymerase I large fragment. The remaining nicks are finally sealed using DNA ligase. If de-

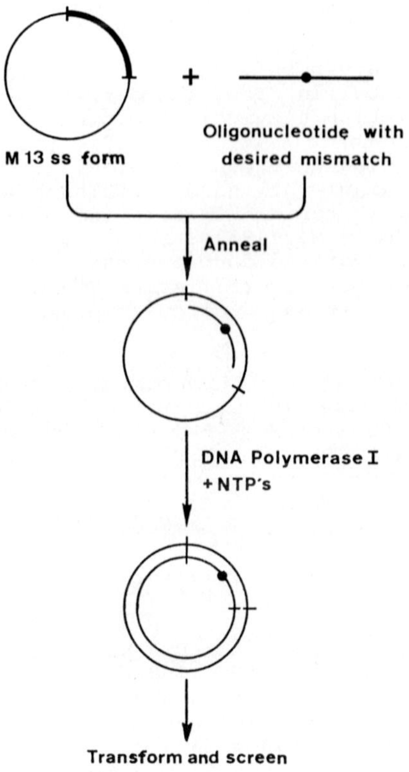

Fig. 11

sirable, this DNA can be purified by a variety of methods, e.g., alkaline sucrose gradient centrifugation (ZOLLER and SMITH 1982). After transformation and selection for the desired mutation its effect can be studied. Using this experimental approach, a tRNA synthetase gene (WINTER et al. 1982) suppressor tRNA's (KUDO et al. 1981, LASKI et al. 1982), phage ØX174 DNA (GILLAM et al. 1980, BAAS et al. 1981), mRNA splice sites (MONDELL et al. 1982), an eukaryotic promotor sequence (GROSSCHEDL et al. 1981), dihydrofolate reductase (VILLAFRANCA et al. 1983) and a signal sequence for secretion across the membrane (INOUYE et al. 1982) have been mutagenized and studied.

Another mechanism reducing the mutation frequency is the mismatch repair system of the transformed *E. coli* strain. Since the mismatch repair system identifies the correct nucleotide in the mismatch by the presence of the *dam* methylation on the parental strand and repairs the opposite nucleotide, the mutation frequency is enhanced when this signal is incorporated in the experimental approach.

For this purpose the DNA strand to be mutagenized is isolated from a *dam*$^-$ mutant of *E. coli*, thus lacking the methylation signal. The M13 vector used to produce the gap at the target site is isolated from a *dam*$^+$ *E. coli* strain. After hybridizing the mismatched oligonucleotide and filling in the remaining gap this DNA is transformed into *E. coli*, which now recognizes the mutation as parental. This protocol leads to a mutation frequency of approximately 40% of the transformants (KRAMER et al. 1982).

b) Mutagenesis After Cloning in Plasmids

If a plasmid is used as a vector for the target DNA, the single-stranded form has to be prepared in the test tube. In order to do this a single nick is introduced, using a restriction enzyme in the presence of ethidium bromide. The strand containing the cleaved phosphodiester bond can then be specifically digested, using an exonuclease, e.g., exonuclease III. The remaining partially single-stranded circular DNA is prepared and may be mutagenized much like the single-stranded phage DNA described above (DALBADIE-MCFARLAND et al. 1982). Sometimes the transformation frequency of the fully restored double-stranded plasmid DNA containing the mismatch is low. Therefore, in some cases it is advisable to prepare the fully double-stranded circular plasmid by alkaline sucrose gradient centrifugation.

This method is frequently used to mutagenize DNA fragments which cannot be stably maintained in single-stranded phages. Recent examples are mutations in the human β-globin gene (WASYLYK et al. 1980), and the deletion of 14 base pairs intervening sequence from the yeast $tRNA^{Tyr}$ gene (WALLACE et al. 1980). Overviews of the above two approaches along with possible perspectives are discussed elsewhere (KATCHALSKI-KATZIR and FREEMAN 1982, SMITH 1982).

4. Mutants Derived from Chemically Synthesized Gene Fragments

In recent years the efficiency of chemical synthesis of DNA has been increased rapidly. For an overview see GASSEN and LANG, 1982. It is now possible to chemically synthesize DNA fragments in a short period of time. Using the well-established methods of recombinant DNA technology, synthetic DNA can easily be inserted into plasmids and this hybrid DNA can be tested in vivo. Up to now it has been still laborious to synthesize long segments of DNA. Therefore, the main work so far in this area focused on rather short segments of DNA with biological activity. Methods are being designed now, however, to facilitate the synthetic approach to complete genes, as will be discussed below.

a) Synthesis of Regulatory Elements

Chemically synthesized DNA has been used to study regulatory signals in the DNA sequence, such as promotors and operators on the transcriptional level and the function of the ribosome binding site of the mRNA on the translational level. This is quite convenient because these signals consist of short DNA sequences with approximately 30-50 base pairs for a promotor, roughly 20-25 base pairs for an operator and some five to eight nucleotides for the ribosome binding site. Plasmids constructed especially for this purpose are used to link the synthetic DNA to indicator genes. The efficiency of regulatory mutants can then be measured in vivo.

The questions addressed with synthetic promotors are the function of the individual base pairs in the consensus sequences (HAWLEY and MCCLURE 1983), the variability of the distance between the consensus sequences, and the involvement of surrounding base pairs in RNA polymerase binding (ROSSI et al. 1983, DOBRYNIN et al. 1980, WINDASS et al. 1982, ROMMENS et al. 1983, AOYOMA et al. 1983, DE HASETH et al. 1983).

The regulation of translation initiation is studied by varying the extent of homology between the mRNA and the ribosomal 16S RNA and the spacer length between the ribosome binding site and the initiation codon ATG (JAY et al. 1982, MATTEUCCI and HEYNECKER 1983).

b) Synthesis of Complete Genes

The ultimate variability of the nucleotide sequence of interest is achieved by a total synthesis of the respective genes. Only a few years back the total synthesis of complete genes was restricted to those with very short reading frames for hormones like angiotensin II (KÖSTER et al. 1975, KÖSTER et al. 1978). Recently, the so far longest synthetic genes consisting of up to 514 base pairs were reported to code for human β-urogastrone (SMITH et al. 1982), human leukocyte interferon (EDGE et al. 1981), human interferon α_2 (EDGE et al. 1983), and human immune interferon (TANAKA et al. 1983). Currently, the methods of chemical DNA synthesis are being developed to a state where the approach of total gene synthesis even becomes a powerful alternative already to the classical isolation procedures for eukaryotic genes (FRANK et al. 1983, MATTHES et al. 1984).

References

AOYOMA, T., TAKANAMI, M., OHTSUKA, E., TANIYAMA, Y., MARUMOTO, R., SATO, H., IKEHARA, M.: Nucleic Acids Res. 11, 5855-5864 (1983).

BAAS, P.D., TEERTSTRA, W.R., VAN MANSFELD, A.D.M., JANSZ, H.S., VAN DER MAREL, G.A., VOENEMAN, G.H., VAN BOOM, J.H.: J. Mol. Biol. 152, 615-639 (1981). - BACKMAN, K., PTASHNE, M., GILBERT, W.: Proc. Natl. Acad. Sci. USA 73, 4174-4178 (1976). - BACKMAN, K., BETLACH, M., BOYER, H.W., YANOFSKI, S.: Cold Spring Harbor Symp. Quant. Biol. 43, 69-76 (1979). - BANNER, C.D.B., MORAN, C.P., LOSICK, R.: J. Mol. Biol. 168, 351-365 (1983). - BATTULA, N., LOEB, L.A.: J. Biol. Chem. 249, 4086-4093 (1974). - BENOIST, C., CHAMBON, P.: Nature 290, 304-310 (1981). - BUSBY, S., KOTLARZ, D., BUC, H.: J. Mol. Biol. 167, 259-274 (1983).

CIAMPI, M.S., MELTON, D.A., CORTESE, R.: Proc. Natl. Acad. Sci. USA 79, 1388-1392 (1982).

DALBADIE-MCFARLAND, G., COHEN, L.W., RIGGS, A.D., MORIN, C., ITAKURA, K., RICHARDS, J.H.: Proc. Natl. Acad. Sci. USA 79, 6409-6413 (1982). - DeHASETH, P.L., GOLDMAN, R.A., CECH, C.L., CARUTHERS, M.H.: Nucleic Acids Res. 11, 773-787 (1983). - DOBRYNIN, V.N., KOROTSKO, V.G., SEVERTSOVA, I.V., BYSTROV, N.S., CHUVPILO, A., KOKOSOV, M.N.; Nucleic Acids Res. 7, 365-375 (1980)

EDGE, M.D., GREENE, A.R., HEATHCLIFFE, G.R., MEACOCK, P.A., SCHUCK, W., SCALON, D.B., ATHINSON, T.C., NEWTON, C.R., MARKHAM, A.F.: Nature 292, 756-762 (1981). - EDGE, M.D., GREENE, A.R., HEATHCLIFFE, G.R., MOORE, V.E., FAULKNER, N.J., CAMBLE, R., PETTER, N.N., TRUEMAN, P., SCHUCH, W., HENNAM, J., ATKINSON, T.C., NEWTON, C.R., MARKHAM, A.F.: Nucleic Acids Res. 11, 6419-6435 (1983). - ENGLUND, P.T., PRICE, S.S., WEIGEL, P.M.: Methods Enzymol. 29, 273-281 (1974). - EVERETT, R.D., CHAMBON, P.: EMBO J. 1, 433-437 (1982).

FRANK, R., HEIKENS, W., HEISTERBERG-MONTSIS, G., BLÖCKER, H.: Nucleic Acids Res. 11, 4365-4377 (1983).

GASSEN, H.G., LANG, A.: Chemical and Enzymatic Synthesis of Gene Fragments. Weinheim: Verlag Chemie 1982. - GILLAM, S., ARTELL, C.R., SMITH, M.: Gene 12, 129-137 (1980). - GIZA, P.E., SCHMIT, D.M., BURR, B.C.: Gene 15, 331-342 (1981). - GREEN, C., TEBBETS, C.: Proc. Natl. Acad. Sci. USA 77, 2455-2459 (1980). - GREENFIELD, L., SIMPSON, L., KAPLAN, D.: Biochim Biophys. Acta 407, 365-375 (1975). - GROSSCHEDL, R., WASLYK, B., CHAMBON, P., BIRNSTIEL, M.L.: Nature 294, 178-180 (1981).

HAWLEY, D.K., McCLURE, R.: Nucleic Acids Res. 11, 2237-2255 (1983). - HAYATSU, H.: Prog. Nucleic Acids Res. Mol. Biol. 16, 75-124 (1976). - HEFFRON, F., BEDINGER, P., CHAMPOUX, J.J., FALKOW, S.: Proc. Natl. Acad. Sci. USA 74, 702-706 (1977. - HEIDECKER, G., MESSING, J., GRONENBORN, B.: Gene 10, 69-73 (1980). - HERRMANN, R., NEUGEBAUER, K., PIRKL, E., ZENTGRAF, H., SCHALLER, H.: Mol. Gen. Genet. 177, 231-241 (1980). - HILLEN, W., GATZ, C., ALTSCHMIED, L., SCHOLLMEIER, K., MEIER, I.: J. Mol. Biol. 169, 707-721 (1983). - HIROSE, S., TAKEUCHI, K., SUZUKI, Y.: Proc. Natl. Acad. Sci. USA 79, 7258-7262 (1982). - HU, N., MESSING, J.: Gene 17, 271-277 (1982).

INOYE, S., SOBERON, X., FRANCESCHINI, T., NAKUMURA, K., ITAKURA, K., INOYE, M.: Proc. Natl. Acad. Sci. USA 79, 3438-3441 (1982).

JAY, E., SETH, A.K., ROMMENS, J., SOOD, A., JAY, G.: Nucleic Acids Res. 10, 6319-6329 (1982).

KALDERON, D., OOSTRA, B.A., ELY, B.K., SMITH, A.E.: Nucleic Acids Res. 10, 5161-5171 (1982). - KATCHALSKI-KATZIR, E., FREEMAN, A.: TIBS, 427-430 (1982). - KÖSTER, H., BLÖCKER, H., FRANK, R., GEUSSENHEIMER, S., KAISER, W.: Hoppe Seyler's Z. Physiol. Chem. 356, 1585-1593 (1975); - Liebig's Ann. Chem. 1978, 839-853 (1978). - KRAMER, W., SCHUGHART, K., FRITZ, H.-J.: Nucleic Acids Res. 10, 6475-6485 (1982). - KUDO, H., LEINEWEBER, M., RAJBHANDARY, U.L.: Proc. Natl. Acad. Sci. USA 78, 4753-4757 (1981). - KUNKEL, T.A., LOEB, L.A.: Science 213, 765-766 (1981).

LACATENA, R.M., CESARENI, G.: Nature 294, 623-626 (1981). - LASKI, F. A., BELAGAJE, R., RAJBHANDARI, U.L., SHARP, P.A.: Proc. Natl. Acad. Sci. USA 79, 5813-5817 (1982). - LEGERSKI, R.Y., HODNETT, Y.L., GRAY, H.B.: Nucleic Acids Res. 5, 1445-1463 (1978). - LINDAHL, T.: Proc. Nucleic Acids Mol. Biol. 22, 135-192 (1979). - LIVNEH, Z.: Proc. Natl. Acad. Sci. USA 80, 237-241 (1983).

MATTEUCCI, M.D., HEYNECKER, H.L.; Nucleic Acids Res. 11, 3113-3121 (1983). - MATTHES, H.W.D., ZENKE, W.M., GRUNDSTROEM, T., STAUB, A., WINTZERITH, M., CHAMBON, P.: EMBO J., 3, 801-805 (1984). MAXAM, A., GILBERT, W.: Methods Enzymol. 65, 499-560 (1980). McENTEE, K., WEINSTOCK, G.M., LEHMAN, S.R.: Proc. Natl. Acad. Sci. USA 76, 2615-2619 (1979). - MESSING, J.: pp. 143-153 in: Recombinant DNA, Third Cleveland Symposium on Macro-Molecules, ed. A. WELTON, Amsterdam: Elsevier, 1982. - MESSING, J., CREA, R., SEEBURG, P.H.: Nucleic Acids Res. 9, 309-321 (1981). - MILLER, J.H.: Experiments in Molecular Genetics. Cold Spring Harbor, NY: Cold Spring Harbor Laboratory Press 1972. - MILLER, L.K., WELLS, R.D.: J. Biol. Chem. 247, 2667-2674 (1972). - MONDELL, C., FISHER, E.F., CARUTHERS, M.H., BERK, A.J.: Nature 295, 380-384 (1982).

NORRIS, K., NORRIS, F., CHRISTIANSEN, L., FIIL, N.: Nucleic Acids Res. 11, 5103-5115 (1983).

PARKER, R.C., WATSON, R.M., VINOGRAD, J.: Proc. Natl. Acad. Sci. USA 74, 851-855 (1977). - PEDEN, K., NATHANS, D.: Proc. Natl. Acad. Sci. USA 79, 7214-7217 (1982). - PIPAS, J.M., ADLER, S.P., PEDEN, K.W.C., NATHANS, D.: Cold Spring Harbor Symp. Quant. Biol. 44, 285-291 (1980). - PUTNEY, S.D., BENKOVIC, S.J., SCHIMMEL, D.R.: Proc. Natl. Acad. Sci. USA 78, 7350-7354 (1981).

RADDING, C.M., BEATTIE, K.L., HOLLOMAN, W.K., WIEGAND, R.C.: J. Mol. Biol. 116, 825-839 (1977). - RAWLINS, D.R., MUZYCKA, N.: J. Virol. 36, 611-616 (1980). - ROMMENS, J., MacKNIGHT, D., PONOROY-CLONEY, L.,

JAY, E.: Nucleic Acids Res. 11, 5921-5939 (1983). - ROSSI, J.J., SOBERON, X., MARUMOTO, Y., McMAHON, J., ITAKURA, K.: Proc. Natl. Acad. Sci. USA 80, 3203-3207 (1983).

SHENK, T.E.: J. Mol. Biol. 113, 503-515 (1977). - SHENK, T.E., RHODES, C., RIGBY, P.W.Y., BERG, P.: Proc. Natl. Acad. Sci. USA 72, 989-993 (1975). - SHIBITA, T., DAS GUPTA, C., CUNNINGHAM, R.P., RADDING, C.M.: Proc. Natl. Acad. Sci. USA 76, 1638-1642 (1979). - SHORTLE, D., BOTSTEIN, D.: Molecular and Cellular Mechanisms of Mutagenesis, eds. J.F. LEMOTT, W.M. GENEROSA, p. 147. New York: Plenum 1982; - Methods Enzymol. 100, 457-468 (1983). - SHORTLE, D., NATHANS, D.: Proc. Natl. Acad. Sci. USA 75, 2170-2174 (1978). - SHORTLE, D., KOSHLAND, D., WEINSTOCK, G.M., BOTSTEIN, D.: Proc. Natl. Acad. Sci. USA 77, 3575-3579 (1980). - SHORTLE, D., DiMAIO, D., NATHANS, D.: Annu. Rev. Genet. 15, 265-294 (1981). - SHORTLE, D., GRISAFI, P., BENKOVIC, S.J., BOTSTEIN, D.: Proc. Natl. Acad. Sci. USA 79, 1588-1592 (1982). - SMITH, J., COOK, E., FOTHERINGHAM, I., PHEBY, S., DERBYSHIRE, R., EATON, M.A.W., DOEL, M., LILLY, D.M., PARDON, J.F., PATEL, T., LEWIS, H., BELL, L.D.: Nucleic Acids Res. 10, 4467-4479 (1982). - SMITH, M.: Nucleic Acids Res., Symposium Series No 7, 387-395 (1980).

TANAKA, S., OSHIMA, T., OHSUYE, K., ONO, T., MIZONO, A., KENO, A., NAKAZATO, H., TSUJIMOTO, M., HIGASHI, N., NOGUCHI, T.: Nucleic Acids Res. 11, 1707-1723 (1983). - TRABONI, C., CILIBERTO, G., CORTESE, R.: EMBO Journal 1, 415-420 (1982).

VILLAFRANCA, J.E., HOWELL, E.E., VOET, D.H.; STROBEL, M.S., OGDEN, R. C., ABELSON, J.N., KRAUT, J.: Science 222, 782-788 (1983).

WALLACE, R.B., JOHNSON, P.F., TANAKA, S., SCHOLD, M., ITAKURA, K., ABELSON, J.: Science 205, 1396-1400 (1980). - WARTELL, R.M., REZNIKOFF, W.S.: Gene 9, 307-319 (1980). - WASYLYK, B., DERBYSHINE, R., GUY, A., MOLKO, D., ROGET, A., TEOULE, R., CHAMBON, P.: Proc. Natl. Acad. Sci. USA 77, 7024-7028 (1980). - WEIHER, H., SCHALLER, H.: Proc. Natl. Acad. Sci. USA 79, 1408-1412 (1982). - WEISSMANN, C., NAGATA, J., TANIGUCHI, T., WEBER, T., MEYER, F.: pp 133-159 in: Genetic Engineering, eds. J.K. SEDLOW, A. HOLLAENDER, Vol. 1. New York: Plenum Press 1979. - WINDASS, J.D., NEWTON, C.R., DeMAEYER-GUIGNARD, J., MOORE, V. E., MARKHAM, A.F., EDGE, M.D.; Nucleic Acids, Res. 10, 6639-6657 (1982). - WINTER, G., FERSHT, A.R., WILKINSON, A.J., ZOLLER, M.J., SMITH, M.: Nature 299, 756-758 (1982).

ZAKOUR, R.A., LOEB, L.A.: Nature 295, 708-710 (1982). - ZOLLER, M.J., SMITH, M.: Nucleic Acids Res. 10, 6487-6499 (1982); - Methods Enzymol. 100, 468-492 (1983).

Dipl. Biol. CHRISTIANE GATZ
Professor Dr. WOLFGANG HILLEN
Institut für Physikalische Biologie
Universität Düsseldorf
Universitätsstraße 1
D-4000 Düsseldorf, FRG

IV. Extrakaryotic Inheritance: Mitochondrial Genetics

By Elke Pratje

1. Introduction

In previous articles of this series the mitochondrial genomes of human, yeast, and plants (MICHAELIS and PRATJE 1982), and DNA plasmids in mitochondria (TUDZYNSKI 1982) have been reviewed.

The present article will focus on the organization and function of mitochondrial genomes of higher plants. Exciting new results concern a possible DNA transfer between mitochondria, chloroplasts, and nuclei. In the third part of this article mitochondrial transcription and RNA processing will be discussed. These studies on RNA synthesis were mainly performed with human and fungal mitochondria. The fourth and last chapter summarizes the progress made in the analysis of nuclear-mitochondrial interactions. In the future this field will probably attract much attention, because it is fundamental for mitochondrial gene expression and mitochondrial biogenesis.

Two proceedings are characterizing the current research on mitochondrial genomes and mitochondrial biogenesis: *Mitochondrial Genes*, edited by Slonimski et al. (1982), and *Mitochondria 1983*, edited by SCHWEYEN et al. (1983). Furthermore, the reader is referred to the following books and review articles: the structure and function of plant genomes (CIFERRI and DURE eds. 1983), isolation of mitochondrial membranes from plant cells (HALL and MOORE eds. 1983), membrane biogenesis (FLEISCHER and FLEISCHER eds. 1983), mitochondrial DNA (GRIVELL 1983), replication of animal mitochondrial DNA (CLAYTON 1982), structural variation of mitochondrial genomes (GRAY 1982, SEDEROFF 1984), and the molecular biology of the mitochondrion (ROSAMOND 1982).

2. The Mitochondrial Genome in Higher Plants

Restriction analysis, renaturation kinetics, and electron microscopy of mitochondrial DNA from higher plants demonstrated high complexities and molecules of various sizes. In general, genome complexities from 200 to 600 kb have been reported (for reviews see LEAVER and GRAY 1982, LEVINGS 1983, MICHAELIS and PRATJE 1982). Within one family, the *Cucurbitaceae*, the mitochondrial genomes vary from 330 kb in the watermelon to 2400 kb in the muskmelon (WARD et al. 1981).

In addition to large linear DNA molecules plant mitochondria have yielded proportions of circular DNA (BRENNICKE and BLANZ 1982, DALE et al. 1983, FONTARNAU and HERNANDEZ-YAGO 1982, KEMBLE et al. 1983a, NEGRUK et al. 1982, POWLING 1982, WEISSINGER 1982). Circular molecules are often heterogeneous in size, 1 to 100 kb. Generally, the frequency of circles is low, about 5%, but can reach values up to 40% in cell cultures.

Progress in Botany, Vol. 46

Physical mapping analysis of the mitochondrial genomes of *Brassica campestris* (PALMER and SHIELDS 1984) and *Zea mays* (LONDSDALE et al. 1983a) was performed by gene cloning. The results of both groups suggest that the genomes are organized on discrete circular molecules which can be converted to each other by recombination, analogous to the integration and excision model for phage lambda. The entire mitochondrial sequence complexity of *Brassica campestris* seems to be integrated in a large circle of 218 kb (PALMER and SHIELDS 1984). This circle includes two copies of an about 2 kb direct repeat separated by 135 and 83 kb of DNA sequences. Intramolecular reciprocal recombination between the two repeated elements creates two smaller circles, 135 and 83 kb in size (Fig. 1). In a similar approach LONSDALE et al. (1983a) have mapped the 570 kb mitochondrial genome of maize with cosmid clones. The *Zea mays* mitochondrial DNA contains more than two direct repeats of about 3 kb. Several small circles can be generated from the large circular molecule by intramolecular recombination. Such a model is supported by restriction analysis of the ribosomal RNA genes in wheat (FALCONET et al. 1984).

Evidence for the existence of a mitochondrial recombination system in higher plants comes also from heteroplasmic fusion experiments of somatic cells. The fusion products contain restriction patterns of mitochondrial DNA different from those of both parents (BOESHORE et al. 1983, NAGY et al. 1983).

Why are plant mitochondrial genomes so complex? Do they contain functions not present in the mitochondrial DNA of other organisms? We know that plant mitochondrial DNA codes for three ribosomal RNA's (rRNA's), a 26S, 18S, and 5S rRNA (BONEN and GRAY 1980, CHAO et al. 1983, STERN et al. 1982). So far a 5S rRNA has not been found in animal and fungal mitochondrial genomes or ribosomes.

Plant mitochondrial DNA also codes for transfer RNA's (tRNA's). Some of the lupine tRNA genes have been cloned and their restriction sites determined (AUGUSTYNIAK et al. 1983). GRAY and SPENCER (1983) sequenced the gene for an initiator methionine tRNA of wheat.

Plant mitochondrial genes coding for proteins were identified by heterologous hybridization using well characterized mitochondrial genes from yeast or bovine as specific probes.

These probes were radioactively labeled by nick translation and hybridized to restriction fragments of maize mitochondrial DNA under various less stringent conditions.

The structural gene for subunit 2 of cytochrome oxidase could be identified by cross-hybridization with the corresponding gene from *Saccharomyces cerevisiae* (FOX and LEAVER 1981). DNA sequence analysis of the maize mitochondrial gene for cytochrome oxidase subunit 2 revealed a centrally located intervening sequence. The respective *Oenothera* gene is not interrupted by an intron as shown by HIESEL and BRENNICKE (1983). These workers identified the *Oenothera* mitochondrial gene by hybridization with the maize DNA probe. The predicted amino acid homology between *Oenothera berteriana* and *Zea mays* subunit 2 is 88%, whereas 47% is found between *Zea mays* and *Saccharomyces cerevisiae*. With a similar approach the genes for apocytochrome b (Cyt b) and cytochrome oxidase subunit 1 (CO I) were identified in maize by LEAVER et al. (1983). The DNA sequences of both maize mitochondrial genes contain no intervening sequences in contrast to the situation in *Saccharomyces cerevisiae*, where complex mosaic genes have been described. In maize the Cyt b gene is composed of 1164 bp, and the CO I gene of 1550 bp.

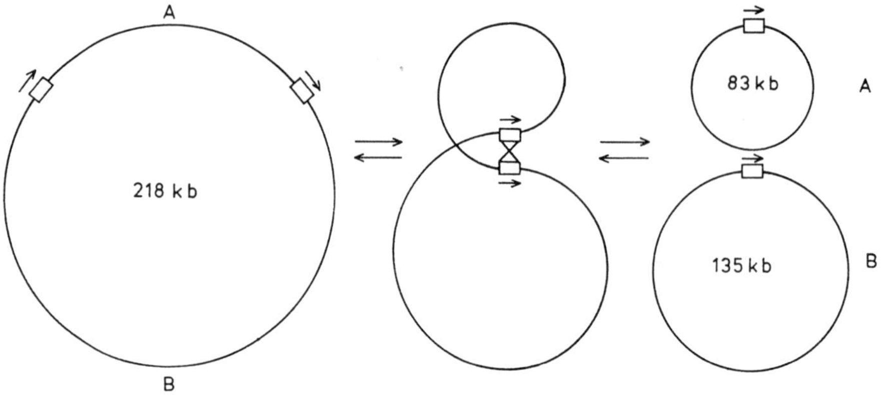

Fig. 1. Model for the organization of the *Brassica campestris* mitochondrial genome. Recombination between the 2 kb repeats (*boxes*) gives either one large cointegrated circle (218 kb) or two smaller circles *A* (83 kb) and *B* (135 kb). Adapted from PALMER and SHIELDS (1984)

The deduced amino acid sequences show homologies of 50% and 67% with the respective sequences from yeast.

The absence of TGA codons in the sequenced plant genes suggests that plant mitochondria do not use the opal codon as tryptophan. This is in contrast to all other mitochondrial translation systems analyzed so far. Nevertheless, plant mitochondria seem to posses a genetic code slightly different from the universal code. The CGG codon may code for tryptophan rather than arginine (FOX and LEAVER 1981, HIESEL and BRENNICKE 1983).

Another approach to identify plant mitochondrial gene products is the analysis of proteins synthesized by isolated mitochondria. Proteins translated on mitochondrial ribosomes seem to be encoded by the mitochondrial genome. No evidence for import of mRNA into mitochondria has been obtained so far. Isolated mitochondria of *Zea mays* were labeled with [^{35}S]-methionine and the translation products were separated on one- or two-dimensional polyacrylamide gels. At least 20 polypeptides can be distinguished (LEAVER et al. 1982). Antibodies against cytochrome oxidase subunit 1 and 2 from yeast were used to identify the respective proteins of maize (DIXON and LEAVER 1982). A maize mitochondrial translation product is extractable with chloroform-methanol and specifically binds [^{14}C]-dicyclohexylcarbodiimide (LEAVER and GRAY 1982). These characteristics suggest that it is the subunit 9 of the oligomycin-sensitive ATPase. In contrast, the homologous polypeptides of the mammalian and *Neurospora* ATPase complexes are encoded in the nuclear genome. ATPase subunit 9 is a mitochondrial gene product in yeast. Isolated plant mitochondria synthesize the α subunit of the F1-ATPase complex as demonstrated in *Vicia faba* (BOUTRY et al. 1983) and *Zea mays* (HACK and LEAVER 1983). This is a very interesting result because all subunits of the soluble F1-ATPase complex are nuclear gene products in lower eukaryotes and mammals.

3. Gene Shuffling

Surprisingly, several recent studies reveal evidence for movement of genes between mitochondria, chloroplasts, and the nucleus. Pieces of chloroplast DNA were found in the mitochondrial genome of *Zea mays*.

STERN and LONSDALE (1982) characterized a 12 kb mitochondrial DNA sequence homologous to chloroplast DNA by restriction analysis and hybridization. Cross-contamination is highly improbable, because the DNA sequence homologous to chloroplast DNA was covalently joined to mitochondrial DNA. In chloroplasts these sequences code for 16S rRNA, $tRNA^{ile}$, and $tRNA^{val}$. It is unknown whether these chloroplast DNA fragments have a function in maize mitochondria.

Mitochondrial DNA sequences hybridizing to the chloroplast structural gene for the large subunit of ribulose-1,5-bisphosphate carboxylase are another example of gene shuffling between chloroplasts and mitochondria (LONSDALE et al. 1983b).

The ribulose-1,5-bisphosphate carboxylase gene of maize mitochondria can be expressed in an *Escherichia coli* transcription-translation system. The in vitro synthesized polypeptide is sufficiently conserved to allow precipitation with antibodies against chloroplast ribulose-1,5-bisphosphate carboxylase from wheat.

Maize with the male sterile cytoplasm cms-S contains sequences of the mitochondrial genome amplified as linear episomes, S1 and S2. These DNA molecules possessing terminal-inverted repeats resemble transposable elements (KIM et al. 1982, LEVINGS and SEDEROFF 1983). Cloned S1 sequences do not only hybridize to mitochondrial but also to nuclear DNA (KEMBLE et al. 1983b).

Evidence for a possible transfer of genetic information from the mitochondria to the nucleus or vice versa has also been discovered in fungi, sea urchin, locust, and mammals. In some strains of *Saccharomyces cerevisiae* nuclear DNA sequences homologous to the mitochondrial genes for var1, cytochrome b, and a replication origin were detected by hybridization and DNA sequencing (FARRELLY and BUTOW 1983). Most probably these mitochondrial sequences found in the yeast nuclear genome are not expressed, because only fragments of the respective genes are present.

Another very interesting example for gene movement is known from *Neurospora crassa*. In this ascomycete the active gene for ATPase subunit 9 lies in the nucleus, but, in addition, a homologous nucleotide sequence is present on the mitochondrial genome (VAN DEN BOOGAART et al. 1982). This mitochondrial gene may be silent, because a translation product was undetectable. In yeast, mitochondrial DNA encodes the respective ATPase subunit. Such a gene is absent from the mitochondrial genomes of human, bovine, and mouse and, therefore, has to be located in the nuclear DNA of these organisms.

Senescence in the fungus *Podospora anserina* is associated with amplification of particular mitochondrial DNA sequences (KOLL et al. 1984, OSIEWACZ and ESSER 1983, WRIGHT and CUMMINGS 1983a). These sequences cover regions of the genes for cytochrome oxidase subunit 1 or 3, respectively. Senescent mitochondria possess the plasmids in high copy number. Moreover, the senescence DNA plasmids can be transposed and integrated into nuclear DNA, suggesting an active gene transfer from mitochondria to the nucleus during senescence (WRIGHT and CUMMINGS 1983b).

Coding sequences for cytochrome oxidase subunit 1 have also been discovered in the chromosomal DNA of the sea urchin, *Strongylocentrotus purpuratus* (JACOBS et al. 1983). DNA complementary to mitochondrial transcripts was cloned and found to hybridize to nuclear DNA. This region contains sequences homologous with the 3'end of the mitochondrial 16S rRNA gene flanked by one complete and one truncated copy of the cytochrome oxidase subunit 1 gene.

Mitochondrial DNA fragments repeated in the nuclear genome of the locust, *Locusta migratoria*, were described by GELLISSEN et al. (1983). These authors isolated clones from a gene library of this insect. Several clones hybridizing to the nuclear genome contain DNA homologous to mitochondrial ribosomal RNA.

Finally, mitochondrial and nuclear DNA share common sequences in human and rat. These sequences contain portions of the mitochondrial genes for ribosomal RNA's (HADLER et al. 1983, TSUZUKI et al. 1983).

All these results substantiate a mechanism transferring genes from one compartment to another.

4. Mitochondrial Transcription and RNA Processing

a) Initiation of Transcription

Both strands of the 16 kb mitochondrial DNA of human are almost completely transcribed. The transcriptional initiation sites (Fig. 2) were identified by in-vitro transcription with isolated homologous RNA polymerase (WALBERG and CLAYTON 1983). In addition, an in-vitro capping reaction with vaccinia virus guanylyltransferase and [α^{32}P] GTP was used (MONTOYA et al. 1982). Guanylyltransferase specifically labels 5'ends of RNA retaining the terminal triphosphate of transcriptional initiation (LEVENS et al. 1981). The initiation site for the light-strand transcript has been mapped within the displacement loop, more precisely at the 5'end of the coding sequence for 7S RNA (MONTOYA et al. 1982, WALBERG and CLAYTON 1983). For heavy-strand transcription two initiation sites were localized at the region coding for the ribosomal RNA's (MONTOYA et al. 1983). Transcription at the first site starts about 25 bp upstream of the $tRNA^{phe}$ gene and terminates at the 3'end of the 16S rRNA gene. Transcription at the second site initiates near the 5'end of the 12S rRNA gene and a polycistronic RNA corresponding to almost the entire heavy strand is produced. Different efficiencies of the two initiation sites could explain the high level of rRNA synthesis. Ribosomal RNA's are about 30 to 60 times more frequent than the other heavy-strand transcripts.

In yeast transcription of the approximate 76 kb mitochondrial genome is more complex (for a review see BORST 1983). Twenty initiation sites were found with guanylyltransferase and 13 of them have been precisely mapped (CHRISTIANSON and RABINOWITZ 1983). Eleven major sites are included in Fig. 3. Initiation sites have been identified in front of the genes for the two ribosomal RNA's, $tRNA^{phe}$, $tRNA^{f\text{-}met}$, $tRNA^{glu}$, cytochrome oxidase subunit I, and ATPase subunit 9. In front of the gene for ATPase subunit 9 transcription can start from two initiation points separated by 80 bp. Furthermore, replication origins rep1 (= ori 3), rep2 (= ori5), and rep3 (= ori2) represent signals for initiation of RNA synthesis. All initiation sites of yeast mitochondrial DNA share the highly conserved oligonucleotide sequence aTATAaGTA (BALDACCI and BERNARDI 1982, CHRISTIANSON et al. 1983, OSINGA et al. 1982). The last A is the first transcribed base. The lower case letters

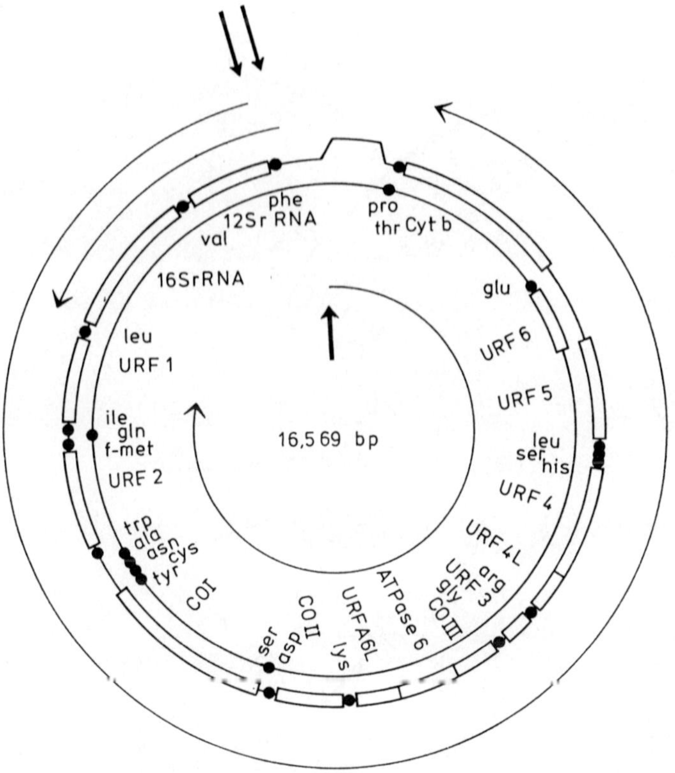

Fig. 2. Map of the human mitochondrial genome (16,569 bp) showing the positions of the three transcriptional initiation sites *(thick arrows)*. Transcription of the heavy strand *(outer ring)* starts at two sites near the region for the large ribosomal RNA and proceeds in a counterclockwise direction. Transcription of the light strand *(inner ring)* initiates at the displacement loop region and proceeds in a clockwise direction. Abbreviations: *COI, COII,* and *COIII* for the three largest subunits of cytochrome oxidase; *Cyt b* for apocytochrome b; *ATPase 6* for subunit 6 of mitochondrial ATPase; *URF* for unassigned reading frame. *Black dots* denote tRNA genes. Data for the map are taken from ANDERSON et al. (1981) and MICHAELIS and PRATJE (1982), data for the transcription are taken from MONTOYA et al. (1982)

indicate positions where variations have been found. These variations occur in genes for less abundant transcripts like phenylalanine, f-methionine, and glutamic tRNAs.

No initiation sites could be localized in front of the yeast genes coding for ATPase subunit 6, cytochrome b, cytochrome oxidase subunit 3, $tRNA^{ser1}$, and $tRNA^{thr2}$. These genes are probably transcribed as multigenic precursor RNA's from the closest upstream initiation sites (CHRISTIANSON et al. 1983, COBON et al. 1982, EDWARDS et al. 1982, 1983, FRONTALI et al. 1982, THALENFELD et al. 1983a, 1983b).

b) RNA Processing

In human mitochondria tRNA sequences within the multigenic transcripts seem to be signals for RNA processing. Almost every human mitochondrial gene is flanked by tRNA genes, as seen in Fig. 2. Palindromic GC-rich

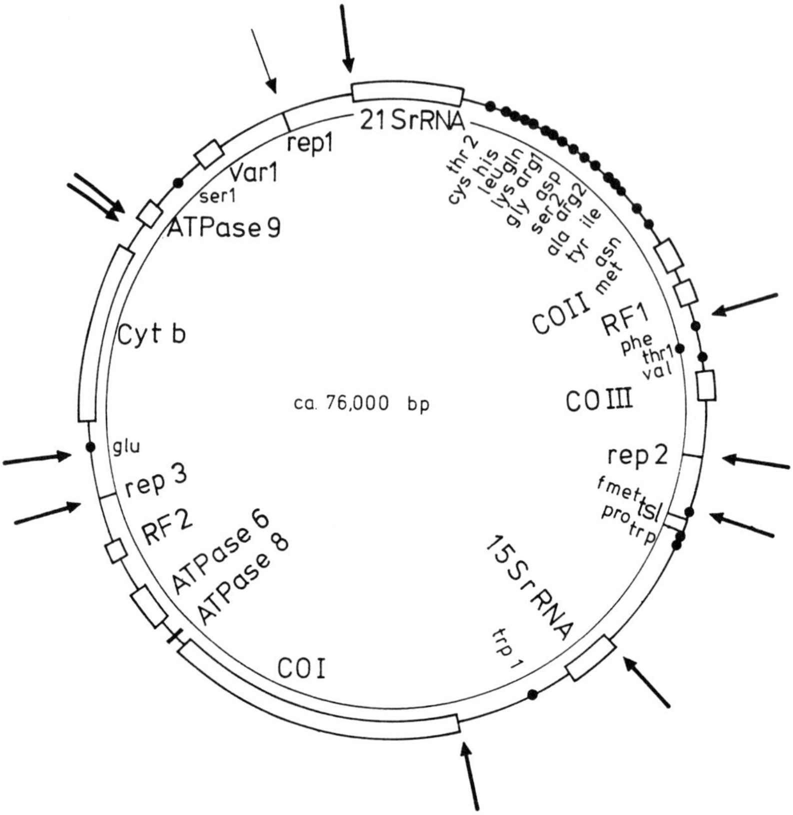

Fig. 3. Map of the mitochondrial genome of *Saccharomyces cerevisiae* (about 76 kb) showing the positions of transcriptional initiation sites *(arrows)*. Most of the sites *(thick arrows)* are located on the positive strand containing almost all genes. One initiation site *(thin arrow)* has been located on the complementary strand near the *rep1* replication origin. Abbreviations: *COI*, *COII*, and *COIII* for the three largest subunits of cytochrome oxidase; *Cyt b* for apocytochrome b; *ATPase 6*, *8*, and *9* for subunits 6, 8, and 9 of mitochondrial ATPase; *Var1* for a polypeptide associated with the small subunit of mitochondrial ribosome; *RF1* and *RF2* for unidentified reading frames; *tsl* for the tRNA synthesis locus; *rep1*, *2*, and *3* for replication origins. *Black dots* denote tRNA genes. Data for the map are taken from GRIVELL (1983) and MICHAELIS and PRATJE (1982), data for the transcriptional initiation sites are compiled from CHRISTIANSON and RABINOWITZ (1983) and TABAK et al. (1983)

sequences have probably the respective function in *Neurospora crassa* mitochondria (YIN et al. 1981).

Factors required for maturation of individual mitochondrial transcripts have been studied in more detail in yeast. Most of these factors must be encoded by nuclear genes since many petite mutants are capable to process their remaining mitochondrial RNA, although they lack large parts of their mitochondrial genome and mitochondrial protein synthesis. Nevertheless, one mitochondrial gene necessary for tRNA processing has been identified (MILLER and MARTIN 1983, UNDERBRINK-LYON et al. 1983). Mutants without this tRNA synthesis locus accumulate high molecular weight tRNA transcripts. The defect of these petite mutants can be restored by mating with petite mutants retaining the tRNA synthesis locus. Since the locus is functional without mitochondrial protein synthesis, its product is probably an RNA. This RNA could have a function similar to the RNA component of RNase P from *Escherichia coli*.

A family of transcripts is often produced from one yeast mitochondrial gene. Five or nine different transcripts contain the coding sequences for the ATPase subunit 6 or cytochrome oxidase subunit 2, respectively (COBON et al. 1982, CORUZZI et al. 1981). Four RNA species hybridize to a DNA probe with the var1 gene (FARRELLY et al. 1982, ZASSENHAUS et al. 1983). The mechanism how primary yeast transcripts are processed is widely unknown. The sequence ATTCTTA, identified near the 3' termini of six different mitochondrial genes may be a candidate for a RNA processing signal (THALENFELD et al. 1983a).

The transcript patterns of the two split genes for cytochrome b and cytochrome oxidase subunit 1 (COI) are extremely complex. Yeast strains with the long forms of these two genes synthesize primary transcripts including either all five cytochrome b or the nine COI introns. More than 60 RNA species, most likely splicing intermediates, can be visualized by hybridizing mitochondrial RNA with a DNA probe containing the gene for cytochrome oxidase subunit 1 (HENSGENS et al. 1983a).

Excision of mitochondrial introns depends on nuclear gene products (BERTRAND et al. 1982, DIECKMANN et al. 1982, DUJARDIN et al. 1983, FAYE and SIMON 1983, McGRAW and TZAGOLOFF 1983, PILLAR et al. 1983) as well as proteins encoded by the introns themselves (ANZIANO et al. 1982, DUJARDIN et al. 1982, WEISS-BRUMMER et al. 1982). The fact that several mitochondrial introns contain long open reading frames in phase with the preceding exons constitutes a unique situation. Studies on numerous mitochondrial intron mutations support the following model. An intron together with the preceding exon encodes an enzyme, called maturase, which catalyzes the excision of the intron itself. Maturases control autocatalytically their synthesis by destroying their own messenger RNA. One intron of the cytochrome b gene of yeast contributes to a splicing function which excises this intron as well as an intron of the transcript for subunit 1 of cytochrome oxidase.

The nucleotide sequences at the splice junctions are variable, a consensus sequence is lacking. In addition to the exon-intron boundaries sequences within the introns are important for splicing. These sequences have been identified by mutations and they are probably necessary to take up a defined secondary RNA structure (BONITZ et al. 1982, BURKE and RAJBHANDARY 1982, CITTERICH et al. 1983, DAVIES et al. 1982, DE LA SALLE et al. 1982, GARRIGA and LAMBOWITZ 1983, HENSGENS et al. 1983b, LAMB et al. 1983, MICHEL et al. 1982, MICHEL and DUJON 1983, NETTER et al. 1982, SCHMELZER et al. 1982, WEISS-BRUMMER et al. 1983, WOLLENZIEN et al. 1983). Some yeast intron sequences are excised and accumulate as covalently closed circular RNA molecules with unknown functions.

5. Nuclear-Mitochondrial Interactions

a) Transport of Proteins into Mitochondria

The biogenesis of functional mitochondria requires a tight collaboration between mitochondrial and nuclear genes. Less than 10% of the protein mass in mitochondria is encoded by mitochondrial DNA, whereas more than 90% is encoded by nuclear DNA. These nuclear-encoded proteins are translated in the cytosol and then transported into mitochondria, either into the matrix, the inner membrane, the intermembrane space, or the outer membrane. Most, but not all proteins destined for the mitochondrial matrix, inner membrane, or intermembrane space are synthesized as larger precursors with N-terminal extended presequences (for reviews see ADES 1982, SCHATZ and BUTOW 1983). The precursors must contain information for targeting these proteins to their submitochondrial locations. The translocation of proteins across the mitochondrial inner membrane is energy-dependent (GASSER et al. 1982), requires a cytosolic protein (ARGAN et al. 1983, MIURA et al. 1983, OHTA and SCHATZ 1984), but is independent of the proteolytic cleavage of the presequences (ZWIZINSKI and NEUPERT 1983). The large precursors destined for the mitochondrial matrix or the inner membrane are converted to their mature forms by a protease located in the matrix (BÖHNI et al. 1983, McADA and DOUGLAS 1982, MIURA et al. 1982).

In contrast, a two-step processing of the presequences has been found for the intermembrane space enzymes, cytochrome c peroxidase and cytochrome b_2 from yeast. Both enzymes are synthesized as larger precursors in the cytosol and transported across the mitochondrial outer membrane. The N-terminal parts of their presequences have to cross the inner membrane to be cloven off by the protease in the matrix. A second proteolytic processing step, probably occurring in the inner membrane, releases the mature enzymes into the intermembrane space (DAUM et al. 1982a,b, REID et al. 1982). A similar two-step maturation process has been documented for cytochrome c_1, a component of the inner mitochondrial membrane (OHASHI et al. 1982, TEINTZE et al. 1982). In the first step the precursor form of cytochrome c_1 is converted to an intermediate form by the matrix-localized protease. This intermediate form has to be combined with heme before maturation can proceed by a second energy-dependent proteolysis.

The precursor forms of apocytochrome c and the ADP/ATP carrier are interesting examples for inner membrane proteins because they do not differ in molecular weight from their mature proteins (HENNIG et al. 1983, ZWIZINSKI et al. 1983). Proteins of the outer membrane are not synthesized as larger precursors and they are integrated into the membrane in an energy-independent step (FREITAG et al. 1982, GASSER and SCHATZ 1983, MIHARA et al. 1982).

Surprisingly, yeast cytochrome oxidase subunit 2, an inner membrane protein encoded by mitochondrial DNA and translated on mitochondrial ribosomes, is also synthesized as an N-terminal extended precursor, although the precursor is not exported out of the mitochondria (PRATJE et al. 1983). Two nuclear mutations have been characterized which prevent maturation of this subunit (MANNHAUPT et al. 1983a). The N-termini of cytochrome oxidase subunit 2 from wild-type and one mutant have been sequenced. A presequence consisting of 15 amino acid residues was found in the mutant protein. The amino acid composition of this presequence differs from known signal sequences. In animal mitochondria the respective subunit is synthesized without an N-terminal extension.

b) Identification of Nuclear Genes for Mitochondrial Functions

Proteins encoded by the nuclear genome are involved in all major mitochondrial functions, e.g., the citrate cycle, the respiratory chain, or the ATP synthesis. Nuclear genes are also required for replication of mitochondrial DNA, transcription of mitochondrial genes, processing of the transcripts, translation of mature mRNA's, and post-translational modification of proteins. In the past, only few nuclear genes coding for mitochondrial proteins have been identified, mainly by classical genetics. The structural gene for cytochrome c is one example. In recent years, two main approaches were used to identify and isolate respective nuclear genes from recombinant DNA libraries: the antibody screening and the complementation of well-defined mutants with cloned wild-type genes.

Antigenically recognizable cytochrome c peroxidase peptides could be detected in *Escherichia coli* cells carrying a hybrid plasmid with the respective yeast structural gene (GOLTZ et al. 1982). DNA sequence analysis of the cloned cytochrome c peroxidase gene established the primary structure of a presequence that comprises 68 amino acid residues (KAPUT et al. 1982). The predicted secondary structure of this presequence is consistent with a transmembrane orientation which is necessary for the two-step processing mechanism, mentioned in the preceding chapter.

Nuclear genes coding for five subunits of the yeast ubiquinol-cytochrome c reductase complex (complex III) have been isolated by VAN LOON et al. (1982, 1983). The authors took advantage of the fact that the levels of mRNA's for these proteins are repressed by glucose and derepressed by a nonfermentable carbon source. Recombinant *Escherichia coli* clones carrying random fragments of yeast DNA were selected, which hybridize more effectively with derepressed than with repressed yeast RNA. The clones containing glucose repressible yeast genes were used to select the respective mRNA's from yeast. These mRNA's were translated in a reticulocyte lysate and the in vitro synthesized proteins could be identified by immunoprecipitation with specific antibodies. The structural gene for one subunit of complex III was reintroduced into yeast cells on a free replicating high copy number plasmid. Despite a 30-fold increase in gene dosage only a 6- to 8-fold increase in protein level was observed. This overproduction does not seem to interfere with mitochondrial functions and has no effect on the synthesis of the other subunits of the complex.

Specific antibodies were used to isolate the genes for the proteolipid subunit of mitochondrial ATPase (VIEBROCK et al. 1982) and the ADP/ATP carrier (ARENDS and SEBALD 1984) from an *Escherichia coli* clone bank carrying *Neurospora crassa* DNA. By DNA sequence analysis the precursor of the *Neurospora crassa* proteolipid was characterized to contain a 66 amino acid residues presequence in addition to the 81 residues of the mature protein. The high polar presequence is probably designed to solubilize the extremely hydrophobic proteolipid for transport into mitochondria. Nucleotide sequence analysis of the structural gene for the ADP/ATP carrier confirmed previous observations that this protein is synthesized without an N-terminal extension. The two *Neurospora crassa* genes contain two introns which have to be excised from the respective primary transcripts.

Complementation of nuclear mutations was applied to isolate the cloned yeast genes for the ADP/ATP carrier (O'MALLEY et al. 1982), subunit V of cytochrome oxidase (CUMSKY et al. 1983), and the β subunit of mitochondrial ATPase (SALTZGABER et al. 1983). The yeast gene for the mitochondrial ATPase β subunit can complement not only a respective mutant of *Saccharomyces cerevisiae*, but also a mutant of *Schizosaccharomyces pombe* lacking the β subunit (BOUTRY and DOUGLAS 1983). In the latter case the *Saccharomyces cerevisiae* β subunit is assembled with the other subunits of *Schizosaccharomyces pombe* into a functional hybrid ATPase complex. This experiment indicates that the transport mechanism and the assembly of the mitochondrial ATPase complex are conserved in these two different yeast species.

Nuclear mutations affecting recombination of mitochondrial DNA have been described by FOURY and KOLODYNSKI (1983). These mutations prevent the integration of tandemly arrayed petite DNA's into the wild-type mitochondrial genome.

Proteins with analogous functions in the cytosol and in mitochondria are generally assumed to differ in structure and to be encoded by separate nuclear genes. Enzyme like malate dehydrogenases, RNA polymerases, or aminoacyl-tRNA synthetases are found as distinct mitochondrial and cytosolic isoenzymes. The finding that single nuclear mutations affect simultaneously, the methylation of mitochondrial as well as cytoplasmic tRNA's could contradict the general assumption (HOPPER et al. 1982). The possibility is now discussed that a single tRNA modification enzyme can function in two cellular compartments.

To elucidate the coordinate interplay between the nuclear and organelle genomes regulatory mutants are valuable tools. Recent publications describe the analysis of nuclear mutations affecting the synthesis of mitochondrial translation products (FOX and REIF 1983, MANNHAUPT et al. 1983b, MICHAELIS et al. 1982). One or several mitochondrially synthesized proteins are altered or absent in such mutants. According to these studies, nuclear genes seem to control nearly all stages of mitochondrial gene expression. So far evidence is lacking that mitochondria send back macromolecular signals to the nucleus.

References

ADES, I.Z.: Mol. Cell. Biochem. 43, 113-127 (1982). - ANDERSON, S., BANKIER, A.T., BARRELL, B.G., DE BRUIJN, M.H.L., COULSON, A.R., DROUIN, J., EPERON, I.C., NIERLICH, D.P., ROE, B.A., SANGER, F., SCHREIER, P. H., SMITH, A.J.H., STADEN, R., YOUNG, I.G.: Nature 290, 457-465 (1981). - ANZIANO, P.Q., HANSON, D.K., MAHLER, H.R., PERLMAN, P.S.: Cell 30, 925-932 (1982). - ARENDS, H., SEBALD, W.: EMBO J. 3, 377-382 (1984). - ARGAN, C., LUSTY, C.J., SHORE, G.C.: J. Biol. Chem. 258, 6667-6670 (1983). - AUGUSTYNIAK, H., BORSUK, P., HIRSCHLER, I., STEPIEN, P.P., BARTNIK, E.: Gene 22, 69-74 (1983).

BALDACCI, G., BERNARDI, G.: EMBO J. 1, 987-994 (1982). - BERTRAND, H., BRIDGE, P., COLLINS, R.A., GARRIGA, G., LAMBOWITZ, A.M.: Cell 29, 517-526 (1982). - BIRKY, C.W.: Science 222, 468-475 (1983). - BOESHORE, M.L., LIFSHITZ, I., HANSON, M.R., IZHAR, S.: Mol. Gen. Genet. 190, 459-467 (1983). - BÖHNI, P.C., DAUM, G., SCHATZ, G.: J. Biol. Chem. 258, 4937-4943 (1983). - BONEN, L., GRAY, M.W.: Nucl. Acids Res. 8, 319-335 (1980). - BONITZ, S.G., HOMISON, G., THALENFELD, B.E., TZAGOLOFF, A., NOBREGA, F.G.: J. Biol. Chem. 257, 6268-6274 (1982). - BORST, P.: CRC Crit. Rev. Biochem. 14, 297 (1983). - BOUTRY, M., DOUGLAS, M.G.: J. Biol. Chem. 258, 15214-15219 (1983). - BOUTRY, M., BRIQUET, M., GOFFEAU, A.: J. Biol. Chem. 258, 8524-8526 (1983). - BRENNICKE, A., BLANZ, P.: Mol. Gen. Genet. 187, 461-466 (1982). - BURKE, J.M., RAJBHANDARY, U.L.: Cell 31, 509-520 (1982).

CHAO, S., SEDEROFF, R.R., LEVINGS, III, C.S.: Plant Physiol. 71, 190-193 (1983). - CHRISTIANSON, T., RABINOWITZ, M.: J. Biol. Chem. 258, 14025-14033 (1983). - CHRISTIANSON, T., EDWARDS, J.C., MUELLER, D.M., RABINOWITZ, M.: Proc. Natl. Acad. Sci. USA 80, 5564-5568 (1983). - CIFERRI, O., DURE, L. (eds.): Structure and Function of Plant Genomes. New York London: Plenum 1983. - CITTERICH, M.H., MORELLI, G., MACINO, G.: EMBO J. 2, 1235-1242 (1983). - CLAYTON, D.A.: Cell 28, 693-705 (1982). - COBON, G.S., BEILHARZ, M.W., LINNANE, A.W., NAGLEY, PH.: Curr. Genet. 5, 97-107 (1982). - CORUZZI, G., BONITZ, S.G., THALENFELD, B.E., TZAGOLOFF, A.: J. Biol. Chem. 256, 12780-12787 (1981). - CUMSKY,

M.G., McEWEN, J.E., KO, CH., POYTON, R.O.: J. Biol. Chem. 258, 13418-13421 (1983).

DALE, R.M.K., WU, M., KIERNAN, M.C.C.: Nucl. Acids Res. 11, 1673-1685 (1983). - DAUM, G., BÖHNI, P.C., SCHATZ, G. : J. Biol. Chem. 257, 13028-13033 (1982a). - DAUM, G., GASSER, S.M., SCHATZ, G.: J. Biol. Chem. 257, 13075-13080 (1982b). - DAVIES, R.W., WARING, R.B., RAY, J. A., BROWN, T.A., SCAZZOCCHIO, C.: Nature 300, 719-724 (1982). - DE LA SALLE, H., JACQ, C., SLONIMSKI, P.P.: Cell 28, 721-732 (1982). - DIECKMANN, C.L., PAPE, L.K., TZAGOLOFF, A.: Proc. Natl. Acad. Sci. USA 79, 1805-1809 (1982). - DIXON, L.K., LEAVER, C.J.: Plant Mol. Biol. 1, 89-102 (1982). - DUJARDIN, G., JACQ, C., SLONIMSKI, P.P.: Nature 298, 628-632 (1982). - DUJARDIN, G., LABOUESSE, M., NETTER, P., SLONIMSKI, P.P.: pp. 233-250 in: Mitochondria 1983, eds. R.J. SCHWEYEN, K. WOLF, F. KAUDEWITZ. Berlin, New York: de Gruyter 1983.

EDWARDS, J.C., LEVENS, D., RABINOWITZ, M.: Cell 31, 337-346 (1982). - EDWARDS, J.C., CHRISTIANSON, T., MUELLER, D., BISWAS, T.K., LEVENS, D., LI, D., WETTSTEIN, J., RABINOWITZ, M.: pp. 69-78 in: Mitochondria 1983, eds. R.J. SCHWEYEN, K. WOLF, F. KAUDEWITZ. Berlin, New York: de Gruyter 1983.

FALCONET, D., LEJEUNE, B., QUETIER, F., GRAY, M.W.: EMBO J. 3, 297-302 (1984). - FARRELLY, F., BUTOW, R.A.: Nature 301, 296-301 (1983). - FARRELLY, F., ZASSENHAUS, H.P., BUTOW, R.A.: J. Biol. Chem. 257, 6581-6587 (1982). - FAYE, G., SIMON, M.: Cell 32, 77-87 (1983). - FLEISCHER, S., FLEISCHER, B. (eds.): Biomembranes, Part K, Methods in Enzymology, Vol. 97. Orlando: Academic Press 1983. - FORTARNAU, A., HERNANDEZ-YAGO, J.: Plant Physiol. 70, 1678-1682 (1982). - FOURY, F., KOLODYNSKI, J.: Proc. Natl. Acad. Sci. USA 80, 5345-5349 (1983). - FOX, T.D., LEAVER, C.J.: Cell 26, 315-323 (1981). - FOX, T.D., REIF, M.K.: pp. 441-448 in: Mitochondria 1983, eds. R.J. SCHWEYEN, K. WOLF, F. KAUDEWITZ. Berlin, New York: de Gruyter 1983. - FREITAG, H., JAMES, M., NEUPERT, W.: Eur. J. Biochem. 126, 197-202 (1982). - FRONTALI, L., PALLESCHI, C., FRANCISCI, S.: Nucl. Acids Res. 10, 7283-7293 (1982).

GASSER, S.M., DAUM, G., SCHATZ, G.: J. Biol. Chem. 257, 13034-13041 (1982). - GASSER, S.M., SCHATZ, G.: J. Biol. Chem. 258, 3427-3430 (1983). - GARRIGA, G., LAMBOWITZ, A.M.: J. Biol. Chem. 258, 14745-14748 (1983). - GELLISSEN, G., BRADFIELD, J.Y., WHITE, B.N., WYATT, G.R.: Nature 301, 631-634 (1983). - GOLTZ, S., KAPUT, J., BLOBEL, G.: J. Biol. Chem. 257, 11186-11190 (1982). - GRAY, M.W.: Can. J. Biochem. 60, 157-171 (1982). - GRAY, M.W., SPENCER, D.F.: FEBS L. 161, 323-327 (1983). - GRIVELL, L.A.: Sci. Am. 248, March, 60-73 (1983).

HACK, E., LEAVER, C.J.: EMBO J. 2, 1783-1789 (1983). - HADLER, H.I., DIMITRIJEVIC, B., MAHALINGAM, R.: Proc. Natl. Acad. Sci. USA 80, 6495-6499 (1983). - HALL, J.L., MOORE, A.L. (eds.): Isolation of Membranes and Organelles from Plant Cells. London: Academic Press 1983. - HENNIG, B., KOEHLER, H., NEUPERT, W.: Proc. Natl. Acad. Sci. USA 80, 4963-4967 (1983). - HENSGENS, L.A.M., ARNBERG, A.C., ROOSENDAAL, E., VAN DER HORST, G., VAN DER VEEN, R., VAN OMMEN, G.-J.B., GRIVELL, L.A.: J. Mol. Biol. 164, 35-58 (1983a). - HENSGENS, L.A.M., BONEN, L., DE HAAN, M., VAN DER HORST, G., GRIVELL, L.A., Cell 32, 379-389 (1983b). - HIESEL, R., BRENNICKE, A.: EMBO J. 2, 2173-2178 (1983). - HOPPER, A. K., FURUKAWA, A.H., PHAM, H.D., MARTIN, N.C.: Cell 28, 543-550 (1982).

JACOBS, H.T., POSAKONY, J.W., GRULA, J.W., ROBERTS, J.W. XIN, J.-H., BRITTEN, R.J., DAVIDSON, E.H.: J. Mol. Biol. 165, 609-632 (1983).

KAPUT, J., GOLTZ, S., BLOBEL, G.: J. Biol. Chem. 257, 15054-15058 (1982). - KEMBLE, R.J., GUNN, R.E., FLAVELL, R.B.: Theor. Appl. Genet. 65, 129-144 (1983a). - KEMBLE, R.J., MANS, R.J., GABAY-LAUGHNAN, S., LAUGHNAN, J.R.: Nature 304, 744-747 (1983b). - KIM, B.D., MANS, R.J., CONDE, M.F., PRING, D.R., LEVINGS, C.S. III: Plasmid 7, 1-14 (1982). - KOLL, F., BEGEL, O., KELLER, A.-M., VIERNY, C., BELCOUR, L.: Curr. Genet. 8, 127-134 (1984).

LAMB, M.R., ANZIANO, P.Q., GLAUS, K.R., HANSON, D.K., KLAPPER, H.J., PERLMAN, P.S., MAHLER, H.R.: J. Biol. Chem. 258, 1991-1999 (1983). - LEAVER, C.J., GRAY, M.W.: Annu. Rev. Plant Physiol. 33, 373-402 (1982). - LEAVER, C.J., FORDE, B.G., DIXON, L.K.: pp. 457-470 in: Mitochondrial Genes, eds. P. SLONIMSKI, P. BORST, G. ATTARDI. Cold Spring Harbor Monograph 12. Cold Spring Harbor, NY: Cold Spring Harbor Laboratory Press 1982 . - LEAVER, C.J., HACK, E., DAWSON, A.J., ISAAC, P.G., JONES, V.P.: pp. 267-283 in: Mitochondria 1983, eds. R.J. SCHWEYEN, K. WOLF, F. KAUDEWITZ. Berlin, New York: de Gruyter 1983. - LEVENS, D., TICHO, B., ACKERMAN, E., RABINOWITZ, M.: J. Biol. Chem. 256, 5226-5232 (1981). - LEVINGS, C.S. III: Cell 32, 659-661 (1983). - LEVINGS, C.S. III, SEDEROFF, R.R.: Proc. Natl. Acad. Sci. USA 80, 4055-4059 (1983). - LONSDALE, D.M., HODGE, T.P., FAURON, C.M.-R., FLAVELL, R.B., in: UCLA Symp. on Molecular and Cellular Biology, Vol. 12, ed. R.B. GOLDBERG. New York: Liss 1983a. - LONSDALE, D.M., HODGE, T.P., HOWE, Chr.J., STERN, D.B.: Cell 34, 1007-1014 (1983b).

MANNHAUPT, G., MICHAELIS, G., PRATJE, E., SCHWEIZER, E., HAWTHORNE, D.C.: pp. 449-454 in: Mitochondria 1983, eds. R.J. SCHWEYEN, K. WOLF, F. KAUDEWITZ. Berlin, New York: de Gruyter 1983a. - MANNHAUPT, G., PRATJE, E., MICHAELIS, G., BEYREUTHER, K., SCHWEIZER, E.: pp. 219-228 in: Endocytobiology II, eds. H.E.A. SCHENK, W. SCHWEMMLER. Berlin, New York: de Gruyter 1983b. - McADA, P.C., DOUGLAS, M.G.: J. Biol. Chem. 257, 3177-3182 (1982). - McGRAW, P., TZAGOLOFF, A.: J. Biol. Chem. 258, 9459-9468 (1983). - MICHAELIS, G., PRATJE, E.: Progr. Bot. 44, 286-296 (1982). - MICHAELIS, G., MANNHAUPT, G., PRATJE, E., FISCHER, E., NAGGERT, J., SCHWEIZER, E.: pp. 311-321 in: Mitochondrial Genes, eds. P. SLONIMSKI, P. BORST, G. ATTARDI. Cold Spring Harbor Monograph 12. Cold Spring Harbor, NY: Cold Spring Harbor Laboratory Press 1982. - MICHEL, F., DUJON, B.: EMBO J. 2, 33-38 (1983). - MICHEL, F., JACQUIER, A., DUJON, B.: Biochimie 64, 867-881 (1982). - MIHARA, K., BLOBEL, G., SATO, R.: Proc. Natl. Acad. Sci. USA 79, 7102-7106 (1982). - MILLER, D.L., MARTIN, N.C.: Cell 34, 911-917 (1983). - MIURA, S., MORI, M., AMAYA, Y., TATIBANA, M.: Eur. J. Biochem 122, 641-647 (1982). - MIURA, S., MORI, M., TATIBANA, M.: J. Biol. Chem. 258, 6671-6674 (1983). - MONTOYA, J., CHRISTIANSON, T., LEVENS, D., RABINOWITZ, M., ATTARDI, G.: Proc. Natl. Acad. Sci. USA 79, 7195-7199 (1982). - MONTOYA, J., GAINES, G.L., ATTARDI, G.: Cell 34, 151-159 (1983).

NAGY, F., LÁZÁR, G., MENCZEL, L., MALIGA, P.: Theor. Appl. Genet. 66, 203-207 (1983). - NEGRUK, V.I., CHERNY, D.I., NIKIFOROVA, I.D., ALEKSANDROV, A.A., BUTENKO, R.G.: FEBS L. 142, 115-117 (1982). - NETTER, P., JACQ, C., CARIGNANI, G., SLONIMSKI, P.P.: Cell 28, 733-738 (1982).

OHASHI, A., GIBSON, J., GREGOR, I., SCHATZ, G.: J. Biol. Chem. 257, 13042-13047 (1982). - OHTA, S., SCHATZ, G.: EMBO J. 3, 651-657 (1984). - O'MALLEY, K., PRATT, P., ROBERTSON, J., LILLY, M., DOUGLAS, M.: J. Biol. Chem. 257, 2097-2103 (1982). - OSIEWACZ, H.D., ESSER, K.: Curr. Genet. 7, 219-223 (1983). - OSINGA, K.A., DE HAAN, M., CHRISTIANSON, T., TABAK, H.F.: Nucl. Acids Res. 10, 7993-8006 (1982).

PALMER, J.D., SHIELDS, C.R.: Nature 307, 437-440 (1984). - PALMER, J.D., SHIELDS, C.R., COHEN, D.B., ORTON, Th.J.: Nature 301, 725-728 (1983). - PILLAR, T., LANG, B.F., STEINBERGER, I., VOGT, B., KAUDEWITZ, F.: J. Biol. Chem. 258, 7954-7959 (1983). - POWLING, A.: Heredity 49, 117-120 (1982). - PRATJE, E., MANNHAUPT, G., MICHAELIS, G., BEYREUTHER, K.: EMBO J. 2, 1049-1054 (1983). - PUSYRIOV, A.T., MAZO, A.M., MICHENKO, A.G., AVDONIA, T.A.: Gene 24, 115-124 (1983).

REID, G.A., YONETANI, T., SCHATZ, G.: J. Biol. Chem. 257, 13068-13074 (1982). - ROSAMOND, J.: Biochem. J. 202, 1-8 (1982).

SALTZGABER, J., KUNAPULI, S., DOUGLAS, M.: J. Biol. Chem. 258, 11456-11470 (1983). - SCHATZ, G., BUTOW, R.A.: Cell 32, 316-318 (1983). - SCHMELZER, C., SCHMIDT, C., SCHWEYEN, R.J.: Nucl. Acids Res. 10, 6797-6808 (1982). - SCHWEYEN, R.J. WOLF, K., KAUDEWITZ, F. (eds.): Mitochondria 1983. Berlin, New York: de Gruyter 1983. - SEDEROFF, R.R., in: Molecular Genetics of Plants. Advances in Genetics, Vol. 22, ed. J.G. SCANDALIOS. New York: Academic Press 1984.- SLONIMSKI, P., BORST, P., ATTARDI, G. (eds.): Mitochondrial Genes. Cold Spring Harbor Monograph 12. Cold Spring Harbor, NY: Cold Spring Harbor Laboratory Press 1982. - STERN, D.B., LONSDALE, D.M.: Nature 299, 698-702 (1982). - STERN, D.B., DYER, T.A., LONSDALE, D.M.: Nucl. Acids Res. 10, 3333-3340 (1982).

TABAK, H.F., OSINGA, K.A., DE VRIES, E., VAN DER BLIEK, A.M., VAN DER HORST, G.T.J., GROOT KOERKAMP, M.J.A., VAN DER HORST, G., ZWARTHOFF, E.C., MACDONALD, M.E.: pp. 79-93 in: Mitochondria 1983, eds. R.J. SCHWEYEN, K. WOLF, F. KAUDEWITZ. Berlin, New York: de Gruyter 1983. - TEINTZE, M., SLAUGHTER, M., WEISS, H., NEUPERT, W.: J. Biol. Chem. 257, 10364-10371 (1982). - THALENFELD, B.E., BONITZ, S.G., NOBREGA, F. G., MACINO, G., TZAGOLOFF, A.: J. Biol. Chem. 258, 14065-14068 (1983a). - THALENFELD, B.E., HILL, J., TZAGOLOFF, A.: J. Biol. Chem. 258, 610-615 (1983b). - TSUZUKI, T., NOMIYAMA, H., SETOYAMA, C., MAEDA, S., SHIMADA, K.: Gene 25, 223-229 (1983). - TUDZYNSKI, P.: Progr. Bot. 44, 297-307 (1982).

UNDERBRINK-LYON, K., MILLER, D.L., ROSS, N.A., FUKUHARA, H., MARTIN, N.C.: Mol. Gen. Genet. 191, 512-518 (1983).

VAN DEN BOOGAART, P., SAMALLO, J., AGSTERIBBE, E.: Nature 298, 187-189 (1982). - VAN LOON, A.P.G.M., VAN EIJK, E., GRIVELL, L.A.: EMBO J. 2, 1765-1770 (1983). - VAN LOON, A.P.G.M., DE GROOT, R.J., VAN EYK, E., VAN DER HORST, G.T.J., GRIVELL, L.A.: Gene 20, 323-337 (1982). - VIEBROCK, A., PERZ, A., SEBALD, W.: EMBO J. 1, 565-571 (1982).

WALBERG, M.W., CLAYTON, D.A.: J. Biol. Chem. 258, 1268-1275 (1983). - WARD, B.L., ANDERSON, R.S., BENDICH, A.J.: Cell 25, 793-803 (1981). - WEISS-BRUMMER, B., RÖDEL, G., SCHWEYEN, R.J., KAUDEWITZ, F.: Cell 29, 527-536 (1982). - WEISS-BRUMMER, B., HOLL, J., SCHWEYEN, R.J., RÖDEL, G., KAUDEWITZ, F.: Cell 33, 195-202 (1983). - WEISSINGER, A.K., TIMOTHY, D.H., LEVINGS, C.S. III, Hu, W.W.L., GOODMAN, M.M.: Proc. Natl. Acad. Sci. USA 79, 1-5 (1982). - WOLLENZIEN, P.L., CANTOR, C.R., GRANT, D.M., LAMBOWITZ, A.M.: Cell 32, 397-407 (1983). - WRIGHT, R.M., CUMMINGS, D.J.: Current Genet. 7, 457-464 (1983a); - Nature 302, 86-88 (1983b).

YAGINUMA, K., KOBAYASHI, M., TAIRA, M., KOIKE, K.: Nucl. Acids Res. 10, 7531-7542 (1982). - YIN, S., HECKMAN, J., RAJBHANDARY, U.L.: Cell 26, 325-332 (1981).

ZASSENHAUS, H.P., FARRELLY, F., HUDSPETH, M.E.S., GROSSMAN, L.I., BUTOW, R.A.: Mol. Cell. Biol. 3, 1615-1624 (1983). - ZWIZINSKI, C., NEUPERT, W.: J. Biol. Chem. 258, 13340-13346 (1983). - ZWIZINSKI, C., SCHLEYER, M., NEUPERT, W.: J. Biol. Chem. 258, 4071-4074 (1983).

ELKE PRATJE
Botanisches Institut
Universität Düsseldorf
D-4000 Düsseldorf 1, FRG

V. Phytopathology

a) Biological Control of Plant Pathogenic Fungi

By Friedhelm Meinhardt

1. Introduction

Genetics of plant pathology is largely focused on the evaluation of recognition systems of the organisms concerned. A comprehensive review covering different aspects of this topic was published quite recently (DAY 1984).

This contribution will therefore concentrate on another important point in studying fungal pathogens, i.e., the search for better ways to minimize economic losses caused by plant diseases. A most promising way to do it is biological control of phytopathogenic fungi by extrachromosomal genetic elements, as will be pointed out below in some detail for fungal pathogens, in which biological control is reported or seems at least possible.

The study of the genetics of phytopathogenic fungi does not only result in benefits which are obtained by the ability to control plant diseases, but also contributes to the understanding of the mutual interactions of genetic material of different organisms "living" together as close as a host and its parasite.

After a brief description of the fungal pathogens, of the damages caused by them, and of the agents which provide the ability to control these harmful organisms, prospects of future research and practical application will be discussed.

2. Survey of Cases Reported

a) *Endothia parasitica*

The chestnut blight fungus *Endothia parasitica* has in North America almost eliminated its host, the American chestnut tree *(Castanea dentata)*. The fungus infects through wounds, and from the point of infection the mycelium continues to grow until the bark of the host is completely occupied. The production of callus tissue by the tree as a response to infection does not really protect it as the fungus easily penetrates the callus (for details see ANAGNOSTAKIS 1982).

A massive epidemic, comparable to that which destroyed the American chestnut population early in this century, seemed to arise also in Europe when the first chestnut cankers were observed in Italy (BIRAGHI 1953).

The detection of trees which resisted the fungus and looked at least apparently healthy led to the discovery of *Endothia* strains which are

Progress in Botany, Vol. 46

less aggressive and were therefore designated "hypovirulent" or H-strains (GRENTE 1965). In addition, these hypovirulent forms were able to cure already existing blight when they were inoculated into diseased trees (GRENTE and SAURET 1969a).

Hypovirulent strains of *E. parasitica* are unstable in culture, and plating of single-spore cultures results in a great variety of morphologically different phenotypes (GRENTE and SAURET 1969b).

Hypovirulence is conferred by a cytoplasmic determinant (BERTHELAY-SAURET 1973, VAN ALFEN et al. 1975). All hypovirulent strains so far known contain dsRNA, which is transmitted along with hypovirulence to formerly virulent strains by anastomosis (DAY et al. 1977, ANAGNOSTAKIS 1981).

A compilation of properties of *E. parasitica* strains that lack or contain dsRNA is given by DAY and DODDS (1979).

H-strains are being used for biological control of blight in European plantations of chestnut trees (GRENTE and BERTHELAY-SAURET 1979). In Europe cytoplasmic hypovirulence has evidently spread naturally, resulting in a balanced control of the disease. In contrast, there is no evidence of natural spread of hypovirulence in North American chestnut populations, although the existence of H-strains containing dsRNA is repeatedly reported (DAY et al. 1977, ANAGNOSTAKIS 1981). This might be due to different host species in Europe *(C. sativa)* and North America *(C. dentata)* or to the absence of natural carriers, such as insects, etc. in American forests.

Efforts were and are undertaken in the United States to develop a biological control system based on hypovirulence. Applying mixtures of H-strains to chestnut cankers (JANES and ELLISTON 1980) reduced canker development, and KUHLMANN (1983) reported that treatment of chestnut blight cankers with compatible H-strains reduced disease in comparison with the appropriate control treatments, but many problems – e.g., the objectives caused by heterogenic incompatibility (ANAGNOSTAKIS and WAGGONER 1981) – sill need to be solved before hypovirulence becomes a potential tool for chestnut blight control.

The statement given by ANAGNOSTAKIS (1982) in her paper on the biological control of chestnut blight is therefore still valid: "To succeed in the United States, we must find or produce stable H-strains of *E. parasitica* that (i) can survive in nature without killing *C. dentata*, (ii) are present in broad ranges of vegetative compatibility, and (iii) are conducive to natural spread."

b) *Rhizoctonia solani*

Rhizoctonia is a widely distributed soil pathogen which causes damping off diseases in numerous plants. It is the causative agent of pocks in potatoes and is aggressive to many cultured and wild-living plants (BAKER 1970). Several isolates of *R. solani* were found to be affected by a degenerative disease called "Rhizoctonia decline" which manifests in a white color, floccose texture, irregular appearance, production of few or no sclerotia and retarded growth of the cultures, whereas healthy cultures are characterized by a brown color, with a mycelium close to the agar surface, rapid growth and the production of numerous sclerotia (CASTANHO and BUTLER 1978a,b).

Three out of 13 isolates of *R. solani* studied by CASTANHO et al. (1978) were found to contain strain-specific dsRNA molecules. These isolates were all diseased compared with the other strains, which looked healthy and did not harbor dsRNA. One diseased nonpathogenic strain was studied in some detail and basidiospore analysis revealed a cytoplasmic inheritance of the disease agent. Fusion between a diseased and a healthy culture derived from a hyphal tip of the diseased strain resulted in the conversion of the healthy into the diseased phenotype, thus indicating that the agent is transmitted by hyphal anastamosis.

The converted formerly healthy and dsRNA lacking isolate was found to contain the same dsRNA pattern as the diseased partner (CASTANHO et al. 1978). Attempts to transfer the agent into other strains belonging to the same or other vegetative compatibility groups (anastomosis groups) failed.

Rhizoctonia decline is therefore due to a transmissible, cytoplasmic factor, presumably dsRNA, and in this respect comparable to the decline of *Endothia parasitica*, although some striking differences exist, for example the unability to transmit the disease to other isolates. Nevertheless one may guess that the exploitation of other known hypovirulent *Rhizoctonia solani* isolates might succeed in disease control at least.

Besides the possible use of dsRNA for biological control of *Rhizoctonia*, the report of a linear plasmid (2.8 kbp in size) in weakly pathogenic isolates of this fungus represents perhaps a new landmark for biological control (HASHIBA et al. 1983). The Japanese workers reported cases evidencing that the decline of pathogenicity is transmissible, because they were able to transform pathogenic isolates into weak pathogenicity in soil experiments. The addition of linear plasmid DNA isolated from weakly pathogenic strains to protoplasts of a normal virulent isolate resulted in abnormal growth characteristics and many of the transformed strains had a decreased ability to parasitize radish plants, thus indicating that plasmid DNA may affect pathogenicity directly.

c) *Ceratocystis ulmi*

Highly pathogenic strains of the ascomycetous fungus *Ceratocystis ulmi* – the causative agent of Dutch elm disease – have brought about severe epidemics of Dutch elm disease throughout North America, Europe and Asia, destroying almost the entire elm population in these areas (BRASIER and GIBBS 1975, BRASIER 1979). For a long period of time one had no idea how to stop or even attenuate the effects of Dutch elm disease. The discovery of a cytoplasmically transmissible disease in one of the aggressive strains from North America, which severely affects the fungus, completely changed the situation (BRASIER 1983).

When isolates of the North American aggressive strain were investigated with a view to determining their incompatibility relationships and so to classify them into vegetative compatibility (v-c) groups, it became evident that one isolate constantly causes abnormal growth in the partners with which it was paired.

Comprehensive analyses indicated that the disease is caused by a transmissible cytoplasmic factor which can be transmitted from mycelium to mycelium without nuclear exchange. BRASIER (1983) has termed it disease or "d"-factor. Transmission of the "d"-factor is possible only within a v-c group, i.e., heterogenic incompatibility prevents exchange of cytoplasms. Spore-germination experiments indicated that the "d"-factor is transmitted to the conidia but not to the ascospores. These results resemble the situation in the coprophilous ascomycete *Podospora anserina*, in which senescence is caused by a mitochondrial plasmid (STAHL et al.

1978, 1980), which is infective and has been identified by sequence analysis as a mobile intron of the cytochrome oxidase gene subunit I (OSIEWACZ and ESSER 1984, OSIEWACZ 1984).

By analogy with other transmissible diseases of fungi (see above) the "d"-factor could be associated with dsRNA. In fact dsRNA was detected in strains of *C. ulmi* (PUSEY and WILSON 1982), it was present in both more aggressive and less aggressive isolates. Since less aggressive isolates generally possessed more segments, these dsRNA, unique to the less aggressive strains, may contribute to lowered pathogenicity, although other factors evidently are instrumental in pathogenesis (BRASIER 1977).

d) *Gaeumannomyces graminis var. tritici*

This ascomycetous fungus causes the "take-all disease", it infects roots of cereals and several grass seedlings. The disease causes at least yellow leaves, but more commonly severe dwarfing, and may result in total loss of grains.

"Take-all decline" occurs after successive crops of wheat and barley. This reduction in the ability to produce disease on susceptible hosts was suggested to be caused by fungal viruses (LAPIERRE et al. 1970), because disease was greatly reduced when mixtures of virus-free normal pathogenic strains and virus-infected strains were applied to wheat plants.

Viruses of *G. graminis var. tritici* were found to contain dsRNA components. Depending on the strain, 1 to 12 segments could be detected ranging in molecular weight from 1 to 3 × MD. The pattern of dsRNA in each strain was different from any other (ALMOND et al. 1977).

Because of the complexity of the *G. graminis* viruses it appears to be rather impossible to use this system for biological control of the take-all disease. In addition, there was no correlation between weak pathogenicity and the presence of viruses (RAWLINSON et al. 1973).

Again, the analysis of DNA has shed new light on disease control in this case. HONEYMAN and CURRIER (1983) described two DNA elements of presumably mitochondrial origin. Virulent isolates of different geographic origin were all found to contain similar elements; however, in an avirulent isolate – obtained as a sector of an original virulent strain – the plasmids could not be detected.

The plasmids are linear, as revealed by restriction analysis and electron microscopy, having molecular weights of 4.8 MD (7.5 kbp) and 5.6 MD (8.5 kbp), respectively.

HONEYMAN and CURRIER also reported that another avirulent strain isolated from nature was lacking these plasmids, and that is why they suggest that these elements are involved in pathogenesis.

To complete the number of phytopathogenic fungal organisms for which cytoplasmically transmissible diseases have been described, three *Helminthosporium* species have to be mentioned: *H. victoriae*, *H. maydis*, and *H. oryzae* (LINDBERG 1959, 1960, 1971). Since the nature of the disease agents remains completely obscure, these fungi and their diseases are not discussed in detail.

3. Discussion

As outlined above, either fungal viruses or linear extrachromosomal DNA molecules are the agents causing disease of the pathogens, eventually leading to hypovirulence of the strains concerned, thus being useful for biological control. Fungal viruses or their nucleic acids (dsRNA) are most prominent among the causative agents of the parasite's disease.

Viruses of plant pathogenic fungi were reviewed by DAY and DODDS (1979). Among all groups of fungi the occurrence of viruses is frequently reported; for further information see a book edited by LEMKE (1979), which provides sufficient information on mycoviruses and which covers a wide array of relevant mycovirus studies.

The most famous among the fungal viruses are probably those associated with the killer character of baker's yeast *Saccharomyces cerevisiae* (for further information see BEVAN and MITCHELL 1979, WICKNER 1979) or *Ustilago maydis* (Day and DODDS 1979), and the ones causing the virus disease of the edible mushroom *Agaricus bisporus* (VAN ZAAYEN 1979).

At least in *Endothia parasitica* and *Rhizoctonia solani* the decrease in growth rate and virulence is associated with the presence of dsRNA, furthermore the dsRNA in *E. parasitica* is in one strain associated with club-shaped virus-like particles (DODDS 1980). *Endothia* is an outstanding example for biological control of a fungal parasite, whereas, systems as efficient as in *E. parasitica* are as yet not established for the other fungi described.

In this context it is of interest that dsRNA might also provide a mechanism for biological control of aflatoxin production by *Aspergillus flavus* and *A. parasiticus* (SCHMIDT et al. 1983).

Testing known inhibitors of protein synthesis inferred an inverse correlation between aflatoxin production and the presence of double-stranded RNA.

One might have the impression that fungal viruses generally cause disease or trigger pathways, but this is by no means what is normally observed. In contrast to most of the mycoviruses described, almost nothing is known about effects on or in their hosts (for details see BUCK 1979, GHABRIAL 1980).

The linear plasmids observed in *Rhizoctonia solani*, which cause abnormal growth of the fungus and a decrease in pathogenicity, can be used for biological control in a way analogous to the dsRNA in this fungus. Since heterogenic incompatibility is apparently instrumental in *Rhizoctonia* – transmission of dsRNA is possible exclusively within derivates of the same isolate – it becomes obvious that for practical application the same difficulties are superimposed and have to be overcome for both dsRNA and extrachromosomal DNA.

Compared to *Rhizoctonia* linear plasmids in *Gaeumannomyces graminis var. tritici* have an inverse effect, because virulence is suggested to be correlated with the presence of the plasmids, whereas it is hypovirulence in *Rhizoctonia*. Nevertheless, the linear plasmids in *G. graminis var. tricici* and *R. solani* open the possibility to investigating pathogenesis on the nucleic acid level.

The occurrence of linear plasmids is also reported for the ergot fungus *Claviceps purpurea* (TUDZYNSKI et al. 1983) which is a parasite of rye; it is not yet clear whether these plasmids are involved in pathogenesis or not.

Linear plasmids are not restricted to phytopathogenic fungi, they occur in a great number of fungi and higher plants (for review and details see TUDZYNSKI 1982, ESSER et al. 1983) and even in prokaryotes, e.g., *Streptomyces* (HIROCHIKA et al. 1984).

Extrachromosomal DNA elements in phytopathogenic organisms remind one of the tumor-inducing (Ti) plasmids of *Agrobacterium tumefaciens*: a defined fragment of the Ti plasmid after transmission to the host cell integrates into the nuclear DNA of the tumor cell (CHILTON et al. 1977, 1980). The capacity to induce tumors is thus correlated with the presence of the Ti plasmid. For further information see a review by HILLE et al. 1983. The detection of linear plasmids in the mykorrizha-forming fungus *Morchella conica* (MEINHARDT and ESSER 1984) supports the opinion that extrachromosomal DNA might be the key for the understanding of communication on the genetic level between different organisms living together in a host-parasite or a symbiontic relation.

Besides their possible significance for interorganismal relations extrachromosomal DNA in fungi represent a potential tool for genetic manipulation of fungi – and perhaps higher plants – via genetic engineering techniques.

References

ALMOND, M.R., BUCK, K.W., RAWLINSON, C.J.: Abstr. 2nd Intern. Mycol. Congr. (Tampa), p. 13 (1977). - ANAGNOSTAKIS, S.: Exp. Mycol. 5, 236-242 (1981); - Science 215, 466-471 (1982). - ANAGNOSTAKIS, S., WAGGONER, P.E.: Phytopathology 71, 1198-1202 (1981).

BAKER, K.F.: pp. 125-148 in: Rhizoctonia Solani, Biology and Pathology, ed. J.R. DARMENTER. University of California Press 1970. - BERTHELAY-SAURET, S.: Annu. Phytopathol. 5, 31 (1973). - BEVAN, E.A., MITCHELL, D.J.: pp. 161-169 in: Viruses and Plasmids in Fungi, ed. P.A. LEMKE. New York, Basel: Dekker 1979. - BIRAGHI, A.: pp. 643-646 in: Reports of 11th Congress of the International Union of Forest Research Organizations. Rome: International Union of Forest Research Organization 1953. BRASIER, C.M.: Trans. Br. Mycol. Soc. 68, 45-52 (1977); - Nature 281, 78-79 (1979); - ibid. 305, 220-223 (1983). - BRASIER, C.M., GIBBS, J.N.: Ann. Appl. Biol. 80, 231-235 (1975). - BUCK, K.W.: pp. 93-160 in: Viruses and Plasmids in Fungi, ed. P.A. LEMKE. New York, Basel: Dekker 1979.

CASTANHO, B., BUTLER, E.E.: Phytopathology 68, 1505-1510 (1978a); - ibid. 68, 1511-1514 (1978b). - CASTANHO, B., BUTLER, E.E., SHEPHERD, R.J.: Phytopathology 68, 1515-1519 (1978). - CHILTON, M.D., DRUMMOND, M.H., MERLO, D.J., SCIAKY, D., MONTOYA, A.L., GORDON, M.P., NESTER, E.W.: Cell 11, 263-271 (1977). - CHILTON, M.D., SAIKI, R.K., YADAV, N., GORDON, M.P., QUETIER, F.: Proc. Natl. Acad. Sci. USA 77, 4060-1064 (1980).

DAY, P.R.: pp. 135-147 in: Cellular Interactions. Encyclopedia of Plant Physiology, N.S. Vol. 17, eds. H.F. LINSKENS, J. HESLOP-HARRISON. BERLIN, Heidelberg, New York: Springer 1984. - DAY, P.R., DODDS, J.A.: pp. 201-238 in: Viruses and Plasmids in Fungi, ed. P.A. LEMKE. New York Basel: Dekker 1979. - DAY, P.R., DODDS, J.A., ELLISTON, J.E., JAYNES, R.A., ANAGNOSTAKIS, S.: Phytopathology 67, 1393-1396 (1977). - DODDS, J.A.: Virology 107, 1-12 (1980).

ESER, K., KÜCK, U., STAHL, U., TUDZYNSKI, P.: Curr. Genet. 7, 239-243 (1983).

GHABRIAL, S.A.: Annu. Rev. Phytopathol. 18, 441-461 (1980). - GRENTE, J.: C.R. Hebd. Seances Acad. Agr. Fr. 51, 1033 (1965). - GRENTE, J., BERTHELAY-SAURET, S.: Biological control of chestnut blight in France, pp. 30-34 in: Proc. Am. Chestnut Symp., eds. W.L. MacDONALD, F.C. CECH, J. LUCHOK, C. SMITH. Morgantown: W. Virginia University Books 1979. - GRENTE, J., SAURET, S.: C.R. Acad. Sci. (Paris), Ser.D, 268, 2347-2350 (1969a); - ibid. 268, 3173-3176 (1969b).

HASHIBA, T., HYAKUMACHI, N., HONMA, Y., MATSUDA, I.: Abstr. Third Int. Mycol. Congr. Tokyo (1983). - HILLE, J., van KAN, J., SCHILPEROORT, R.A.: pp. 223-227 in: Molecular Genetics of the Bacteria-Plant Interaction, ed. A. PÜHLER. Berlin, Heidelberg, New York: Springer 1983. - HIROCHIKA, H., NAKAMURA, K., SAKAGUCHI, K.: Embo J. 3, 761-776 (1984). - HONEYMAN, A.L., CURRIER, T.C.: Abstr. 83rd Annu. Meeting of the American Society for Microbiology, N. Orleans (1983).

JANES, R.A., ELLISTON, J.E.: Phytopathology 70, 453-456 (1980).

KUHLMANN, E.G.: Phytopathology 73, 1030-1034 (1983).

LAPIERRE, H., LEMAIRE, J.M., JONAN, B., MOLIN, G.: C.R. Acad. Sci. (Paris), Ser. D, 271, 1833-1836 (1970). - LEMKE, P.A. (ed.): Viruses and Plasmids in Fungi. New York, Basel: Dekker 1979. - LINDBERG, G.D.: Phytopathology 49, 29-32 (1959); - ibid. 50, 457-460 (1960); - ibid. 61, 900-901 (1971).

MEINHARDT, F., ESSER, K.: Curr. Genet. 8, 15-18 (1984).

OSIEWACZ, H.D.: Bibliotheca Mycologica, Bd. 94. Vaduz: Cramer 1984. - OSIEWACZ, H.D., ESSER, K.: Curr. Genet. 8, 299-305 (1984).

PUSEY, L.P., WILSON, C.L.: Phytopathology 72, 423-428 (1982).

RAWLINSON, C.J., HORNBY, D., PEARSON, V., CARPENTER, J.M.: Ann. Appl. Biol. 74, 197-209 (1973)

SCHMIDT, F.R., DAVIS, N.D., DIENER, U.L., LEMKE, P.A.: Bio/Technology, (Nature) 1, 794-795 (1983). - STAHL, U., LEMKE, P.A., TUDZYNSKI, P., KÜCK, U., ESSER, K.: Mol. Gen. Genet. 162, 341-343 (1978). - STAHL, U., KÜCK, U., TUDZYNSKI, P., ESSER, K.: Mol. Gen. Genet. 178, 639-646 (1980).

TUDZYNSKI, P.: Progr. Bot. 44, 297-307 (1982). - TUDZYNSKI, P., DÜVELL, A., ESSER, K.: Curr. Genet. 7, 145-150 (1983).

VAN ALFEN, N.K., JAYNES, R., ANAGOSTAKIS, S., DAY, P.R.: Science 189, 890-891 (1975). VAN ZAAYEN, A.: pp. 239-324 in: Viruses and Plasmids in Fungi, ed. P.A. LEMKE. New York, Basel: Dekker 1979.

WICKNER, R.B.: Plasmid 2, 305-322 (1979).

Dr. FRIEDHELM MEINHARDT
Lehrstuhl für Allgemeine Botanik
Ruhr-Universität
Postfach 102148
D-4630 Bochum 1, FRG

D. Taxonomy

I. Systematics and Evolution of the Algae

By MICHAEL MELKONIAN

1. General Aspects

This report covers books, reviews and original publications on the systematics and evolution of algae, including prokaryotic algae, which have been published mainly during the period 1982/1983. The number of papers in this field has again increased and only about 30% of the publications are included in this review. The scope of this review is again fairly broad taking into consideration the various approaches which lead to significant conclusions on the systematics and evolution of this important and diverse assembly of organisms.

For teaching purposes and as a reference guide the most useful single publication on algae is certainly the book: *Selected Papers in Phycology II* (ROSOWSKI and PARKER 1982, eds.). In addition to 77 reprinted papers on phycological topics (a selection of important papers published during the last 10 years) it includes a number of original contributions and bibliographies (some contributions are listed in the appropriate chapters of this report.

Other more general books on algae include LOBBAN and WYNNE (eds., 1981): *The Biology of Seaweeds*, DRING (1982): *The Biology of Marine Plants*, DARLEY (1982): *Algal Biology: A Physiological Approach*, JOHANSEN (1981): *Coralline Algae, a First Synthesis*, and DAWES (1981): *Marine Botany*. A new review series: *Progress in Phycological Research* (ROUND and CHAPMAN eds., 1982) has been launched and some of the reviews of Vol. 1 of this series are discussed in later chapters. Students of "Endocytobiology" will find the recent compilation of papers published as *Endocytobiology II* (SCHENK and SCHWEMMLER eds., 1983) very useful. Additional information on problems surrounding endosymbiosis can be found in various review articles (e.g., CAVALIER-SMITH 1982a, GRAY and DOOLITTLE 1982, HARRINGTON and THORNLEY 1982, KREMER 1982a, WALLACE 1982, TAYLOR 1983a). The evolution of flagella and flagella beat patterns are discussed by CAVALIER-SMITH (1982b) and SLEIGH (1981), and hypotheses on the evolution of flagellated algal cells are presented by CHADEFAUD (1982a, 1982b).

2. Cyanobacteria (Blue-Green Algae)

a) Books, Reviews

The most important book on cyanobacteria published during the last 2 years is that by CARR and WHITTON (eds., 1982). Another book deals specifically with marine cyanobacteria (HUMM and WICKS 1980). LANG and WALSBY (1982) present an introduction to the group and provide a bibliography.

Progress in Botany, Vol. 46

b) Nomenclature

The question of whether the taxonomy of cyanobacteria (or Cyanophyceae) should be treated according to the rules of the bacteriological or the botanical nomenclatorial system (cf. Progr. Bot. 44, 316) has again been controversially discussed (FRIEDMANN and BOROWITZKA 1982, RIPPKA and COHEN-BAZIRE 1983). The different views were summarized at a symposium at the XIII. International Botanical Congress and a compromise was proposed (FRIEDMANN and BOROWITZKA 1982): cyanobacteria (blue-green algae) may be described following either the Botanical or the Bacteriological Code and the published names are valid. When following the Botanical Code it is recommended that all efforts should be made to obtain a living pure culture which should be deposited in a permanently established culture collection. If a new taxon is described according to the Bacteriological Code, and a culture is designated as a type, a sample of the culture should be preserved and deposited in a permanent institution (herbarium) together with photomicrographs or drawings. Subsequently, some changes have been made in the Code of Botanical Nomenclature toward a compromise as outlined above.

c) Taxonomy and Biology of Cyanobacteria

Taxonomic studies in the cyanobacteria are relatively rare compared to the flood of publications on the physiology and biochemistry of selected cyanobacteria. This may be partly due to the unsettled nomenclatorial problems (see above). Following the traditional approach of studying cyanobacteria from natural samples GOLUBIC and colleagues continue to investigate endolithic cyanobacteria and their geological activities (GOLUBIC 1983, LUKAS and GOLUBIC 1983). In a culture study of 16 strains of *Anabaena* the following morphological criteria have been found to be useful as taxonomic characters: cell dimension, position of the akinete in the trichome, and shape of the terminal cell (STULP and STAM 1982). These characters are apparently stable over a wide range of temperatures and light intensities (STULP 1983).

The occurrence of cyanophycin inclusion bodies may be taxonomically significant in cyanobacteria (LAWRY and SIMON 1982). Formation of cyanophycin inclusion bodies can be induced by various environmental conditions (e.g., in *Spirulina* by sulfur deficiency), but they are not formed in two strains of *Synechococcus*. In other strains of *Synechococcus* differing in their relative GC-content (cf. Progr. Bot. 44, 316), they are however present.

In their study on different *Oscillatoria* strains KLAVENESS and SKULBERG (1982) investigated the possibility of using in-vivo spectrophotometry as a simple tool in cyanobacterial taxonomy. They find that *O. bornetii* strains differ considerably from *O. agardhii/rubescens* strains, the latter being a heterogeneous assemblage with respect to pigment content. In his study on morphological variation of *O. agardhii* and *O. rubescens* GONZALEZ (1981), however, concludes that "coloration" has no taxonomical significance in distinguishing species and advocates using "cell width" as a distinguishing character. Electrophoretic patterns of phycochromoproteins of cyanobacteria may be useful for identification of species and strains (SCHENK and KUHFITTIG 1983). Functional phycobilisomes have been assembled in vitro from allophycocyanin of *Nostoc* sp. and phycoerythrin-phycocyanin of *Fremyella diplosiphon* (CANAANI and GANTT 1982). It has not been possible to obtain hybrid phycobilisomes between allophycocyanin from Nostoc and phycoerythrin-phycocyanin complexes from *Phormidium persicinum* or *Porphyridium sordidum* (CANAANI and GANTT 1982).

In a study of autotrophic picoplankton in the eastern tropical Pacific it has again been found that a substantial portion (20% - 80% of CO_2 fixation; 70% of ribulose-bisphosphate carboxylase) of the autotrophic production is associated with particles passing a 1 µm screen (LI et al. 1983, cf. Progr. Bot. 44, 316). In this size range unicellular cyanobacteria are a significant component of picoplankton (JOHNSON and SIEBURTH 1982).

The symbiosis between marine sponges and cyanobacteria has recently attracted considerable interest (WILKINSON 1983a, 1983b, BERTHOLD et al. 1982). Cyanobacterial symbioses are significant for the nutrition of many sponges which evolved increased surface areas to capture incident light (WILKINSON 1983b). WILKINSON (1983a) speculates that the sponge/cyanobacterial symbiosis is a long-established one (since the Precambrian?).

d) *Prochloron*

Prochloron is a prokaryotic organism which lacks phycobilins but contains chlorophyll a and b and occurs only as an ectosymbiont of colonial ascidians (cf. Progr. Bot. 44, 318; and reviews by LEWIN 1981, CHAPMAN and TRENCH 1982). *Prochloron* has received much attention since it was raised to phylum rank by LEWIN, especially in relation to the endosymbiosis theory and the possibility that *Prochloron* might be regarded as similar to the progenitors of the green chloroplast. Recent data obtained from analysis of 16S rRNA and 5S rRNA of *Prochloron* strains collected from different ascidian hosts do, however, not support views that *Prochloron* might be related to the progenitors of green chloroplasts (SEEWALDT and STACKEBRANDT 1982, MacKAY et al. 1982, STACKEBRANDT 1983). In fact, the partial sequence analysis of both rRNA-types reveals that *Prochloron* clusters with the cyanobacteria and not with the chloroplasts of Euglenophyceae, green algae and higher plants. STACKEBRANDT (1983) presents evidence that the evolutionary clock of 16S rRNA is fairly "isochronic" in *Prochloron*, cyanobacteria and chloroplasts. These data indicate that (a) *Prochloron* should be classified with the cyanobacteria (although remaining in a somewhat isolated position) and not in a separate phylum, (b) *Prochloron* most likely is not related to the immediate progenitors of green chloroplasts. Sequence determination data of 5S rRNA and of ferredoxin from *Synechococcus lividus* and spinach chloroplasts led to the conclusion that *Synechococcus* is more related to higher plant chloroplasts than *Prochloron* (DELIHAS et al. 1982). These data correlate well with a recent suggestion that cyanelles and green chloroplasts share a common ancestry and originated by a single endosymbiotic event (MELKONIAN 1983a).

The disposition of DNA has been studied in *Prochloron* by means of the fluorochrome DAPI (COLEMAN and LEWIN 1983). DNA is present as a network of 15 - 50 aggregates scattered at the periphery of the cell, very similar to the arrangement found in some unicellular cyanobacteria (TSCHERMAK-WOESS and SCHÖLLER 1982). The murein type of *Prochloron* is identical to that found in cyanobacteria and many gram-negative bacteria (STACKEBRANDT and KANDLER 1982). Lipid composition in *Prochloron* has been investigated by MURATA and SATO (1983) who find typical cyanobacterial lipids. PATTERSON and WITHERS (1982) have attempted laboratory cultivation of *Prochloron* and showed that the organisms is a tryptophan auxotroph.

3. Rhodophyceae

a) Reviews

Reviews on the Rhodophyceae include: morphology and classification (KRAFT 1981, DIXON 1982a), life history (WEST and HOMMERSAND 1981), and systematics and biology (KRAFT and WOELKERLING 1981). An introduction to the group is presented by NORRIS and KUGRENS (1982) (marine Rhodophyceae) and by OTT and SOMMERFELD (1982) (freshwater Rhodophyceae; extensive bibliography!). GOFF (1982) reviews parasitic red algae.

b) Morphology and Taxonomy

In reviewing papers on the systematics of Rhodophyceae one notes again the many papers on new taxa and on life histories of selected species (cf. Progr. Bot. 44, 318-320). It appears that most of the changes in the taxonomy of this group take place at the subfamily level (inclusion or transfer of genera and species), whereas the taxonomy of larger systematic categories appears to be relatively settled (see however PUESCHEL and COLE 1982). It is interesting to note that the majority of the new descriptions in the Rhodophyceae come from the southern hemisphere (e.g., HUISMAN and KRAFT 1982, 1983, KRAFT and NIN-THEIN 1983).

The current taxonomic status of genera in the Delesseriaceae is discussed by WYNNE (1983). Subdivision of the family into two subfamilies (Delesserioideae, Nitophylloideae) with a greater number of "tribes" ("groups") is indicated. The *Audouinella (Acrochaetium-Rhodochorton)* complex has been reviewed by WOELKERLING (1983). A great taxonomic confusion (with 24 different classification schemes and 1007 nomenclatural combinations!) is evident in this complex. Generic attributes are unclear but the type of sexual cycle may provide a stable basis to delineate genera (four different types of sexual cycles occur in the complex). MOE and SILVA (1983) continue their studies on Antarctic Ceramiaceae (cf. Progr. Bot. 44, 320) with two members of the tribe Ptilotea (*Georgiella* and *Plumariopsis*). Different types of tetrasporangial conceptacle development have been described in two tribes of the Corallinaceae (Mastophoroideae and Lithophylloideae) and may be taxonomically significant (TOWNSEND 1981, TURNER and WOELKERLING 1982). As a first contribution to a systematic study of *Bangia* in North America (cf. Progr. Bot. 44, 319) COLE et al. (1983) have provided karyotypes of *Bangia* populations from 88 sites along the coast of British Columbia. The morphology of postfertilization stages has been examined in members of the Solieriaceae (Gigartinales; GABRIELSON and HOMMERSAND 1982a). The Solieriaceae now contain three different tribes (Solierieae, Areschougieae trib. nov., Agardhielleae trib. nov., GABRIELSON and HOMMERSAND 1982b).

c) Life History and Polymorphism

A review of the life histories in the Florideophyceae with special reference to the Nemaliales is presented by DIXON (1982b). During the last 2 years the life history of various Rhodophyceae has been investigated in culture. Most life histories are of the "Polysiphonia-type" (GOFF 1981, GUIRY and MAGGS 1982, MAGGS and GUIRY 1982, WEST and GUIRY 1982) but some are very different: DE CEW and WEST (1982) have reinterpreted the sexual life history of *Rhodophysema* in the light of the recently described life history of *Palmaria palmata* (cf. Progr. Bot. 44, 320). In *R. elegans* there is no carposporophyte and no free-living tetrasporophyte. Structures on tetrasporangial crusts, previously believed to be "hair cells", are shown to be carpogonia. The fertilized carpogonium divides into a stalk cell and a tetrasporangium (the stalk cell produces successive tetrasporangia). The vegetative tissue of the crust is gametophytic with carpogonia and spermatia produced on the same crust. MAGNE (1982) tries to explain the newly discovered life histories of *Palmaria palmata* and *Rhodophysema elegans* as derivatives from the "Polysiphonia-type". He interprets these life histories as a result of regressive evolution and raises the question whether the stalk cell in *Rhodophysema* is of tetrasporophytic or carposporophytic origin.

DE CEW and WEST (1981) report that three species of *Farlowia* (Dumontiaceae) have a life history in which upright gametophytes alternate with a crustose tetrasporophyte resembling *Haematocelis* and/or *Cruoriopsis*. Carpospores of two *Gloiosiphonia* species germinate to form *Cruoriopsis*-like crustose tetrasporophytes (DE CEW et al. 1981).

Hybridization experiments of rhodophycean taxa from different geographic regions provide important information for species delimitation and for separating genetically different strains from phenotypic variation (RUENESS and RUENESS 1982, WEST et al. 1983). Studies on the genetics of *Gracilaria tikvahiae* are continued by PATWARY and VAN DER MEER (1982).

d) Cellular Organization

Two areas of ultrastructure research in relation to the systematics of the Rhodophyceae appear to be most promising at present: studies on mitosis/cell division and those on pit plugs (cf. Progr. Bot. 44, 320-321).

SCHORNSTEIN and SCOTT (1982) investigated mitosis/cytokinesis in the unicellular *Porphyridium purpureum* and SCOTT (1983) studied mitosis in *Batrachospermum* (Nemaliales). In all genera of Rhodophyceae studied to date both an intranuclear spindle composed of chromosomal and non-chromosomal microtubules and a metaphase plate of chromosomes are present. Also NAO's (nucleus associated organelles) are present at the division poles. The nuclear envelope in all genera is intact during metaphase-anaphase except at the division poles. In *Porphyridium* and *Batrachospermum* a single large gap is present, whereas numerous small fenestrations are present in the "Polysiphonia-type" algae studied. On the other hand, polar rings (PR) (although of a special type) and perinuclear envelope (PER) are present in *Batrachospermum* as in higher Florideophyceae. Features of mitosis in *Batrachospermum* are therefore somewhat intermediate between *Porphyridium* and *Polysiphonia*, as would be expected from the ancestral position of the Nemaliales within the Florideophyceae. It is clear that many more studies on rhodophycean mitosis/cytokinesis are needed in a phylogenetic approach. The chromatin of *Porphyridium aerugineum* has a nucleosome structure and is therefore not fundamentally different from that of higher eukaryotes (BARNES et al. 1982).

A survey of pit plug structure in 63 species representing 34 families of Rhodophyceae has been published (PUESCHEL and COLE 1982, cf. Progr. Bot. 44, 320). The main conclusions: the validity of the Palmariales is confirmed (two-layered pit plug caps). The Hildenbrandiaceae (one) cap layer) and the Corallinaceae (two-layered plug caps) should be removed from the Cryptonemiales (no plug caps) and raised to ordinal status. In a similar way the Bonnemaisoniaceae (incl. Naccariaceae, no plug caps), the Gelidiaceae (one cap layer), and the Batrachospermaceae (incl. Lemaneaceae and Thoreaceae; dome-shaped elaboration of the two-layered plug caps) should be removed from the Nemaliales (two-layered plug caps) and raised to ordinal status.

e) Chemotaxonomy

Chemotaxonomic investigations in the Rhodophyceae deal with phycoerythrins (GLAZER et al. 1982), carrageenan patterns (McCANDLESS et al. 1982, 1983), cell-wall composition (GRETZ et al. 1982), quaternary ammonium compounds (BLUNDEN et al. 1982), and carotenoids (BROWN and McLACHLAN 1982).
In a survey of phycoerythrins from representatives of all orders of Bangiophyceae and Florideophyceae GLAZER et al. (1982) distinguish five types of phycoerythrins

(B-phycoerythrins types I and II and R-phycoerythrins types I, II, and III) on the basis of absorption, excitation, and fluorescence emission spectra. They conclude that phycoerythrins are probably not useful as chemotaxonomic markers at familial, ordinal and class level, but may be useful at generic and species level (cf. Progr. Bot. 44, 320).

Cell-wall composition has been studied in the generic phase of *Bangia atropurpurea* (GRETZ et al. 1982). The major wall components are mannans and galactans with no cellulose found (cf. Progr. Bot. 44, 320). Protein is a significant wall component (10.6% of total wall weight) with hydroxyproline (0.3% of wall weight) present (absence of hydroxyproline was reported for cell walls of *Porphyra tenera*, cf. Progr. Bot. 44, 320).

f) *Cyanidium*

The uncertain taxonomic status of thermophilic phycobilin-containing eukaryotic algae (*"Cyanidium"*) has been discussed and reviewed by BROCK (1978), SECKBACH et al. (1981, 1983), LEWIN and GIBBS (1982), and KREMER (1982, 1983). Despite some 600 - 700 papers which have been published on *"Cyanidium"*, there is still no consensus among researchers as to its taxonomic and phylogenetic status. In general *"Cyanidium"* is either regarded as a primitive red alga or an endocyanome (and then placed into the Glaucophyceae sensu KIES). SECKBACH and colleagues (1981, 1983), however, believe that *"Cyanidium"* is a "bridge alga" directly linking Cyanobacteria with Rhodophyceae.

"Cyanidium" is not a single taxon but consists of at least three different genera (*Cyanidium*, *Cyanidioschyzon* and *Galdieria*; MEROLA et al. 1982). These genera have recently been placed into a separate class within the "Rhodophyta", the Cyanidiophyceae (MEROLA et al. 1982). DE LUCA and MORETTI (1983) have found floridoside and isofloridoside in the three genera, thus confirming earlier results presented by NAGASHIMA and FUKUDA (1981) for *C. caldarium*. These results again contrast with those of KREMER (1982) who from $[^{14}C]$-labeling experiments concludes that floridosides are absent from *C. caldarium* as compared to *Rhodella violacea*. It may be that floridoside is not labeled in short-time experiments under the experimental conditions chosen. Interestingly, REED (1983) has shown that *Cyanidium* produces significant amounts of floridoside only when the organism is subjected to salt stress. In an interesting study STEINMÜLLER et al. (1983) have shown that in addition to the α- and β-subunits of phycocyanin and allophycocyanin also both subunits of ribulose-1,5-bisphosphate carboxylase (RuBPCase) are synthesized in an in-vitro system by non-polyadenylated mRNA from *Cyanidium caldarium* and *Porphyridium aerugineum* as mature forms. This contrasts with the situation in green algae and higher plants (but see results on *Cyanophora*, Sect. 14) and indicates that both subunits of RuBPCase are synthesized by the plastid.

In conclusion present evidence suggests that *Cyanidium* is related to unicellular Rhodophyceae, but it cannot readily be incorporated into existing taxa. This view, however, does not preclude the possibility that *Cyanidium* is an endocyanome, since if the endosymbiosis hypothesis is accepted, then of course the plastids of the Rhodophyceae are also former endosymbionts and, in fact, the differences between the Rhodophyceae and the Glaucophyceae are somewhat blurred in such genera as *Cyanidium* and *Glaucosphaera*. To the reviewer it appears possible that the Rhodophyceae, Glaucophyceae and green algae had a common ancestor and that their plastids originated by a single endosymbiotic event (see also Sect. 14 on Glaucophyceae).

4. Dinophyceae

a) Reviews

DODGE (1982) in his book on the marine dinoflagellates of the British Isles gives descriptions and illustrations of 250 species including a key to the genera. The economically important problem of red tides is covered in the recent book by YENTSCH and TAYLOR (1982). An introduction to the Dinophyceae with a bibliography (172 references) is presented by DODGE and STEIDINGER (1982). DODGE (1983) also discussed the changing views on the systematics of the Dinophyceae and in one of his conclusions he indicates that the Prorocentrales should be regarded as an advanced group with a fairly recent history and not as an ancestral order in the Dinophyceae (cf. Progr. Bot. 44, 322). Special reviews deal with chromosome structure and mitosis (SPECTOR and TRIEMER 1981), basic chromatin proteins (RIZZO 1981), evolution (BUJAK and WILLIAMS 1981), flagella structure (MOESTRUP 1982), the amphiesma (MORRILL and LOEBLICH 1983), cysts (DALE 1983, SARJEANT 1982), and toxins (BADEN 1983).

b) Cellular Organization

One of the most unusual features of the dinoflagellate cell, the nucleus, has again received much attention (cf. Progr. Bot. 44, 322). SPECTOR and TRIEMER (1981) present a new diagrammatic model of the dinoflagellate chromonema involving a central core of 9 nm fibrils and an outer and inner helix of 2.5 nm fibrils. SOYER (1981) describes the presence of intranuclear filaments (5 - 6 nm diameter) in *Prorocentrum micans* using ultrarapid freezing and cryosubstitution. If no artifact, these filaments could represent actin filaments and might be involved in chromosome rearrangements. Nuclei have been isolated from the Florida red tide dinoflagellate *Ptychodiscus brevis* (RIZZO et al. 1982). Nucleosomes are absent as in other dinoflagellates (cf. Progr. Bot. 44, 322). While typical histones are absent, two histone-like proteins were detected (12,000 and 13,000 Mr). Further studies on histone-like proteins in dinoflagellates may help to dermine the evolutionary history of histones and nucleosomes.

The ultrastructure of mitosis has been studied in two further marine dinoflagellates (TRIEMER 1982, FRITZ and TRIEMER 1983). In a very interesting study TRIEMER (1982) shows that mitosis in the colorless *Oxyrrhis marina* is different not only from other dinoflagellates, but from all other organisms investigated to date. Dense plaques develop within the nuclear envelope at prophase and give rise to an intranuclear spindle. Some microtubules associate with the chromosomes, while others extend across the nucleus. Intranuclear cytoplasmic channels are absent. *Oxyrrhis* shares several typical characteristics with other dinoflagellates and should therefore be retained within the Dinophyceae, but probably in a new subgroup. Because of the presence of basic proteins and its intranuclear spindle TRIEMER (1982) regards *Oxyrrhis* as more advanced than the autotrophic thecate dinoflagellates. In some interesting speculations possible evolutionary steps of the mitotic spindle from extranuclear to intranuclear via an *Oxyrrhis*-type are discussed and the euglenoid mitotic apparatus is viewed as being derived from an *Oxyrrhis*-like ancestor. Mitosis in *Prorocentrum minimum* is similar to that of other dinoflagellates (intranuclear cytoplasmic invagination containing microtubules; microtubules in contact with kinetochore-like structures which attach through the channel membrane with chromosomes) and does not justify the separation of the Desmophyceae from other dinoflagellates (FRITZ and TRIEMER 1983).

c) Life History and Cysts

A prominent area of research in dinoflagellate life histories is the study of parasitic and phagotrophic forms (cf. Progr. Bot. 44, 324). Ameboid and helizoid life stages appear to be more widespread in sessile dinoflagellates than previously anticipated (POPOVSKY and PFIESTER 1982). Four consecutive developmental stages can be distinguished in the life history of the phagotrophic ectoparasite *Paulsenella* (DREBES and SCHNEPF 1982). The dinoflagellate *Caryotoma bernardi*, parasitic in the endoplasm of the radiolarian *Thallassicolla*, is recognized as a member of the Oodinida on the basis of life history and ultrastructure (HOLLANDE and CORBEL 1982).

Symbiodinium microadriaticum which is symbiotic in numerous marine invertebrates ("zooxanthellae") has been intensively studied in recent years (cf. Progr. Bot. 44, 324). Different strains of *S. microadriaticum* have been distinguished on the basis of differences in isoenzyme patterns, morphology and infectivity (reviews by TRENCH 1981, TRENCH et al. 1981). In addition, strains were found to differ in periodicity of diel motility (FITT et al. 1981), fatty acid and sterol composition (BISHOP and KENRICK 1980, KOKKE et al. 1981). CHANG and TRENCH (1982) now report different patterns of peridinin-chlorophyll a proteins in different strains of *S. microadriaticum* which are stable under different environmental conditions. TAYLOR (1983b) reports on possible free-living *S. microadriaticum*.

An excellent survey of the study of living dinoflagellate cysts has been presented by DALE (1983). The stratigraphic literature and cyst terminology (review: SARJEANT 1982) are excluded from his review, but cyst morphology, its role in dinoflagellate life histories, and the ecological and classificatory impact of cysts are adequately covered.

d) Taxonomy

Thecal plate patterns are the most useful single character in the taxonomy of the Dinophyceae (cf. Progr. Bot. 44, 323-324). A detailed comparison of thecal plate homology between *Ceratium cornutum* and four other members of the Peridiniales (belonging to the genera *Peridinium* and *Gonyaulax*) reveals a possible common type of pattern with four apical plates [HAPPACH-KASAN 1982, in contrast to the views of TAYLOR (cf. Progr. Bot. 44, 322)]. The apical pore structures of 15 marine genera were studied by SEM and classified into several groups (DODGE and HERMES 1981). A new terminology is proposed for the three types of apical plates that may be involved (pore plate, cover plate, canal plate).

Amphidinium cryophilum sp. nov. has been investigated by WEDEMAYER et al. (1982) and WILCOX et al. (1982). This study is a fine example of modern taxonomy including natural samples and cultures as well as light microscopy, SEM and thin sections. On the basis of a study of 56 clonal cultures of the heterotrophic dinoflagellate *Crypthecodinium cohnii* from 54 geographical sites BEAM and HIMES (1982) identify 28 sibling species. They conclude that sympatric speciation has contributed substantially to the genetic diversification within *C. cohnii* s.l. Electrophoresis of soluble enzymes (MDH, GDH) has been successfully used to demonstrate differences in different species of freshwater *Peridinium* (HAYHOME and PFIESTER 1983) and also between the sibling species of *Crypthecodinium cohnii* s.l. (BEAM et al. 1982).

5. Cryptophyceae

A summary of the taxonomy, ultrastructure and phylogeny of the Cryptophyceae is given by OAKLEY and SANTORE (1982). This small class of algae continues to raise interest because of its numerous unique features (cf. Progr. Bot. 44, 324-326). The chloroplast and the nucleomorph have been the subject of a number of recent studies. DWARTE and VESK (1983) investigated thylakoids of two cryptomonads (*Chroomonas* sp.; *Cryptomonas* sp.) with the freeze-fracture technique. The thylakoids of both strains are organized into stacks of typically two thylakoids. Evaluation of particle sizes in the different membrane faces suggests that there is partial segregation of photosystem II into stacked membranes and photosystem I into unstacked membranes. These results indicate that the cryptomonad thylakoid is more similar to the thylakoid of other chlorophyll c-containing plants (DWARTE and VESK 1982) than to the thylakoid of the red algae or cyanobacteria.

The ultrastructure of nucleomorph division has been the subject of two studies (McKERRACHER and GIBBS 1982, MORRALL and GREENWOOD 1982). The single nucleomorph divides into two after basal body replication, but generally before chloroplast division. Nucleomorph division is accomplished by an invagination of the nucleomorph envelope. The inner envelope membrane invaginates first, followed later by the outer membrane. No microtubules are involved as also evidenced by studies with colchicine. Colchicine inhibits nuclear division, but not division of the nucleomorph (McKERRACHER and GIBBS 1982). The cluster of globules usually present in the nucleomorph is also divided into two groups by the infurrowing nucleomorph envelope. McKERRACHER and GIBBS (1982) have proposed that the dense globules are viral in nature. Although nucleomorph division appears to be unlike any known form of nuclear division, it does not invalidate the hypothesis that it is a vestigial nucleus of a eukaryotic endosymbiont (cf. Progr. Bot. 44, 325), since the endosymbiotic nucleus of *Peridinium balticum* (presumably of chrysophyte affinities) also divides amitotically (TIPPIT and PICKETT-HEAPS 1976). This question must remain unsettled unless DNA is unequivocally demonstrated in the nucleomorph or isolated from nucleomorphs.

The flagellar root system of two marine species of *Cryptomonas* has been investigated by GILLOTT and GIBBS (1983). Although the flagellar root system in these autotrophic cryptomonads is basically similar to that of the leucoplast-containing *Chilomonas* (cf. Progr. Bot. 44, 326) it is much reduced in complexity and some flagellar roots present in *Chilomonas* are apparently lacking in the two *Cryptomonas* sp. Even with respect to the two "species" of *Cryptomonas* there is considerable difference in the detailed structure of the flagellar root system. GILLOTT and GIBBS (1983) reach the conclusion that the flagellar root system in the Cryptophyceae does not resemble that observed in any other algal group. These authors also do not favor the concept that a phylogenetic link exists between the cryptomonads and the zooflagellate groups Oxymonadida and Trichomonadida (cf. Progr. Bot. 44, 326).

Another ultrastructural study deals with the presence of flagellar and body scales in a *Chroomonas* sp. (of the mesostigmatica-group) and in a *Cryptomonas* sp. (SANTORE 1983), extending the previous observations on *Hemiselmis* (cf. Progr. Bot. 44, 326) to other cryptophycean genera. A comparative ultrastructural study of three species of *Hemiselmis* justifies the separation of *Hemiselmis* from other cryptophytes at the family level (i.e., Hemiselmidaceae, SANTORE 1982).

The endosymbiotic cryptomonad in the ciliate *Mesodinium rubrum* is apparently a complete cell, since some nuclei previously assigned to the ciliate may in fact rather belong to the endosymbiont (GRAIN et al. 1982). An interesting ectosymbiotic association between *Cryptomonas (curvata?)* and bacteria of the genus *Caulobacter* has been studied by KLAVENESS (1982). The *Cryptomonas* cells possess a transverse constriction caused by the bacteria which is lost when axenic cultures of the alga are established. The taxonomic validity of species of *Cryptomonas* described on the basis of transverse constrictions may therefore be questioned.

6. Prymnesiophyceae

An introduction to the class and a bibliography (114 references) has been presented by BONEY and GREEN (1982) dealing with taxonomy, ecology, life history, physiology, and biochemistry. The present taxonomy (cf. Progr. Bot. 44, 326) recognizes four orders (Pavlovales, Isochrysidales, Coccosphaerales, Prymnesiales). The Pavlovales (summary by GREEN 1980) with three genera constitute a natural group. With respect to the other three orders the system is rather artificial. This is exemplified by the morphological similarity of unmineralized scales and coccoliths between genera belonging to all three orders. In a thoughtful discussion of the problems encountered in the systematics of coccolithophorids NORRIS (1983) stresses the importance of cultures, life-history data and the effect of environmental factors on morphologcial variability of coccoliths. As in other flagellate algal classes scale characters have contributed invaluable information for the identification of various taxa (cf. Progr. Bot. 44, 327).

New taxa in natural samples of Galapagos nanoplankton have been identified on the basis of their scale structure (MANTON 1982, 1983, MANTON and OATES 1983). Based on a comparative study of coccolith structure in the two genera *Papposphaera* and *Pappomonas*, the inclusion of the two genera in a restructured family Deflandriaceae (including fossil coccoliths) is recommended (NORRIS 1983).

Descriptions of prymnesiophycean taxa based on light and electron microscope studies of cultures have recently been presented (*Prymnesium:* GREEN et al. 1982, BILLARD 1983; *Pleurochrysis:* GAYRAL and FRESNEL 1983a; *Platychrysis:* GAYRAL and FRESNEL 1983b). A second type of body scale has now been discovered in four species of *Prymnesium* (GREEN et al. 1982, BILLARD 1983) raising questions about its delimitation from *Chrysochromulina*. *Chrysochromulina* (which usually has two or more types of scales) is nevertheless distinguished from *Prymnesium* by several prominent features: a very long coiling haptonema, highly elaborated nonmineralized organic scales, homodynamic apically inserted flagella which are posteriorly oriented during rapid movement, and probably lack of cysts.

There is a continuing interest in intracellular coccolith formation and biocalcification associated with coccoliths (reviews by DE JONG et al. 1983 and WESTBROEK et al. 1983). A Ca^{2+}-binding polysaccharide has been isolated and characterized from coccoliths of *Emiliana huxleyi* (BORMAN et al. 1982). *Emiliana huxleyi* has gained additional prominence since it has been shown to be responsible for extensive patches of water giving strong reflectance of visible light along the margin of the north-west European continental shelf scanned each year with the Coastal Zone Color Scanner of the Nimbus 7 satellite (HOLLIGAN et al. 1983).

7. Eustigmatophyceae

An introduction to the Eustigmatophyceae and a bibliography is given by HIBBERD (1982) covering systematics, ultrastructure, physiology and ecology.

Probably the only original work during 1982/1983 is the investigation on keto-carotenoids in two species of the marine *Nannochloropsis* (ANTIA and CHENG 1982). In this study astaxanthin and astacene are found for the first time in the class.

8. Tribophyceae (Xanthophytes)

OTT (1982) reviews taxonomy, ultrastructure, physiology and parasitic associations in the Tribophyceae. In a review of flagellar root systems in the Tribophyceae MOESTRUP (1982) indicates that the root system of the *Vaucheria* spermatozoid is considerably different from that of other Tribophycean genera (notably *Ophiocytium*, *Bumilleria*) and instead shows some similarities to that of the Fucophyceae, but at present any phylogenetic conclusions would be premature.

9. Raphidophyceae

HEYWOOD (1982) presents a detailed review of the class and provides a bibliography.

The discussion about the problematic taxonomic status of *Olisthodiscus luteus* continues (cf. Progr. Bot. 44, 329). MOESTRUP (1982) considers *Olisthodiscus* to be a member of the Raphidophyceae (in contrast to GIBBS et al. 1980). This view is mainly based on the flagellar root system of *Olisthodiscus* which shows striking similarities to the root system of *Chatonella* and *Gonyostomum*, two members of the Raphidophyceae. In a preliminary communication HARA (1981) has claimed that the organism usually referred to as *Olisthodiscus luteus* and available under this name in culture collections is, in fact, *Heterosigma inlandica*. It is clear that more work is needed to settle this taxonomic problem. *Olisthodiscus luteus* continues to be a very successful model organism for the study of extranuclear DNA (ALDRICH et al. 1982) and has recently been used to isolate relatively pure fractions of nuclei (RIZZO and BURGHARDT 1983).

10. Chrysophyceae

Two recent reviews on the Chrysophyceae have been prepared by KRISTIANSEN (1982) and by KRISTIANSEN and TAKAHASHI (1982). In the first the emphasis is on the systematics and classification, while the second covers taxonomy, ultrastructure, physiology, life history and ecology, and includes a bibliography (170 references). KRISTIANSEN (1982) distinguishes two subclasses within the Chrysophyceae, the Chrysophycidae (approximately 800 species) and the Dictyochophycidae (the silicoflagellates with only 3 species). For classification of the Chrysophycidae into orders KRISTIANSEN (1982) gives preference to the classical system of PASCHER (as modified by FOTT 1971) over the system of BOURRELLY (1981, cf. Progr. Bot. 44, 328), although the artificial nature of the former is recognized. The current status of the Pedinellales within the Chrysophycidae is extremely uncertain and the isolated position of the order is for example reflected in the peculiar flagellar root system (MOESTRUP 1982, for review).

ANDERSEN and MULEY (1983) studied 34 strains representing 25 species of Chrysophyceae for the occurrence of chlorophylls c_1 and c_2. Most strains contained both types of chlorophyll c, however, all strains studied of the genera *Mallomonas* and *Synura* (12 strains) contained only chlorophyll c_1 but lacked c_2. This situation is apparently unique in the chromophyte algae. In addition to their peculiar chlorophyll c distribution the genera *Synura* and *Mallomonas* (and probably other genera of the Mallomonadaceae) are distinguished by additional characters from other members of the Chrysophyceae. These features lead ANDERSEN and MULKEY (1983) to the conclusion that the Mallomonadaceae may be more distinct than is suggested by present classification as a family in the Ochromonadales (e.g., KRISTIANSEN 1982).

In the Mallomonadaceae a greater number of taxonomic studies have been performed in the last 2 years based in principle on structural features of the silicious scales (e.g., ASMUND et al. 1982, DÜRRSCHMIDT 1983, CROOME and TYLER 1983, PREISIG and HIBBERD 1982a, 1982b). With the exception of some *Paraphysomonas* species (PREISIG and HIBBERD 1982a, 1982b, 1983) the organisms in question have not been cultured, often even living cells have not been seen and the descriptions are based solely on electron microscopy of the siliceous scales. The detailed study of PREISIG and HIBBERD (1982a, 1982b) has shown that the type species of *Spiniferomonas* (originally believed to be pigmented in contrast to the colorless *Paraphysomonas*) lacks plastids and therefore has been attributed to *Paraphysomonas*. The analysis of scale structure within taxa of *Paraphysomonas* allowed the erection of 11 species groups within the genus.

The siliceous resting cysts (statospores, stomatocysts) produced by Chrysophyceae (cf. Progr. Bot. 44, 329) are currently receiving considerable attention as a potentially important group of microfossils (ADAM 1980, CARNEY and SANDGREN 1982, SANDGREN 1983a). In an experimental study SANDGREN (1983b) investigated structure and variability of statospores from six different populations of *Dinobryon cylindricum* produced under defined conditions. The results suggest that some features of cyst morphology are identical in all populations, while other characters show distinctive variations. It is suggested that statospores as common freshwater microfossils may improve the accuracy of paleoecological interpretations.

11. Diatomophyceae

a) Books, Reviews

An overview of current diatom research is presented in the Proceedings of the Sixth Symposium on Recent and Fossil Diatoms (ROSS 1981). Topics range from taxonomy, cellular organization to ecology with emphasis on frustule characters. Another compilation of papers is the Festschrift NILS FOGED (HAKANSSON and GERLOFF 1982). Here emphasis is on floristics and ecology. FOGED (1981) himself presents a monograph of the diatoms of Alaska, where he gives an account of 987 taxa with 45 described as new to science. Two new volumes of the diatom journal Bacillaria (Vol. 5 and Vol. 6, 1982, 1983) have appeared. An introduction to the class and a bibliography (197 papers) is given by AMSPOKER and CZARNECKI (1982).

Other useful reviews include ROUND (1981a, 1981b) on the phylogenetic relationships within silicified algae and the origin of diatoms (cf. Progr. Bot. 44, 329-330), HARGRAVES and FRENCH (1983) on diatom resting spores, CRAWFORD (1981) on the siliceous components of the diatom cell wall and their morphological variation, SULLIVAN and VOLCANI (1981) on the role of silicon in the cellular metabolism of diatoms, and HERTH and SCHNEPF (1982) on β-chitin fibril formation in centric diatoms.

b) Cellular Organization

Two important papers deal with valve morphogenesis and development of valve symmetries in different pennate diatoms. PICKETT-HEAPS (1983) has investigated the relationship of cell organelles to valve morphogenesis in three species of *Nitzschia* with different valve symmetries ("nitzschioid" and "hantzschioid", cf. Progr. Bot. 44, 331). Unexpectedly the observations revealed considerable variations in the ultrastructural systems associated with valve morphogenesis (i.e., silicalemma, and "Microtubule Center") in the different species. An interesting difference between the investigated species of *Nitzschia* compared to *Hantzschia* is that in *Nitzschia* the fibulae in one daughter cell grow from the valve surface to the girdle surface (as in *Hantzschia*) whereas in the other daughter cell the fibulae grow from the girdle surface to the valve surface. An important taxonomic difference between *Nitzschia* and *Hantzschia* may therefore depend on whether the longitudinal axis of the forming valves moves in the same or opposite directions across the cell during morphogenesis.

MANN (1983) presents a diagrammatic scheme of the cryptic lateral polarity of the valve in raphe-bearing diatoms and distinguishes the two types of frustules as "cis-type" (both valves have the same orientation) and "trans-type" (both valves have opposite orientations). Based on the study of 45 taxa several conclusions were drawn: (a) different populations of the same species exhibit same behavior, (b) some species only produce cis-frustules (e.g., *Cymbella* spp., *Gomphonema* spp., some *Navicula* species), (c) in other taxa both cis- and trans-trustules are found, no taxon reveals only trans-frustules. Although the taxonomic significance of cis/trans symmetry of frustules in raphe-bearing diatoms remains to be established, the genera exhibiting only cis-frustules share other characteristics. MANN (1983) states that in dividing raphid diatoms both new valves have the same orientation and therefore of the six possible modes of division only three seem to be present in nature and the relative proportions of these explain the ratios of cis/trans-symmetries observed in the different taxa. With respect to the "nitzschioid" symmetry the results by PICKETT-HEAPS (1983) and MANN (1983) taken together indicate that here also initially daughter valves have the same orientation (cis-type; exemplified by direction of growth of the fibulae!) and only later due to the movement of "Microtubule Centers" into opposite directions does the typical "nitzschioid" ("pseudo-trans") symmetry arise.

c) Taxonomy

Numerous taxonomic studies mainly based on SEM observations of frustule characters have been carried out during the last 2 years. The most important single contribution appears to be the monographic treatment of the Cymatosiraceae, a new family of diatoms by HASLE et al. (1983). The reason for regarding the Cymatosiraceae as a family of centric diatoms lies in the structure of the auxospore envelope (presence of spinescent scales, transverse perizonium of closed bands) and the presence of flagellated male gametes in four of the nine genera. The Cymatosiraceae are clearly distinguished from other families of centric diatoms by some peculiar features of frustule structure (e.g., low apical elevation each with an ocellulus).

On the basis of SEM investigations of frustules other new taxa of diatoms have been described (e.g., COX 1982, HENDEY 1982, ROUND 1982, HASLE 1983, FRYXELL and HASLE 1983). FRYXELL and MEDLIN (1981) demonstrate that chain formation in the marine diatom genus *Chaetoceros* can occur in markedly different ways demonstrating parallel evolution of an apparently advantageous character. SEM studies of the vegetative cell, auxospore and initial cell of *Cerataulus laevis* by EHRLICH et al. (1982a, 1982b) document a fimbriate edge to the valve mantle, thus possibly supporting the new hypothesis for the evolution of diatoms from ancestral forms covered by fimbriate scales (cf. Progr. Bot. 44, 329, 330). Studies on *Rhoicosphenia* (MANN 1982a, 1982b) and *Cocconeis* (HOLMES et al. 1982) support the hypothesis that monoraphid diatoms derive from the biraphid series and that the araphid genera are merely modified biraphid forms. The species concept in diatoms is discussed in papers by MANN (1982c) on key characters in the classification of *Nitzschia* species and by GEISSLER (1982) on the effect of environmental parameters on frustule morphology of *Stephanodiscus hantzschii*. Auxospores continue to play a prominent role in taxonomic studies of diatoms. A review of auxospore envelopes has been provided by VON STOSCH (1982) with emphasis on anisometric auxospore growth. In centric species with anisometric auxospores only the first isometric developmental stage is contained in a scale-bearing envelope (as in isodiametric auxospores, cf. Progr. Bot. 44, 329-330), subsequent anisometric growth is supported by a secondary cover of silica rings originating inside and coalescent with the scaly layer (termed "properizonium"). The properizonia occur in two types: one with closed bands throughout and the other with a majority of open hoops. VON STOSCH (1982) concludes that the latter type of properizonium in the biradiate Centrales may be homologous to the transverse perizonia of pennate diatoms and that the Biddulphiaceae could therefore be the family from which the Pennales may have evolved (the *Rhabdonema* auxospore envelope with its scales bears a striking resemblance to those of the Biddulphiaceae).

12. Fucophyceae (Phaeophyceae)

Recent research on Fucophyceae is summarized in several review articles: WYNNE (1982a, 1982b) reports on morphology and classification, PEDERSEN (1982) on life histories. In addition CLAYTON (1981a) treats the Fucophyceae from an Australasian viewpoint. An introduction to the class with a bibliography (255 references) is provided by WYNNE (1982c). Reproduction and early development in *Fucus* is reviewed by EVANS et al. (1982). A monograph on European Sphacelariaceae has been prepared by PRUD'HOMME VAN REINE (1982).

Life history studies based on cultures continue to play an important role in contemporary taxonomic studies of Fucophyceae (cf. Progr. Bot. 44, 332). A review of life history studies in the Ectocarpales is presented by CLAYTON (1982).

The importance of parthenogenesis in the life history of many Fucophyceae is becoming apparent. CLAYTON (1981b) investigated the origin of asexually reproducing blades of complanate *Scytosiphon* and demonstrated that these blades arise parthenogenetically, developing directly from a small percentage of gametes formed by gametophytic blades. The seasonal occurrence of *Scytosiphon* plants in Southern Australia may be explained by considering the sequence of phases in the life history together with the effects of environmental factors on reproduction. MOE and HENRY (1982) studied the antarctic endemic *Acroseira mirabilis* both in situ and in culture. The life history is interpreted according to a fucalean scheme with a free-living sporophyte bearing a reduced cytological gametophyte. The biflagellate swarmers develop parthenogenetically into a plant identical to the parent. The authors suggest that *Acroseira* be placed into a separate order *Acroseirales*. Additional life history studies using cultures of Fucophyceae include *Arthrocladia villosa* (MÜLLER and MEEL 1982; sexuality is suppressed and replaced by apomeiosis and parthenogenesis, the life history supports inclusion of *Arthrocladia* into the Desmarestiales), and *Analipus japonicus* (NELSON and COLE 1981, NELSON 1982a, 1982b; this crustose brown alga is referred to the monotypic family Heterochordariaceae within the Ectocarpales and the problematic status of the Ralfsiales is discussed).

The taxonomy, dispersal and colonization of *Sargassum muticum*, an invasive brown alga, which since its accidental introduction from Japan has spread along the entire coast of North America and is now rapidly invading Western Europe, is subject of intensive study (FLETCHER 1980, NORTON and FETTER 1981, DEYSHER and NORTON 1982, NORTON 1983). In a new approach statistical analysis of morphogenetic measurements has been applied to taxonomic problems in *Fucus serratus* and *F. vesiculosus* (MARSDEN et al. 1983a, 1983b). Discriminant analysis of 12 continuous morphological variables can separate closely related populations of *F. serratus* and *F. vesiculosus*.

Hybridization experiments have been performed in the genera *Laminaria* and *Saccorhiza* (COSSON and OLIVARI 1982, BOLTON et al. 1983). Successful hybridization between Pacific and North Atlantic representatives of *Laminaria* are reported by BOLTON et al. (1983). It is concluded that the genus *Laminaria* probably originated in the Pacific and with a few species invaded the forming North Atlantic fairly recently in geological times. Studies dealing with phenotypic plasticity of Fucophyceae are also prominent fields of research (cf. Progr. Bot. 44, 332). Formation of blades, colorless hairs and plurilocular sporangia may be influenced by light, temperature and nitrogen in *Desmotrichum undulatum* (LOCKHART 1982). In *Macrocystis integrifolia* blade morphology and growth characteristics from geographically isolated populations when grown under the same culture conditions over time became increasingly similar, only frond elongation and the rate of frond initiation appeared to be genotypically fixed (DRUEHL and KEMP 1982). FRIES (1982) showed that vanadium enhances growth and in addition stimulates branching in axenic cultures of *Fucus spiralis*.

At the level of cellular organization studies have been carried out on the ultrastructure of swarmers in the Laminariales (HENRY and COLE 1982a, 1982b). In the laminarialean zoospore absence of an eyespot and presence of elaborate adhesion vesicles are regarded as an evolutionary advancement. *Chorda* zoospores and spermatozoids, however, contain an eyespot and the genus represents the most primitive extant member of the Laminariales (HENRY and COLE 1982b). The ultrastructure of *Heribaudiella fluviatilis*, a freshwater brown alga, has been studied by PUESCHEL and STEIN (1983) who clearly demonstrate its fucophycean structural characteristics.

Investigations on pheromones of Fucophyceae (cf. Progr. Bot. 44, 333) continue to yield results of potential phylogenetic significance. A sperm attractant has been identified in another fucoid *Ascophyllum nodosum* and named finavarrene (MÜLLER et al. 1982a). The molecular size of finavarrene is identical to that of related substances in other Fucophyceae in which the attractants are all C_{11} olefines, *Fucus* with a C_8 molecule being the exception. *Ascophyllum* and *Fucus*, however, share the straight-chain character of their attractants, whereas all pheromones of the subclass "Phaeophycidae" are monocyclic cyclopentane or cycloheptane derivatives. Interestingly, finavarrene is also found in *Spermatochnus paradoxus* (order Chordariales) and in vegetative thalli of *Dictyopteris* (order Dictyotales).

Pheromones do not only act as sperm attractants but may in some species also affect discharge of gametes from the antheridium. Identification and bioassy of three compounds secreted by eggs of *Desmarestia aculeata* and *D. viridis* showed that only one, named desmarestene (another C_{11} olefine), was active in releasing gametes, the other two, ectocarpene and viridene, were much less effective (MÜLLER et al. 1982b). It is interesting to note that egg secretions of *Laminaria* and *Cutleria* also contain three compounds, only one of which has pheromone activity. In both cases ectocarpene was an inactive by-product. Ectocarpene, however, is the active pheromone in the genus *Ectocarpus*. MÜLLER (1981) has speculated that the occurrence of ectocarpene in the more advanced members of the Fucophyceae represents the conservation of an ancestral character, its function in those genera may be related to interspecific competition phenomena.

13. Euglenophyceae

Most of the current research in the Euglenophyceae is concerned with biochemistry, molecular and cell biology of two strains of *Euglena gracilis* and is extensively summarized in volumes III and IV of the book *The Biology of Euglena* (BUETOW ed. 1982a, 1982b). Research on other euglenophycean genera is very limited and so are the studies on the systematics and evolution of this algal class. LEEDALE et al. (1982) have presented an introduction to the class including a bibliography.

In his review on flagellar structure in algae MOESTRUP (1982) concludes that the Euglenophyceae may be related to the green algae by two non-plastid characters: (a) the presence of a multilayered structure associated with a flagellar root, (b) the presence of disc-shaped mitochondrial cristae which may be similar to those of some naked green flagellates. Although this is an attractive hypothesis there are not yet enough data to substantiate it.

The first marine euglenophyte to be studied for its pigment composition is *Eutreptiella gymnastica* (BJÖRNLAND 1982). The combination of diatoxanthin, diadinoxanthin and neoxanthin, together with chlorophylls a and b, seems to be characteristic for the Euglenophyceae; α-carotene, zeaxanthin, antheroxanthin and the ketocarotenoids astaxanthin, canthaxanthin and echinenone are absent from *E. gymnastica*. PAGNI et al. (1981), however, report the presence of α-carotene from isolated stigmata of *Euglena gracilis*, none of the seven stigma carotenoids was a ketocarotenoid.

In a survey on the occurrence of intracellular bacteria in the Euglenophyceae SUREK and MELKONIAN (1983) found that about 5% of the species examined contained intracellular bacteria. In an axenic culture system (*Euglena mutabilis*/intracellular bacteria) it was shown that numbers of bacteria per algal cell remained constant during prolonged culture. It was concluded that the association between Euglenophyceae and intracellular bacteria represents a recently acquired endosymbiosis and that present-day green Euglenophyceae are able to phagocytose suitable bacteria.

14. Glaucophyceae

This recently resurrected class of algae (cf. Progr. Bot. 44, 334-335) is receiving considerable interest because of the peculiar nature of their photosynthetic organelles, the cyanelles. Recent reviews of the cyanelles have been prepared by TRENCH (1982a, 1982b). As more data accumulate it becomes obvious that the cyanelle is not an endosymbiotic cyanobacterium but instead much more resembles present-day chloroplasts. A peculiar feature of a cyanelle, however, is the presence of a peptidoglycan cell wall between the two envelope membranes (AITKEN and STANIER 1979). In fact, a recent investigation by GIDDINGS et al. (1983) using the freeze-fracture technique indicates that the whole boundary of the cyanelle is very similar to that of a cyanobacterium: the inner envelope membrane of a cyanelle (from *Cyanophora*) resembles the plasma membrane of a cyanobacterium, the outer envelope membrane is strikingly similar to the lipopolysaccharide layer of cyanobacteria. The latter observation is particularly interesting since it has previously been assumed that the cyanelle is enclosed in a "host vacuole" (SCHNEPF and BROWN 1971). It seems more likely now that the "host vacuole" of cyanelles is in fact a lipopolysaccharide layer (does this also explain the phylogeny of the outer envelope membrane in red algal and green algal chloroplasts ?). The genome of the cyanelle of *Cyanophora paradoxa* has been further investigated (BOHNERT et al. 1982, BOHNERT and LÖFFELHARDT 1982, BOHNERT et al. 1983). The size of cyanelle DNA (cyDNA) has now been determined to be 126.5 Kbp for the "Pringsheim-strain" and 138 Kbp for the "Kies-strain" (BOHNERT et al. 1983). The cyDNA is circular and contains two inverted repeat regions each of ca. 10 Kpb (similar to chloroplast DNA of *Marchantia polymorpha*; OHYAMA et al. 1983). A partial physical map of the cyDNA is presented by BOHNERT et al. (1983) and shown to be very similar to chloroplast DNA. Discussing these results BOHNERT et al. (1983) conclude that "the cyanelles are a, probably ancient form of plastid". Of great interest is the observation by HEINHORST and SHIVELY (1983) that the genes for both subunits of ribulose-bisphosphate carboxylase (RubPCase) are located on the cyanelle genome as shown by heterologous hybridization. These data correspond well with recent results on the location of RubPCase genes in *Cyanidium* and *Porphyridium* (see Sect. 3).

FLOENER et al. (1982) and FLOENER and BOTHE (1982) studied metabolic activities in *Cyanophora paradoxa* with emphasis on the relationships between the cyanelle and its "host". Assimilatory nitrate reduction proceeds as in green plants and is different from the pathway in cyanobacteria. In addition KCN-sensitive respiration could not be detected in cyanelles (respiratory cytochrome oxidase is lacking.

More information on "cytoplasmic compartments" of Glaucophyceae should be very useful in relating the glaucophycean "host" cell with other algal classes, notably the Rhodophyceae (see Sect. 3) and the green algae. From a comparison between the flagellar apparatus of *Cyanophora paradoxa* and that of some scaly green flagellates it is concluded that the Glaucophyceae and the green algae presumably had a common ancestor (MELKONIAN 1983a).

15. Green Algae

a) Books, Reviews

A number of books and reviews have appeared on green algae: For the new edition of the *Süßwasserflora von Mitteleuropa* ETTL (1983) has compiled information on the "Phytomonadina". In the second part of Vol. 1 of *Die Desmidiaceen Mitteleuropas* RUZICKA (1981) covers the genera *Actinotaenium*, *Tetmemorus*, *Euastrum*, and *Micrasterias*. Sections 3 and 4 of Part IV of the *Synopsis of North American Desmids* (PRESCOTT et al. 1981, PRESCOTT et al. 1983) deal with the genera *Cosmarium* (Sect. 3), *Arthrodesmus*, *Spinocosmarium*, *Xanthidium*, and *Staurastrum* (Sect. 4). VAN DEN HOEK (1982) provides a taxonomic revision of the American species of *Cladophora*. KOMAREK and FOTT (1983) treat the "Chlorococcales" for *Das Phytoplankton des Süßwassers*.

Shorter reviews on the systematics of the green algae include SILVA (1982) and VAN DEN HOEK (1981). Introduction and bibliographies to several "classes" of green algae have been written (Prasinophyceae: NORRIS 1982, Chlorophyceae: MELKONIAN and ICHIMURA 1982, Charophyceae: GRANT and SAWA 1982). Reviews on special topics include KESSLER (1982a) on the chemotaxonomy of the "Chlorococcales", COLEMAN (1983) on resting spores and akinetes in green algae, MOESTRUP (1982) on green algal flagella, KOCHERT (1982) on sexual processes in the Volvocales, and MELKONIAN (1982) on evolutionary aspects of the flagellar apparatus.

b) Ultrastructure and Phylogeny

Ultrastructural aspects of green algae (especially the motile, flagellated cell, mitosis/cytokinesis and cell surface features) continue to provide important information on (1) classification of green algae at the class and order level, (2) evolutionary relationships between the various green algal groups. These aspects have been partially summarized in recent reviews (MOESTRUP 1982, MELKONIAN 1982) and will be further discussed in two forthcoming books to appear in 1984 (cf. Progr. Bot. 48). From original publications that appeared during 1982/1983 the following conclusions may be drawn: (a) The flagellar apparatus of motile cells of the Ulvophyceae sensu Stewart and Mattox (including siphonalean green algae: ROBERTS et al. 1982) is characterized by a peculiar displacement of its basal bodies into the 11/5 o'clock direction (ROBERTS et al. 1982, O'KELLY and FLOYD 1983). (b) Basal bodies of the flagellar apparatus in the Chlorophyceae sensu Stewart and Mattox, if displaced, are always displaced into the opposite direction (1/7 o'clock; MELKONIAN and BERNS 1983). These differences present another major distinction between these two "classes" of green algae. (c) A group of green algae including the genera *Trebouxia*, *Pseudotrebouxia*, *Friedmannia*, *Pleurastrum*, and *Microthamnion* has been shown to form a natural assemblage on the basis of their motile cell ultrastructure (MELKONIAN and BERNS 1983). In this group, as in the Ulvophyceae sensu Stewart and Mattox, basal bodies are displaced into the 11/5 o'clock direction. The final taxonomic status of these genera is unclear at present, but they cannot be retained within the traditional orders of the Chlorophyceae sensu Stewart and Mattox. (d) The flagellar apparatus of the zoospore of *Coleochaete pulvinata* has been investigated by SLUIMAN (1983) who shows that in addition to the previously known unilateral MLS-containing flagellar root there exists a second smaller root composed of three microtubules, both roots associated with the same basal body. SLUIMAN (1983) provides convincing evidence that the zoospore of *Coleochaete* exhibits a great similarity to the bryophyte spermatozoid thus providing further evidence for charophycean ancestry of the archegoniate land plants. (e) The flagellar apparatus of the prasinophyte *Mesostigma viride* has been reinvestigated and shown to exhibit a cruciate flagellar root system (MELKONIAN 1983b, cf. Progr. Bot. 44, 336) with two opposite flagellar roots containing a typical multilayered structure (MLS). (f) Gametes of the trentepohliacean *Cephaleuros virescens* have finally been shown to exhibit a cruciate microtubular flagellar system with MLS-remnants and an 11/5 o'clock displacement of their basal bodies (CHAPMAN and HENK 1983). This study further documents the unique position of the Trentepohliales. (g) In studies on the flagellar apparatus of colonial Volvocales it has been shown that the flagellar apparatus, which is basically *Chlamydomonas*-like, undergoes dramatic rearrangements during ontogenesis, probably as an adaptation to the colonial habit and special type of flagellar motion (HOOPS and FLOYD 1982a, 1982b, HOOPS and FLOYD 1983). (h) The zoospore of *Urospora penicilliformis* shows several unique features but is related in ultrastructure to other Ulvophyceae sensu Stewart and Mattox (SLUIMAN et al. 1982). Its mitosis and cytokinesis (LOKHORST and STAR 1983) is also unusual, involving, in the interpretation of the authors, a modified phycoplast.

c) Systematics of Lower Categories

Numerous taxonomic studies have appeared over the last 2 years at the level of the lower categories. Due to limitation of space only few of them are included here.

In scaly green flagellates ("Prasinophyceae") taxonomy is mostly based on the structure of organic scales (e.g., MELKONIAN and ROBENEK 1981, McFADDEN et al. 1982, PENNICK 1983), although other ultrastructural features (e.g., pyrenoids: HORI et al. 1982) are increasingly used for taxonomic purposes. JOHNSON and SIEBURTH (1982) have found a very small nonflagellate prasinophyte (related to *Mantoniella?)* in oceanic picoplankton. Fossil phycomata of prasinophytes from silurian formations have been described by COLBATH (1983).

"*Chlamydomonas*-type organisms": Interbreeding between two species of *Chlamydomonas (C. reinhardii* and *C. smithsii)* leads to fertile offspring and indicates that morphologically defined species in *Chlamydomonas* may not correspond to genetically defined species (BELL and CHAIN 1983). A comparative analysis of matrix polysaccharides in colonial Volvocales as compared with *Chlamydomonas* suggests that there is an increase in the amount of sulfate with evolutionary advancement (CRAYTON 1982).

In an important study HOFFMAN (1983) describes a new genus within the Sphaeropleales, *Atractomorpha*. The organism differs from the other genus of the order *(Sphaeroplea)* by (a) vegetative growth (single, coenocytic cells), (b) formation of biflagellate zoospores, (c) anisogamous sexual reproduction. The study indicates that the Sphaeropleales have probably evolved from coenocytic, unicellular ancestors of chlorococcalean affinities.

Chlorococcales: Review: ETTL and KOMAREK (1982). *Nanochlorum eucaryotum* is one of the smallest eukaryotic cells to be found in seawater picoplankton (WILHELM et al. 1982, JOHNSON and SIEBURTH 1982). Studies on the taxonomy of 28 *Scenedesmus* strains from culture collections have shown that they belong to six species, an additional species "*Chlorella fusca*" (Utex 343) is transferred to *Scenedesmus* as *S. fuscus* (HEGEWALD 1982). Clear distinguishing characteristics have been found between the genera *Ankistrodesmus* and *Selenastrum* (KOMAREK and GONZALEZ 1982). In addition to traditional morphological characters also other criteria have been used in the taxonomy of certain chlorococcalean genera (occurrence of polyamines: HEGEWALD and KNEIFEL 1982; cell-wall structure and composition: LOOS and MEINDL 1982, DEASON 1983; utilization of organic carbon and nitrogen compounds: KESSLER 1982b). In addition numerical taxonomy and cluster analysis are gaining more popularity (MARVAN 1982, KOMAROMY 1982). Differences which arise in distinguishing members of the "Chlorococcales" from members of the "Chlorosarcinales" have been discussed by TSCHERMAK-WOESS (1982).

"Ulvophyceae": Here as in other marine macroalgae new genera and species continue to be described from natural samples or cultures (e.g., SEARS and BRAWLEY 1982, EISEMAN and EARLE 1983). Two taxonomically very difficult genera of the Ulvales, *Ulva* and *Enteromorpha*, have recently received considerable attention (KOEMAN and VAN DEN HOEK 1982a, 1982b, HOEKSEMA and VAN DEN HOEK 1983, PHILLIPS and CLAYTON 1983). In addition, a monograph of the genus *Halimeda* has been prepared (HILLIS-COLINVAUX 1980). KORNMANN and SAHLING (1983) studied the life history of *Chlorocystis cohnii* and showed that zygotes developed into *Codiolum*-like cells. *Chlorocystis* is therefore transferred from the "Chlorococcales" to a new order Chlorocystidales (probably in the Ulvophycean line of evolution).

Charophyceae sensu Stewart and Mattox: Taxonomy of Zygnematales follows traditional lines and HOSHAW (1983) shows the usefulness of glycerine-preserved specimens for present-day research since the specimens can be subjected to some modern methods of analysis (scanning electron microscopy, cytofluorometry). Investigations into the chemical composition of cell walls in zygnematalean algae yield interesting results of taxonomic significance (DE FRIES et al. 1983, MIX et al. 1982). Putative placental transfer cells surrounding zygotes of the green alga *Coleochaete orbicularis* have been identified on the basis of localized wall ingrowths, location, and time of development (GRAHAM and WILCOX 1983). These cells resemble placental transfer cells of embryophytes and therefore provide further evidence for the close phylogenetic relationship between *Coleochaete* and simple archegoniate land plants.

References

ADAM, D.P.: Geol. Surv. Profess. Paper (Wash. DC) 1175, 242 pp (1980). - AITKEN, A., STANIER, R.Y.: J. Gen. Microbiol. 112, 219-233 (1979). - ALDRICH, J., GELVIN, S., CATTOLICO, R.A.: Plant Physiol. 69, 1189-1195 (1982). - AMSPOKER, M.C., CZARNECKI, D.B.: Bacillariophyceae: introduction and bibliography, pp. 712-718. In: see ref. ROSOWKSKI and PARKER (eds.) 1982. - ANDERSEN, R.A., MULKEY, T.J.: J. Phycol. 19, 289-294 (1983). - ANTIA, N.J., CHENG, J.Y.: Br. Phycol. J. 17, 39-50 (1982). - ASMUND, B., CRONBERG, G., DÜRRSCHMIDT, M.: Nord. J. Bot. 2, 383-395 (1982).

BADEN, D.G.: Int. Rev. Cytol. 82, 99-143 (1983). - BARNES, K.L., CRAIGIE, R.A., CATTINI, P.A., CAVALIER-SMITH, T.: J. Cell Sci. 57, 151-160 (1982). - BEAM, C.A., HIMES, M.: J. Protozool. 29, 8-15 (1982). - BEAM, C.A., HIMES, M., DAGGETT, P., NERAD, T.A.: J. Protozool. 29, 494 (1982). - BELL, R.A., CAIN, J.R.: Phycologia 22, 243-247 (1983). - BERTHOLD, R.J., BOROWITZKA, M.A., MACKAY, M.A.: Phycologia 21, 327-335 (1982). - BILLARD, C.: Phycologia 22, 141-151 (1983). - BISHOP, C.A., KENRICK, J.R.: Lipids 15, 799-804 (1980). - BJØRNLAND, T.: Phytochemistry 21, 1715-1719 (1982). - BLUNDEN, G., GORDON, S.M., McLEAN, W.F.H., GUIRY, M.D.: Bot. Marina 25, 563-567 (1982). - BOHNERT, H.J., LÖFFELHARDT, W.: FEBS Lett. 150, 403-406 (1982). - BOHNERT, H.J., CROUSE, E.J., POUYET, J., MUCKE, H., LÖFFELHARDT, W.: Eur. J. Biochem. 126, 381-388 (1982). - BOHNERT, H.J., MICHALOWSKI, C., KOLLER, B., DELIUS, H., MUCKE, H., LÖFFELHARDT, W.: The cyanelle genome from Cyanophora paradoxa, pp. 433-448. In: see ref. SCHENK and SCHWEMMLER (eds.) 1983. - BOLTON, J.J., GERMANN, I., LÜNING, K.: Phycologia 22, 133-140 (1983). - BONEY, A.D., GREEN, J.C.: Prymnesiophyceae (Haptophyceae): introduction and bibliography, pp. 705-711. In: see ref. ROSOWKSI and PARKER (eds.) 1982. - BORMAN, A.H., DE JONG, E.W., HUIZINGA, M., KOK, D.J., WESTBROEK, P., BOSCH, L.: Eur. J. Biochem. 129, 179-183 (1982). - BOURRELLY, P.: Les Algues d'Eau Douce: Initiation a la Systematique. II Les Algues jaunes et brunes, Chrysophycees, Pheophycees, Xanthophycees et Diatomees, 517 pp. Paris: Boubee 1981. - BROCK, T.D.: Thermophilic Microorganisms and Life at High Temperatures, 465 pp. New York, Heidelberg, Berlin: Springer 1978. - BROWN, L.M., McLACHLAN, J.: Phycologia 21, 9-16 (1982). - BUETOW, D.E. (ed.): The Biology of Euglena, Vols. III, IV, 363 pp. New York: Academic Press 1982a, 1982b. - BUJAK, J.P., WILLIAMS, G.L.: Can. J. Bot. 59, 2077-2087 (1981).

CANAANI, O., GANTT, E.: Proc. Natl. Acad. Sci. USA 79, 5277-5281 (1982). - CARNEY, H.J., SANDGREN, C.D.: Hydrobiologia 55, 1-8 (1982). - CARR, N.G., WHITTON, B.A. (eds.): The Biology of Cyanobacteria, 700 pp. Oxford: Blackwell 1982. - CAVALIER-SMITH, T.: Biol. J. Linn. Soc. 17, 289-306 (1982a); - The evolutionary origin and phylogeny of eukaryotic flagella, pp. 465-493. In: Prokaryotic and Eukaryotic Flagella, W.B. AMOS, J.G. DUCKETT (eds.). Cambridge: University Press 1982b. - CHADEFAUD, M.: Cryptogam. Algol. 3, 147-164 (1982a); - ibid. 3, 241-256 (1982b). - CHANG, S.S., TRENCH, R.K.: Proc. R. Soc. Lond. B 215, 191-210 (1982). - CHAPMAN, D.J., TRENCH, R.L.: Prochlorophyceae: introduction and bibliography, pp. 656-658. In: see ref. ROSOWKSI and PARKER (eds.) 1982. - CHAPMAN, R.L., HENK, M.C.: Am. J. Bot. 70, 1340-1355 (1983). - CLAYTON, M.N.: Phaeophyta, pp. 104-137. In: Marine Botany: An Australasian Perspective. Melbourne: Longman 1981a; - Phycologia 20, 358-364 (1981b); - Bot. Marina 25, 111-116 (1982). - COLBATH, G.K.: Phycologia 22, 249-265 (1983). - COLE, K.M., HYMES, B.J., SHEATH, R.G.: J. Phycol. 19, 136-145 (1983). - COLEMAN, A.W.: The roles of resting spores and akinetes in chlorphyte survival, pp. 1-21. In: Survival Strategies of the Algae, G.A. FRYXELL (ed.). Cambridge: University Press 1983. - COLEMAN, A.W., LEWIN, R.A.: Phycologia 22, 209-212 (1983). - COSSON, J., OLIVARI, R.: C.R. Acad. Sci. Paris 295, 381-384 (1982). - COX, E.J.: Br. Phycol. J. 17, 147-168 (1982). - CRAYTON, M.A.: J. Phycol. 18, 336-344 (1982). - CRAWFORD, R.M.: The siliceous components of the diatom cell wall and their morphological variation, pp. 129-156. In: Silicon and Siliceous Structures in Biological Systems, T.L. SIMPSON, B.E. VOLCANI (eds.). Berlin, Heidelberg, New York: Springer: 1981. - CROOME, R.L., TYLER, P.A.: Br. Phycol. J. 18, 357-365 (1983).

DALE, B.: Dinoflagellate resting cysts, pp. 69-136. In: Survival Strategies of the Algae, G.A. FRYXELL (ed.). Cambridge: University Press 1983. - DARLEY, W.M.: Algal Biology: A Physiological Approach, 168 pp. Oxford: Blackwell 1982. - DAWES, C.J.: Marine Botany, 628 pp. New York: Wiley 1981. - DEASON, T.R.: J. Phycol. 19, 248-251 (1983). - DE CEW, T.C., WEST, J.A.: Phycologia 20, 342-351 (1981); - ibid. 21, 67-74 (1982). - DE CEW, T.C., WEST, J.A., GANESAN, E.K.: Phycologia 20, 415-423 (1981). - DE FRIES, P.J.R., SIMONS, J., VAN BEEM, A.P.: Acta Bot. Neerl. 32, 25-28 (1983). - DE JONG, E.W., VAN DER WAL, P., BORMAN, A.H., DE VRIND, J.P.M., VAN EMBURG, P., WESTBROEK, P., BOSCH, L.: Calcification in Coccolithophorids, pp. 291-301. In: Biomineralization and Biological Metal Accumulation, P. WESTBROEK, E.W. DE JONG (eds.). Dordrecht: Reidel 1983. - DELIHAS, N., ANDRESINI, W., ANDERSEN, J., BERNS, D.: J. Mol. Biol. 162, 721-727 (1982). - DE LUCA, P., MORETTI, A.: J. Phycol. 19, 368-369 (1983). - DEYSHER, L., NORTON, T.A.: J. Exp. Mar. Biol. Ecol. 56, 179-195 (1982). - DIXON, P.S.: Rhodophycota, pp. 61-79. In: Synopsis and Classification of Living Organisms, S.P. PARKER (ed.). New York: McGraw Hill 1982a; - Bot. Marina 25, 611-621 (1982b). - DODGE, J.D.: Marine Dinoflagellates of the British Isles, 310 pp. London: H.M.S.O. 1982; - Br. Phycol. J. 18, 335-356 (1983). - DODGE, J.D., HERMES, H.B.: Phycologia 20, 424-430 (1981). - DODGE, J.D., STEIDINGER, K.A.: Dinophyceae: introduction and bibliography, pp. 691-697. In: see ref. ROSOWSKI and PARKER (eds.) 1982. - DREBES, G., SCHNEPF, E.: Helgol. Meeresunters. 35, 501-515 (1982). - DRING, M.A.: The Biology of Marine Plants, 199 pp. London: Arnold 1982. - DRUEHL, L.D., KEMP, L.: Can. J. Bot. 60, 1409-1413 (1982). - DÜRRSCHMIDT, M.: Plant Syst. Evol. 143, 175-196 (1983). - DWARTE, D., VESK, M.: Micron. 13, 325-326 (1982). - DWARTE, D., VESK, M.: Protoplasma 117, 130-141 (1983).

EHRLICH, A., CRAWFORD, R.M., ROUND, F.E.: Br. Phycol. J. 17, 195-203 (1982a); - ibid. 17, 205-214 (1982b). - EISEMAN, N.J., EARLE, S.A.: Phycologia 22, 1-6 (1983). - ELLIS-ADAM, A.C.: Acta Bot. Neerl. 32, 1-23 (1983). - ETTL, H.: Chlorophyta I, Phytomonadina, pp. 1-807. In: Süßwasserflora von Mitteleuropa, H. ETTL, J. GERLOFF, H. HEYNIG, D. MOLLENHAUER (eds.). Stuttgart: Fischer 1983. - ETTL, H., KOMAREK, J.: Arch. Hydrobiol. Suppl. 60, 345-374 (1982). - EVANS, L.V., CALLOW, J.A., CALLOW, M.E.: Progr. Phycol. Res. 1, 67-110 (1982).

FITT, W.K., CHANG, S.S., TRENCH, R.K.: Bull. Mar. Sci. 31, 436-443 (1981). - FLETCHER, R.L.: Bot. Marina 23, 425-432 (1980). - FLOENER, L., BOTHE, H.: Planta 156, 78-83 (1982). - FLOENER, L., DANNEBERG, G., BOTHE, H.: Planta 156, 70-77 (1982). - FOGED, N.: Diatoms in Alaska. Bibliotheca Phycologica, Vol. 55, 317 pp. Vaduz: Cramer 1981. - FOTT, B.: Algenkunde, 581 pp. Jena: Fischer 1971. - FRIEDMANN, E.I., BOROWITZKA, L.J.: Taxon 31, 673-683 (1982). - FRIES, L.: Planta 154, 393-396 (1982). - FRITZ, L., TRIEMER, R.E.: J. Protozool. 30, 437-444 (1983). - FRYXELL, G.A., HASLE, G.R.: Polar Biol. 2, 53-62 (1983). - FRYXELL, G.A., MEDLIN, L.K.: Cryptogam. Algol. 2, 3-29 (1981).

GABRIELSON, P.W., HOMMERSAND, M.H.: J. Phycol. 18, 31-45 (1982a); - ibid. 18, 45-58 (1982b). - GAYRAL, P., FRESNEL, J.: Protistologica 1983, 245-261 (1983a); - Phycologia 22, 29-45 (1983b). - GEIßLER, U.: Beih. Nova Hedwigia 73, 211-247 (1982). - GIBBS, S.P., CHU, L.L., MAGNUSSEN, C.: Phycologia 19, 173-177 (1980). - GIDDINGS, T.H.Jr., WASMANN, C., STAEHELIN, L.A.: Plant Physiol. 71, 409-419 (1983). - GILLOTT, M.A., GIBBS, S.P.: Can. J. Bot. 61, 1964-1978 (1983). - GLAZER, A.N., WEST, J.A., CHAN, C.: Biochem. Syst. Ecol. 10, 203-215 (1982). - GOFF, L.J.: Phycologia 20, 397-406 (1981); - Progr. Phycol. Res. 1, 289-370 (1982). - GOFF, L.J., HOMMERSAND, M.H.: J. Phycol. 18, 295-304 (1982). - GOLUBIC, S.: Stromatolites, fossil and recent: a case history, pp. 1-14. In: Biomineralization and Biological Metal Accumlation, P. WESTBROEK, E.W. WONG (eds.). Dordrecht: Reidel 1983; - GONZALEZ, F.T.: Nova Hedwigia 35, 167-191 (1981). - GRAHAM, L., WILCOX, L.W.: Am. J. Bot. 70, 113-120 (1983). - GRAIN, J., PUYTORAC, P. DE, GROLIERE, C.-A. LA: Protistologica 18, 7-21 (1982). - GRANT, M.C., SAWA, T.: Charophyceae: introduction and bibliography, pp. 754-759. In: see ref. ROSOWSKI and PARKER (eds.) 1982. - GRAY, M.W., DOOLITTLE, W.F.: Microbiol. Rev. 46, 1-42 (1982). - GREEN, J.C.: Br. Phycol. J. 15, 151-191 (1980). - GREEN, J.C., HIBBERD, D.J., PIENAAR, R.N.: Br. Phycol. J. 17, 363-382 (1982). - GRETZ, M.R., SOMMERFELD, M.R., ARONSON, J.M.: Bot. Marina 25, 529-539 (1982). - GUIRY, M.D., MAGGS, C.A.: Br. Phycol. J. 17, 215-228 (1982).

HAKANSSON, H., GERLOFF, J. (eds.): Diatomaceae III. Festschrift Nils Foged. Beih. Nova Hedwigia 73. Vaduz: Cramer 1982; - HAPPACH-KASAN, C.: Arch. Protistenk. 125, 181-207 (1982). - HARA, Y.: XIII Int. Bot. Congr. Sydney, 183 (Abstr.) Sydney: Australian Academy of Sciences 1981. - HARGRAVES, P.E., FRENCH, F.W.: Diatom resting spores: significance and strategies, pp. 49-68. In: Survival Strategies of the Algae, G.A. FRYXELL (ed.). Cambridge: University Press 1983. - HARRINGTON, A., THORNLEY, A.L.: J. Mol. Evol. 18, 287-292 (1982). - HASLE, G.R.: J. Phycol. 19, 220-229 (1983). - HASLE, G.R., STOSCH, H.A. von, SYVERTSEN, E.E.: Bacillaria 6, 9-156 (1983). - HAYHOME, B.A., PFIESTER, L.A.: Am. J. Bot. 70, 1165-1172 (1983). - HEGEWALD, E.: Arch. Hydrobiol. Suppl. 60, 375-406 (1982). - HEGEWALD, E., KNEIFEL, H.: Arch. Hydrobiol. Suppl. 60, 489-499 (1982). - HEINHORST, S., SHIVELY, J.M.: Nature 304, 373-374 (1983). - HENDEY, N.I.: Beih. Nova Hedwigia 73, 275-280 (1982). - HENRY, E.C., COLE, K.M.: J. Phycol. 18, 550-569 (1982a); - ibid. 18, 570-579 (1982b). - HERTH, W., SCHNEPF, E.: Chitin-fibril formation in algae, pp. 185-206. In: Cellulose and other Natural Polymer Systems, R. MALCOLM BROWN, Jr. (ed.). New York: Plenum 1982. - HEYWOOD, P.: Raphidophyceae (Chloromonadophyceae): introduction and bibliography, pp. 719-722. In: see ref. ROSOWSKI and PARKER (eds.) 1982. - HIBBERD, D.J.: Eustigmatophyceae: introduction and bibliography, pp. 728-730. In: see ref. ROSOWSKI and PARKER (eds.) 1982. - HILLIS-COLINVAUX, L.: Adv. Mar. Biol. 17, 1-327 (1980). - HOEK, C.: Chlorophyta: morphology and classification, pp. 86-132. In: The Biology of Seaweeds, C. LOBBAN, M.J. WYNNE (eds.). London: Blackwell 1981; - A taxonomic revision of the American species of Cladophora (Chlorophyceae) in the North Atlantic Ocean and their geographic distribution, 236 pp. Amsterdam: North Holland 1982. - HOEKSEMA, B.W., HOEK, C.: Bot. Marina 26, 65-86 (1983). - HOFFMAN, L.R.: J. Phycol. 19, 76-86 (1983). - HOLLANDE, A., CORBEL, J.-C.: Protistologica 28, 123-133 (1982). - HOLLIGAN, P.M., VIOLLIER, M., HARBOUR, D.S., CAMUS, P., CHAMPAGNE-PHILIPPE, M.: Nature 304, 339-343 (1983). - HOLMES, R.W., CRAWFORD, R.M., ROUND, F.E.: Phycologia 21, 370-381 (1982). - HOOPS, H.J., FLOYD, G.L.: J. Phycol. 18, 455-462 (1982a); - ibid. 18, 462-466 (1982b); - J. Cell Sci. 63, 21-41 (1983). - HORI, T., NORRIS, R.E., CHIHARA, M.: Bot. Mag. Tokyo 95, 49-61 (1982). - HOSHAW, R.W.: Am. J. Bot. 70, 145-153 (1983). - HUISMAN, J.M., KRAFT, G.T.: J. Phycol. 18, 177-192 (1982); - Phycologia 22, 285-294 (1983). -

HUMM, H.J., WICKS, S.R.: Introduction and Guide to the Marine Blue-Green Algae, 194 pp. New York: Wiley 1980.

JOHANSEN, H.W.: Coralline Algae, a First Synthesis, 239 pp. Boca Raton: CRC Press 1981. - JOHNSON, P.W., SIEBURTH, J. McN.: J. Phycol. 18, 318-327 (1982).

KESSLER, E.: Progr. Phycol. Res. 1, 112-135 (1982a); - Arch. Microbiol. 132, 103-106 (1982b). - KLAVENESS, D.: Nord. J. Bot. 2, 183-188 (1982). - KLAVENESS, D., SKULBERG, O.M.: Nord. J. Bot. 2, 91-96 (1982). - KOCHERT, G.: Progr. Phycol. Res. 1, 235-256 (1982). - KOEMAN, R.P.T., VAN DEN HOEK, C.: Arch. Hydrobiol. Suppl. 63, 279-330 (1982a); - Cryptogam. Algol. 3, 37-70 (1982b). - KOKKE, W.C.M.C., FENICAL, W., BOHLIN, L., DJERASSI, C.: Comp. Biochem. Physiol. B 68, 281-287 (1981). - KOMAROMY, ZS.P.: Arch. Hydrobiol. Suppl. 60, 432-438 (1982). - KOMAREK, J., FOTT, B.: Chlorophyceae (Grünalgen), Ordnung: Chlorococcales. In: Das Phytoplankton des Süsswassers, Vol. 7, C. HUBER-PESTALOZZI (ed.). Stuttgart: Schweizerbart 1983. - KOMAREK, J., GONZALEZ, A.C.: Arch. Hydrobiol. Suppl. 63, 259-277 (1982). - KORNMANN, P., SAHLING, P.-H.: Helgol. Meeresunters. 36, 1-65 (1983). - KRAFT, G.T.: Rhodophyta: morphology and classification, pp. 6-51. In: The Biology of Seaweeds, C.S. LOBBAN, M.J. WYNNE (eds.). Oxford: Blackwell 1981; - KRAFT, G.T., MIN-THEIN, U.: Phycologia 22, 171-183 (1983). - KRAFT, G.T., WOELKERLING, WM.J.: Rhodophyta: systematics and biology, pp. 61-103. In: Marine Botany, an Australasian Perspective, M. CLAYTON, R. KING (eds.). Melbourne: Longman 1981. - KREMER, B.P.: Naturwissenschaften 69, 428-435 (1982a); - Br. Phycol. J. 17, 51-61 (1982); - Cyanidium caldarium: Rhodophyte, Cyanome or transitional species, pp. 963-970. In: see ref. SCHENK and SCHWEMMLER (eds.) 1983. - KRISTIANSEN, J.: Chromophycota, pp. 81-84. In: Synopsis and Classification of Living Organisms, S.P. PARKER (ed.). New York: McGraw Hill 1982. - KRISTIANSEN, J., TAKAHASHI, E.: Chrysophyceae: introduction and bibliography, pp. 698-704. In: see ref. ROSOWSKI and PARKER (eds.) 1982.

LANG, N.J., WALSBY, A.E.: Cyanophyceae: introduction and bibliography, pp. 647-655. In: see ref. ROSOWSKI and PARKER (eds.) 1982. - LAWRY, N.H., SIMON, R.D.: J. Phycol. 18, 391-399 (1982). - LEEDALE, G.F., SCHIFF, J.A., BUETOW, D.E.: Euglenophyceae: introduction and bibliography, pp. 687-690. In: see ref. ROSOWSKI and PARKER (eds.) 1982. - LEWIN, R.A.: The Prochlorophytes, pp. 257-266. In: The Procaryotes. M.P. STARR, H. STOLP, H.G. TRÜPER, A. BARLOWS, H.G. SCHLEGEL (eds.). Berlin, Heidelberg, New York: Springer 1981. - LEWIN, R.A., GIBBS, S.P.: Algae of uncertain taxonomic position: introduction and bibliography, pp. 659-662. In: see ref. ROSOWSKI and PARKER (eds.) 1982. - LI, W.K.W., SUBBA RAO, D.V., HARRISON, W.G., SMITH, J.C., CULLEN, J.J., IRWIN, B., PLATT, T.: Science 219, 292-295 (1983). - LOBBAN, C.S., WYNNE, M.J. (eds.): The Biology of Seaweeds, 800 pp. Oxford: Blackwell 1981. - LOCKHART, J.C.: Phycologia 21, 264-272 (1982). - LOEBLICH, A.R., FINE, K.E.: Proc. Biol. Soc. Wash. 90, 388 (1977). - LOKHORST, G.M., STAR, W.: Protoplasma 117, 142-153 (1983). - LOOS, E., MEINDL, D.: Planta 156, 270-273 (1982). - LUKAS, K.J., GOLUBIC, S.: J. Phycol. 19, 129-136 (1983).

MacKAY, R.M., SALGADO, D., BONEN, L., STACKEBRANDT, E., DOOLITTLE, W.F.: Nucl. Acids Res. 9, 2963-2970 (1982). - MAGGS, C.A., GUIRY, M.D.: Bot. Marina 25, 589-599 (1982). - MAGNE, F.: Cryptogam. Algol. 3, 265-271 (1982). - MANN, D.G.: J. Phycol. 18, 162-176 (1982a); - ibid. 18, 264-274 (1982b); - Pl. Syst. Evol. 141, 143-152 (1982c); - Ann. Bot. 52, 573-581 (1983). - MANTON, I.: Bot. Marina 25, 163-169 (1982); - ibid. 26, 15-22 (1983). - MANTON, I., OATES, K.: Bot. Marina 26, 517-525 (1983). - MARSDEN, W.J.N., EVANS, L.V., CALLOW, J.A.: Bot. Marina 26,

383-392 (1983a); - ibid. 26, 527-531 (1983b). - MARVAN, P.: Arch. Hydrobiol. Suppl. 60, 420-431 (1982). - McCANDLESS, E.L., WEST, J.A., GUIRY, M.D.: Biochem. Syst. Ecol. 10, 275-284 (1982); - ibid. 11, 175-182 (1983). - McFADDEN, G.I., MOESTRUP, Ø., WETHERBEE, R.: Phycologia 21, 103-111 (1982). - McKERRACHER, L., GIBBS, S.P.: Can. J. Bot. 60, 2440-2452 (1982). - MELKONIAN, M.: Taxon 31, 255-265 (1982); - Evolution of green algae in relation to endosymbiosis, pp. 1003-1008. In: see ref. SCHENK and SCHWEMMLER (eds.) 1983a; - Br. Phycol. J. 18, 206 (1983b). - MELKONIAN, M., BERNS, B.: Protoplasma 114, 67-84 (1983). - MELKONIAN, M., ICHIMURA, T.: Chlorophyceae: introduction and bibliography, pp. 747-753. In: see ref. ROSOWSKI and PARKER (eds.) 1982. - MELKONIAN, M., ROBENEK, H.: Phycologia 20, 365-376 (1981). - MEROLA, A., CASTALDO, R., DE LUCA, P., GAMBARDELLA, R.: G. Bot. Jt. 116, 189-195 (1982). - MIX, M., LORCH, D., MANSHARD, E.: Mitt. Inst. Allg. Bot. Hamburg 18, 5-22 (1982). - MOE, R.L., HENRY, E.C.: Phycologia 21, 55-66 (1982). - MOE, R.L., SILVA, P.C.: Br. Phycol. 18, 275-298 (1983). - MOESTRUP, Ø.: Phycologia 21, 427-528 (1982). - MORRALL, S., GREENWOOD, A.D.: J. Cell Sci. 54, 311-328 (1982). - MORRILL, L.A., LOEBLICH, A.R. III.: Int. Rev. Cytol. 82, 151-180 (1983). - MÜLLER, D.G.: The role of olefinic hydrocarbons in sexual reproduction of marine brown algae, pp. 57-70. In: Proc. 10th Intern. Seaweed Sympos., T. LEVRING (ed.). Berlin: de Gruyter 1981; - MÜLLER, D.G., MEEL, H.: Br. Phycol. J. 17, 419-425 (1982). - MÜLLER, D.G., GASSMANN, G., MARNER, F.-J., BOLAND, W., JAENICKE, L.: Science 218, 1119-1120 (1982a). - MÜLLER, D.G., PETERS, A., GASSMANN, G., BOLAND, W., MARNER, F.-J., JAENICKE, L.: Naturwissenschaften 69, 290-292 (1982b). - MURATA, N., SATO, N.: Plant cell Physiol. 24, 133-138 (1983).

NAGASHIMA, H., FUKUDA, I.: Phytochemistry 20, 439-442 (1981). - NELSON, W.A., COLE, K.: Phycologia 20, 435-437 (1981); - Bot. Marina 25, 357-369 (1982a); - Br. Phycol. J. 17, 311-320 (1982b). - NORRIS, J.N., KUGRENS, P.: Marine Rhodophyceae: introduction and bibliography, pp. 663-670. In: see ref. ROSOWSKI and PARKER (eds.) 1982. - NORRIS, R.E.: Prasinophyceae: introduction and bibliography, pp. 740-746. In: see ref. ROSOWSKI and PARKER (eds.) 1982; - Phycologia 22, 161-169 (1983). - NORTON, T.A.: J. Mar. Biol. Ass. U.K. 63, 181-193 (1983). - NORTON, T.A., FETTER, R.: J. Mar. Biol. Ass. U.K. 61, 929-940 (1981).

OAKLEY, B.R., SANTORE, U.J.: Cryptophyceae: introduction and bibliography, pp. 682-686. In: see ref. ROSOWKSI and PARKER (eds.) 1982. - OHYAMA, K., YAMANO, Y., FUKIZAWA, H., KOMANO, T., YAMAGISHI, H., FUJIMOTO, S., SUGIURA, M.: Mol. Gen. Genet. 189, 1-9 (1983). - O'KELLY, C.J., FLOYD, G.L.: J. Phycol. 19, 153-164 (1983). - OTT, D.W.: Tribophyceae (Xanthophyceae): introduction and bibliography, pp. 723-727. In: see ref. ROSOWKSI and PARKER (eds.) 1982. - OTT, F.D., SOMMERFELD, M.: Freshwater Rhodophyceae: introduction and bibliography, pp. 671-681. In: see ref. ROSOWSKI and PARKER (eds.) 1982.

PAGNI, P.G.S., WALNE, P.L., PAGNI, R.M.: Phycologia 20, 431-434 (1981). - PATTERSON, G.M., WITHERS, N.W.: Science 217, 1934-1935 (1982). - PATWARY, M.U., VAN DER MEER, J.P.: Can. J. Bot. 60, 2556-2564 (1982). - PEDERSEN, P.M.: Phaeophyta: Life histories, pp. 194-217. In: The Biology of Seaweeds, C.S. LOBBAN, M.J. WYNNE (eds.). Oxford: Blackwell 1982. - PENNICK, N.C.: Arch. Protistenk. 127, 1-7 (1983). - PHILLIPS, J.A., CLAYTON, M.N.: Aust. J. Bot. 31, 11-18 (1983). - PICKETT-HEAPS, J.: J. Phycol. 19, 269-281 (1983). - POPOVSKY, J., PFIESTER, L.A.: Arch. Protistenk. 125, 115-127 (1982). - PREISIG, H.R., HIBBERD, D.J.: Nord. J. Bot. 2, 397-420 (1982a); - ibid. 2, 601-638 (1982b); - ibid. 3, 695-723 (1983). - PRESCOTT, G.W., CROASDALE, H.T., VINYARD, W.C., BICUDO, C.E.M.: A Synopsis of North American Desmids Part IV. Desmidiaceae: Placodermae, Section 3, 720 pp. Lincoln: University Press

1981. - PRESCOTT, G.W., BICUDO, C.E.M., VINYARD, W.C.: A Synopsis of North American Desmids. Part II. Desmidiaceae: Placodermae, Section 4, 700 pp. Lincoln: University Press 1983. - PRUD'HOMME VAN REINE, W.F.: A taxonomic Revision of the European Sphacelariaceae (Sphacelariales, Phaeophyceae). 294 pp. Leiden: Brill 1982. - PUESCHEL, C.M., COLE, K.M.: Am. J. Bot. 69, 703-720 (1982). - PUESCHEL, C.M., STEIN, J.R.: J. Phycol. 19, 109-125 (1983).

QUADER, H., VAN KEMPEN, R., STELZER, R., ROBINSON, D.G.: Eur. J. Cell Biol. 30, 283-287 (1983).

REED, R.H.: Phycologia 22, 351-354 (1983). - RIPPKA, R., COHEN-BAZIRE, G.: Ann. Microbiol. B 134, 21-36 (1983). - RIZZO, P.J.: BioSystems 14, 433-443 (1981). - RIZZO, P.J., BURGHARDT, R.C.: J. Phycol. 19, 348-351 (1983). - RIZZO, P.J., JONES, M., RAY, S.M.: J. Protozool. 29, 217-222 (1982). - ROBERTS, K.R., STEWART, K.D., MATTOX, K.R.: J. Phycol. 18, 498-508 (1982). - ROSOWSKI, J.R., PARKER, B.C. (eds.): Selected papers in Phycology II, 866 pp. Lawrence: Phycol. Soc. America 1982. - ROSS, R. (ed.): Proc. 6th Symposium on Recent and Fossil Diatoms, 487 pp. Koenigstein: Koeltz 1981. - ROUND, F.E.: BioSystems 14, 483-486 (1981a); Morphology and phyletic relationships of the silicified algae and the archetypal diatom-monophyly or polyphyly, pp. 97-128. In: Silicon and Silicon Structures in Biological Systems, T.L. SIMPSON, B.E. VOLCANI (eds.). Berlin, Heidelberg, New York: Springer 1981b; - Arch. Protistenk. 125, 323-329 (1982). - ROUND, F.E., CHAPMAN, D.J. (eds.): Progress in Phycological Research, Vol. 1. Amsterdam: Elsevier 1982; - RUZICKA, J.: Die Desmidiaceen Mitteleuropas. Bd. 1, Lf. 2, pp. 293-736. Stuttgart: Schweizerbart 1981; - RUENESS, M., RUENESS, J.: Phycologia 21, 137-144 (1982).

SANDGREN, C.D.: Survival strategies of chrysophyte flagellates: reproduction and the formation of resistent resting spores, pp. 23-40. In: Survival Strategies in the Algae, G. FRYXELL (ed.). Cambridge: University Press 1983a; - J. Phycol. 19, 64-70 (1983b). - SANTORE, U.J.: Br. Phycol. J. 17, 81-99 (1982); - ibid. 18, 239-248 (1983). - SARJEANT, W.A.S.: Can. J. Bot. 60, 922-945 (1982). - SCHENK, H.E.A., KUHFITTIG, G.: Biochem. Syst. Ecol. 11, 163-174 (1983). - SCHENK, H.E.A., SCHWEMMLER, W. (eds.): Endocytobiology II, 1071 pp. Berlin: de Gruyter 1983; - SCHNEPF, E., BROWN, R.M.Jr.: On relationships between endosymbiosis and the origin of plastids and mitochondria, pp. 299-322. In: Origin and Continuity of Cell Organlles, J. REINERT, H. URSPRUNG (eds.). Berlin, Heidelberg, New York: Springer 1971. - SCHORNSTEIN, K.L., SCOTT, J.: Can. J. Bot. 60, 85-97 (1982). - SCOTT, J.: Protoplasma 118, 56-70 (1983). - SEARS, J.R., BRAWLEY, S.H.: Am. J. Bot. 69, 1450-1461 (1982). - SECKBACH, J., HAMMERMAN, I.S., HANANIA, J.: Ann. N.Y. Acad. Sci. 361, 409-425 (1981). - SECKBACH, J., FREDRICK, J.F., GARBARY, D.J.: Auto- or exogenous origin of transitional algae: an appraisal, pp. 947-962. In: see ref. SCHENK and SCHWEMMLER (eds.) 1983. - SEEWALDT, E., STACKEBRANDT, E.: Nature 295, 618-620 (1982). - SILVA, P.: Chlorophyceae, pp. 133-161. In: Synopsis and Classification of Living Organisms, ed. S.P. PARKER. New York: McGraw Hill 1982. - SLEIGH, M.A.: BioSystems 14, 423-431 (1981). - SLUIMAN, H.J.: Protoplasma 115, 160-175 (1983). - SLUIMAN, H.J., ROBERTS, K.R., STEWART, K.D., MATTOX, K.R., LOKHORST, G.M.: J. Phycol. 18, 1-12 (1982). - SOYER, M.-O.: BioSystems 14, 299-304 (1981). - SPECTOR, D.L., TRIEMER, R.E.: Biosystems 14, 289-298 (1981). - STACKEBRANDT, E.: A phylogenetic analysis of Prochloron, pp. 921-931. In: Endocytobiology II, H.E.A. SCHENK, W. SCHWEMMLER (eds.). Berlin: de Gruyter 1983. - STACKEBRANDT, E., KANDLER, O.: Zbl. Bakt. Hyg. Abt. 1 Orig. 3, 354-357 (1982). - STEINMÜLLER, K., KALING, M., ZETSCHE, K.: Planta 159, 308-313 (1983). - STULP, B.K.: Arch. Hydrobiol. Suppl. 63, 165-176 (1983). - STULP, B.K., STAM, W.T.: Arch. Hydrobiol. Suppl. 63, 35-52 (1982). - SULLIVAN, C.W., VOLCANI, B.E.: Silicon in the cellular metabolism of diatoms, pp. 15-42. In: Silicon and Silicon Structures in Biological Systems. Berlin, Heidelberg, New York: Springer 1981. - SUREK, B., MELKONIAN, M.: Intracellular bacteria in the Euglenophyceae: prolonged axenic culture of an algal-bacterial system, pp. 475-486. In: see ref. SCHENK and SCHWEMMLER (eds.) 1983.

TAYLOR, F.J.R.: Int. Rev. Cytol. Suppl. 14, 1-28 (1983a); - Possible free-living Symbiodinium microadriaticum (Dinophyceae) in tide pools in southern Thailand, pp. 1009-1014. In: see ref. SCHENK and SCHWEMMLER (eds.) 1983b. - TIPPIT, D.H., PICKETT-HEAPS, J.D.: J. Cell Sci. 21, 273-289 (1976). - TOWNSEND, R.A.: Phycologia 20, 407-414 (1981). - TRENCH, R.K.: Pure Appl. Chem. 53, 819-835 (1981); - Progr. Phycol. Res. 1, 257-288 (1982a); - Cyanelles, pp. 55-76. In: On the Origins of Chloroplasts, J.A. SCHIFF (ed.). New York: Elsevier/North Holland 1982b. - TRENCH, R.K., COLLEY, N.J., FITT, W.K.: Chem. Ber. 94, 568-577 (1981). - TRIEMER, R.E.: J. Phycol. 18, 399-411 (1982). - TSCHERMAK-WOESS, E.: Pl. Syst. Evol. 139, 295-301 (1982). - TSCHERMAK-WOESS, E., SCHÖLLER, A.: Plant Syst. Evol. 140, 207-223 (1982). - TURNER, J.A., WOELKERLING, W.M.J.: Phycologia 21, 218-235 (1982).

VON STOSCH, H.A.: Bacillaria 5, 127-156 (1982).

WALLACE, D.C.: Microbiol. Rev. 46, 208-240 (1982). - WEDEMAYER, G.J., WILCOX, L.W., GRAHAM, L.E.: J. Phycol. 18, 13-17 (1982). - WEST, J.A., GUIRY, M.D.: Bot. Marina 25, 205-211 (1982). - WEST, J.A., HOMMERSAND, M.H.: Rhodophyta: life histories, pp. 133-193. In: The Biology of Seaweeds, C.S. LOBBAN, M.J. WYNNE (eds.). Oxford: Blackwell 1981. - WEST, J.A., GUIRY, M.D., MASUDA, M.: Further investigations on the genetic affinities and life history patterns of the red alga Gigartina, pp. 137-166. In: Proc. joint China-U.S. Phycol. Sympos., C.K. TSENG (ed.). Peking: Science Press 1983. - WESTBROEK, P., DE JONG, E.W., VAN DER WAL, P., BORMAN, T., DE VRIND, J.P.M., VAN EMBURG, P.E., BOSCH, L.: Calcification in Coccolithophoridae - wasteful or functional?, pp. 291-299. In: Environmental Biogeochemistry, Ecol. Bull. (Stockholm) 35 (1983). - WILCOX, L.W., WEDEMAYER, G.J., GRAHAM, L.E.: J. Phycol. 18, 18-30 (1982). - WILHELM, C., EISENBEIS, G., WILD, A., ZAHN, R.: Z. Naturforsch. 37c, 107-114 (1982). - WILKINSON, C.: Phylogeny of bacterial and cyanobacterial symbionts in marine sponges, pp. 993-1002. In: see ref. SCHENK and SCHWEMMLER (1983a); - Science 219, 410-412 (1983b). - WOELKERLING, W.M.J.: Phycologia 22, 59-92 (1983). - WYNNE, M.J.: Phaeophyta: morphology and classification, pp. 52-85. In: The Biology of Seaweeds, S.C. LOBBACH, M.J. WYNNE (eds.). Oxford: Blackwell 1982a; - Phaeophyceae: introduction and bibliography, pp. 731- 739. In: see ref. ROSOWSKI and PARKER (eds.) 1982b; - Phaeophyceae, pp. 115-125. In: Synopsis and Classification of Living Organisms, S.P. PARKER (ed.). New York: McGraw-Hill 1982c; - Bot. Marina 26, 437-450 (1983).

YENTSCH, C.M., TAYLOR, D.L.: Red Tides, 250 pp. Boston: Harvard Univ. Press 1982.

Priv.-Doz. Dr. MICHAEL MELKONIAN
Botanisches Institut der
Westfälischen Wilhelms Universität
Schloßgarten 3
D-4400 Münster, FRG

II. Taxonomy and Phylogeny of Fungi

By WALTER GAMS and WALTER JÜLICH

1. General Considerations, Phylogeny, and Ecology

We welcome the publication of the 7th edition of AINSWORTH and BISBY's *Dictionary of Fungi* (HAWKSWORTH et al. 1983) which reflects the considerable progress achieved in mycology over the past 12 years. We also wish to mention the much improved 4th edition of MÜLLER and LOEFFLER's *Mykologie*(1982) which now presents a strongly modernized classification of the fungi. In order to provide a pragmatic framework, we adopt here a compromise between these two works (for ascomycetes also ERIKSSON 1982 and M. BARR 1983 and for basidiomycetes KHAN and KIMBROUGH 1982) as basis for the systematic arrangement. The full integration of lichens in the fungal system cannot, however, be adopted for the present report.

Speciation in fungi was the topic of BURNETT's (1983) presidential address to the British Mycological Society. Numerous examples of genetic differentiation at or below species level were reviewed. Genetic separation need not be allopatric, and sympatric sibling species are drawing more and more attention. Recent examples are the separation of *Armillaria mellea* (MARXMÜLLER 1982; ROMAGNESI and MARXMÜLLER 1983) and *Heterobasidion annosum* into several host-specific microspecies (CHASE and ULLRICH 1983, WORRALL et al. 1983). (See also Uredinales).

Studying haploid mitotic fruiting in *Polyporus ciliatus*, PRILLINGER and SIX (1983) emphasized that the sexual cycle is controlled genetically by mating type factors on the one hand, and by a multitude of different genes determining fruitbody formation on the other. Similar conclusions were reached by MEINHARDT and ESSER (1983) while surveying genetic aspects of sexual differentiation in fungi. PRILLINGER (1982a) reviewed the occurrence of haploid apomixis and amphithallism among basidiomycetes, which lead to fruiting without previous sexual reaction. Forcibly discharged meiotic basidiospores are regarded as derived from mitotic ballistospores, and haploid apomixis as a primitive character. Heterothallism is believed to have evolved from homothallism along different lines within the fungal kingdom (PRILLINGER 1982b) as suggested by KNIEP in the 1920's. Some cases of homothallism are, however, derived from a heterothallic condition. JAHRMANN and PRILLINGER (1983) found a yeast phase in the homobasidiomycete *Asterophora lycoperdoides*; in order to interpret this surprising observation, they reviewed various old and new ideas of fungal phylogeny; if it has a phylogenetic implication at all, this yeast phase would be a sign of atavism. Chitinous fungi may have been derived from primarily heterotrophic unicellular organisms; primitive fungi might have adapted to terrestrial life in a coccoid phase.

[1] As in our previous contribution, the second author deals with the taxonomy of basidiomycetes and the senior author with remaining groups.

Progress in Botany, Vol. 46

Nomenclature. The problems caused by the changes in Art. 13 ICBN (VOSS et al. 1983) noted in our previous report were discussed by KORF (1982 a, b), PETERSEN (1983b) and RAUSCHERT (1983); GAMS and KUYPER (1984) tried to solve problems encountered by GAMS (1984) who compiled an index to the sanctioned names. The fate of names adopted by FRIES before the "Systema" was traced by PETERSEN (1983a).

Ultrastructure. An atlas of fungal morphology illustrated mainly by SEM illustrations was published by TSUNEDA (1983). Architecture and chemistry of fungal cell walls was reviewed by COLE (1981). The use of frozen-hydrated specimens considerably improved accuracy of surface details in SEM pictures (READ et al. 1983). An atlas of wall structures and budding mechanisms in yeasts was published by GARRISON and ARNOLD (1981) (for terminology of cell aggregates cf. also JAHRMANN and PRILLINGER, 1983).

Chemotaxonomy. 5S rRNA was sequenced for a few basidiomycetes by WALKER and DOOLITTLE (1982) and the results were recalculated by TEMPLETON (1983); the findings support a fundamental separation in doliporous and non-doliporous basidiomycetes but do not object against a monophyletic origin of the basidiomycetes. Digestion of DNA by restriction enzymes and analysis of the resulting fragments by gel electrophoresis or hybridization with radioactively labeled RNA was applied for the first time to fungi but with species of *Aspergillus* it did not yield plausible results (KOZLOWSKI and STEPIEN 1982). Coenzyme-Q systems showed considerable heterogeneity in yeasts and yeast-like fungi (YAMADA et al. 1982, 1983). Consistent results were obtained at the genus level, with exceptions in *Trichosporon* (YAMADA et al. 1982).

The study of cytochrome absorption spectra is another approach which supports previous conclusions on the existence of two distinct groups of species within *Kluyveromyces* (FIOL and CLAISSE 1982). Analysis of glucuronic acid content in cold-water extracts of fruitbodies of diverse asco- and basidiomycetes showed unusually high percentages in the Phallales and relatively low levels in the ascomycetes examined (TSUCHIHASHI et al. 1982). Sterols were analyzed in some species of Oomycota and Hyphochytridiomycota (WARNER et al. 1983); the preferential utilization of cycloartenol over lanosterol in all taxa studied may indicate an ancestral affinity to photosynthetic organisms. The abundant production of ergosterol by *Zoophagus* (Pythiaceae) is unusual, casting doubts on its classification (WARNER et al. 1982). Sterol content in two species of Endogonales differs from that of other Zygomycetes (BEILBY 1980, BEILBY and KIDLEY 1980). An unusual sterol, brassicasterol, was found in *Taphrina* and *Protomyces*, in addition to ergosterol in some species of *Taphrina* (VAN EIJK and ROEIJMANS, 1982). Large quantitative differences in composition of cellular fatty acids allowed four groups of yeasts to be distinguished (MOSS et al. 1982). TLC analysis of secondary metabolite profiles (some pigments, some mycotoxins) in *Penicillium* yielded taxonomic criteria that may be useful in deciding about delimitation of critical species (FRISVAD and FILTENBORG 1983). The levels of the naphthoquinone pigments (toxins) xanthomegnin and viomellein allowed two groups of isolates of *Penicillium viridicatum* to be distinguished (CIEGLER et al. 1981). In contrast with previous unsatisfactory results using isoenzyme analysis for fungal taxonomy, JONES and NOBLE (1982) obtained a good correlation with morphological classification in some Dermatophytes. Pectic zymograms seem to be particularly promising as a taxonomic tool in *Sclerotinia* (CRUIKSHANK 1983a) and *Botrytis* (CRUIKSHANK 1983b). BERNIER et al. (1983) succeeded in differentiating between aggressive and nonaggressive isolates of *Ceratocystis ulmi*, using five particular enzymes. The highest specificity of serological techniques is achieved by means of monoclonal antibodies; IANNELLI et al. (1982, 1983) even were able to differentiate between formae speciales of *Fusarium oxysporum* by applying at least five hybridomas in one analysis. Analysis of amino acids in total fungal protein of some yeasts and yeast-like fungi led to the crude distinction of eight somewhat heterogeneous groups (KOCKOVÁ-KRATOCHVÍLOVÁ et al. (1981). (More work on DNA and serology is cited under "Endomycetales").

Fungal ecology and geography. From many years' surveys of plant-parasitic fungi in Australia and New Zealand, J. WALKER (1983) noted unusual deficiencies in the Australian and New Zealand flora for species of *Synchytrium* and *Physoderma*, Peronosporaceae, *Taphrina*, Erysiphales, *Ravenelia* on *Acacia*, and relatively few rusts occurring on Myrtaceae and Proteaceae. These taxa are mostly represented by species occurring on plants introduced from the northern hemisphere. Hence these distribution patterns may have something to do with the plant-geographic history of these countries.

The studies on fungal endophytes noticed in our previous contribution are continued for Araceae, Bromeliaceae and Orchidaceae (PETRINI and DREYFUSS 1982), evergreen shrubs (PETRINI et al. 1982), and *Abies alba* (GOURBIÈRE 1983). Most of the endophytes of coniferous foliage utilize xylan and pectin, are lipolytic and grow on gallic acid, fewer consume cellulose and remaining hemicelluloses; thus different endophytes might coexist in a host by partitioning the available resoures (CARROLL and PETRINI 1983). Parasitic microfungi on trees in Western Canada were compiled by FUNK (1981). HAWKSWORTH (1983) keyed out 218 lichen-forming and lichen-inhabiting fungi. (More ecological groups under "Hyphomycetes")

Medical mycology. Our previous critical remarks on the taxonomic side of medical mycology can be attenuated: an ambitious work edited by HOWARD (1983) contains some critical reviews on the major fungal categories that contain human pathogens, and surveys of morphology, ultrastructure and conidiogenesis.

2. Fungus-like PROTOCTISTA

a) Myxomycota

α) *Dictyosteliomycetes* and *Acrasiomycetes*. BONNER (1982) reviewed evolutionary strategies under the headings: size, division of labor, early signs of differentiation, compound fruiting bodies, regulation, stalkless migration, microcysts and macrocysts and isolating mechanisms (acrasins); the acrasids seem to be polyphyletic and the occurrence of at least eight different acrasins that prevent mutual interference in the dictyostelids is still difficult to explain. *Dictyostelium caveatum* WADDELL (1982), however, overcomes the physiological barrier and preys on various other dictyostelids. Numerous collections of the orange-colored acrasid *Pocheina* (=*Guttulina*) were compared and a new species had to be described (OLIVE et al. 1983). *Fonticula alba*, an organism distinct from both dictyostelid and acrasid slime moulds, produces spores within the stalk necks of broadly conical fruiting bodies (SEM and TEM study by DEASEY 1982).

β) *Myxomycetes*. An abridged version by FARR of the MARTIN and ALEXOPOULOS monograph from 1969 appeared (MARTIN et al. 1983), in which the classification is brought up to date at the generic level, the 1969 plates are reprinted. The use of polarized light to study birefringence in capillitium, spore walls and peridium yielded criteria which are useful at family level and led to some changes in classification and a new key to the families of the Liceales (NANNENGA-BREMEKAMP 1982). A major taxonomic-floristic contribution covers the genus *Diderma* with 16 species occurring in Belgium (BUYCK 1982). In the genus *Tubifera*, peridial and spore ornamentations were used to establish a modified key to species (NELSON et al. 1982). The operculum of *Trichia crateriformis* consisting of a single layer of spore-like bodies (SEM by FREDERICK et al. 1983), is unique among the Myxomycetes. The unusual capillitium, consisting of moniliform threads in *Listerella*, was studied by SEM (ELIASSON and GILERT 1982). Amebal and plasmodial mitoses in *Echinostelium minutum* are different at the light-microscopic level (HINCHEE and HASKINS 1981). The microtubular rootlets of the flagellar apparatus of *Ceratiomyxa* are more similar to protostelids than to Myxomycetes (SPIEGEL 1981). In *Physarum polycephalum* the nucleotide sequence of 5S ribosomal RNA was very distinct from that of the cellular slime mould *Dictyostelium discoideum* and closer to that of Protozoa and Metazoa than to those from fungi (KOMIYA and TAKEMURA 1981). A new genus, *Semimorula*, seems to connect the Myxomycetes with the Protostelids (TEM and SEM by HASKINS et al. 1983). A considerable number of taxa were added to the Myxomycete flora of Japan since EMOTO's monograph (NANNENGA-BREMEKAMP and YAMAMOTO 1983, including a key to the species of *Macbrideola*).

γ) *Plasmodiophoromycetes*. The zoospores of *Polymyxa graminis* are unique in possessing two kinds of rootlets extending from each kinetosome to the plasmalemma; they are thus closer to Protozoa than to any other zoosporic fungi (D.J.M. BARR and ALLAN 1982). The haploid chromosome number of *Plasmodiophora brassicae* was established to be 20 by means of serial thin sections of pachytene nuclei (BRASELTON 1982). The obligate biotrophic endoparasite *Woronina pythii* was found to infect 15 species of *Pythium* but none of *Phytophthora* or various Saprolegniaceous genera (DYLEWSKI and MILLER 1983). The molecular weight of ribosomal RNA of three genera of the Plasmodiophorales was found to be 1.25 and 0.66 x 10^6 for larger and smaller molecules, respectively, and thus quite distinct from Myxomycetes and Acrasiomycetes or other higher fungal taxa. An evolutionary origin from protozoan ancestors cannot be excluded on the available data (FRASER and BUCZACKI 1983). The resting spore wall of *Plasmodiophora brassicae* has chitin and protein as major constituents (MOXHAM and BUCZACKI 1983).

b) Labyrinthulomycota

Detailed studies of the fine structure of *Ulkenia amoeboidea* by RAGHU KUMAR (1982 a, b) of the vegetative thallus, ameboid stage and zoospores are consistent with previous reports suggesting a classification within protists with unknown affinities.

c) Oomycota (and other zoosporic fungi in general)

The flagellar rootlet system consisting of microtubules and finely fibrillar or electron-opaque material is particularly conservative among zoosporic fungi second to the flagellar axoneme, the kinetosome and structure of the nuclear membrane; therefore it is a valuable taxonomic criterion (D.J.M. BARR 1981); selected examples indicate that Thraustochytridiales, Hyphochytridiomycetes and Oomycetes probably evolved as three distinctive lines from chrysophycean ancestors. Flagellar structures in zoosporic plant pathogens were also reviewed by LANGE and OLSON (1983) and D.J.M. BARR (1983), reconfirming that the Mastigomycotina are an artificial group and that the Chytridiomycotina might be included in the Eumycota.

α) *Saprolegniales*. The genera *Achlya*, *Dictyuchus* and *Brevilegnia* differ not only in zoospore release mechanism but also in sporangial wall thickness (decreasing in the above order (ARMBRUSTER and JOHNSON 1983). Ultrastructure of sporangiogenesis in these three genera was studied in detail by ARMBRUSTER (1982a, b). MIL'KO (1983) keyed out and documented 33 taxa of *Achlya*. Three species were retained in a revision of *Apodachlya* by JACOBS (1982). The finding of three new marine *Phytophthora* species prompted a survey of all marine Pythiaceae known (GERRETTSON-CORNELL and SIMPSON 1984). The zoospores of *Haliphthoros milfordensis* differ from those of other Oomycetes by mitochondria grouped in the anterior part of the cell around the nucleus, while the posterior end of the zoospores is filled with numerous fibrous vesicles (OVERTON et al. 1983).

β) *Peronosporales*. A volume on biology, taxonomy, ecology and pathology of *Phytophthora* (ERWIN et al. 1983) contains accounts of taxonomic criteria (WATERHOUSE et al.), variability (ERWIN), and new taxonomic criteria (GALLEGLY). Problems, mainly concerning speciation were reviewed in this volume by BRASIER, but the taxonomic situation remains unsatisfactory.

Isolates obtained from *Hevea* in Sri Lanka (mainly *Ph. meadii*) were studied in all details concerning colony morphology, micromorphology, cytology and compatibility by DANTANARAYANA et al. (1984). HO (1983) and BRASIER (in ERWIN et al. 1983) compared ontogeny of paragynous and amphigynous antheridia, the latter being considered as more advanced. CONSTANTINESCU and NEGREAN (1983) published a check-list of Romanian Peronosporaceae (*Peronospora* with 88, *Plasmopara* with 10, *Pseudoperonospora* with 4 species plus four monotypic genera) and Albuginaceae (6 species). As Romania has the richest known flora of these families and some taxonomic decisions had to be taken in critical cases, this survey is of wider importance. SKALICKY (1983) revised *Peronospora* species occurring on Rosaceae and recognized 11 species.

d) Chytridiomycota

Most research was concerned with ultrastructure (TEM): Flagellar rootlet structures suggest that Chytridiales and Spizellomycetales present two major lines from which the Monoblepharidales and Blastocladiales, respectively, evolved (D.J.M. BARR 1981). Zoospores of *Caulochytrium protostelioides* (POWELL 1981) and resting body formation in *Chytriomyces hyalinus* (PECHAK and MILLER 1981) were studied with TEM. Zoospore structure of the anaerobic gut inhabitant *Neocallimastix frontalis* brings it near to the Spizellomycetales in spite of the multiple flagella (HEATH et al. 1983). Studies on morphology and development of the fungal endoparasites *Rozella* (Olpidiaceae) and *Rozellopsis* (Olpidiopsidaceae) were summarized by HELD (1981), who noted a morphogenetic cooperation between parasite and host determining the shape of the former. Within the recently described order Spizellomycetales, the Urophlyctaceae were distinguished as a family by HADAR (1982). The polyflagellate anaerobic rumen protist *Neocallimastix* is associated to the Spizellomycetales in a separated family, because of its flagellar rootlet structure (HEATH et al. 1983). Within the Blastocladialean genus *Coelomomyces* interspecific hybrids were experimentally produced that yielded a functional sporophyte but inviable gametophytes (FEDERICI 1982); meiosis was shown to occur in the resistant sporangia and ten synaptonemal complexes were observed (WHISLER et al., 1983).

3. FUNGI

ZYGOMYCOTA

a) Zygomycetes

α) *Mucorales*. The asexual structures of representatives of the sections of *Mortierella* were studied with SEM and TEM by ANSELL and YOUNG (1983 a); some new details of spore ornamentation were discovered; *M. indohii* is one of the few species of the genus with invested zygospores and these were investigated by ANSELL and YOUNG (1983b). Criteria used to distinguish species of *Rhizopus* were evaluated by SEVIOUR et al. (1983) with emphasis of spore structures as observed with SEM. Taxonomic conclusions on speciation in this difficult genus were drawn by SCHIPPER (1984), who recognized the species groups *stolonifer* and *microsporus* (each with two species) in addition to the variable species *Rh. oryzae*, based on temperature relationships, mating behavior and SEM of sporangiospores.

SEM and TEM studies were carried out on zygospores of *Cokeromyces* (JEFFRIES and YOUNG 1983), sporangioles of *Choaneophora* (HIGHAM and COLE 1982) and of *Cunningham* (LUNN and SHIPTON 1983); the latter authors combined *C. elegans* and *C. echinulat* as varieties of one species, despite their sexual incompatibility. An intermediate isolate was compared in detail with the two recognized species of *Radiomyces* by KITZ et al. (1982). The variation of the new species *Syncephalis leadbeateri* was studied on various host fungi and compared with its relatives by BAWCUTT (1983) yielding criteria that permit a sound classification of the genus. VON ARX (1982b) shuffled around some families and genera of Mucorales, but his conclusions still await verification by experts.

β) *Dimargaritales*. Ultrastructure of the complex septal pores in *Tieghemiomyces* agrees with that of other representatives of the order (BRAIN et al. 1982).

γ) *Entomophthorales* . The great innovations of classification reviewed in our previous report are becoming consolidated (HUMBER 1983); arguments and taxonomical and phylogenetical criteria were further detailed by BEN-ZE'EV and KENNETH (1982a), who arranged the genera of the three recognized families in a kind of cladistic diagram. A key to species according to the old system by WATERHOUSE and BRADY (1982) still has its practical usefulness. The genus *Erynia* (syn. *Zoophthora*), with three subgenera, was revised according to modern criteria by BEN-ZE'EV and KENNETH (1982b), while *Strongwellsea* was recognized as distinct from this complex by HUMBER (1982). Formation of azygospores (resting spores) in *Conidiobolus obscurus* was studied by means of SEM and TEM by LATGÉ et al. (1982).

δ) *Endogonales*. *Endogone pisiformis* was reported to grow from surface-sterilized sporocarps in axenic cultures on a variety of conventional media (BERCH and FORTIN 1983); it did not form mycorrhizae with any of the plants tested. Six new species were added to the endomycorrhizal genus *Glomus*, and a modified synoptic key to the Endogonaceae known in Florida was presented by SCHENCK and SMITH (1982). The ectendomycorrhiza partner of pines described as a possible member of the Endogonaceae, *Complexipes moniliformis* C. WALKER (1979), turned out to be an anamorphic ascomycete (DANIELSON 1982).

b) Trichomycetes

In *Smittium culisetae* zygospores were obtained for the first time, and studies on trichospore germination confirmed their homology with monosporic sporangia (WILLIAMS 1983). Five new species of Harpellales were described from aquatic Diptera larvae and many more new taxa are to be expected (LICHTWARDT 1984).

ASCOMYCOTA

ERIKSSON (1982) and M. BARR (1983) present further suggestions for an arrangement of classes, subclasses and orders of ascomycetes, but without pretending to give a phylogenetic scheme. While most other authors tend to place the Endomycetes remote from other ascomycetes, ERIKSSON leaves them in a fourth group of the Euascomycetes together with various orders of pseudoprototunicate ascomycetes, and BARR removes the Taphrinomycota to the Dicaryomycotera and leaves the remaining orders of Endomycota in the Monocaryomycotera.

a) Endomycetes (Hemiascomycetes)

Taphrinales and *Protomycetales*. VON ARX et al. (1982) confirmed previous findings that *Taphrina* and *Protomyces* have ascomycete-type cell walls and are thus unrelated to the basidiomycetes *Microstroma* and *Exobasidium*. Cultures of *Taphrina* and *Protomyces* were indistinguishable on the basis of carotenoid and sterol production (VAN EIJK and ROEIJMANS 1982). A thorough TEM study of mitosis in yeast cells of several *Taphrina* species reconfirms their ascomycete affinity; an invagination of the nuclear envelope below the NAO during prophase observed in two species is an unusual feature similar to that observed in the red alga *Porphyridium*, whilst the occurrence of a pore in the nuclear envelope at the NAO center in two other species is also unique, suggesting that the genus may be polyphyletic (HEATH et al. 1982).

Endomycetales. The yeast compendium by BARNETT et al. (1983) is more than a revision of the *Guide to identifying yeasts* by these authors (1979); it contains tables of the distribution of 83 physiological characters, several dichotomous keys to species according to morphological and physiological characters and succinct descriptions of 469 species. In addition, BELIN (1981) compiled the more than 100 new species and the redispositions published since LODDER (1970). KURTZMAN et al. (1983) reviewed work on nucleic acid as a tool in yeast taxonomy and concluded that molecular techniques have provided a more reliable guide for discerning relationships than the traditional morphological and physiological criteria. Using DNA base sequence complementarity, the status of some isolates so far distinguished as varieties was revised by KURTZMAN (1984): Isolates with 94% homology were considered identical, others with 19% homology as specifically distinct, and only a homology of 41% is still regarded as reflecting varietal difference.

Immunoprecipitation was applied in 13 problematic cases of perfect and imperfect yeasts,but it confirmed previous taxonomic conclusions (MONTROCHER 1982). According to a light-microscopic study, meiosis and nuclear fusions in *Saccharomyces rosei* take place in the bud cell, followed by the migration of the diploid nucleus back into the mother cell, unlike other species of the genus (supporting present classification of this fungus in *Torulaspora* as *T. delbrueckii*) (FERREIRA 1982). Isoenzyme studies on species of *Kluyveromyces* support the distinction of 13 species (SIDENBERG and LACHANCE 1983), in contrast with the previously reported findings from DNA hybridization (FIOL and PONCET 1980) and from cytochrome studies (FIOL and CLAISSE, 1982), who recognized three main types. Sexual hybridization and protoplast fusion applied to species of *Schizosaccharomyces* suggested synonymy of *S. pombe* and *S. malidevorans*; this species is related to *S. octosporus* but more distant from *S. japonicus* (SIPICZKI et al. 1982); these findings are at variance with JOHANNSEN's findings (1981, noted in our last report), who recognized five distinct species.

b) Ascomycetes

General considerations. CHADEFAUD (1982a–c) attempted to synthesize his and his followers' ontogenetic studies on ascomata: The "gynocarp" (sexual ascoma initial), consisting of carpocentrum and an envelope, gives rise to discoid or globular ascomata; a discoid gynocarp like that of *Dothiora* is comparable to the conidioma of *Coryneum* (1982a). Five categories of admittedly polyphyletic discocarpous ascomata were distinguished, among which the Pezizales type would be the most evolutionarily advanced; a diagram of possible evolution of ascus structures is included (1982b). Among the very heterogeneous pyrenocarpous ascomycetes, a major line is drawn in the conventional way between Ascoloculares and Ascohymeniales, while the Plectascales are considered as a product of retrogressive evolution from one or both groups.

Further studies on ascoma development and ultrastructure have been carried out concerning crozier formation in *Byssochlamys* (with SEM, SAUTTER and HOCK 1982), ascus formation in *Eleutherascus* (TEM study continued by STEFFENS and JONES 1983). Detailed studies on centrum development were carried out on *Gnomonia comari* (HUANG and LUTTRELL 1982) and *Diaporthe phaseolorum* (JENSEN 1983) with a Diaporthaceous structure and *Gelasinospora reticulispora* (JENSEN 1982) with a Sordariaceous structure. While in previously studied *Tuber* species ascospores are delimited singly as in the Hemiascomycetes, those of *T. magnatum* were found to be delimited by two membranes which originate from the ascus plasmalemma (BERTA and FUSCONI 1983). In the asci of typical Loculoascomycetes an electron-transparent zone develops between the layers of the exoascus and the endoascus following local lysis of the latter (PARGUEY-LEDUC and JANEX-FAVRE 1982) as shown in species of *Leptosphaeria, Pleospora, Sporormia* and *Massaria*. A similar structure is seen in *Leptosphaeria senegalensis* (PARGUEY-LEDUC et al. 1982). A three-layered structure of the ectotunica and homogeneous, thick endotunica was also described for *Hysterographium fraxini* (Dothideales), the structure of which shows similarities with the *Placynthium* type (Peltigerales), linking the pyrenomycetous bitunicate ascus structure to that of the Lecanorales (BELLEMÈRE and HAFELLNER 1982 b). Centrum development in *Didymella bryoniae* is typical of the Pleosporales (SKARSHAUG 1981). Meiosis in *Cochliobolus heterostrophus* was studied with light microscopy by GUZMAN et al. (1982) who observed eight chromosomes.

Floristic contributions: Collections representing 16 lignicolous pyrenomycete genera in southern Germany were described by HILBER et al. (1983) and coprophilous pyrenomycetes in arctic Canada by BOOTH (1982). The studies of muscicolous ascomycetes were continued by DOBBELER (1981) dealing with fungi growing on *Dawsonia*.

Taxonomy

Two major attempts of a classification of the orders by ERIKSSON (1982) and M. BARR (1983) have been mentioned above.

Laboulbeniomycetidae

Important floristic-taxonomic studies were published on species found in Finland and adjacent parts of the U.S.S.R. (HULDÉN 1983), Madagascar, the Comores and the Mascarenes (BALAZUC 1982), Caribic countries (ROSSI and CESARE ROSSI 1982), and Borneo (SUGIYAMA and YAMAMOTO 1982).

Ascomycetidae

α) *Pezizales* (incl. *Tuberales*). Previous studies on *Peziza* in south-eastern France were summarized by DONADINI (1981) who also provided a key to all species found in France; the subgenus *Pachyella* with gelatinous ectal excipulum is treated separately. PFISTER (1982) provided a nomenclatural revision for SEAVER's classic book on North-American operculate cup fungi. The ultrastructure of vegetative and generative structures in some Pezizales and *Rhytisma* was studied by KAMALETDINOVA and VASSILYEV (1982).

The genus *Scutellinia* was revised for the Soviet Union by KULLMAN (1982); its section *Trechispora* was monographed in more detail by DONADINI (1983). In the complex of globose-spored Pezizaceae two species with prominent pustulate ascospore sculpture form a distinct genus, *Scabropezia* (DISSING and PFISTER 1981). Eight species of *Pulvinula* were described from the north-west Himalayas (KAUSHAL 1982); seven species of *Coprotus* were described for Norway (AAS 1983). A cytological study revealed details of the isomorphic gametangia of two species of *Ascodesmis* (RANALLI et al. 1982). The new species *Ascobolus ferrugineus* was studied culturally and cytologically by RANALLI and FORCHIASSIN (1981). *Hydnotria michaelis* was removed from the Tuberaceae and placed in the Helvellaceae because of its 4-nucleate ascospores (BERTHET 1982). Two species of the desert truffle genus *Tirmania* were redescribed by ALSHEIKH and TRAPPE from Kuwait (1983).

β) *Helotiales*. Genera revised taxonomically include: *Chloroscypha*, 5 species endophytic in Cupressaceae (PETRINI 1982); *Hysterostegiella* characterized by pointed paraphyses, apothecia with hairs and an operculum, thus obscuring the familial circumcription of the Dermateaceae, Leotiaceae and Hyaloscyphaceae (HEIN 1983); genera of the *Hysteropezizella* complex in which the length of paraphysis cells was used as a taxonomic criterion (HEIN 1982); *Naeviella*, a member of this complex with an unusual bilateral symmetry of the ascospores due to a laterally protuberant germ pore (NANNFELDT 1982), and *Nimbomollisia* and *Discocurtisia* lacking germ pores (NANNFELDT 1983); species of *Monilinia* growing on Ericaceae with various degrees of host specificity (BATRA 1983); and long-spored, mainly neotropical, lignicolous species of *Lachnum* (=*Dasyscyphus*) (HAINES and DUMONT 1984). The critical discomycete flora of Macaronesia was continued by a revision of the Sclerotiniaceae (KOHN, 1982). A SEM study of ascoma development in *Sclerotinia sclerotiorum* showed the presence of an amorphous membrane protecting the young asci (TU 1982). The ascus of *Dactylospora* is functionally unitunicate but has an unusual opening mechanism; this non-lichenized genus appears intermediate between Helotiales and Lecanorales, though too highly evolved to be considered a common ancestor (BELLEMÈRE and HAFELLNER 1982 a).

γ) *Phacidiales*. DICOSMO et al. (1983 a) redefined the family Phacidiaceae by an ascoma roof wall consisting of columnar cells. Accordingly, of the 24 genera previously classified in this family, only *Lophophacidium* and *Phacidium* are retained there. They are associated with *Apostrasseria* and *Ceuthospora* anamorphs, respectively.

δ) *Diaporthales*. A comprehensive revision of the Gnomoniaceae with emphasis on European material and cultural studies was provided by MONOD (1983). Of the 214 species considered, 37 are new. When 89 species of this family were compared numerically, the present generic classification appeared unnatural, although it is still convenient for identification (MONOD and ZIEGLER 1983).

ε) *Sordariales*. *Apiosordaria* was redefined to include *Echinopodospora* and *Lacunospora*; it now contains 15 species (KRUG et al. 1983). Conversely, *Melanospora* and similar fungi were revised and redefined in a restricted sense, and seven similar genera were recognized by CANNON and HAWKSWORTH (1982). Ascus shape and germ pores of the ascospores as primary taxonomic criteria in the Chaetomiaceae (VON ARX et al., 1984) have been emphasized previously; the dextrinoid ascospore walls of the *Chaetomium bostrychodes* group are a useful taxonomic criterion too (CARTER and MALLOCH, 1982).

ζ) *Diatrypales*. Studies of *Selenosporella*-like anamorphs in *Eutypa*, *Eutypella* (GLAWE and ROGERS 1982a), *Diatrype* and *Diatrypella* (GLAWE and ROGERS 1982b) revealed both percurrently and sympodially proliferating conidiogenous cells in closely related taxa.

η) *Xylariales*. *Hypoxylon* is represented with 15 species in southern Germany (ENDERLE 1982). The new genus *Helicogermslitia*, though having non-amyloid asci, may belong to the Xylariaceae (HAWKSWORTH and LODHA 1983).

ϑ) *Hypocreales*. *Nectria* and related genera are richly represented in the neotropics and 30 spp. (+ 2 of related genera) were treated systematically for Panama by SAMUELS and DUMONT (1982), including anamorph connections. The phragmosporous Nectriaceae were revised by ROSSMAN (1983); 42 species were treated in *Nectria*, and this genus can no longer be restricted to species with two-celled ascospores; only five species remain in *Calonectria*, where *Cylindrocarpon* anamorphs are a major criterion. Both of these important contributions contain synoptic keys to the species treated.

ι) *Clavicipitales*. A beautifully illustrated monograph of *Cordyceps* and allied genera was published by KOBAYASHI and SHIMIZU (1984), following the same authors' (1982) monograph of *Torrubiella*. Various groups of *Cordyceps* and their anamorphs were studied: species from Ghana by SAMSON et al. (1982), from the Galápagos Islands by EVANS and SAMSON (1982), species occurring on neotropical Formicidae by EVANS and SAMSON (1983, 1984); and species occurring in Denmark by LAESSØE (1982).

κ) *Halosphaeriaceae*. The classification of *Corollospora* was revised based on previous electron-microscopic studies by JONES et al. (1983a). Ultrastructural studies on spore appendages were carried out for two further species by JONES et al. (1983 b).

λ) *Dothideales*. A survey of bitunicate ascomycetes and their anamorphs is presented by SIVANESAN (1984). The north-American Pleomassariaceae were revised and classified in three genera by M. BARR (1982). The genus *Pleospora* sensu Wehmeyer was heterogeneous and its species were distributed by CRIVELLI (1983) over 12 genera according to ascoma morphology and biology; dematiaceous anamorphs are characteristic of some groups but most species are homothallic and do not produce an anamorph.

The segregation of *Pseudocochliobolus* from *Cochliobolus* is not justified when all criteria are carefully considered (ALCORN 1983a). *Hypobryon* is a remarkable bryophilic genus with hypophyllous ascomata which perforate the moss leaf to discharge ascospores on the upper surface (DOBBELER 1983). Further revisionary work concerns *Asterinema* (FARR 1983), *Balladynopsis* and related genera (SIVANESAN 1981), Australian leaf-inhabiting fungi of *Semifissispora* (SWART 1982a) and *Pachysacca* (SWART 1982b); in the sooty mould genus *Trichomerium*, REYNOLDS (1982) retained only one species with numerous synonyms and excluded some taxa after screening a vast amount of material.

μ) *Erysiphales*. BRAUN continued his monographic studies on *Microsphaera*, *Erysiphe* (1982a,b), *Podosphaera* (1984) and the anamorph-genus *Oidium* (1982c). Some new taxa and combinations in *Erysiphe* were reported from China by ZHENG and CHEN (1981a,b), and in *Microsphaera* by YU and LAI (1981a,b).

ν) *Eurotiales* (and associated cleistocarpous families). 33 species of *Eupenicillium* and related sclerotigenous *Penicillium* were revised by STOLK and SAMSON (1983), adopting rather broad species concepts; advances in the knowledge of teleomorphs affect the taxonomy of the anamorphs considerably. *Emericella* was revised by CHRISTENSEN and STATES (1982). The medically important genus *Petriellidium* was revised and the name found to be a later synonym of *Pseudallescheria* by McGINNIS et al. (1982). The puzzling genus *Monascus*, important in production of oriental fermented food, was revised by HAWKSWORTH and PITT (1983), who retained three species. *Elaphomyces* species found in Nordic countries were revised and keyed out by KERS (1981).

ξ) The *Ascosphaerales* were formalized as a distinct order by SKOU (1982) and characterized by the possession of spore cysts; they may be classified between the Protomycetales and Eurotiales, and a relationship might exist with *Amorphotheca*. The species of Ascosphaerales were keyed out by ROSE et al. (1984).

FUNGI IMPERFECTI (Deuteromycota)

a) General Considerations and Teleomorph-Anamorph Connections

SUBRAMANIAN (1983) and COOKE (1983) attempted to classify deuteromycetes into orders and families, based mainly on conidiogenesis, but thus united quite heterogeneous elements in single taxa. Numerous papers are concerned with new anamorph-teleomorph connections: PUNITHALINGAM (1981) obtained several Sphaeropsidales from ascospore-isolates. Acervuloid to pycnidial conidiomata of *Ephelis* type, but with polyblastic conidiogenesis, were found in *Myriogenospora* (Clavicipitaceae) (RYKARD et al. 1982). Connections between various pyrenomycetes and hyphomycetes were established by SIVANESAN (1983), MÜLLER and SAMUELS (1982a,b) and SUBRAMANIAN and SEKAR (1980). Because of a *Hyalorhinocladiella* anamorph, *Klasterskya* was reclassified in the Ophiostomataceae by MINTER (1983). As expected, *Cristulariella* was found to have a Sclerotiniaceous teleomorph, viz. in the new genus *Grovesinia* (CLINE et al. 1983). The aero-aquatic genus *Pseudaegerita* is connected with *Hyaloscypha* (ABDULLAH and WEBSTER 1983), the marine *Anguillospora marina* with *Lindra* (Halosphaeriaceae) (NAKAGIRI and TUBAKI 1983 a), and the nematode parasite *Harposporium anguillulae* with *Atricordyceps* SAMUELS (1983).

b) Coelomycetes

A detailed analysis of pycnidial types and conidiophore structures in lichens by VOBIS (1980) should be carefully considered in coelomycete taxonomy. Generic revisions of note include: *Strasseria* and two related genera by NAG RAJ (1983), and *Cytospora* by GVRITISHVILI (1982).

c) Hyphomycetes

SUBRAMANIAN (1983) published a noteworthy book on taxonomy and biology of the hyphomycetes; taxonomy occupies a minor portion but is efficiently placed in a historical context. The author again expounds his views on conidiogenesis, some of which are controversial. The book complements the *Biology of Conidial Fungi* by COLE and KENDRICK (1981, noted previously). Modes of conidiogenesis continue to be discussed (e.g. COLE and SAMSON, in HOWARD 1983). MINTER et al. (1982, 1983a,b) emphasized the need for describing the phases of conidium formation step by step (as suggested already by KENDRICK in 1971); an advance in these lengthy papers is the relegation of the term enteroblastic to a mechanism of wall formation and not to a category of conidiogenesis.

Conidiogenesis in *Fusarium* is typically phialidic (TEM by SCHNEIDER and SEAMAN, 1982). In the apparently phialidic *Cylindrocladium crotalariae* conidium formation ceases after the first conidium because the phialides lose their nucleus (JONES, 1981). The taxonomic importance of rhexolytic secession was realized by VON ARX (1982 b) when he decided to expand the genus *Dicyma* with species of *Hansfordia*, *Basifimbria*, *Gonytrichella* and *Puciola*.

α) *Ecological groups*: Litter of *Laurus nobilis* is a rich substrate for interesting hyphomycetes (KIRK 1982b). Numbers of new taxa were described in the aquatic hyphomycetes (DESCALS and WEBSTER 1982a,b). Aquatic hyphomycetes with unbranched conidia were reviewed for Norway by BRÅTHEN (1982). The aero-aquatic bubble-trap conidia of some genera were examined in detail by MICHAELIDES and KENDRICK (1982). In freshwater swamps in various parts of the U.S.A. CRANE and SCHOKNECHT (1982) found several interesting hyphomycete genera. As endophytic hyphomycetes are becoming important as causal agents of toxicity in forage grasses, the taxonomic distinction between *Acremonium coenophialum*, the fescue endophyte, and the anamorph of *Epichloe typhina* was described by MORGAN-JONES and GAMS (1982). Entomogenous hyphomycetes were studied from the Galápagos Islands by EVANS and SAMSON (1982), for more references see under "Clavicipitales". Some species of *Hirsutella* and one of *Ramularia* parasitizing mites were described by MINTER et al. (1983c) and two new species of *Beauveria* by SAMSON and EVANS (1982 b). The number of known *Termitaria* parasitizing termites was raised from two to five by KIMBROUGH and LENZ (1982). Another termite parasite is *Mattirolella*, which was redescribed by KIMBROUGH and THORNE (1982). A new helicosporous parasite of spiders was described as *Clathroconium* by SAMSON and EVANS (1982). An unusual parasite of ciliated protozoans was named *Meria harposporioides* by BARRON and SZIJARTO (1982).

β) *Imperfect yeasts*. DE HOOG and SMITH (1983) emphasized the numerous connections between yeast anamorphs and hyphomycetes, and the necessity for morphological, ultrastructural and biochemical comparisons in both. Supposed teleomorph connections in 19 species were verified by electrophoretic comparison of ten enzymes of sugar metabolism by YAMAZAKI and KOMAGATA (1982). Killer strains of nine yeasts are an efficient tool for identifying these species (POLONELLI et al. 1983).

γ) *Dematiaceous hyphomycetes*. ALCORN (1983b) reviewed and supported the present generic concepts of *Drechslera*, *Bipolaris* and *Exserohilum*. Four species of *Bipolaris* parasitizing grass ovaries were critically distinguished by ALCORN (1982). A critical revision of *Alternaria* species is published in fractions by SIMMONS (1982 a, b); four species of this genus occurring on sesame seeds were critically compared by YU et al. (1982). DEIGHTON (1983) continued to revise *Cercospora* and related genera. In the same complex VON ARX (1983) keyed out and discussed 23 accepted anamorph-genera connected with *Mycosphaerella*, and made several new combinations. Over 200 taxa of *Cercospora* can be distinguished in Burma (THAUNG 1984). Critical generic revisions were published for *Spadicoides* and *Diplococcium* in Czechoslovakia by HOLUBOVÁ-JECHOVÁ (1982), four genera of hyphopodiate sooty moulds from North America by HUGHES (1983), the pleo-anamorphic genus *Triadelphia* by CONSTANTINESCU and SAMSON (1982), and the sporodochial genus *Koorchaloma* with *Kananascus* teleomorphs by NAG RAJ (1984). Whether genera of apparently closely related anamorphs of *Chaetosphaeria* are to be distinguished primarily on conidiophore branching or on minor details of conidiogenesis (DICOSMO et al. 1983b), seems of minor importance. The generic delimitation of *Ramichloridium*, *Veronaea* and related fungi was discussed by DE HOOG et al. (1983 b, with key to recognized species). In the course of a revision of *Dactylaria* and related genera, a mycoparasite of *Erysiphe*, which liberates its blastoconidia in pairs, was described as *Dissoconium*, and *Cordana* with 8 species was revised (DE HOOG et al. 1983a). Considerable numbers of new taxa were found by KIRK (1982a) in Devon.

δ) *Mucedinaceous hyphomycetes*. Pleo-anamorphic genera pose problems of classification and nomenclature. An extreme example of the anatomic system of naming each state separately is given by BUFFIN and HENNEBERT (1984) with *Cylindrodendrum* and the very similar synanamorph named *Cylindrocarpon*. GAMS (1982) and CONSTANTINESCU and SAMSON (1982) would rather admit several forms of sporulation in one genus that has been defined accordingly. Another example of pleomorphism is given in *Engyodontium*, in which both polyblastic and phialidic (*Verticillium*) conidiogenesis may occur jointly (GAMS et al. 1984). *Thozetella* is a sporodochial genus with different morphology in vivo and in vitro and therefore problematic from the viewpoint of species differentiation (SUTTON and COLE 1983). A segregation of the *Monilia*-like anamorphs of *Neurospora* from *Monilia* (anamorphs of *Monilinia*), long overdue, was carried out by VON ARX (1982 a) as *Chrysonilia* for *Neurospora* anamorphs. Little-differentiated phialidic anamorphs often elude taxonomic distinctions on morphological grounds, a borderline case being *Phialemonium* W. GAMS and McGINNIS (1983). A taxonomic revision of the principally basidiomycetous genus *Sporotrichum* (only three species retained) by STALPERS (1984) created order among many names given to unrelated hyphomycetes. The stilbellaceous genus *Didymostilbe* was revised by CRANE and HEWINGS (1982). A survey on taxonomy and biology of *Verticillium* species by FILIPPOV et al. (1980) contains a very useful amount of documentation, though a critical revision is not yet available.

In *Aspergillus*, the *ochraceus* group was revised by CHRISTENSEN (1982). In *Penicillium* a great diversity of taxonomic views emerged following the book by PITT (1981, noted previously) : a conservative narrow species concept in a beautifully illustrated atlas by RAMÍREZ (1982), a much broader species concept based on teleomorph taxa by STOLK and SAMSON (1983). Concerning secondary metabolites see under "Chemotaxonomy". Fractions of a taxonomic revision are also coming from Japan by ABE et al. (1982, 1983a–c); these authors as well as MARTÍNEZ et al. (1982) illustrated conidial ornamentation with SEM. Considerable advance and convergence of opinions can be reported for *Fusarium*. Use of "carnation-leaf agar" (FISCHER et al. 1982) helps to obtain optimally differentiated conidia within a short time, it is therefore now widely used. Three equally well-illustrated handbooks appeared within a short time (GERLACH and NIRENBERG 1983, NELSON et al. 1983, and BURGESS and LIDDELL 1983), which help in species identification. The most refined species concept adopted at Berlin is also defended by GERLACH in a book (NELSON et al., 1981) dealing mainly with diseases and biology of *Fusarium*. Some controversy still exists in naming species of section *Liseola* (GAMS 1982), where KUHLMAN (1982) demonstrated sexual incompatibility between four taxa so far distinguished morphologically thus supporting their specific distinction. (Concerning serology see under "Chemotaxonomy".)

4. BASIDIOMYCOTA

a) General Aspects

A reevaluation of the basidiomycetes by KHAN and KIMBROUGH (1982) was based upon septal and basidial structures. According to pore cap structures, four classes of basidiomycetes were distinguished: (1) the Teliomycetes (Uredinales, Septobasidiales, Exobasidiales) with simple, single-pored, nonflared walls; (2) the Hemibasidiomycetes (Ustilaginales, Tilletiales, Tremellales) with septal walls slightly flared or with dolipores with parenthesomes; (3) Phragmobasidiomyetes (Auriculariales, Dacrymycetales, Brachybasidiales, Tulasnellales) often with simple, nonperforate pore caps; and (4) the Holobasidiomycetes, almost all of which have multiperforate complex pore caps.

In a book dealing with evolution, cytology, function and development of basidia and basidiocarps (WEELS and WELLS 1982), OBERWINKLER surveyed the importance of basidium morphology in the phylogeny of basidiomycetes; he preferred to retain the two classes, Hetero- and Homobasidiomycetes. McLAUGHLIN discussed ultrastructure and cytochemistry of basidium and basidiospore development, mainly of while THIELKE dealt with the meiotic divisions in the basidium. Differentiation in the basidiomycetes was reviewed by MANACHÈRE et al. (1983) mainly from a physiological point of view. Geometrical relationships of basidia and basidiospores were defined mathematically by WASSER and BERGER (1980, 1982). A flora of the European taxa of Aphyllophorales, Gasteromycetes and Heterobasidiomycetes (excluding rusts and smuts) was published by JÜLICH (1984).

Ultrastructure of the pore apparatus in basidiomycetes was also studied by other authors: In *Hypholoma fasciculare* (Strophariaceae, Agaricales) parenthesomes with an outer cap of fibrillar material were found, as opposed to a granular cap being found in other fungi. Fibrillar as well as granular outer caps thus occur within one family. Pore caps of *Tremella* and *Pseudohydnum* were studied by PATRIGNANI et al. (1983). The marine and terrestrial *Rhodosporidium capitatum* was found to have a primitive type of dolipore similar to that of *Tilletia*, *Filobasidium* and *Filobasidiella*; other species of *Rhodosporidium* have a simple pore (NAKAGIRI and TUBAKI, 1983 b).

Mycorrhiza and ecology . The capacity of basidiomycetes to form ectomycorrhiza was found to be correlated with the possession of spores with a close-pore hilum, no mycorrhiza being formed by Coprinaceae, Agaricaceae, Strophariaceae and some other families, all having an open-pore hilum (MILLER 1982). The ultrastructure of mycorrhizal associations was reviewed by SCANNERINI and BONFANTE-FASOLO (1982).

Two volumes on ecology and coenology of macrofungi (mainly Agaricales in grasslands and moist heathlands) were written by ARNOLDS (1981, 1982), part one dealing with synecology, part two with autecology and taxonomy.

b) Phragmobasidiomycotina (Heterobasidiomycetes)

α) *Miscellaneous taxa* . A flora of Polish jelly fungi was written by WOJEWODA (1981). JULICH (1983 b) published a key to the hymenomycetes parasitizing other fungi; this group also contains species lacking fruit-bodies of their own, growing inside the hymenium or between the hyphae of other fungi. The poroid genus *Aporpium*, revised by SETLIFF and RYVARDEN (1982), was supposed to exemplify such a parasitic origin. The puzzling genus *Syzygospora* and its relatives were placed in a new family independently by OBERWINKLER and BANDONI (1982a) as Carcinomycetaceae and by JULICH (1982 b) as Syzygosporaceae. Other new families distinguished include the Eocronartiaceae (JULICH 1982b) and Sebacinaceae (WELLS and OBERWINKLER 1982). A synopsis of the three poroid genera of Tremellaceae appeared from BANDONI et al. (1982). A taxonomic survey of the gasteroid, auricularioid Heterobasidiomycetes was written by OBERWINKLER and BANDONI (1982b). This small group of genera with nonforcibly discharged spores was distributed over three families (incl. Chionosphaeraçeae, fam. nov.) and placed in a new order, Atractiellales.

Interesting new genera include *Sterigmatosporidium*, the heterothallic teleomorph of the yeast *Sterigmatomyces*, producing a dikaryotic mycelium with clamps and sexual spores in ramified whorls (KRAEPELIN and SCHULZE 1982); *Cystofilobasidium* (Filobasidiaceae, OBERWINKLER et al. 1983), *Stilbotulasnella* (Tulasnellaceae, BANDONI and OBERWINKLER 1982), characterized by synnemata and blastoconidia, and *Trimorphomyces* (Tremellaceae), a parasite of *Arthrinium* with blastoconidia conjugating in pairs (OBERWINKLER and BANDONI 1983). JULICH (1982 a) studied two clavarioid genera with auriculariaceous basidia, *Paraphelaria* (South-east Asia) and *Aphelariopsis* (South America and South-east Asia), demonstrating the enormous plasticity of the Heterobasidomycetes.

The species of *Rhodosporidium* fall into two groups, one with 2–4 cells in the promycelium, lacking inositol assimilation and starch formation and with simple septa, the other with one-celled promycelium, capable of inositol assimilation, starch formation and possessing primitive dolipores (NAKAGIRI and TUBAKI 1983b). Species of the red yeast genera *Sporidiobolus*, *Rhodosporidium* and *Sporobolomyces* were analyzed for micromorphology, teliospore formation, ultrastructure of mitosis and biochemical characters by DE HOOG (1982) and coworkers; a considerable variation was noticed in colony characters and chemical composition, pointing to a heterogeneity of *Rhodosporidium*, *Rh. infirmo-miniatum* being close to the Filobasidiaceae. The ultrastructure and taxonomic position of *Graphiola phoenicis* , a biotrophic parasite of palms, was studied independently by OBERWINKLER et al. (1982) and COLE (1983). The fungus, formerly classified as myxomycete, discomycete, rust, pyrenomycete, smut and hyphomycete, belongs to the Heterobasidiomycetes and was placed in a separate order, Graphiolales. Chains of fertile cells produce two-celled spores which finally secede and form yeast cells.

In *Exobasidium* BLANZ and OBERWINKLER (1983) found an unusually wide range in G+C content, from 39.1 to 68.1 mol %, which allowed the arrangement of the taxa into ten different groups. Spore germination and conidium development in *Auricularia* and the Dacrymycetales were described by INGOLD (1982 and 1983, respectively).

β) *Uredinales* . The genera of rust fungi were surveyed according to present knowledge by CUMMINS and HIRATSUKA (1983). SAVILE and URBAN (1982) reviewed the evolution and ecology of the extremely complex rust species *Puccinia graminis*, the specific delimitation of which is not yet fully understood. Coevolution of this species with grasses and its alternate host, *Berberis* spp., is supposed to have started in relatively undisturbed Eurasian forest and "forest steppe" ecosystems, allowing regular transfer of rust biotypes between grasses and preventing a complete speciation of rust strains. Taxonomic implications of the coevolution of Uredinales with their plant hosts were shown by McCAIN and HENNEN (1982) in the small autoecious rust genus *Cumminsiella*: two species with hygroscopic pedicels were known to grow on *Berberis*, five species with nonhygroscopic pedicels on *Mahonia*; a new species was found in Brazil on *Berberis* with nonswelling pedicels, supporting the view of some authors that *Mahonia* is only a subgenus of *Berberis*.

Further studies deal with the Uredinales of the German Democratic Republic (BRAUN, 1982d); axenic monokaryotic cultures of *Coleosporium tussilaginis* (DEML et al., 1982a,b); pycnidial structure and pycnosporogenesis in *Gymnosporangium fuscum* (METZLER 1981); ultrastructure of aeciospore ornamentation in *Cronartium quercuum* (MIMS and ROBERSON 1983); a flora of the Uredinales on legumes and Compositae in North America (CUMMINS 1982); the surface structure of aeciospores of Uredinales (SATO and SATO 1982); the classification of *Melampsoridium* based on the position of the urediniospore germ pores (KANEKO and HIRATSUKA 1981); and a very detailed study of the ultrastructure of meiosis of *Puccinia malvacearum* (O'DONNELL and McLAUGHLIN, 1981).

γ) *Ustilaginales*. Four noteworthy papers deal with siderophore formation in anther smuts of Caryophyllaceae (DEML and OBERWINKLER 1982a), characterization of *Ustilago violacea* from *Saponaria officinalis* and the transfer of the smuts of Caryophyllaceae to a separate genus, *Microbotryum* (DEML and OBERWINKLER 1982b), the genus *Schroeteria* (VANKY, 1981), and the genus *Thecaphora* on Compositae in North America (DURÁN 1982).

c) Holobasidiomycotina

α) *Aphyllophorales* . FIASSON and DAVID (1983) analyzed micro- and macroscopic characters of European poroid Hymenochaetaceae taximetrically. Some of the current generic concepts based partly on differences in hyphal systems were found to be incorrect. Studies of several enzymatic activities showed especially the *Phellinus* complex to be heterogeneous (FIASSON and BERNILLON 1983); a subgeneric rearrangement was proposed. The tropical wood-rotting genus *Amauroderma* (Ganodermataceae), especially wide-spread in South America, was monographed by FURTADO (1981) and CORNER (1983), the latter also included observations on *Ganoderma*. In *Ganoderma* before the normal basidiospores "proterospores" are formed, which are morphologically different from normal basidiospores, have no complex spore walls and germinate readily (NUSS 1982); proterospores occur also in various other basidiomycetes as a kind of dimorphic meiospores, comparable with dimorphic mitospores of Deuteromycetes.

JULICH and STAR (1983a) studied the ultrastructure of basidiospores of *Beenakia*, a small hydnoid genus from South America, in which ornamentation is comparable with that found in *Gomphus* and *Ramaria*. The gomphoid complex characterized i.a. by warted, cyanophilous spores, therefore includes not only clavarioid, but also hydnoid taxa. The group is placed either in the Aphyllophorales or in a separate order, Gomphales. Another part of the large genus *Ramaria*, the subgenus *Echinoramaria*, was monographed by PETERSEN (1981) and type specimens of numerous species described by COKER were revised by PETERSEN (1982).

Further studies were concerned with *Epithele* and related genera (BOIDIN and LANQUETIN 1983a), *Botryohypochnus* and *Botryobasidium* from Gabon (BOIDIN and GILLES 1982), *Vuilleminia* (BOIDIN and LANQUETIN 1983b), affinities between *Heterobasidion* and *Bonderzewia* (GLUCHOFF-FIASSON et al. 1983), a revision of *Hydnochaete* (Hymenochaetaceae). by RYVARDEN (1982), a revision of the type specimens of polypores described by PATOUILLARD (RYVARDEN 1983), a world monograph of *Coniophora* (GINNS, 1982), a flora of the Hydnaceae and Steccherinaceae of Poland (DOMAŃSKI 1981), and a revision of ten species of the *Ganoderma lucidum* complex from Argentina (BAZZALO and WRIGHT 1982). *Skeletocutis* was monographed by DAVID (1982) with an expanded genus concept, to include 15 (6 new) species. The genus *Coriolus* was revised for Argentina including cultural details (RAJCHENBERG 1982). Pink Corticiaceae causing turf grass diseases were studied by STALPERS and LOERAKKER (1982), who segregated *Limonomyces* from *Laetisaria* because of multinucleate, clampless cells and lack of laccase.

β) *Agaricales* s.l. The fifth edition of MOSER's (1983) well-known key to the Agaricales s.l. and Boletales (English translation of the 4th edition by PHILLIPS 1983) appeared, and also the third part of the British Fungus Flora dealing with the Bolbitiaceae (WATLING 1982). An impressive treatise of the Agaricales s.l. of the Lesser Antilles was furnished by PEGLER (1983), containing also some agaricoid taxa of Aphyllophorales and Cantharellales. A symposium on arctic and alpine mycology presented papers on taxonomy, distribution, ecology and physiology of higher fungi, mainly Agaricales (LAURSEN and AMMIRATI 1980).

PEGLER and YOUNG (1981) illustrated the basidiospore structure of Russulaceae (SEM, TEM), distinguishing genera with ballisto- and statismospores. HILBER (1982) published a comprehensive and well-illustrated revision of the European taxa of *Pleurotus* , concentrating especially on the *P. eryngii* complex. Part I appeared of an extensive revision of the North American species of *Clitocybe* (BIGELOW 1982), for Europe the genus was revised by CLÉMENÇON (1984a). In order to germinate basidiospores of *Laccaria* , FRIES (1983) placed an actively growing mycelium of the same species of *Laccaria* next to the spores; growing hyphae of *L. laccata* showed the typical homing reaction toward the germinating spores, the resulting conjugation being either viable or lethal; this, together with mating tests, showed that the 14 isolates tested represent four intersterile groups. But altogether eleven species were distinguished for Europe (CLÉMENÇON 1984b). Further studies deal with *Cortinarius* subgen. *Cortinarius* in northern Europe (BRANDRUD 1983); the genus *Xerula* (DORFELT 1981a,b, 1982), the anatomy of the

Lentinus hymenophore (PEGLER and YOUNG 1983), a biological species concept in the *Collybia dryophila* group (VILGALYS and MILLER 1983), *Armillaria* species from Australia (KILE and WATLING 1983), nomenclature and species differentiation in *Armillaria* in Europe (WATLING et al. 1982, MARXMULLER 1982, ROMAGNESI and MARXMULLER 1983), and the occurrence of stable diploid strains in this genus (ANDERSON and ULLRICH 1982). The genus *Phlebopus* was revised on world scale by HEINEMANN and RAMMELOO (1982). Two authors studied the *Omphalina* complex: LANGE (1981) included the genera *Gerronema*, *Phaeotellus*, *Rickenella*, and *Haasiella* in one genus, *Omphalina*, but CLÉMENÇON (1983) divided the same group into five genera, including *Chrysomphalina* gen. nov. The distinction of three species and two varieties in *Flammulina* (BAS 1983) was based on spore statistics and cultural interfertility tests.

γ) "*Gasteromycetes*". Ultrastructure: In five genera of Phallales, BURK et al. (1982) found the spores to be completely smooth under SEM, confirming light-microscopic observations. JULICH (1982c) illustrated the spore ornamentation (SEM) of *Mycenastrum* (with pits) and *Pyrenogaster* (with warts); in *Queletia*, the spores have rather long, complex, conical spines (JULICH 1983b). In *Pyrenogaster* JULICH and STAR (1983b) found a distinct hyaline "endospore", which is comparable with the structure found in spores of *Geastrum*, but differs from the situation in *Bovista* and *Lycoperdon*, where a hyaline "endospore" is either absent or very thin. This study confirms the relationship between *Geastrum* and *Pyrenogaster*. In a revision of *Calostoma* for North America, spore ultrastructure and peridial wall details were taken into account (CASTRO-MENDOZA et al. 1983).

Further studies deal with *Lycoperdon* in southern Europe (DEMOULIN 1983), *Mesophellia* in Australia (BEATON and WESTE 1983), cytology of the basidiocarp of *Sphaerobolus* (DYKSTRA 1982), and the Galeropsidaceae found west of the Rocky Mountains (SINGER and PONCE DE LEON 1982).

Acknowledgements. We wish to acknowledge the assistance of Mr. K.A. Seifert in the correction of the text, and the supply of references by several colleagues of the Centraalbureau voor Schimmelcultures.

References

AAS, O.: Nord. J. Bot. 3, 253-259 (1983). – ABDULLAH, S. K., WEBSTER, J.: Trans. Br. Mycol. Soc. 80, 247-254 (1983). – ABE, S., IWAI, M., SUGIHARA, T.: Trans. Mycol. Soc. Jpn 23, 149-163 (1982). – ABE, S., IWAI, M., YOSHIOKA, H.: Trans. Mycol. Soc. Jpn 24, 39-51 (1983a). – ABE, S., IWAI, M., TANAKA, H.: Trans. Mycol. Soc. Jpn 24, 95-108 (1983b). – ABE, S., IWAI, M., AWANO, M.: Trans. Mycol. Soc. Jpn 24, 109-120 (1983c). – ALCORN, J. L.: Mycotaxon 15, 20-48 (1982); – ibid. 16, 353-379 (1983a); – ibid. 17, 1-86 (1983b). – ALSHEIKH, A. M., TRAPPE, J. M.: Trans. Br. Mycol. Soc. 81, 83-90 (1983). – ANDERSON, J. B., ULLRICH, R. C.: Can. J. Bot. 60, 432-439 (1982). – ANSELL, P. J., YOUNG, T. W. K.: Nova Hedwigia 36, 309-339. '1982' (1983a); – Mycologia 75, 64-69 (1983b). – ARMBRUSTER, B. L.: Mycologia 74, 433-459 (1982a); – ibid. 74, 975-999 (1982b). – ARMBRUSTER, B. L., JOHNSON, T. W. jr.: Can. J. Bot. 61, 545-550 (1983). – ARNOLDS, E.: Ecology and Coenology of Macrofungi in Grasslands and Moist Heathlands in Drenthe, the Netherlands. Bibliotheca. Mycol. Vol. 83, 407 pp. Lehre: Cramer. 1981; – Bibliotheca. Mycol. Vol. 90, 501 pp., 8 col. plates. Lehre: Cramer. 1982. – ARX, J. A. von: Sydowia 34, 13-29. '1981' (1982a); – ibid. 35, 10-26 (1982b); – Proc. K. Ned. Akad. Wet., Ser. 3, 85, 21-28 (1982c); – ibid. 86, 15-54 (1983). – ARX, J. A. von, VAN DER WALT, J.

P., LIEBENBERG, N. V. D. M.: Mycologia 74, 285-296 (1982). – ARX, J. A. von, DREYFUSS, M., MÜLLER, E.: Persoonia 12, 169-179 (1984).

BALAZUC, J.: Bull. Mens. Soc. Linn. Lyon 51, 6-27 (1982). – BANDONI, R. J., OBERWINKLER, F.: Can. J. Bot. 60, 1875-1879 (1982). – BANDONI, R. J., OBERWINKLER, F., WELLS, K.: Can. J. Bot. 60, 998-1003 (1982). – BARNETT, J. A., PAYNE, R. W., YARROW, D.: Yeasts, Characteristics and – Identification. 811 pp. Cambridge: University Press. 1983. – BARR, D. J. M.: Bio Systems 14, 359-370 (1981); – 43-83 in: see ref. BUCZACKI 1983. – BARR, D. J. M., ALLAN, P. M. E.: Can. J. Bot. 60, 2496-2504 (1982). – BARR, M.: Mycotaxon 15, 349-383 (1982); – Mycologia 75, 1-13 (1983). – BARRON, G. L., SZIJARTO, E.: Can. J. Bot. 60, 1031-1034 (1982). – BAS, C.: Persoonia 12, 51-66 (1983). – BATRA, L. R.: Mycologia 75, 131-152 (1983). – BAWCUTT, R. A.: Trans. Br. Mycol. Soc. 80, 219-230 (1983). – BAZZALO, M. E., WRIGHT, J. E.: Mycotaxon 16, 293-325 (1982). – BEATON, G., WESTE, G.: Trans Br. Mycol. Soc. 80, 209-218 (1983). – BEILBY, J. B.: Lipids 15, 949-952 (1980). – BEILBY, J. B., KIDBY, D. K.: Lipids 15, 375-378 (1980). – BELIN, J. M.: Can. J. Microbiol. 27, 1235-1251 (1981). – BELLEMÈRE, A., HAFELLNER, J.: Crypt. Mycol. 3, 71-93 (1982a); – ibid. 3, 261-295 (1982b). – BEN-ZE'EV, I., KENNETH, R. G.: Mycotaxon 16, 393-455 (1982a); – ibid. 16, 456-475 (1982b). – BERCH, S. M., FORTIN, J. A.: Can. J. Bot. 61, 899-905 (1983). – BERNIER, L., JENG, R. S., HUBBES, M.: Mycotaxon 17, 456-472 (1983). – BERTA, G., FUSCONI, A.: Trans. Br. Mycol. Soc. 80, 201-207 (1983). – BERTHET, P.: Bull. Trimest. Soc. Mycol. Fr. 98, 75-79 (1982). – BIGELOW, H. E.: Nova Hedwigia, Beih. 72, 280 pp. (1982). – BLANZ, P., OBERWINKLER, F.: System. Appl. Microbiol. 4, 199-206 (1983). – BOIDIN, J., GILLES, G.: Mycotaxon 14, 280-304 (1982). – BOIDIN, J., LANQUETIN, P.: Mycotaxon 16, 461-499 (1983a); – Bull. Trimest. Soc. Mycol. Fr. 99, 269-279 (1983b). – BONNER, J. T.: Am. Naturalist 119, 530-552 (1982). – BOOTH, T.: Can. J. Bot. 60, 1115-1125 (1982). – BRAIN, A. P. R., JEFFRIES, P., YOUNG, T. W. K.: Mycologia 74, 173-181 (1982). – BRANDRUD, T. E.: Nord. J. Bot. 3, 577-592 (1983). – BRASELTON, J. P.: Can. J. Bot. 60, 403-408 (1982). – BRÅTHEN, I.: Nord. J. Bot. 2, 575-582 (1982). – BRAUN, U.: Mycotaxon 14, 369-374 (1982a); – ibid. 15, 121-137 (1982b); – Zentralbl. Mikrobiol. 137, 138-152 (1982c); – Feddes Repert. 93, 213-333 (1982d); – Mycotaxon 19, 369-374 (1984). – BUCZACKI, S. T. (ed.): Zoosporic Plant Pathogens. London: Academic Press. 1983. – BUFFIN, N., HENNEBERT, G. L.: Mycotaxon 19, 323-341 (1984). – BURGESS, L. W., LIDDELL, C. M.: Laboratory Manual for Fusarium Research. 162 pp. Univ. Sydney. 1983. – BURK, W. R., FLEGLER, S. L., HESS, W. M.: Mycologia 74, 166-168 (1982). – BURNETT, J. H.: Trans. Br. Mycol. Soc. 81, 1-14 (1983). – BUYCK, B.: Bull. Jard. Bot. Natl. Belg. 52, 165-210 (1982).

CANNON, P. F., HAWKSWORTH, D. L.: Bot. J. Linn. Soc. 84, 115-160 (1982). – CARROLL, G., PETRINI, O.: Mycologia 75, 53-63 (1983). – CARTER, A., MALLOCH, D.: Can. J. Bot. 60, 1249-1252 (1982). – CASTRO-MENDOZA, E., MILLER, O. K., jr., STETLER, D. A.: Mycologia 75, 36-45 (1983). – CHADEFAUD, M.: Crypt. Mycol. 3, 1-9 (1982a); – ibid. 3, 103-144 (1982b); – ibid. 3, 199-235 (1982c). – CHASE, T.E., ULLRICH, R.C.: Mycologia 75, 825-831 (1983). – CHRISTENSEN, M.: Mycologia 74, 210-225 (1982). – CHRISTENSEN, M., STATES, J. S.: Mycologia 74, 226-235 (1982). – CIEGLER, A., LEE, L. S., DUNN, J. J.: Appl. Environ. Microbiol. 42, 446-449 (1981). – CLÉMENÇON, H.: Z. Mykol. 48, 195-237 (1983); – Beih. Z. Mykol. 5(1) (1984a); – Z. Mykol. 50: 3-12 (1984b). – CLINE, M. N., CRANE, J. L., CLINE, S. D.: Mycologia 75, 988-994 (1983). – COLE, G. T.: Microbiology 1981, 227-231. Am. Soc. Microbiol. (1981); – Mycologia 75, 93-116 (1983). – CONSTANTINESCU, O., NEGREAN, G.: Mycotaxon 16, 537-556 (1983). – CONSTANTINESCU, O., SAMSON, R. A.: Mycotaxon 15, 472-486 (1982). – COOKE, W. B.: Revista Biol. 12, 279-296 (1983). – CORNER,

E. J. H.: Nova Hedwigia Beih. 75, 182 pp. (1983). – CRANE, J. L., HEWINGS, A. D.: Mycotaxon 16, 133-140 (1982). – CRANE, J. L., SCHOKNECHT, J. D.: Can. J. Bot. 60, 369-378 (1982). – CRIVELLI, P. G.: Über die heterogene Ascomycetengattung Pleospora; Vorschlag für eine Aufteilung. Diss. ETH, Zürich, 213 pp. (1983). – CRUICKSHANK, R. H.: Trans. Br. Mycol. Soc. 80, 117-119 (1983a); – Abstr. 4th Int. Congr. Pl. Path., Melbourne, 708 (1983b). – CUMMINS, G. B.: Rust Fungi on Legumes and Composites in North America. Tucson: Univ. Arizona Press. 1982. – CUMMINS, G. B., HIRATSUKA, Y.: Illustrated Genera of Rust Fungi. Rev. ed. 152 pp. St. Paul, Minnesota: Am. Phytopath. Soc. 1983.

DANIELSON, R. M.: Can. J. Bot. 60, 7-18 (1982). – DANTANARAYANA, D. M., PERIES, O. S., LIYANAGE, A. de S.: Trans. Br. Mycol. Soc. 82, 113-126 (1984). – DAVID, A.: Natur. Can. 109, 235-272 (1982). – DEASEY, M. C.: Mycologia 74, 607-613 (1982). – DEIGHTON, F. C.: Mycol. Pap. 151 (1983). – DEML, G., BAUER, R., OBERWINKLER, F.: Phytopathol. Z. 103, 149-155 (1982a); – ibid. 104, 39-45 (1982b). – DEML, G., OBERWINKLER, F.: Zentralbl. Bakt. Hyg., Abt. 1, Orig. C 3, 475-477 (1982a); – Phytopathol. Z. 104, 345-356 (1982b). – DEMOULIN, V.: Revista Biol. 12, 65-70 (1983). – DESCALS, E., WEBSTER, J.: Trans. Br. Mycol. Soc. 78, 405-437 (1982a); – ibid. 79, 45-64 (1982b). – DICOSMO, F., NAG RAJ, T. R., KENDRICK, B.: Can. J. Bot. 61, 31-44 (1983a). – DICOSMO, F., BERCH, S., KENDRICK, B.: Mycologia 75, 949-973 (1983b). – DISSING, H., PFISTER, D. H.: Nord. J. Bot. 1, 102-108 (1981). – DÖBBELER, P.: Mitt. Bot. Staatssamml. München 17, 393-474 (1981); – Nova Hedwigia 37, 1-7 (1983). – DONADINI, J.-C.: Le genre Peziza dans le Sud-Est de la France. Marseille: Lab. Chimie gén. 119 pp. + 11 plates 1981; – Docum. Mycol. 13(49), 9-37 (1983). – DOMAŃSKI, S.: Flora of Poland. Fungi. Vol. 13, 94 pp. Warszawa: Państwowe Wydawnictwo Naukowe. 1981. – DÖRFELT, H.: Feddes Repert. 92, 255-291 (1981a); – ibid. 92, 631-674 (1981b); – Flora, Jena 172, 533-561 (1982). – DURÁN, R.: Can. J. Bot. 60, 1512-1522 (1982). – DYKSTRA, M. J.: Mycologia 74, 44-53 (1982). – DYLEWSKI, D. P., MILLER, C. E.: Mycologia 75, 412-422 (1983).

EIJK, G. W. van, ROEIJMANS, H. J.: Antonie van Leeuwenhoek 48, 257-264 (1982). – ELIASSON, U., GILERT, E.: Nord. J. Bot. 2, 249-255 (1982). – ENDERLE, M.: Z. Mykol. 48, 141-164 (1982). – ERIKSSON, O.: Mycotaxon 15, 203-248 (1982). – ERWIN, D. S., BARTNICKI-GARCIA, S., TSAO, P. H. (eds.): Phytophthora, its Biology, Taxonomy, Ecology, and Pathology. St. Paul, Minnesota: Am. Phytopath. Soc. 1983. – EVANS, H. C., SAMSON, R. A.: Can. J. Bot. 60, 2325-2333 (1982); – Trans. Br. mycol. Soc. 79, 431-453 (1983); – ibid. 82, 127-150 (1984).

FARR, M. L.: Mycologia 75, 1036-1043 (1983). – FEDERICI, B. A.: Mycologia 74, 555-562 (1982). – FERREIRA, J. D.: Antonie van Leeuwenhoek J. Microbiol. 48, 1-24 (1982). – FIASSON, J.-L., BERNILLON, J.: Bull. Trimest. Soc. mycol. Fr. 99, 93-107 (1983). – FIASSON, J.-L., DAVID, A.: Can. J. Bot. 61, 442-451 (1983). – FILIPPOV, V. V., ANDREEV, L. N., BAZILINSKAYA, N. V.: Fitopatogennye Griby Roda Verticillium. Moskva: Iz'vo Nauka, 1980. – FIOL, J. B., CLAISSE, M.: Mycopathologia 78, 177-184 (1982). – FIOL, J. B., PONCET, S.: Mycopathologia 70, 13-23 (1980). – FISCHER, N. L., BURGESS, L. W., TOUSSOUN, T. A., NELSON, P. E.: Phytopathology 72, 151-153 (1982). – FRASER, R. S. S., BUCZACKI, S. T.: Trans. Br. Mycol. Soc. 80, 107-112 (1983). – FREDERICK, L., ROTH, I. L., SIMONS, R.: Mycologia 75, 46-52 (1983). – FRIES, N.: Mycologia 75, 221-227 (1983). – FRISVAD, J. C., FILTENBORG, O.: Appl. Environ. Microbiol. 46, 1301-1310 (1983). – FUNK, A.: Parasitic Microfungi of Western Trees. 190 pp. Victoria B.C.: Can. Forestry Serv. 1981. – FURTADO, J. S.: Mem. N. Y. Bot. Gard. 34, 109 pp. (1981).

GAMS, W.: Mycotaxon 15, 459-464 (1982); - ibid. 19, 219-270 (1984). - GAMS, W., McGINNIS, M. R.: Mycologia 75, 977-987 (1983). - GAMS, W., KUYPER, T. W.: Mycotaxon 20, 619-631 (1984). - GAMS, W., SAMSON, R. A., de HOOG, G. S.: Persoonia 12: 135-147 (1984). - GARRISON, R. G., ARNOLD, W. N. in: Yeast Cell Envolopes: Biochemistry, Biophysics and Ultrastructure. ed. W. N. ARNOLD, Vol. 1, 5-24. Boca Raton: C.R.C. Press, 1981. - GERLACH, W., NIRENBERG, H.: Mitt. Biol. Bundesanst. Land- Forstwirtsch. Berl.Dahlem 209, 406 pp. (1983). - GERRETTSON-CORNELL, L., SIMPSON, J.: Mycotaxon 19, 453-470 (1984). - GINNS, J.: Opera Bot. 61, 61 pp. (1982). - GLAWE, D. A., ROGERS, J. D.: Mycotaxon 14, 334-346 (1982a); - Can. J. Bot. 60, 245-251 (1982b). - GLUCHOFF-FIASSON, K., DAVID, A., DEQUATRE, B.: Cryptog. Mycol. 4, 135-143 (1983). - GOURBIÈRE, F.: Bull. Trimest. Soc. Mycol. Fr. 99, 203-215 (1983). - GUZMAN, D., GARBER, R. C., YODER, O. C.: Can. J. Bot. 60, 1138-1141 (1982). - GVRITISHVILI, M. N.: (Fungi of the Genus Cytospora in the USSR).Tbilisi: Iz'vo Sabchota Sakartvelo, 214 pp. 1982.

HADAR, T.: C. R. Hebd. Séanc. Acad. Sci. Paris, Sér. 3, 294, 329-334 (1982). - HAINES, J. H., DUMONT, K. P.: Mycotaxon 19, 1-39 (1984). - HASKINS, E. F., McGUINNESS, M. D., BERRY, C. S.: Mycologia 75, 153-158 (1983). - HAWKSWORTH, D. L.: Lichenologist 15, 1-44 (1983). - HAWKSWORTH, D. L., LODHA, B. C.: Trans. Br. Mycol. Soc. 81, 91-96 (1983). - HAWKSWORTH, D. L., PITT, J. I.: Aust. J. Bot. 31, 51-61 (1983). - HAWKSWORTH, D. L., SUTTON, B. C., AINSWORTH, G. C.: Ainsworth & Bisby's Dictionary of the Fungi, 7th ed. Kew: Commonw. Mycol. Inst. 1983. - HEATH, I. B., ASHTON, M.-L., RETHORET, K., HEATH, M. C.: Can. J. Bot. 60, 1696-1725 (1982). - HEATH, I. B., BAUCHOP, T., SKIPP, R. A.: Can. J. Bot 61, 295-307 (1983). - HEIN, B.: Willdenowia 12, 293-302 (1982); - Nova Hedwigia 38, 669-702 (1983). - HEINEMANN, P., RAMMELOO, J.: Mycotaxon 15, 384-404 (1982). - HELD, A. A.: Bot. Rev. 47, 451-515 (1981). - HIGHAM, M. T., COLE, K. M.: Can. J. Bot. 60, 2313-2324 (1982). - HILBER, O.: Die Gattung Pleurotus. Bibliotheca Mycologica Vol. 87, 448 pp., 25 plates. Lehre: J. Cramer 1982. - HILBER, O., HILBER, R., ENDERLE, M.: Mitt. Ver. Naturw. Math. Ulm 32, 77-199 (1983). - HINCHEE, A. A., HASKINS, E. F.: Protoplasma 18, 361-367 (1981). - HO, H. H.: Mycopathologia 83, 119-123 (1983). - HOLUBOVÁ-JECHOVÁ, V.: Folia Geobot. Phytotaxon., Praha, 17, 295-327 (1982). - HOOG, G. S. de (ed.): Stud. Mycol. 22, 74 pp. (1982). - HOOG, G. S. de, SMITH, M. Th., in: Yeasts in human environment, 9th Int. Symp. on Yeasts, Smolenice, 22-23 (1983). - HOOG, G. S. de, OORSCHOT, C. A. N. van, HIJWEGEN, T.: Proc. K. Ned. Akad. Wet., Ser. C, 86, 197-206 (1983a). - HOOG, G. S. de, RAHMAN, M. A., BOEKHOUT, T.: Trans. Br. Mycol. Soc. 81, 485-490 (1983b). - HOWARD, D. H. (ed.): Fungi pathogenic for humans and animals. Part. A. 652 pp. New York, Basel: Dekker. 1983. - HUANG, L. H., LUTTRELL, E. S.: Am. J. Bot. 69, 421-431 (1982). - HUGHES, S. J.: Can. J. Bot. 61, 1727-1767 (1983). - HULDÉN, L.: Karstenia 23, 31-136 (1983). - HUMBER, R. A.: Mycotaxon 15, 167-184 (1982); - 98-108 in: Fungus-insect relationships. eds. Q. WHEELER, M. BLACKWELL. New York: Columbia Univ. Press 1983.

IANNELLI, D., CAPPARELLI, R., CRISTINZIO, G., MARZIANO, F., SCALA, F., NOVIELLO, C.: Mycologia 74, 313-319 (1982). - IANNELLI, D., CAPPARELLI, R., MARZIANO, F., SCALA, F., NOVIELLO, C.: Mycotaxon 17, 523-532 (1983). - INGOLD, C. T.: Trans. Br. Mycol. Soc. 78, 161-190 (1982); - ibid. 81, 563-571 (1983).

JACOBS, R. P. W. M.: Antonie van Leeuwenhoek J. Microbiol. 48, 389-399 (1982). - JAHRMANN, H. J., PRILLINGER, H.: Z. Mykol. 49, 195-235 (1983). - JEFFRIES, P., YOUNG, T. W. K.: Mycologia 75, 509-517 (1983). - JENSEN, J. D.: Mycologia 74, 724-737 (1982); - ibid. 75, 1074-1091 (1983). - JONES, E. B. G., JOHNSON, R. G., MOSS, S. T.: Bot. J. Linn.

Soc. 87, 193-212 (1983a). – JONES, E. B. G., MOSS, S. T., CUOMO, V.: Trans. Br. Mycol. Soc. 80, 192-200 (1983b). – JONES, J. P.: Mycologia 73, 923-930 (1981). – JONES, M. G., NOBLE, W. C.: J. Gen. Microbiol. 128, 1101-1108 (1983). – JÜLICH, W.: Persoonia 11, 397-406 (1982a); – ibid. 11, 421-428 (1982b); – Int. J. Mycol. Lichenol. 1, 61-74 (1982c); – ibid. 1, 169-174 (1983a); – ibid. 1, 189-203 (1983b); – Die Nichtblätterpilze, Gallertpilze und Bauchpilze (Aphyllophorales, Heterobasidiomycetes, Gastromycetes). Kleine Kryptogamenflora II b/1, 626 pp. Stuttgart: Fischer 1984. – JÜLICH, W., STAR, W.: Persoonia 12, 67-74 (1983a); – Int. J. Mycol. Lichenol. 1, 175-182 (1983b).

KAMALETDINOVA, F. I., VASSILYEV, A. E.: Tsitologiya Diskomitsetov. Alma-Ata: Iz'vo Nauka, Kazakh. SSR. 1982. – KANEKO, S., HIRATSUKA, N.: Trans. Mycol. Soc. Jpn 22, 463-473 (1981). – KAUSHAL, R.: Mycotaxon 16, 117-122 (1982). – KERS, L. E.: Nord. J. Bot. 1, 795-800 (1981). – KHAN, S. R., KIMBROUGH, J. W.: Mycotaxon 15, 103-120 (1982). – KILE, G. A., WATLING, R.: Trans. Br. Mycol. Soc. 81, 129-140 (1983). – KIMBROUGH, J. W., LENZ, M.: Bot. Gaz. 143, 262-272 (1982). – KIMBROUGH, J. W., THORNE, B. L.: Mycologia 74, 201-209 (1982). – KIRK, P. M.: Trans. Br. Mycol. Soc. 78, 55-74 (1982a); – ibid. 78, 293-303 (1982b). – KITZ, D. J., EMBREE, R. W., CAZIN, J., jr.: Mycologia 74, 110-116 (1982). – KOBAYASI, Y., SHIMIZU, D.: Bull. natn. Sci. Mus., Tokyo, Ser. B, 8, 43-78 (1982); – Iconography of Vegetable Wasps and Plant Worms. Osaka: Hoikusha 1984. – KOCKOVÁ-KRATOCHVÍLOVÁ, A., MARKOVIČ, O., SLÁVIKOVÁ, E.: Folia Microbiol. 26, 221-227 (1981). – KOHN, L. M.: Mycotaxon 16, 1-34 (1982). – KOMIYA, H., TAKEMURA, S.: J. Biochem. Tokyo 90, 1577-1582 (1981). – KORF, R. P.: Mycotaxon 14, 476-490 (1982a); – Mycologia 74, 250-255 (1982b); – Mycotaxon 16, 341-352 (1983). – KOZLOWSKI, M., STEPIEN, P. P.: J. Gen. Microbiol. 128, 471-476 (1982). KRAEPELIN, G., SCHULZE, U.: Antonie van Leeuwenhoek J. Microbiol. 48, 471-483 (1982). – KRUG, J. C., UDAGAWA, S., JENG, R. S.: Mycotaxon 17, 533-549 (1983). – KUHLMAN, E. G.: Mycologia 74, 759-768 (1982). – KULLMAN, B. B.: A Revision of the genus Scutellinia (Pezizales) in Soviet Union. 158 pp. Tallinn: Inst. Zool. Bot. 1982. – KURTZMAN, C. P.: Mycotaxon 19, 271-279 (1984). – KURTZMAN, C. P., PHAFF, H. J., MEYER, S. A.: 139-166 in: Fundamental and applied aspects of yeast genetics. eds. J. F. T. SPENCER, D. M. S. SPENCER, A. R. W. SMITH. Heidelberg, New York: Springer 1983.

LAESSØE, T.: Svampe (6), 73-83 (1982). – LANGE, L., OLSON, L. W.: 1-42 in: see ref. BUCZACKI (1983). – LANGE, M.: Nord. J. Bot. 1, 691-696 (1981). – LATGÉ, J. P., PRÉVOST, M. C., PERRY, D. F., REISINGER, O.: Can. J. Bot. 60, 413-431 (1982). – LAURSEN, G. A., AMMIRATI, J. F. (eds.): Arctic and Alpine Mycology. 559 pp. – Seattle, London: Univ. Washington Press. 1980. – LICHTWARDT, R. W.: Mycotaxon 19, 529-550 (1984). – LODDER, J. (ed.): The Yeasts, a Taxonomic Study. 1385 pp. Amsterdam: North Holland. – LUNN, J. A., SHIPTON, W. A.: Trans. Br. Mycol. Soc. 81, 303-312 (1983).

MANACHÈRE, G., ROBERT, J.-C., DURAND, R., BRET, J.-P., FÈVRE, M.: 481-514 in: Fungal Differentiation, ed. J. E. Smith. New York, Basel: Dekker 1983. – MARTIN, G. W., ALEXOPOULOS, C. J., FARR, M. L.: The Genera of Myxomycetes. 561 pp. Iowa City: Univ. Iowa Press 1983. – MARTÍNEZ, A. T., CALVO, M. A., RAMÍREZ, C.: Antonie van Leeuwenhoek J. Microbiol. 48, 245-255 (1982). – MARXMÜLLER, H.: Bull. Trimest. Soc. Mycol. Fr. 98, 87-142 (1982). – McCAIN, J. W., HENNEN, J. F.: Syst. Bot. 7, 48-59 (1982). – McGINNIS, M. R., PADHYE, A. A., AJELLO, L.: Mycotaxon 14, 94-102 (1982). – MEINHARDT, F., ESSER, K.: 537-557 in: Fungal Differentiation. ed. J. E. Smith. New York, Basel: Dekker 1983. – METZLER, B.: Z. Mykol. 47, 271-280 (1981). – MICHAELIDES, J., KENDRICK, B.: Mycotaxon 14, 247-260 (1982). – MILLER, O. K.: Can. J.

Bot. 61, 909-916 (1982). – MIL'KO, A. A.: Mikol. Fitopatol. 17, 289-296 (1983). – MIMS, C. W., ROBERSON, R. W.: Mycologia 75, 400-411 (1983). – MINTER, D. W.: Trans. Br. Mycol. Soc. 80, 162-163 (1983). – MINTER, D. W., KIRK, P. M., SUTTON, B. C.: Trans. Br. Mycol. Soc. 79, 75-93 (1982); – ibid. 80, 39-66 (1983a). – MINTER, D. W., SUTTON, B. C., BRADY, B. L.: Trans. Br. Mycol. Soc. 81, 97-107 (1983b). – MINTER, D. W., BRADY, B. L.,, HALL, R. A.: Trans. Br. Mycol. Soc. 81, 455-471 (1983c). – MONOD, M.: Beih. Sydowia 9, 315 pp. (1983). – MONOD, M., ZIEGLER, P.: Mycol. Helv. 1, 101-124 (1983). – MONTROCHER, R.: Cell. Molec. Biol. 28, 1-14 (1982). – MORGAN-JONES, G., GAMS, W.: Mycotaxon 15, 311-318 (1982). – MOSER, M.: Die Röhrlinge und Blätterpilze (Polyporales, Boletales, Agaricales, Russulales). Kleine Kryptogamenflora II b/2, 533 pp. Stuttgart: Fischer 1983. – MOSS, C. W., SHINODA, T., SAMUELS, J. W.: J. Clin. Microbiol. 16, 1073-1079 (1982). – MOXHAM, S. E., BUCZACKI, S. T.: Trans. Br. mycol. Soc. 80, 297-304 (1983). – MÜLLER, E., LOEFFLER, W.: Mykologie, 4. Aufl. 366 pp. Stuttgart: Thieme 1982. – MÜLLER, E., SAMUELS, G. J.: Sydowia 35, 150-154 (1982a); – ibid. 35, 155-161 (1982b).

NAG RAJ, T. R.: Can. J. Bot. 61, 1-30 (1983); – Mycotaxon 19, 167-212 (1984). – NAKAGIRI, A., TUBAKI, K.: Mycologia 75, 487-497 (1983a); – Can. J. Bot. 61, 1898-1905 (1983b). – NANNENGA-BREMEKAMP, N. E.: Proc. K. Ned. Akad. Wet., C, 85, 541-562 (1982). – NANNENGA-BREMEKAMP, N. E., YAMAMOTO, Y.: Proc. K. Ned. Akad. Wet., C, 86, 207-241 (1983). – NANNFELDT, J. A.: Sydowia 35, 162-175 (1982); – Mycologia 75, 292-310 (1983). – NELSON, P. E., TOUSSOUN, T. A., COOK, R. J. (eds): Fusarium, diseases, biology and taxonomy. 457 pp. University Park, PA: Pennsylvania St. Univ. Press, 457 pp. 1981. – NELSON, P. E., TOUSSOUN, T. A., MARASAS, W. F. O.: Fusarium Species, an Illustrated Manual for Identification. 193 pp. University Park, PA: Pennsylvania St. Univ. Press, 1983. – NELSON, R. K., SCHEETZ, R. W., ALEXOPOULOS, C. J.: Mycologia 74, 541-548 (1982). – NUSS, I.: Plant Syst. Evol. 141, 53-79 (1982).

OLIVE, L. S., STOIANOVITCH, C., BENNETT, W. E.: Mycologia 75, 1019-1029 (1983). – OBERWINKLER, F., BANDONI, R. J.: Nord. J. Bot. 2, 501-516 (1982a); – Can. J. Bot. 60, 1726-1750 (1982b); – Syst. appl. Microbiol. 4, 105-113 (1983). – OBERWINKLER, F., BANDONI, R. J., BLANZ, P., DEML, G., KISIMOVA-HOROVITZ, L.: Plant Syst. Evol. 140, 251-277 (1982). – OBERWINKLER, F., BANDONI, R. J., BLANZ, P., KISIMOVA-HOROVITZ, L.: Syst. Appl. Microbiol. 4, 114-122 (1983). – O'DONNELL, K. L., McLAUGHLIN, D. J.: Protoplasma 108, 225-244, 245-263, 265-288 (1981). – OVERTON, S. V., THARP, T. P., BLAND, C. E.: Can. J. Bot. 61, 1165-1177 (1983).

PARGUEY-LEDUC, A., JANEX-FAVRE, M. C.: Can. J. Bot. 60, 1222-1230 (1982). – PARGUEY-LEDUC, A., JANEX-FAVRE, M. C., ANDRIEU, S., LACOSTE, L., TRAORE, F.: Ann. Parasitol. 57, 179-195 (1982). – PATRIGNANI, G., PELLEGRINI, S., GEROLA, F. M.: Caryologia 36, 19-26 (1983). – PECHAK, D. G., MILLER, C. E.: Micron 12, 215-216 (1981). – PEGLER, D. N.: Agaric Flora of the Lesser Antilles. Kew Bull., Add. Ser. IX, 668 pp., 27 col. plates. 1983. – PEGLER, D. N., YOUNG, T. W. K.: World Pollen Spore Flora 10, 35 pp. (1981); – Trans. Br. Mycol. Soc. 80, 469-482 (1983). – PERRY, D. F., TYRRELL, D., DeLYZER, A. J.: Mycologia 74, 549-554 (1982). – PETERSEN, R. H.: Ramaria Subgen. Echinoramaria. Bibliotheca Mycol. Vol. 70, 261 pp., 4 col. plates. Lehre, Cramer. 1981; – Sydowia 35, 176-205 (1982); – Mycotaxon 17, 87-147 (1983a); – Cryptog. Mycol. 4, 221-230 (1983b). – PETRINI, O.: Sydowia 35, 206-222 (1982). – PETRINI, O., DREYFUSS, M.: Sydowia 34, 135-148 (1982). – PETRINI, O., STONE, J., CARROLL, F. E.: Can. J. Bot. 60, 789-796 (1982). – PFISTER, D. H.: Occas. Pap. Farlow Herb. 17, 32 pp. (1982). – PHILLIPS, R.: Keys to Agarics and Boleti (transl. from MOSER, 1978), London (see Bull. Br.

Mycol. Soc. 17, 64, 1983). – POLONELLI, L., ARCHIBUSACCI, C., SESTITO, M., MORACE, G.: J. Clin. Microbiol. 17, 774-780 (1983). – POWELL, M. J.: Am. J. Bot. 68, 1074-1089 (1981). – PRILLINGER, H.: Z. Mykol. 48, 275-296 (1982a); – Z. Mykol. 48, 297-324 (1982b). – PRILLINGER, H., SIX, W.: Plant Syst. Evol. 141, 341-371 (1983). – PUNITHALINGAM, E.: Mycol. Pap. 149, 60 pp. (1981); – Nova Hedwigia 36, 151-198 (1982).

RAGHU KUMAR, S.: Can. J. Bot. 60, 1092-1102 (1982a); – ibid. 60, 1103-1114 (1982b). – RAJCHENBERG, M.: Bol. Soc. Argent. Bot. 21, 17-57 (1982). – RAMÍREZ, C.: Manual and Atlas of the Penicillia. 874 pp. Amsterdam: Elsevier. 1982. – RANALLI, M. E., FORCHIASSIN, F.: Physis B. Aires, Sec. C, 38, 43-55 (1981). – RANALLI, M. E., FORCHIASSIN, F., MERCURI, O. A.: Nova Hedwigia 35, 417-432 (1982). – RAUSCHERT, S.: Boletus 7, 21-38 (1983). – READ, N. D., PORTER, R., BECKET, A.: Can. J. Bot. 61, 2059-2078 (1983). – REYNOLDS, D. R.: Mycotaxon 14, 189-220 (1982). – ROMAGNESI, H., MARXMÜLLER, H.: Bull. Trimest. Soc. Mycol. Fr. 99, 301-321 (1983). – ROSE, J. B., CHRISTENSEN, M., WILSON, W. T.: Mycotaxon 19, 41-55 (1984). – ROSSI, W., CESARE ROSSI, M. G.: Can. J. Bot. 60, 306-309 (1982). – ROSSMAN, A. Y.: Mycol. Pap. 150, 164 pp. (1983). – RYKARD, D. M., LUTTRELL, E. S., BACON, C. W.: Mycologia 74, 648-654 (1982). – RYVARDEN, L.: Mycotaxon 15, 425-447 (1982); – Occas. Papers Farlow Herb. 18, 39 pp. (1983).

SAMSON, R. A., EVANS, H. C.: Can. J. Bot. 60, 1577-1580 (1982a); – J. Invertebr. Pathol. 39, 93-97 (1982b). – SAMSON, R. A., EVANS, H. C., HOEKSTRA, E. S.: Proc. K. Ned. Akad. Wet., Ser. C 85, 589-606 (1982). – SAMUELS, G. J.: N. Z. J. Bot. 21, 171-176 (1983). – SAMUELS, G. J., DUMONT, K. P.: Caldasia 13, 379-423 (1982). – SATO, T., SATO, S.: Trans. Mycol. Soc. Jpn 23, 51-63 (1982). – SAUTTER, C., HOCK, B.: Mycologia 74, 256-264 (1982). – SAVILE, D. B. O., URBAN, Z.: Preslia 54, 97-104 (1982). – SCANNERINI, S., BONFANTE-FASOLO, P.: Can. J. Bot. 61, 917-943 (1982). – SCHENCK, N. C., SMITH, G. S.: Mycologia 74, 77-92 (1982). – SCHIPPER, M. A. A.: Stud. Mycol. 25, 34 pp. (1984). – SCHNEIDER, E. F., SEAMAN, W. L.: Trans. Br. Mycol. Soc. 79, 283-290 (1982). – SETLIFF, E. C., RYVARDEN, L.: Can. J. Bot. 60, 1004-1011 (1982). – SEVIOUR, R. J., PITT, E. E., McCLURE, S., PYLE, J.: Can. J. Bot. 61, 2374-2383 (1983). – SIDENBERG, D. G., LACHANCE, M.-A.: Int. J. Syst. Bacteriol. 33, 822-828 (1983). – SIMMONS, E. G.: Mycotaxon 14, 17-43 (1982a); – ibid. 14, 44-57 (1982b). – SINGER, R., PONCE DE LEON, P.: Mycotaxon 14, 82-90 (1982). – SIPICZKI, M., KUCSERA, J., UŁASZEWSKI, S., Zsolt, J.: J. Gen. Microbiol. 128, 1989-2000 (1982). – SIVANESAN, A.: Mycol. Pap. 146, 38 pp. (1981); – Trans. Br. mycol. Soc. 81, 313-332 (1983); – The Bitunicate Ascomycetes and their Anamorphs. Vaduz: J. Cramer 1984. – SKALICKY, V.: Folia Geobot. Phytotaxon. 18, 71-101 (1983). – SKARSHAUG, A. J.: Am. J. Bot. 68, 1096-1103 (1981). – SKOU, J. P.: Mycotaxon 15, 487-499 (1982). – SPIEGEL, F. W.: J. Elisha Mitchell Scient. Soc. 97, 183-189 (1981). – STALPERS, J. A.: Stud. Mycol. 24, 105 pp. (1984). – STALPERS, J. A., LOERAKKER, W. M.: Can. J. Bot. 60, 529-537 (1982). – STEFFENS, W. L., JONES, J. P.: Can. J. Bot. 61, 1599-1617 (1983). – STOLK, A. C., SAMSON, R. A.: Stud. Mycol. 23, 149 pp. (1983). – SUBRAMANIAN, C. V.: Hyphomycetes, Taxonomy and Biology. 502 pp. London: Academic Press, 1983. – SUBRAMANIAN, C. V., SEKAR, G.: Kavaka 8, 73-77 (1980). SUGIYAMA, K., YAMAMOTO, H.: Trans. Mycol. Soc. Jpn 23, 119-130 (1982). – SUTTON, B. C., COLE, G. T.: Trans. Br. Mycol. Soc. 81, 97-107 (1983). – SWART, H. J.: Trans. Br. Mycol. Soc. 78, 259-264 (1982a); – ibid. 79, 261-269 (1982b).

TEMPLETON, A. R.: Nature, Lond. 303, 731-732 (1983). – THALER, I., GAILHOFER, M.: Phyton, Horn 22, 137-142 (1982). – THAUNG, M. M.: Mycotaxon 19, 425-452 (1984). – TSUNEDA, A.: Fungal Morphology and Ecology - mostly scanning electron microscopy. 320 pp. Tottori Mycol. Inst.

1983. – TSUSHIHASHI, H., NUNOMURA, K., YADOMAE, T., MIYAZAKI, T.: Trans. Mycol. Soc. Jpn 23, 21-28 (1982). – TU, J. C.: Can. J. Bot. 60, 173-178 (1982).

VANKY, K.: Sydowia 34, 157-166 (1981). – VILGALYS, R., MILLER, O. K.: Mycologia 75, 707-722 (1983). – VOBIS, G.: Bau und Entwicklung der Flechten-Pycnidien und ihrer Conidien. Vaduz: Cramer, Vaduz. 1980. – VOSS, E. G. et al.: International Code of Botanical Nomenclature. Regnum Veget. 111, 472 pp. (1983).

WADDELL, D. R.: Nature, Lond. 298, 464-466 (1982). – WALKER, C.: Mycotaxon 10, 99-104 (1979). – WALKER, J.: Aust. J. Bot. Suppl. Ser., No. 10: 89-136 (1983). – WALKER, W. F., DOOLITTLE, W. F.: Nature, Lond. 299, 723-724 (1982). – WARNER, S. A., SOVOCOOL, G. W., DOMNAS, A. J.: Phytochemistry 21, 2135-2136 (1982); – Mycologia 75, 285-291 (1983). – WASSER, S. P., BERGER, E. G.: Mycotaxon 10, 351-364 (1980); – Ukrain. Bot. Zhurn. 38, 67-95 (1982). – WATERHOUSE, G. M., BRADY, B. L.: Bull. Br. Mycol. Soc. 16, 113-143 (1982). – WATLING, R.: British Fungus Flora 3. Bolbitiaceae: Agrocybe, Bolbitius and Conocybe. 139 pp. Edinburgh: R. Bot. Garden. 1982. – WATLING, R., KILE, G. A., GREGORY, N. M.: Trans. Br. Mycol. Soc. 78, 271-285 (1982). – WELLS, K., OBERWINKLER, F.: Mycologia 74, 325-331 (1982). – WELLS, K., WELLS, E. K. (eds.): Basidium and Basidiocarp. Berlin, Heidelberg, New York: Springer. 1982. – WHISLER, H. C., WILSON, C. M., TRAVLAND, L. B., OLSON, L. W., BORKHARDT, B., ALDRICH, J., THERRIEN, C. D., ZEBOLD, S. L.: Exp. Mycol. 7, 319-327 (1983). – WILLIAMS, M. C.: Mycologia 75, 251-256 (1983). – WOJEWODA, W.: Mala flora grzybów, Tom 2: Basidiomycetes, Tremellales, Auriculariales, Septobasidiales. 408 pp. Warszawa, Kraków: Państwowe Wydawnictwo Naukowe. 1981. – WORRALL, J. J., PARMETER, J. R., jr., COBB, F. W. jr.: Phytopathology 73, 304-307 (1983).

YAMADA, Y., NAKAZAWA, E., KONDO, K.: J. Gen. Appl. Microbiol. Tokyo 28, 355-358 (1982). – YAMADA, Y., OHISHI, T., KONDO, K.: J. Gen. Appl. Microbiol. Tokyo 29, 51-57 (1983). – YAMAZAKI, M., KOMAGATA, K.: J. Gen. Appl. Microbiol. Tokyo 28, 119-138 (1982). – YU, S., MATHUR, S. B., NEERGAARD, P.: Trans. Br. Mycol. Soc. 78, 447-458 (1982). – YU, Y., LAI, Y.: Acta Microbiol. Sin. 21, 7-14 (1981a); – ibid. 21, 15-22 (1981b).

ZHENG, R., CHEN, G.: Acta Microbiol. Sin. 21, 23-30 (1981a); – Sydowia 34, 214-325 (1981b).

Professor Dr. WALTER GAMS
Centraalbureau voor Schimmelcultures
Oosterstr. 1
P.O. Box 273
NL 3740 AG Baarn

Dr. WALTER JÜLICH
Rijksherbarium
Schelpenkade 6
P.O. Box 9514
NL 2300 RA Leiden

III. Systematics of Lichenized Fungi

By JOSEF HAFELLNER

By reason of the large number of publications which appeared in the last 2 years and due to the little space available, only strictly systematic publications and articles dealing with problems that might be of interest for systematic studies are included.

1. Morphology of the Mycobiont

In systematics the never-ending conflict of distinguishing between taxonomic units and convergent thallus organizations always played an important role. AHTI (1982b) tries a morphological interpretation of cladoniiform thalli in lichens and states that the *Cladonia* habitus does not indicate a taxonomic unit but can be found in Cladoniaceae, Heterodeaceae, Cladiaceae, Baeomycetaceae and Siphulaceae and therefore cladoniiform lichens are a life-form group. He also gives a definition of a podetium as a lichenized, stem-like portion of the thallus, bearing the hymenial discs and sometimes conidiomata in a fruticose apothecium, and so the term is restricted to lichenized stipes out of generative tissue. A fertile *Cladonia* podetium consequently is interpreted as polydiscoid (!).

Another example of analogous thallus differentiation finds POELT (1983c) in three discocarpous but not closely related genera (*Almbornia*, *Speerschneidera*, *Seirophora*) all of them with scleroplectenchymatic cords in the medulla, which should have a water-preserving function.

Once the principle had been recognized that a fungus may form two different morphotypes, TØNSBERG and HOLTAN-HARTWIG (1983) reported further phycotype pairs in *Nephroma*, *Peltigera* and *Lobaria*, and RENNER and GALLOWAY (1982) in *Pseudocyphellaria*. They proposed also the new terminus "phycosymbiodeme" for this phenomenon. RENNER and GALLOWAY find that chlorosymbiodemes always are formed secondarily by the mycobiont of the cyanosymbiodeme.

With cytological aspects of the triple symbiosis in *Peltigera aphthosa* deals HONEGGER (1982b).

Investigating the ultrastructure of the mycobiont two obligately sterile lichenized fungi were found to be Ascomycetes: *Chiodecton sanguineum* (WITHROW and AHMADJIAN 1983) and *Lichenothrix riddlei* (TSCHERMAK-WOESS et al. 1983). The ultrastructure of *Coenogonium implexum* is investigated by MEIER and CHAPMAN (1983). As in other areas of biology, SEM-studies of lichens become more and more frequent. Mostly included in taxonomic treatments (see Sect. 6), there are also some with only morphological contents. ANGLESEA et al. (1983) investigate the cortical structure and alga cell division in the thallus tips of *Usnea floridana*, and ANGLESEA et al. (1982) describe a method of pretreatment of lichen cortices with an ordinary washing-powder for SEM investigation. JACKSON (1981) studies the lichen pruina crystals and finds the crystallized substance embedded in a matrix material secreted by the hyphae.

The variability of vegetative and generative parts of lichens is dealt with by JAHNS et al. (1982). The structure of lichen plectenchyma is strongly influenced by their function, whereas certain ontogenetic principles cannot be changed. The authors warn of an over-emphasizing of vegetative structures without a knowledge of their ontogeny. LARSON (1983) deals with the morphological variation in a concrete case (*Ramalina menziesii*). An attempt to classify the different types of discocarps (lichenized and non-lichenized) is undertaken by CHADEFAUD (1982). KUROKAWA (1981) reminds us that more attention should be paid to apothecial size in taxonomic studies.

For TEM investigations of different ascus types see Sect. 6. A detailed study of the formation of phialidic conidia directly from muriform ascospores in asci of *Brigantiaea leucoxantha* : HAFELLNER and BELLEMERE (1983).

2. Morphology of the Phycobiont

In the last few years scientists became interested in seasonal alterations of the ultrastructure of lichenized algae. In *Trebouxia* (out of *Bryoria capillaris* and *Hypogymnia physodes*) HOLOPAINEN (1982) thinks they are caused by climate and differences in metabolism between summer and winter. BERGMAN and HUSS-DANELL (1983) examine *Stigonema* in the cephalodia of *Stereocaulon paschale* in winter and autumn and find besides some biochemical differences haustoria only in degenerating cells of the alga. On the other hand, BERGMAN and HÄLLBOM (1982) observe distinctive differences between isolated and lichenized algal cells (*Nostoc* out of *Peltigera canina*) concerning some organelles such as thylakoids and phycobilisomes.

The type of haustoria penetrating *Scytonema* trichoms in *Dictyonema* is characteristic for the genus (TSCHERMAK-WOESS 1983).

3. Ontogenetic Studies

Attempts at an artificial reestablishment of lichens were made by AHMADJIAN and JACOBS (1982) who were successful with a synthetic development of *Usnea strigosa*.

JAHNS and OTT (1983) observed the development of fruitbodies of *Cladonia* and *Baeomyces* for 2 years. While thallus and apothecia of *Baeomyces* finish their developmental cycle in about 2 years, podetia of *Cladonia* need up to 4 years (1 - 2 years for the apothecia). The development is clearly influenced by the microclimate of the habitats. Decomposition of lichen thalli needs about 13 months (WETMORE 1982b).

Ontogenetic studies of fruitbodies are mostly made in connection with taxonomic ones (see Sect. 6). The ontogenetic development of the perithecia of *Porina byssophila* shows features closely resembling *Pyrenula nitida* (JANEX-FAVRE 1981). JANEX-FAVRE (1982) also investigates the pycnidia of *Parmelia acetabulum* and finds similarities with those of *Umbilicaria cinereorufescens*.

4. Chemistry

Chemosystematic studies are reviewed in Sect. 6. CULBERSON and JOHNSON (1982) propose the substitution of diethyl ether in solvent system B of their standardized thin-layer chromatographic method with methyl tert.-buthyl ether. Investigating *Xanthoparmelia* species, ELIX (1982) concludes that some groups of metabolites (e.g., several aliphatic acids and amino acid derivatives) are taxonomically insignificant. The biosequential patterns for the formation of depsides, depsidones and dibenzofurans in *Cladonia* are clarified by HUOVINEN and AHTI (1983) who find *Cladonia* species mainly either with depsides and depsidones of the orcinol series or with β-orcinol depsides or with β-orcinol depsidones.

An influence of the alga upon the secondary metabolism of the fungus is shown by RENNER (1982) who tests cephalodia in search of secondary metabolites.

Growth and production of secondary compounds are affected by genetics and environment (C. CULBERSON et al. 1983). Four clones of *Cladonia cristatella* were grown in a phytotron. Growth was affected by clone and decreased at lower temperature and at higher light intensities. The clones had significantly different capacities for the production of compounds, while the concentration of secondary metabolites from the barbatic acid pathway increased at lower temperatures, and those of the didymic acid pathway changed little. HAMADA (1982) finds higher levels of the medullary depsidone salazinic acid in *Ramalina siliquosa* specimens growing on dark substrates and in areas with higher mean temperatures.

Anthraquinons are widely distributed in lichenized fungi. Parietin is now also reported in the genus *Parmelia* s. lat. (KROG and SWINSCOW 1982).

Cytochemical ultrastructure of *Peltigera canina*: BOISSIERE (1982).

5. Phylogeny and Evolution in Lichenized Fungi

After the 13th International Botanical Congress at Sydney several short versions of contributions to symposia on lichenology held there appeared in J. Hattori Bot. Lab. 52, some of them dealing with the evolution of lichenized fungi.

KROG (1982a) discusses the characters regarded as important for classification of alectoroid, usnoid, hypogymnoid and parmelioid genera as well as their distributional trends and she states that the segregation of the Alectoriaceae may be necessary but that there is no necessity to segregate Hypogymniaceae and Usneaceae from the Parmeliaceae.

MAYRHOFER (1982) reports on many different spore types he has found in Physciaceae, as well as on the distribution of these types within the family. Several lines of parallel evolution become evident, for all groups *Rinodia*-like lichens must be the common ancestors (e.g., *Rinodia* - *Physcia*, *Phaeophyscia*, *Heterodermia*; *Rinodina* - *Physconia* - *Anaptychia* - *Tornabaea*; *Rinodina* - *Diploicia* - *Dirinaria*, *Pyxine*).

Knowledge of cladoniiform lichens is discussed by AHTI (1982a) including several phylogenetic reconstructions of earlier authors (see also Sect. 1).

The evolutionary trends within the genus *Pertusaria* are summarized by DIBBEN (1982), who regards poriform *Pertusariae* with polycarpic verrucae and double-walled spores as the most primitive ones. He assumes that the ancestral *Pertusariae* soon derived the ability to produce both orcinol and β-orcinol-type derivates.

There is a rather large number of lichenicolous fungi restricted to *Peltigera*, a host genus with a similar number of species on all major land masses, which is interpreted as further circumstantial evidence for the ancestry of Peltigerineae (HAWKSWORTH 1982e), an opinion already expressed earlier by other lichenologists after having analyzed the ascus type.

Pseudoparmelia caperata, a widely distributed lichen with an exceptionally broad amplitude of ecological tolerance, is found in eastern North America primarily epiphytically, replaced on non-calcareous rocks by *Pseudoparmelia baltimorensis*, but both are found on tree boles with intermediate ecology. The preference of substrate should be a result of intense competition in the past (W. CULBERSON and CULBERSON 1982).

Further comments on the species pair problem: TEHLER (1982). The secondary species should be treated as formae.

6. Taxonomy

Two introductory textbooks dealing with the different aspects of lichenology appeared in the last 2 years: HALE (1983a) proposes a system for lichenized fungi, which follows mainly ERIKSSON's (1982a). DOLL (1982) published a text which may be useful for beginners, in which simplified keys taken from POELT's keys are included.

The well-known and useful *Ainsworth and Bisby's Dictionary of the Fungi* appeared in a new edition (HAWKSWORTH et al. 1983). It is the most complete listening of generic names of lichen-forming fungi available at the moment. For each genus additional information, such as authors, systematic position, number of species and distribution, is given and many general entries are included.

It becomes more and more usual to incorporate the lichenized fungi in the system of non-lichenized ones, so consequently done by ERIKSSON (see Prog. Bot. 44, 379-381). He now proposes several changes in family and order classification of lichens (ERIKSSON 1982a,b) and he gives an improved outline with some considerable alterations (ERIKSSON 1983). Further improvements have to be expected in near future. This trend is now taken in consideration also in textbooks for teaching at university level. In a new edition of Strasburger's famous textbook on botany the lichens are considered as a "type of organization" of eukaryotic organisms (BRESINSKY 1983) and lichenized fungi are included too in an outline of the fungal system proposed by MÜLLER and LÖFFLER (1982). A key to orders and higher taxa of ascomycetes is given by BARR (1983). HONEGGER (1982c) finds additional criteria for a more natural classification of lichenized fungi, such as ascospore delimitation, phycobiont cell-wall types and storage products. She states that natural groupings within the Lecanorales as well as other groups of lichenized ascomycetes might be characterized by an almost complete correspondence in ascus structure and function. The strategy of all known ascus types is to bring the ascospores up to the hymenial surface, but the means may be very different.

Several ascus types have been studied in detail. The *Pertusaria*-type is functionally unitunicate and is characterized by a specific disintegration of the outer wall layers at the moment of dehiscense (HONEGGER 1982a). CHADEFAUD's idea that the Baeomycetaceae and *Leotia* are closely related is confirmed by means of ascus fine-structure investigation. The Baeomycetaceae belong to the Helotiales (HONEGGER 1983). *Catolechia wahlenbergii* and *Epilichen scabrosus* have the same ascus type which shows similarities with the *Rhizocarpon*-type (BELLEMERE and HAFELLNER 1983). A detailed study on the ontogeny and structure of asci of the *Teloschistes*-type and the bilocular spore type of Teloschistaceae (the equatorial thickening is of an endosporal nature, the spore septation is delayed but conventional): BELLEMERE and LETROUIT-GALINOU (1982). *Bombyliospora* *Brigantiaea* and *Letrouitia* (*Bombyliospora domingensis*-group) have peculiar ascus types. *Bombyliospora* is congeneric with *Megalospora*. *Brigantiaea* is a genus not closely related but with analogies in spore septation to *Lopadium*. *Letrouitia* is related with *Teloschistes* and is characterized by a remarkable infrageneric spore evolution (HAFELLNER and BELLEMERE 1982a,b,c). *Dactylospora*, rich in lichenicolous species, has a peculiar ascus type very different from *Buellia* (BELLEMERE and HAFELLNER 1982).

Some of the larger monographic works published in the last 2 years should be mentioned first: ARVIDSSON (1982 - monograph of *Coccocarpia*), COPPINS (1983b - *Micarea* in Europe), HAFELLNER (1983d - monograph of *Letrouitia*; this is the *Bombyliospora domingensis*-group), HENSSEN and JAMES (1982 - monograph of *Steinera*), SIPMAN (1983 - monograph of Megalosporaceae: *Megalospora* incl. *Bombyliospora*, *Austroblastenia* gen. nov., *Megaloblastenia* gen. nov.), TEHLER (1983 - monograph of *Dirina* and *Roccellina*).

The following compilation is based on the system of ERIKSSON (1983). M means monograph, revision; K means key; C means contribution.

Lecanorales

Acarosporaceae: K to *Acarospora* in Great Britain: HAWKSWORTH (1983c). M, K of *Sporostatia* in U.R.S.S.: GOLUBKOVA (1982). Brigantiaeaceae: Description of the family: HAFELLNER and BELLEMERE (1982b). C to *Brigantiaea* in New Zealand: HAFELLNER (1983e). Cladoniaceae: Outline of a new infrageneric division of *Cladonia* in seven groups correlated to the main medullary components: HUOVINEN and AHTI (1983). *Cladonia chlorophaea*-group in southeastern Australia: ARCHER (1983). *Cladonia chlorophaea* - *fimbriata* complex in western Washington: HENNINGS (1983). C to single or several *Cladonia* species: AHTI (1983b), ASPERGES (1981), BARENDREGT et al. (1982), W. CULBERSON et al. (1982). Coccocarpiaceae: M, K of *Coccocarpia*, 21 species are recognized: ARVIDSSON (1982). M, K of the subantarctic genus *Steinera*, including a study of ascocarp ontogeny, five taxa are recognized: HENSSEN and JAMES (1982). Collemataceae (see also Lichinaceae): M, K of the *Leptogium azureum*-group in Europe: JØRGENSEN and JAMES (1983). M, K of *Leptogium* in India: SHARMA (1982). Heppiaceae (see also Lichinaceae): M, K of *Heppia* and *Peltula* in Tenerife: SWINSCOW (1982). Lecanoraceae s. ampl.: M, K of *Bryonora* gen. nov. for the *Lecanora castanea*-group, seven taxa are recognized: POELT (1983b). M of the subarctic *Lecanora contractula*-group, *Arctopeltis* gen. nov. for an umbilicate Lecanoraceae which is regarded as a convergent form to *Rhizoplaca*: POELT (1983a). M, K of *Squamarina* in Scandinavia: TIMDAL (1983). M, K of the corticolous species of *Haematomma* in Australia: ROGERS (1982c). Revision of Australian lichens described as *Lecanora* and *Placodium* by MÜLLER-ARGOVIENSIS: ROGERS (1982b). C to *Lecanora*: SEDELNIKOVA (1982), BRODO (1982), CRESPO and LLIMONA (1981). Lecideaceae s. ampl.: M, K of *Lecidea* in Japan, 17 species are recognized: INOUE (1982). M, K of *Huilia* in Japan, 11 species are recognized: INOUE (1983a,b,c). *Farnoldia* gen. nov. of Huiliaceae for all but the type species of *Melanolecia*, *Carbonea* gen. nov. for the *Lecidea vorticosa*-group: HERTEL (1983). Remarks to several species of *Lecidea* coll.: HERTEL (1982b). Supraspecific units in Lecideaceae, circumscribed morphologically have clear parallels in ecological characters. An ecological difference may indicate a possible taxonomic separation: WIRTH (1983). M, K of *Roccellinastrum* comprising four southern hemispheric species: HENSSEN et al. (1982). M, K of *Micarea* in Europe, 45 species are recognized, many species have anamorphic forms: COPPINS (1983b). C, K to *Rhizocarpon* species living on *Pertusaria*: POELT and HAFELLNER (1983). M, K of *Lopadium* in southwestern India: PATWARDHAN and MAKHIJA (1981a). Lichinaceae: C to *Lempholemma* in India: AKHTAR (1981). The cyanophilous lichens in hot deserts of Israel with remarks and a K to 30 taxa of Lichinaceae, Heppiaceae and Collemataceae: MARTON and GALUN (1981). Megalosporaceae: Description of the family: HAFELLNER and BELLEMERE (1982a). M, K of the family, *Megalospora* with 25 species, *Austroblastenia* gen. nov. and *Megaloblastenia* gen. nov., with a rather broad specific concept: SIPMAN (1983). C to *Megalospora*: MAKHIJA and NAGARKAR (1981). Pannariaceae: C to *Psoroma*: HENSSEN (1983), HENSSEN et al. (1983). Parmeliaceae s. ampl.: M, K of *Punctelia* gen. nov. with subgen. *Punctelia* (*Parmelia borreri*-group) and subgen. *Flavopunctelia* (*Parmelia flaventior*-group), separation of *Parmelia* s. str. mainly because of the different shape of pseudocyphellae and the chemistry of the medulla: KROG (1982b). C to *Punctelia*: SERUSIAUX (1983a). Segregation of *Foraminella* gen. nov. (*Parmeliopsis hyperopta-ambigua* group) from *Parmeliopsis* (*P. aleurites* group) mainly because of different shape of conidia and ascospores: S. MEYER (1982). C to *Parmeliopsis* in Norway: NORDNES (1982a). K to *Parmelia* coll. in Great Britain: JAMES (1982). M, K of *Parmelia* in southern Australia, a study comprising 78 species: FILSON (1982b). C to *Xanthoparmelia* in Australia: KUROKAWA (1982). C to *Xanthoparmelia* from southern and central Africa: KNOX and BRUSSE (1983). M, K of *Xanthoparmelia* and *Melanoparmelia* in India: AWASTHI (1981). C to *Hypotrachyna* in South Africa: KNOX (1982). C to *Hypotrachyna* and *Parmelina*: SKOREPA (1983). C to *Parmotrema*: EGAN (1982a), KROG and SWINSCOW (1983). C to *Parmelia*: HALE (1982), GOWARD and AHTI (1983). Three chemical strains of *Pseudevernia* in Norway: HALVORSEN and BENDIKSEN (1982). C to *Hypogymnia*: PIKE and HALE (1982). C to *Cetraria*: SHIBUICHI and YOSHIDA (1982). C to *Usnea*: W. CULBERSON et al. (1983). M, K of *Bryoria* in New Zealand: JØRGENSEN and GALLOWAY (1983). Physiaceae: M, K of *Phaeophyscia* in East Africa: MOBERG (1983b). Species with a dual lower cortex should be retained in *Physcia*: HALE (1983b). C to *Phaeophyscia*: TRASS (1981). M, K of *Pyxine* in Europe: MOBERG (1983a). M, K of *Pyxine* in India: AWASTHI (1980). M, K of *Buellia* in India: SINGH and AWASTHI (1981). Synopsis of saxicolous species of *Rinodina* with blue epithecia: SHEARD (1982). M, K of saxicolous species of *Rinodina* in New Zealand: MAYRHOFER (1983). C to *Rinodina*: COPPINS (1983a). Ramalinaceae: Revision of the *Ramalina farinacea*-complex in the tropics: STEVENS (1983b). C to *Ramalina*: W. CULBERSON and CULBERSON (1981), KROG (1983), MOLHO et al. (1981), STEVENS (1982, 1983a). Stereocaulaceae: M, K of *Stereocaulon* in Sweden: CARLIN and CARLIN-SILVANG (1982). M, K of *Stereocaulon* in Colombia: BOEKHOUT (1982). Umbilicariaceae: C to the Umbilicariaceae on the Iberian peninsula: SANCHO (1982). C to *Umbilicaria*: NAGARKAR and PATWARDHAN (1981a), TOPHAM et al. (1982).

Teloschistales

Teloschistaceae: M, K of *Caloplaca* sect. *Xanthoriella* (*Xanthoria lobulata*-group). A difference concerning the thickness of the septum between living and dead ascospores is described: STEINER and POELT (1982). *Seirophora* gen. nov. with a peculiar thallus construction: POELT (1983c). C to *Caloplaca*: EGEA and LLIMONA (1983), HAFELLNER (1983a). C to *Fulgensia*: FOLLMANN and POELT (1981). Letrouitiaceae: Description of the family and the genus *Letrouitia* (*Bombyliospora domingensis*-group): HAFELLNER and BELLEMERE (1982c). M, K of *Letrouitia* comprising 15 species: HAFELLNER (1983d).

Gyalectales

Clathroporina is not present in Europe, the two European species are shown to belong to *Belonia*: JØRGENSEN et al. (1983).

Pertusariales

In the European species of *Pertusaria* 23 main compounds are present and 21 chemotypes are found. In comparison with North America, European species are poorer in xanthones and p-depsides of the orcinol type with long aliphatic side chains: HANKO (1983).

Peltigerales

Lobariaceae: M, K of *Lobaria*, *Pseudocyphellaria* and *Sticta* in India and Nepal: MAMTA and AWASTHI (1982). C to *Pseudocyphellaria* in New Zealand. Chemical profiles for all described *Pseudocyphellaria* species of New Zealand: GALLOWAY et al. (1983). C to *Pseudocyphellaria*: RENNER and GALLOWAY (1982), GALLOWAY (1983). Nephromataceae: Phytochemical study of South American *Nephroma* species: RENNER et al. (1982). Peltigeraceae: K to *Peltigera* in Great Britain: VITIKAINEN (1982), M, K of *Peltigera* in India and Nepal: AWASTHI and JOSHI (1982).

Graphidales

Graphidaceae: C to *Graphis* in India: NAGARKAR and PATWARDHAN (1982). Asterothyriaceae s. ampl.: C to *Gyalideopsis*: VEZDA (1983b).

Caliciales

M, K of Caliciales in Costa Rica, comprising 26 species in ten genera: TIBELL (1982). C to *Chaenotheca*: TIBELL (1983).

Opegraphales

Opegraphaceae: K to *Opegrapha* in Great Britain: PENTECOST and COPPINS (1983). Roccellaceae: M, K of *Dirina* and *Roccellina*, 38 taxa are recognized. The genera are defined by means of the shape of the hypothecium, chemistry and type of thallus cortex. Species pairs are regarded as states of a single species: TEHLER (1983).

Verrucariales

C to *Placidiopsis*: BREUSZ (1983). C to *Macentina*: DÖBBELER and VEZDA (1982).

Pyrenulales

Discussion of a possible relationship between Pyrenula and Acrocordia: ERIKSSON (1982c). C to *Pyrenula*: A. SINGH (1983). C to *Polythelis*, a genus with eu- and distoseptate ascospores: HAWKSWORTH (1983b). M, K of *Anthracothecium* in southwest India: PATWARDHAN and MAKHIJA (1980). C to *Anthracothecium*: A. SINGH (1982a, b).

Dothideales

C to *Pocsia*: VEZDA (1982).

C to several unrelated species: AWASTHI and UPRETI (1981), HOUMEAU and ROUX (1982a), KASHIWADANI and GRADSTEIN (1982), PATWARDHAN and MAKHIJA (1981b).

Lichenicolous fungi

K to lichenicolous fungi occurring in Great Britain: HAWKSWORTH (1983a). Lichenicolous fungi can live as parasites, saprophytes or parasymbionts: HAWKSWORTH (1982f). C to British lichenicolous fungi, with *Everniicola* gen. nov.: HAWKSWORTH (1982g). The new family Odontotremataceae is described for several genera including the lichenicolous *Skyttea*, *Spilomela* and *Lethariicola*: HAWKSWORTH and SHERWOOD (1982). The new family Dactylosporaceae is described because of the peculiar ascus type of the type genus: BELLEMERE and HAFELLNER (1982). K to lichenicolous species of *Nectriella*: HAWKSWORTH (1982d). The monotypic *Lichenostigma* gen. nov. is a common lichenicolous fungus on several corticolous Parmeliaceae in central Europe: HAFELLNER (1983c). Revision of *Pleoscutula*, one of the known species is a *Catillaria*: HAFELLNER (1983b). C to diverse lichenicolous fungi: HAWKSWORTH (1982b,c).

7. Floristics

A well illustrated field guide to Swedish macrolichens and selected microlichens with small distribution maps for Scandinavia was published by MOBERG and HOLMASEN (1982). The flora contains keys to the species of the macrolichens but lacks a key to the genera. SCHUBERT (1983) is the author of the lichenological part of a flora of lower plants growing in Germany. The keys often end at the genus. Only the macrolichens are keyed out down to species. The lichen flora of Poland is continued with an issue containing the Arthoniaceae, Opegraphaceae, Lecanactidaceae and Roccellaceae (NOWAK (1983).

Europe

Central Europe: KALB (1983). Svalbard, Bockfjord region on Spitsbergen: HAFELLNER (1982). Norway, a detailed flora of the island of Vega in Nordland; 668 species are reported with remarks on many species: DEGELIUS (1982). Data on several species: TIMDAL (1982), TØNSBERG and OVSTEDAL (1982). *Cladonia polycarpoides* found in Norway: NORDNES (1982b). Finland, checklist of fungi and lichens of the Koillismaa biological province: ULVINEN et al. (1981). Denmark, report on the vitality of the Danish epiphytic lichen flora, over 100 species reported: SØCHTING and RAMKAER (1982). British Isles, Carlow County: SEAWARD (1982); Birmingham region: LINDSAY (1981); Warwickshire: LINDSAY (1980); Slepe Heath: GIAVERINI (1982); Ludlow in Shropshire: GILBERT and LAMBLEY (1982); Kirkaldy District: COPPINS (1982b); Calcicolous lichens from Ben Alder Range, Scotland: GILBERT et al. (1982); Rossdhu Park, Dunbartonshire: GILBERT and MITCHELL (1981); Berwickshire: COPPINS (1982a). France, Normandie: LEROND (1981); *Cetrelia* in Autunois: BEGUINOT (1983); Massif de la Vanoise: MEYER et al. (1982); Cantal: HOUMEAU and ROUX (1982c). A site for foliicolous lichens in Aveyron: FOUCAULT et al. (1982). Records for the southwestern part of France: HOUMEAU and ROUX (1982b). Island of Port-Cros: RONDON-SEIDENBINDER (1983); Provence: ROUX (1982). Germany, Thüringen: MEINUNGER (1983). C.S.S.R., *Cyphelium sessile* in the C.S.S.R.: LISICKA (1981). Austria, an attempt to an almost complete flora (excl. algae) of a mountain area in Styria: MAURER et al. (1983); Salzburg: HEISELMAYER and TÜRK (1979); Upper Austria: TÜRK and WITTMANN (1983). Poland, Czerkies: BYSTREK (1980). Terricolous lichens of Kielecko Sandomierska Upland: CIESLINSKI (1981); Polish Tatra Mountains: KORNAS (1981). Several *Ramalina* species in Poland: SULMA and BYSTREK (1982). European part of U.S.S.R., Vilsandi State Reserve, Estonia: RANDLANE (1981); flora of Latvia; 464 species reported: PITERANS (1982). Lichens in mountain beech forests of Ukrainia: NAVROTSKA (1982). Spain, Turo de Sant Mateu: TORRES and HLADUN (1982). Lichens on old bones: HLADUN and GOMEZ-BOLEA (1982). *Omphalina umbellifera* in Spain: HLADUN (1982b). Silicicolous lichens in the higher parts of Montseny: HLADUN (1981, 1982a). Foliicolous lichens in Cataluna: GOMEZ and HLADUN (1982). Lichens on wood in decomposition in the woods of Moixero: ALVARO and HLADUN (1983). Casa de Campo de Madrid: CRESPO and BUENO (1982); Pico del Lobo, Segovia: BARRENO and RICO (1982). Silicicolous lichens of the Sierra del Cabo de Palos: EGEA and LLIMONA (1982); Sierra Nevada: CASARES and LLIMONA (1982). Silicicolous lichens of the alpine zone of Sierra Nevada: EGEA et al. (1982). Calcicolous lichens of Granada: CASARES and LLIMONA (1983); Malaga: SEAWARD (1983). Epiphytic lichens in the Sierra de Bejar and the Pena de Francia: MARCOS LASO (1983); Barranco de Linares: GARCIA ROWE et al. (1982).

Asia

Turkey, catalog of 256 species: VERSEGHY (1983). Asian part of U.S.S.R., Parmeliaceae in the Ural Mountains: RJABKOVA (1982a). Baeomycetaceae, Cladoniaceae and Stereocaulaceae in the Ural Mountains: RJABKOVA (1982b). Foliicolous lichens from Colchis, Transkaukasia, some new species: VEZDA (1983a). *Stereocaulon* on Czukotka peninsula: DOMBROVSKAJA and MAKAROVA (1982); Arctic Sibiria: ANDREEV and MAKAROVA (1982); Jamal peninsula: ANDREEV (1982); Tian-schan: BREDINKA (1982a,b). Mongolia, flora of Mongolia; 657 species reported: GOLUBKOVA (1981). China, first part of a catalog: ZHAO et al. (1982); Changbai Shan, NE China: HERTEL and ZHAO (1982). India, pyrenocarpous lichens from NE India: NAGARKAR and PATWARDHAN (1981b). Some macrolichens from NE India: PATWARDHAN and NAGARKAR (1982). Microlichens of Manipur: K. SINGH (1981). Japan, catalog of macrolichens of Japan and adjacent regions: IKOMA (1983).

Africa

Canary Islands, short introduction to several contributions, the results of an excursion of BLS: BRIGHTMAN (1982). Alpine microlichens: FOLLMANN (1982). Many records: TOPHAM and WALKER (1982). *Alectoria* and *Bryoria*: HAWKSWORTH (1982a). Terricolous lichens of Tenerife: GILBERT (1982). *Umbilicaria krempelhuberi*: TOPHAM (1982). Taoro Park, Tenerife: TOPHAM and SWINSCOW (1982). Zimbabwe, foliicolous lichens: SERUSIAUX (1983b).

North America

Caloplaca tominii in the Americas: THOMSON (1982b). *Tholurna dissimilis* in western North America: OTTO (1983). Greenland, epiphytic lichens; 106 species reported: ALSTRUP (1982); Central E Greenland: HANSEN (1982); NW Greenland: HANSEN (1983a). Additions to the known flora: HANSEN (1983b). Canada, Bathurst Inlet Region, Northwest Territories: THOMSON and SCOTTER (1983). *Cavernularia hultenii* in Newfoundland and Labrador: HOISINGTON and MAASS (1982); Newfoundland: AHTI (1983a). USA, Swan Valley, Montana: McCUNE (1982). Lichens in a *Pinus ponderosa* forest in SE Montana: EVERSMAN (1982); Zion National Park, Utah: NASH and SIGAL (1981). Macrolichens on junipers in Boulder County, Colorado: PEARD (1983); Ohio: SHOWMAN (1981); Voyageurs National Park, Minnesota; 405 species recorded: WETMORE (1982a). Key to the lichens in the intertidal region of northeastern US: TAYLOR (1982). Checklist of Louisiana lichens: TUCKER (1981). *Xanthoparmelia barbatica* found in America: EGAN (1982b).

South America

Contribution to the foliicolous lichens: FERRARO (1983). Venezuela, additions to the flora: LOPEZ-FIGUEIRAS (1982, 1983). Columbia, key to the genera of macrolichens in the Paramo zone: SIPMAN and AGUIRRE (1982). Brazil, Guaiba, Rio Grande do Sul: OSORIO et al. (1982); Santa Maria, Rio Grande do Sul: OSORIO and FLEIG (1983); Torres, Rio Grande do Sul: OSORIO and FLEIG (1982); São Jeronimo, Rio Grande do Sul: OSORIO et al. (1983). Uruguay: OSORIO (1983a); Rio de la Plata coast: OSORIO (1983b). Argentina, Misiones: OSORIO (1981); Entre Rios: OSORIO (1982).

Australia

Checklist of Australian lichens: FILSON (1983). Western Australia: RICHARDSON and RICHARDSON (1982); Arid Australia: ROGERS (1982a). New records: ELIX and STREIMANN (1982). New Zealand, Cuvier Island: HAYWARD et al. (1981); Rakitu Island: HAYWARD and HAYWARD (1982).

Antarctica

Comments on the lichen flora of continental Antarctica: FILSON (1982a); Vestfjella, Dronning Mauds Land: ØVSTEDAL (1983).

8. Phytosociology and Ecology

Only some selected publications are reviewed of this subject. A portable instrument for microclimatic measurements of lichen habitats: SCHUSTER et al. (1983).

A cyclic development of mosses and *Baeomyces rufus* is observed by JAHNS (1982). The vegetation rich in terricolous lichens in SE Greenland: DANIELS (1982). Lichen vegetation in the high Arctic: THOMSON (1982a). Lichen rich pine forest vegetation in Finland: OKSANEN and AHTI (1982). Composition and succession of lichen communities in an inner coastal dune area in Sweden: MAGNUSSON (1982). Composition of some calcicolous lichen communities in W Ireland: McCARTHY (1983). Distribution of several macrolichens on a boulder scree and their adaptations to the microclimatic conditions: JAHNS and FRITZLER (1983). Microclimate and a maritime lichen cleft community: McCARTHY (1982). Host specificity and ecology of macrolichens in Swedish spruce forests: ESSEEN (1982). *Lobarion pulmonariae* in S Spain: MARCOS and NAVARRO (1982). The minimum area of *Physconietum griseae* is about 150 - 200 cm^2: RIEUX and ROUX (1982). Water relations in the epiphytic communities of *Oropogonetum loxensis* and *Ramalinetum cactacearum* in a fog oasis in Chile: REDON and LANGE (1983). Phytosociological study of epiphytic lichens in S Brazil: ZANETTE et al. (1981).

9. Chorology

A first volume of the atlas of the lichens of the British Isles was published by SEAWARD and HITCH (1982). It contains maps for 176 species and comments for each of them. The known distribution of 882 species in the Ukrainian Carpathian Mountains can also be seen in an atlas: MAKAREVIC et al. (1983). First results of mapping of lichens in Upper Austria: TÜRK et al. (1982).

The major geological events of the Pacific region are reflected in several distribution patterns of lichens in this area. There are characteristic disjunctions as well as endemic taxa in Australia, Hawaii, New Caledonia and New Zealand: JØRGENSEN (1983). The distribution patterns of macrolichens of E Iowa show great affinity to the eastern North American floristic region: SCHUTTE (1983).

Map of *Umbilicaria aprina*: CRESPO and SANCHO (1982), *Umbilicaria propagulifera*: SEAWARD et al. (1983), *Erioderma* spec. div.: MAASS (1983). Maps may also be included in taxonomic and floristic publications.

10. Herbaria and Bibliographies

An annotated list of lichen exsiccata conserved in Munich (M): HERTEL (1982a). BOULY de LESDAIN and VOUAUX material found in Sydney (NSW): HAWKSWORTH and JAMES (1983).

Bibliography of lichenology in CSSR 1945 - 1980: VANA (1982).

References

AHMADJIAN, V., JACOBS, J.B.: J. Hattori Bot. Lab. 52, 393-399 (1982). - AHTI, T.: J. Hattori Bot. Lab. 52, 331-341 (1982a); - Lichenologist 14, 105-113 (1982b); - Exploration of the lichen flora, pp. 319-360. In: Biogeography and Ecology of the Island of Newfoundland, G.R. SOUTH (ed.). The Hague: Junk 1983 (1983a); - Ann. Bot. Fennici 20, 1-7 (1983b). - AKHTAR, P.: Biol. Mem. 6, 78-80 (1981). - ALSTRUP, V.: Bryologist 85, 64-73 (1982). - ALVARO, I., HLADUN, N.L.: Collectanea Bot. 14, 19-25 (1983). - ANDREEV, M.P.: Novit. Syst. Plant. Non-Vascul. 19, 111-117 (1982). - ANDREEV, M.P., MAKAROVA, I.I.: Novit. Syst. Plant. Non-Vascul. 19, 117-121 (1982). - ANGLESEA, D., VELTKAMP, C., GREENHALGH, G.N.: Lichenologist 14, 29-38 (1982). - ANGLESEA, D., GREENHALGH, G.N., VELTKAMP, C.J.: Lichenologist 15, 73-80 (1983). - ARCHER, A.W.: Bryologist 86, 251-253 (1983). - ARVIDSSON, L.: Opera Bot. 67, 1-96 (1982). - ASPERGES, M.: Cryptogamie Bryol. Lichenol. 2, 349-358 (1981). - AWASTHI, D.D.: Phytomorphology 30, 359-379 (1980); - Indian J. For. 4, 198-204 (1981). - AWASTHI, D.D., JOSHI, M.: Kavaka 10, 47-62 (1982). - AWASTHI, D.D., UPRETI, D.K.: Curr. Sci. 50, 821-823 (1981).

BARENDREGT, A., VAN DEN DRIES, P.J.L., SIPMAN, H.J.M.: Acta Bot. Neerl. 31, 491-494 (1982). - BARR, M.E.: Mycologia 75, 1-13 (1983). - BARRENO, E., RICO, V.J.: Collectanea Bot. 13, 265-277 (1982). - BEGUINOT, J.: Bull. Soc. Hist. Nat. Autun 104, 9-12 (1983). - BELLEMERE, A., HAFELLNER, J.: Cryptogamie Mycol. 3, 71-93 (1982); - Cryptogamie Bryol. Lichenol. 4, 1-36 (1983). - BELLEMERE, A., LETROUIT-GALINOU, M.A.: Cryptogamie Bryol. Lichenol. 3, 95-137 (1982). - BERGMAN, B., HÄLLBOM, L.: Can. J. Bot. 60, 2092-2098 (1982). - BERGMAN, B., HUSS-DANELL, K.: Lichenologist 15, 181-190 (1983). - BOEKHOUT, T.: J. Hattori Bot. Lab. 53, 483-511 (1982). - BOISSIERE, M.-C.: Lichenologist 14, 1-27 (1982). - BREDINKA, L.I.: Novit. Syst. Plant. Non-Vascul. 19, 122-128 (1982a); - ibid. 19, 128-130 (1982b). - BRESINSKY, A.: Niedere Pflanzen, pp. 551-757. In: D. von DENFFER, H. ZIEGLER, F. EHRENDORFER, A. BRESINSKY, Lehrbuch der Botanik für Hochschulen, 32. Aufl. Stuttgart, New York: Fischer 1983. - BREUSZ, O.: Plant Syst. Evol. 142, 247-250 (1983). - BRIGHTMAN, F.H.: Lichenologist 14, 57-61 (1982). - BRODO, I.M.: Bryologist 84, 521-526 (1982). - BYSTREK, J.: Ann. Univ. Mariae Curie-Sklod., Sect. C, 35, 53-64 (1980).

CARLIN, G., CARLIN-SILVANG, U.: Svensk Bot. Tidskr. 76, 289-307 (1982). - CASARES, M., LLIMONA, X.: Acta Bot. Malacitana 7, 5-10 (1982); - Collectanea Bot. 14, 221-230 (1983). - CHADEFAUD, M.: Cryptogamie Mycol. 3, 103-144 (1982). - CIESLINSKI, S.: Fragm. Florist. Geobot. 27, 527-540 (1981). - COPPINS, B.J.: Lichenologist 14, 189-193 (1982a); - Lichens, pp. 129-134. In: The Wildlife and Antiquities of Kirkcaldy District, G.H. BALLANTYNE (ed.). Kirkcaldy: Kirkcaldy Nat. Soc. 1982b; - Lichenologist 15, 147-150 (1983a); - Bull. Brit. Mus. Natl. Hist., Bot. Ser. 11/2, 17-214 (1983b). - CRESPO, A., BUENO, A.G.: Lazaroa 4, 327-356 (1982). - CRESPO, A., LLIMONA, X.: Anal. Jardin Bot. Madrid 38, 25-28 (1981). - CRESPO, A., SANCHO, L.G.: Lazaroa 4, 357-360 (1982). - CULBERSON, C.F., JOHNSON, A.: J. Chromatogr. 238, 483-487 (1982). - CULBERSON, C.F., CULBERSON, W.L., JOHNSON, A.: Biochem. Syst. Ecol. 11, 77-84 (1983). - CULBERSON, W.L., CULBERSON, C.F.: Occ. Papers Farlow Herb. 16, 37-41 (1981); - Syst. Bot. 7, 158-169 (1982). - CULBERSON, W.L., CULBERSON, C.F., JOHNSON, A.: Mycologia 74, 662-667 (1982). - CULBERSON, W.L., CULBERSON, C.F., FISCUS, S.A.: Bryologist 86, 254-256 (1983).

DANIELS, F.J.A.: Meddel. Grønland Biosci. 10, 1-78 (1982). - DEGELIUS, G.: Acta Reg. Soc. Sci. Litt. Gothoburg., Bot. 2, 1-127 (1982). -

DIBBEN, M.: J. Hattori Bot. Lab. 52, 343-355 (1982). - DÖBBELER, P., VEZDA, A.: Mitt. Bot. München 18, 1-8 (1982). - DOLL, R.: Die Flechten. Neue Brehm-Bücherei, Band 40. Wittenberg: Ziemsen 1982. - DOMBROVSKAJA, A.V., MAKAROVA, I.I.: Novit. Syst. Plant. Non-Vascul. 19, 144-148 (1982).

EGAN, R.S.: Bryologist 85, 79-83 (1982a); - ibid. 85, 129-130 (1982b). - EGEA, J.M., LLIMONA, X.: Acta Bot. Malacitana 7, 11-38 (1982); - Collectanea Bot. 14, 265-269 (1983). - EGEA, J.M., LLIMONA, X., CASARES, M.: Collectanea Bot. 13, 295-312 (1982). - ELIX, J.A.: J. Hattori Bot. Lab. 52, 407-415 (1982). - ELIX, J.A., STREIMANN, H.: J. Hattori Bot. Lab. 51, 69-97 (1982). - ERIKSSON, O.: Mycotaxon 15, 203-248 (1982a); - Systema Ascomycetum 1, 1-16 (1982b); - Mycotaxon 15, 189-202 (1982c); - Systema Ascomycetum 2, 1-37 (1983). - ESSEEN, P.-A.: Wahlenbergia 7, 73-80 (1982). - EVERSMAN, S.: Bryologist 85, 204-213 (1982).

FERRARO, L.I.: Bonplandia 5, 21-29 (1983). - FILSON, R.B.: J. Hattori Bot. Lab. 53, 357-360 (1982a); - Austr. J. Bot. 30, 511-582 (1982b); - Checklist of Australian Lichens. Nat. Herb. Victoria: 1983. - FOLLMANN, G.: Philippia 5, 33-39 (1982). - FOLLMANN, G., POELT, J.: Philippia 4, 370-378 (1981). - FOUCAULT, B., SERUSIAUX, E., van HALUWYN, C.: Cryptogamie Bryol. Lichenol. 3, 73-76 (1982).

GALLOWAY, D.J.: Lichenologist 15, 307-308 (1983). - GALLOWAY, D.J., JAMES, P.W., WILKINS, A.L.: Lichenologist 15, 135-145 (1983). - GARCIA-ROWE, J., SILVESTRE, S., LLIMONA, X.: Collectanea Bot. 13, 313-317 (1982). - GIAVERINI, R.J.: Proc. Dorset Nat. Hist. Arch. Soc. 103, 107-110 (1982). - GILBERT, O.L.: Lichenologist 14, 90-91 (1982). - GILBERT, O.L., LAMBLEY, P.W.: Lichenologist 14, 185-188 (1982). - GILBERT, O.L., MITCHELL, J.: Glasgow Naturalist 20, 123-132 (1981). - GILBERT, O.L., FOX, B.W., PURVIS, O.W.: Lichenologist 14, 165-174 (1982). - GOLUBKOVA, N.S.: Biol. Res. Prirod. Usl. Mongol. Nar. Resp. 16, 1-201 (1981); - Novit. Syst. Plant. Non-Vascul. 19, 131-143 (1982). - GOMEZ, A., HLADUN, N.-L.: Collectanea Bot. 13, 319-322 (1982). - GOWARD, T., AHTI, T.: Ann. Bot. Fennici 20, 9-13 (1983).

HAFELLNER, J.: Phyton Austria 22, 23-50 (1982); - Herzogia 6, 283-287 (1983a); - ibid. 6, 289-298 (1983b); - ibid. 6, 299-308 (1983c); - Nova Hedwigia 35, 645-729 (1983d); - Lichenologist 15, 263-266 (1983e). - HAFELLNER, J., BELLEMERE, A.: Nova Hedwigia 35, 207-235 (1982a); - ibid. 35, 237-261 (1982b); - ibid. 35, 263-312 (1982c); - ibid. 38, 169-186 (1983). - HALE, M.E.: Mycotaxon 16, 162-164 (1982); - The Biology of Lichens, 3. ed. London, Melbourne, Baltimore: Arnold 1983a; - Lichenologist 15, 157-160 (1983b). - HALVORSEN, R., BENDIKSEN, E.: Nord. J. Bot. 2, 371-380 (1982). - HAMADA, N.: Can. J. Bot. 60, 383-385 (1982). - HANKO, B.: Bibl. Lichenol. 19, 1-296 (1983). - HANSEN, E.S.: Meddel. Grønland Biosci. 9, 1-33 (1982); - Mycotaxon 18, 175-180 (1983a); - ibid. 18, 483-494 (1983b). - HAWKSWORTH, D.L.: Lichenologist 14, 75-76 (1982a); - ibid. 14, 83-86 (1982b); - ibid. 14, 131-137 (1982c); - Nova Hedwigia 35, 755-762 (1982d); - J. Hattori Bot. Lab. 52, 323-329 (1982e); - ibid. 52, 357-366 (1982f); - Notes R. Bot. Gard. Edinburgh 40, 375-397 (1982g); - Lichenologist 15, 1-44 (1983a); - ibid. 15, 151-156 (1983b); - Brit. Lich. Soc. Bull. 52, 42-44 (1983c). - HAWKSWORTH, D.L., JAMES, P.W.: Cryptogamie Bryol. Lichenol. 4, 169-173 (1983). - HAWKSWORTH, D.L., SHERWOOD, M.A.: Mycotaxon 16, 262-264 (1982). - HAWKSWORTH, D.L., SUTTON, B.C., AINSWORTH, G.C.: Ainsworth and Bisby's Dictionary of the Fungi (including the lichens). 7. ed. Kew Surrey: Commonwealth Mycol. Inst. 1983. - HAYWARD, B.W., HAYWARD, G.C.: Tane, J. Auckland Univ. Field Club 28, 125-133 (1982). - HAYWARD, B.W., WRIGHT, A.E., HAYWARD, G.C.: Tane, J. Auckland Univ. Field Club 27, 13-16 (1981). - HEISELMAYER, P., TÜRK, R.: Florist. Mitt. Salzburg 6, 3-23 (1979). - HENNINGS, C.J.: Bryologist 86, 64-73 (1983). -

HENSSEN, A.: Mycotaxon 18, 97-111 (1983). - HENSSEN, A.M., JAMES, P.W.: Bull. Brit. Mus. Nat. Hist., Bot. Ser. 10/4, 227-256 (1982). - HENSSEN, A., VOBIS, G., RENNER, B.: Nord. J. Bot. 2, 587-599 (1982). - HENSSEN, A., RENNER, B., MARTON, K.: Mycotaxon 18, 29-48 (1983). - HERTEL, H.: Mitt. Bot. München 18, 297-340 (1982a); - ibid. 18, 443-454 (1982b); - ibid. 19, 441-447 (1983). - HERTEL, H., ZHAO, C.-F.: Lichenologist 14, 139-152 (1982). - HLADUN, N.L.: Folia Bot. Miscell. 2, 39-40 (1981); - Aportación a la flora, morfología y vegetación de los líquenes silicícolas de la parte alta del Montseny (Cataluña). Barcelona: Universitat, Centre de Publicacions 1982a; - Collectanea Bot. 13, 323-324 (1982b). - HLADUN, N.L., GOMEZ-BOLEA, A.: Folia Bot. Miscell. 3, 17-19 (1982). - HOISINGTON, B.L., MAASS, W.S.G.: Bryologist 85, 122-125 (1982). - HOLOPAINEN, T.H.: Ann. Bot. Fennici 19, 39-52 (1982). - HONEGGER, R.: Lichenologist 14, 205-217 (1982a). - J. Hattori Bot. Lab. 52, 379-391 (1982b); - ibid. 52, 417-429 (1982c); - Lichenologist 15, 57-71 (1983). - HOUMEAU, J.M., ROUX, C.: Bull. Soc. Bot. Centre-Ouest, N.S. 13, 263-272 (1982a); - ibid. 13, 273-277 (1982b); - ibid. 13, 278-280 (1982c). - HUOVINEN, K., AHTI, T.: Ann. Bot. Fennici 19, 225-234 (1983).

IKOMA, Y.: Macrolichens of Japan and adjacent regions. Tottori City 1983. - INOUE, M.: J. Sci. Hiroshima Univ., Ser. B, Div. 2, 18, 1-55 (1982). - J. Jap. Bot. 58, 113-128 (1983a); - ibid. 58, 161-173 (1983b); - ibid. 58, 225-236 (1983c).

JACKSON, D.W.: Scan. Electr. Micr. 3, 279-284 (1981). - JAHNS, H.M.: Lichenologist 14, 261-265 (1982). - JAHNS, H.M., FRITZLER, E.: Herzogia 6, 243-270 (1983). - JAHNS, H.M., OTT, S.: Herzogia 6, 201-241 (1983). - JAHNS, H.M., PFEIFER, K., SCHUSTER, G.: Ber. Dtsch. Bot. Ges. 95, 313-322 (1982). - JAMES, P.: Brit. Lich. Soc. Bull. 51, 27-36 (1982). - JANEX-FAVRE, M.C.: Cryptogamie Bryol. Lichenol. 2, 253-275 (1981); - ibid. 3, 337-349 (1982). - JØRGENSEN, P.M.: Aust. J. Bot. Suppl. Ser. 10, 43-66 (1983). - JØRGENSEN, P.M., GALLOWAY, D.J.: N. Z. J. Bot. 21, 335-340 (1983). - JØRGENSEN, P.M., JAMES, P.W.: Lichenologist 15, 109-125 (1983). - JØRGENSEN, P.M., VEZDA, A., BOTNEN, A.: Lichenologist 15, 45-55 (1983).

KALB, K.: Herzogia 6, 71-83 (1983). - KASHIWADANI, H., GRADSTEIN, S.R.: Misc. Bryol. Lichenol. 9, 79-81 (1982). - KNOX, M.D.E.: Bryologist 85, 118-121 (1982). - KNOX, M.D.E., BRUSSE, F.A.: J. S. Afr. Bot. 49, 143-160 (1983). - KORNAS, J.: Zesz. Nauk Uniw. Jagiellonsk. Pr. Bot. 9, 99-104 (1981). - KROG, H.: J. Hattori Bot. Lab. 52, 303-311 (1982a); - Nord. J. Bot. 2, 287-292 (1982b); - Lichenologist 15, 305-307 (1983). - KROG, H., SWINSCOW, T.D.V.: Lichenologist 14, 98 (1982); - ibid. 15, 127-130 (1983). - KUROKAWA, S.: Hikobia, Suppl. 1, 439-444 (1981); - Bull. Natl. Sci. Mus. Tokyo, Ser. B. 8, 35-42 (1982).

LARSON, D.W.: Am. J. Bot. 70, 668-681 (1983). - LINDSAY, D.C.: Lichens, pp. 232-243. In: A Fungus Flora of Warwickshire, M.C. CLARK (ed.). London: British Mycol. Soc. 1980; - Proc. Birmingham Natl. Hist. Soc. 24, 125-152 (1981). - LEROND, M.: Act. Mus. Rouen 1981, 3-299 (1981). - LISICKA, E.: Zpravy Cesk. Bot. Spol. Praha 16, 58-60 (1981). - LOPEZ-FIGUEIRAS, M.: Phytologia 51, 423-429 (1982); - ibid. 53, 454-459 (1983).

MAASS, W.S.G.: Nord. J. Bot. 3, 567-576 (1983). - MAGNUSSON, M.: Lichenologist 14, 153-163 (1982). - MAKAREVIC, M.F., NAVROCKAJA, I.L., JUDINA, I.L.: Atlas geograficeskogo rasprostranenija lisainikov v ukrainskich Karpatach. Kiev: Naukova Dumka 1982 (1983). - MAKHIJA, U., NAGARKAR, M.B.: Curr. Sci. 50, 823-824 (1981). - MAMTA, J., AWASHTI, D.D.: Biol. Mem. 7, 165-190 (1982). - MARCOS LASO, B.: Stud. Bot. Salamanca 2, 123-128 (1983). - MARCOS, B., NAVARRO, F.: Stud. Bot. Salamanca 1, 59-64

(1982). - MARTON, K., GALUN, M.: Israel J. Bot. 30, 125-155 (1981). - MAURER, W., POELT, J., RIEDL, J.: Mitt. Abt. Bot. Landesmus. Joanneum Graz 11/12, 1-104; Flechten und Flechtenparasiten, 62-71 (1983). - MAYRHOFER, H.: J. Hattori Bot. Lab. 52, 313-321 (1982); - Lichenologist 15, 267-282 (1983). - McCARTHY, P.M.: Lichenologist 14, 195-197 (1982); - ibid. 15, 231-248 (1983). - McCUNE, B.: Bryologist 85, 13-21 (1982). - MEIER, J.L., CHAPMAN, R.L.: Am. J. Bot. 70, 400-407 (1983). - MEINUNGER, L.: Herzogia 6, 51-59 (1983). - MEYER, M., ROUX, C., SOLLIER, J.: Bull. Soc. Linn. Provence 33, 51-63 (1982). - MEYER, S.L.F.: Mycologia 74, 592-598 (1982). - MOBERG, R.: Lichenologist 15, 161-167 (1983a); - Nord. J. Bot. 3, 509-516 (1983b). - MOBERG, R., HOLMASEN, I.: Lavar. En fälthandbok. Stockholm: Rahm & Stenström 1982. - MOLHO, D.M., BODO, B., CULBERSON, W.L., CULBERSON, C.F.: Bryologist 84, 396-398 (1981). - MÜLLER, E., LOEFFLER, W.: Mykologie, 4. Aufl. Stuttgart: Thieme 1982.

NAGARKAR, M.B., PATWARDHAN, P.G.: Curr. Sci. 50, 1075-1076 (1981a); - Biovigyanam 7, 131-136 (1981b); - ibid. 8, 125-131 (1982). - NASH, T.H. III, SIGAL, L.L.: J. Arizona-Nev. Acad. Sci. 16, 46-50 (1981). - NAVROTSKA, I.L.: Ukr. Bot. Zh. 39, 43-48 (1982). - NORDNES, J.: Nord. J. Bot. 2, 381-382 (1982a); - Blyttia 40, 187-189 (1982b). - NOWAK, J.: Porosty (Lichenes) Tom. 1, Czesc 1. Warszawa, Krakow: Panstwowe Wydawnictwo Naukowe 1983.

OKSANEN, J., AHTI, T.: Ann. Bot. Fennici 19, 275-301 (1982). - OSORIO, H.S.: Comun. Bot. Mus. Hist. Nat. Montevideo 4/63, 1-18 (1981); - ibid. 4/64, 1-8 (1982); - Mycotaxon 18, 81-85 (1983a); - Phytologia 54, 279-282 (1983b). - OSORIO, H.S., FLEIG, M.: Mycotaxon 14, 347-350 (1982); - Phytologia 53, 138-140 (1983). - OSORIO, H.S., HOMRICH, M.H., FLEIG, M.: Phytologia 51, 479-483 (1982). - OSORIO, H.S., BAGINSKI, L.C., PINHEIRO, L.P.: Phytologia 53, 194-196 (1983). - OTTO, G.F.: Bryologist 86, 263-265 (1983). - ØVSTEDAL, D.O.: Cryptogamie Bryol. Lichenol. 4, 217-226 (1983).

PATWARDHAN, P.G., MAKHIJA, U.: Kavaka 8, 17-27 (1980); - Ind. J. Bot. 4, 20-26 (1981a); - Biovigyanam 7, 121-129 (1981b). - PATWARDHAN, P.G., NAGARKAR, M.B.: Curr. Sci. 51, 527-528 (1982). - PEARD, J.L.: Bryologist 86, 244-250 (1983). - PENTECOST, A., COPPINS, B.: Brit. Lich. Soc. Bull. 53, 27-35 (1983). - PIKE, L.H., HALE, M.E.: Mycotaxon 16, 157-161 (1982). - PITERANS, A.V.: Lishainiki Latvii. Riga 1982. - POELT, J.: Int. J. Mycol. Lichenol. 1, 143-160 (1983a); - Nova Hedwigia 38, 73-111 (1983b); - Flora 174, 439-445 (1983c). - POELT, J., HAFELLNER, J.: Herzogia 6, 309-321 (1983).

RANDLANE, T.: Folia Crypt. Estonica 15, 2-5 (1981). - REDON, J., LANGE, O.L.: Flora 174, 213-243 (1983). - RENNER, B.: J. Hattori Bot. Lab. 52, 367-377 (1982). - RENNER, B., GALLOWAY, D.J.: Mycotaxon 16, 197-231 (1982). - RENNER, B., HENSSEN, A., GERSTNER, E.: Z. Naturf., Sect. C, 37, 739-747 (1982). - RICHARDSON, R.M., RICHARDSON, D.H.S.: West. Aust. Herb. Res. Notes 7, 17-20 (1982). - RIEUX, R., ROUX, C.: Ecol. Medit. 8, 153-165 (1982). - RJABKOVA, K.A.: Novit. Syst. Plant. Non-Vascul. 19, 149-154 (1982a); - ibid. 19, 154-164 (1982b). - ROGERS, R.W.: J. Hattori Bot. Lab. 53, 351-355 (1982a); - Muelleria 5, 31-34 (1982b); - Lichenologist 14, 115-129 (1982c). - RONDON-SEIDENBINDER, M.: Cah. Nat., Bull. N.P., n. s. 38, 97-109 (1983). - ROUX, C.: Bull. Soc. Bot. Centre-Ouest, N. S., 13, 210-228 (1982).

SANCHO, L.G.: Collectanea Bot. 13, 339-349 (1982). - SCHUBERT, R.: Flechten - Lichenophyta, pp. 524-626. In: Exkursionsflora für die Gebiete der DDR und der BRD, R. SCHUBERT, H.H. HANDKE, H. PANKOW (eds.), Band 1, Niedere Pflanzen - Grundband. Berlin: VEB Volk und Wissen 1983. - SCHUSTER, G., HEROLD, K., JAHNS, H.M.: Herzogia 6, 183-200 (1983). - SCHUTTE, J.A.: Bryologist 86, 117-120 (1983). - SEAWARD, M.R.D.: Ir. Nat. J. 20, 442-444 (1982); - Nova Hedwigia 37, 325-345 (1983). - SEAWARD, M.R.D., HITCH, C.J.B.: Atlas of the Lichens of the British Isles, Vol. 1. Cambridge: Institute of Terrestrial Ecology 1982. - SEAWARD, M.R.D., BYLINSKA, E.A., TOPHAM, P.B.: Nova Hedwigia 38, 703-716 (1983). - SEDELNIKOVA, N.V.: Novit. Syst. Plant. Non-Vascul. 19, 165-166 (1982). - SERUSIAUX, E.: Nord. J. Bot. 3, 517-520 (1983a); - Lichenologist 15, 283-287 (1983b). - SHARMA, L.R.: Ind. J. Bot. 5, 174-177 (1982). - SHEARD, J.W.: Bryologist 85, 88-95 (1982). - SHIBUICHI, H., YOSHIDA, K.: J. Jap. Bot. 57, 296-298 (1982). - SHOWMAN, R.E.: Ohio J. Sci. 81, 178-179 (1981). - SINGH, A.: Feddes Repert. 93, 67-69 (1982a); - Nova Hedwigia 36, 237-239 (1982b); - ibid. 38, 223-226 (1983). - SINGH, K.P.: Geophytology 11, 242-256 (1981). - SINGH, S.R., AWASTHI, D.D.: Biol. Mem. 6, 169-196 (1981). - SIPMAN, J.J.M.: Bibl. Lichenol. 18, 1-241 (1983). - SIPMAN, H.J.M., AGUIRRE, J.: Caldasia 13/64, 605-634 (1982). - SKOREPA, A.C.: Phytologia 53, 445-446 (1983). - SØCHTING, U., RAMKAER, K.: Nord. J. Bot. 2, 171-181 (1982). - STEINER, M., POELT, J.: Plant Syst. Evol. 140, 151-177 (1982). - STEVENS, G.N.: Lichenologist 14, 39-45 (1982); - ibid. 15, 99-102 (1983a); - ibid. 15, 213-229 (1983b). - SULMA, T., BYSTREK, J.: Acta Mycol. Warszawa 18, 119-121 (1982). - SWINSCOW, T.D.V.: Lichenologist 14, 76-83 (1982).

TAYLOR, R.M.: NMFS Circular 446, I-III, 1-23 (1982). - TEHLER, A.: Taxon 31, 708-714 (1982); - Opera Bot. 70, 1-86 (1983). - THOMSON, J.W.: J. Hattori Bot. Lab. 53, 261-364 (1982a); - Bryologist 85, 251 (1982b). - THOMSON, J.W., SCOTTER, G.W.: Bryologist 86, 14-22 (1983). - TIBELL, L.: Lichenologist 14, 219-254 (1982); - ibid. 15, 131-134 (1983). - TIMDAL, E.: Blyttia 40, 179-185 (1982); - Lichenologist 15, 169-179 (1983). - TØNSBERG, T., HOLTAN-HARTWIG, J.: Nord. J. Bot. 3, 681-688 (1983). - TØNSBERG, T., ØVSTEDAL, D.O.: Blyttia 40, 59-63 (1982). - TOPHAM, P.B.: Lichenologist 14, 86-87 (1982). - TOPHAM, P.B., SWINSCOW, T.D.V.: Lichenologist 14, 91-94 (1982). - TOPHAM, P.B., WALKER, F.J.: Lichenologist 14, 61-75 (1982). - TOPHAM, P.B., SEAWARD, M.R.D., BYLINSKA, E.A.: Lichenologist 14, 47-52 (1982). - TORRES, E., HLADUN, N.-L.: Collectanea Bot. 13, 381-383 (1982). - TRASS, H.: Folia Crypt. Estonica 15, 1-2 (1981). - TSCHERMAK-WOESS, E.: Plant Syst. Evol. 143, 109-115 (1983). - TSCHERMAK-WOESS, E., BARTLETT, J., PEVELING, E.: Plant Syst. Evol. 143, 293-309 (1983). - TUCKER, S.C.: Proc. Louisiana Acad. Sci. 44, 58-70 (1981). - TÜRK, R., WITTMANN, H.: Linzer Biol. Beitr. 14, 127-139 (1983). - TÜRK, R., WITTMANN, H., PILSL, P.: Stapfia 10, 121-137 (1982).

ULVINEN, T., OHENOJA, E., AHTI, T., ALANKO, P.: Oulanka Rep. 2, 1-64; lichens p. 48-63 (1981).

VANA, J.: Folia Geobot. Phytotax. 17, 1-19 (1982). - VERSEGHY, K.P.: Studia Bot. Hung. 16, 53-65, 1982 (1983). - VEZDA, A.: Folia Geobot. Phytotax. 17, 387-392 (1982); - ibid. 18, 45-70 (1983a); - Mitt. Bot. München 19, 151-161 (1983b). - VITIKAINEN, O.: Brit. Lich. Soc. Bull. 50, 28-36 (1982).

WETMORE, C.M.: Bryologist 84, 482-491 (1982a); - Lichenologist 14, 267-271 (1982b). - WIRTH, V.: Ber. Dtsch. Bot. Ges. 96, 103-115 (1983). - WITHROW, K., AHMADJIAN, V.: Mycologia 75, 337-339 (1983).

ZANETTE, V.C., AGUIAR, L.W., MARTAU, L., ARANJO MARIATH, J.E., OSORIO, H.: Iheringia, Ser. Bot. 0/28, 107-140 (1981). - ZHAO, J.D., XU, L.W., SUN, Z.M.: Prodromus lichenum sinicorum. Beijing, China: Science Press 1982.

Doz. Mag. Dr. JOSEF HAFELLNER
Institut für Botanik
Karl-Franzens-Universität
Holteigasse 6
A-8010 Graz, Austria

IV. Systematics of the Bryophytes

By WOLFGANG FREY

1. Fundamental Perspectives

In recent years the pace of bryological research has quickened and the dissemination of new findings has increased, creating a scientific climate comparable only to that at the beginning of this century. Fundamental and comprehensive publications dealing with almost all branches of bryology have revealed the wide scope of this subject and depicted the problems currently under investigation; ideas have been put forward and questions have been raised which illuminate future lines of inquiry in research.

The beginning of this new era is marked by the article by SMITH (1978) entitled *Cytogenetics, biosystematics and evolution in Bryophyta*. Here, in an "outline phyletic scheme of the Bryopsida based on chromosome numbers", he impressively demonstrates that the Musci are a class containing numerous polyploid taxa (gametophytes). There is considerable evidence that the Hepaticae, too – apart from important exceptions such as *Takakia* with a basic number x = 4,5 – are diploid ("paleodiploidy") and in few cases polyploid (SCHUSTER 1981a). Only with these facts in mind can we understand the 400-million-year evolution of these plants and their present-day dominance in some macroniches. No longer can the bryophytes be regarded as the one plant group which has arrived at a "dead end" in evolution and has become fixed, with no potential for further development. On the contrary, they are a highly successful group, which shows evolutionary potential comparable to that of the angiosperms.

The evolution of the terrestrial plants in the Upper Silurian followed two distinct pathways. The first was an "attempt" by nature to conquer the land with the haploid gametophyte and to develop vegetation units. This was done by the bryophytes and Anthocerotae, in which the "haploid" stage is the actual vegetative-productive phase. The second was an attempt to conquer the land with the diploid sporophyte. The decisive breakthrough was achieved with the seed plants, especially the angiosperms. The selection of "paleodiploid" and "polyploid" gametophytes in the bryophytes, which are functionally haploid, served to "buffer" the group against the environment and enabled it to survive and evolve over the course of 400 million years. In spite of all the reduction that must be attributed to adaptation to ecological macroniches, this group exhibits an immense and fascinating diversity. The bryophytes show as wide a range of adaptations to habitat as the angiosperms; they possess a high vegetative plasticity and a similar range of types of life cycle, from short-lived ephemerals to long-lived perennials, i.e., dendroids. Many are able to withstand long periods of unfavorable conditions. They are a group at an advanced stage of development, with many taxa (ANDERSON 1980, FREY 1981, SMITH 1979, VITT 1981a, VRIES et al. 1983).

Progress in Botany, Vol. 46

2. Systematics

a) Floras

The increasing number of floras demonstrates a worldwide increase in bryologic research. Above all in China there is an intensive study of bryophytes (compare Sect. 2c).

MARGADANT and DURING (1982), LANDWEHR (1978, 1980): Netherlands; FRAHM and FREY (1983): Federal Republic of Germany; WATSON (1981): Great Britain; JAHNS (1980): Europe; SJÖGREN (1978): Azores; BACHURINA and PARTYKA (1979): Ukraine. Volumes 3 - 5 on hepaticae of the northern part of the U.S.S.R. were published by SHLYAKOV (1980, 1981, 1982). Thus, there is a richly illustrated extensive flora on the northern part of the U.S.S.R. as far southward as the arid steppes. Conspectus of the moss flora of Mongolia: ABRAMOVA and ABRAMOV (1983). In North America SCHUSTER (1980a) concludes a leafy hepatic treatment, except for the Jubulaceae, from the Cephaloziellaceae through the Lejeuneaceae. CRUM and ANDERSEN (1981) publish the *Mosses of the Eastern North America*. IRELAND's manual (1982) covers the moss flora of the maritime provinces of Canada. One of the most fascinating moss floras is that of MAGILL (1981), of South Africa south of Namibia. The first fascicle (Sphagnaceae-Grimmiaceae) contains excellent figures. This flora, important as far as the arid regions of the southern hemisphere are concerned, promises to become an outstanding work. GANGULEE (1978, 1980) finished his flora of the *Mosses of Eastern India* with fascicles 7 and 8. CHOPRA and KUMAR (1981): West Himalayas; JOHNSON (1980): Singapore and Malaysia; CHOE (1980): Korea; MANUEL (1981a): Malaysia and Singapore; CATCHESIDE (1980): South Australia; MILLER et al. (1983): Polynesia. Two floras (GAO and CHANG 1981, CHANG 1978) and an index (GAO and CHANG 1983) demonstrate the bryological activities in China.

b) Systematics Above the Family Level, Important Lists

New ideas concerning the relationships among the Bryopsida are set forth by VITT (1981b). The Bryopsida are divided into three subclasses: Polytrichidae, Tetraphidae, Bryidae. VITT suggests that the direction of peristome evolution has been one of reduction. The plesiomorphous state in the arthrodontous mosses is diplolepiduous, with endostome segments opposite the exostome teeth. From this type is derived the apomorphous peristome type. Likewise, alternately toothed diplolepiduous types are also derived from the opposite diplolepiduous peristome. Thus, the haplolepiduous peristome has evolved from a diplolepiduous ancestral type. To accord with the above evolutionary considerations, VITT divides the Bryidae into 14 orders, positioned rather differently from what was done in earlier systems: Archidiales, Funariales, Splachnales, Orthotrichales, Bryales, Hypnobryales, Isobryales, Hookeriales, Encalyptales, Buxbaumiales, Pottiales, Dicranales, Fissidentales, Seligeriales, Grimmiales.

Here the peristome teeth are given an extremely important status. It remains to be seen whether this great emphasis upon one trait will stand up to criticism. One should at least point out that it is not consistent with the paleobotanic data, the notion of the divergence of bryophytes through time, and with the comparative data on the gametophyte.

An extremely meticulous study of the systematics of the Musci has been written by WALTHER (1983). His arrangement of the units is based on classical considerations; like VITT (1981b), however, he neglects to discuss the relationships between them. The Orthotrichales are treated separately from the Isobryales. The relationships among the very heterogeneous Isobryales remain unclear. Detailed studies of the systematics and evolution of this order can be expected to result in a substantial revision of our knowledge. FRAHM and FREY (1983) point out the connections between the large related groups of the Musci. Fissidentales, Dicranales, Encalyptales, Pottiales and Grimmiales form one unit (Dicrananae, the Eubryales according to FLEISCHER form two related circles linked with one another only by the double "bryoide" peristome (Bartramianae, Bryanae). The pleurocarpous orders are combined in the superorder Hypnanae. The Funarianae form a complete unit. MILLER (1979) presents the various systems since FLEISCHER *(Musci of the flora of Buitenzorg)*. Due to the differing evaluation of the peristome the interpretation of the family relationships presents insurmountable difficulties. IWATSUKI (1981) emphasizes in his general survey on the taxonomy of mosses that generic revisions of some large families are urgently needed, especially within the families with hypnaceous peristomes (i.e., Amblystegiaceae, Hypnaceae).

The *Nomina generica Hepaticarum* (GROLLE 1983a) certainly are a basis for every taxonomic work on Hepaticae, the list of hepatics and mosses of Europe and the Azores (GROLLE 1983b, CORLEY et al. 1981) a basis for taxonomic work in Europe. SCHUSTER (1979) and ENGEL (1982) give a conjectural phylogenetic classification of the Hepaticae, which is certain to be presented in detail in the *New Manual of Bryology* by SCHUSTER (1983).

c) Systematics on the Family Level and Below (M = revisions and monographs)

Allisoniaceae: *Sandeothallus* gen. nov. *Allisonia*, *Calycularia*, *Sandeothallus*: SCHUSTER (1982). Amblystegiaceae: *Pictus scoticus*, gen. nov. in Scotland. For well-explored Europe a great discovery: TOWNSEND (1982). *Drepanocladus crassicostatus*, sp. nov., which is described as new from North American collections of living populations and from fossils of Pleistocene and Holocene age: JANSSENS (1983). Anthocerotaceae: M of the *Nothothylas* species of Japan and adjacent areas: HASEGAWA (1979). Bryaceae: Variations between different clones isolated from temperate, tropical and polar populations of *Bryum argenteum*. The clones examined are morphologically distinct from each other in cultivation (protonema, shoot arrangement, leaf morphology): LONGTON (1981). M of the propaguliferous species of *Pohlia* in North America with uniform sporophyte characters but distinct propagula: SHAW (1981a). Besides the propagula the ecology could define difficult *Pohlia* species: SHAW (1981b). M of the neotropical Bryoideae, 350 species: OCHI (1980, 1981). Bryoideae in southern South America: OCHI (1982a). Bryoideae in Australasia and migration of species into this region: OCHI (1982b). Calymperaceae: M of the tropical West African *Calymperes*-species: EDWARDS (1980). M of *Mytthyridium* in Oceania: NOWAK (1980). Key for the American *Calymperes*- and *Syrrhopodon*-species: REESE (1983). *Calymperopsis* and *Syrrhopodon* species of China: LIN (1979). Conspectus of the African species of *Syrrhopodon*: ORBAN (1981). Cryphaceae: M of the family in Mexico, Central America and the Caribbean: MANUEL (1981b). Family concept: MANUEL (1982). *Dendrocryphaea*, a genus with antipodal distribution in South America: GRIFFIN et al. (1982). Dicranaceae: M of *Campylopus* in North America: FRAHM (1980a). Range of variations of vegetative characters in the *Campylopus cavifolius* complex: FRAHM (1980b). *Chorisodontium* in southern South America and Indian Ocean: FRAHM (1981a). Taxonomy of South American *Campylopus* species: FRAHM (1981b). Infrageneric classification in the genus *Campylopus*, subgenera *Campylopus*, *Thysanomitrion*: FRAHM (1983a). M of *Pilopogon*. Evolution of species from an ancestor as result of isolation by continental drift: FRAHM (1983b). *Campylopus* subgen. *Pseudocampylopus* with afromontane distribution: De SLOOVER (1983). *Kingiobryum paramicola* in the Paramos of Columbia, taxonomy and ecology: ZANDER and CLEEF (1982). Ditrichaceae: M of *Ditrichum* in Australia, New Zealand and Pacific Ocean: SEPPELT (1982a,b). Encalyptaceae: M of Encalyptaceae with an excellent part on the anatomy of the species: HORTON (1982a, 1983). Present research all over the world: HORTON (1982b). *Encalypta* in the West Himalayas: KUMAR (1980). Entodontaceae: Generic revision of Entodontaceae, mainly based on peristome characters; only *Entodon*, *Mesonodon*, *Erythrodontium* and *Pylaisiobryum* are retained in Entodontaceae, the other genera are transferred to different pleurocarpous families: BUCK (1980). M of *Entodon* in China: RENLIANG (1983). Erpodiaceae: First survey on *Erpodium* and *Aulocopilum* in Brazil: VITAL (1980). Fabroniaceae: M of *Fabronia* in South America: BUCK (1983). Fissidentaceae: *Fissidens viridulus* complex in England: CORLEY (1980). M of *Fissidens* in China: IWATSUKI (1980). M of the section *Pachylomidium* of *Fissidens*, inclusive the *Fissidens bryoides* complex: BRUGGEMAN-NANNENGA (1982). Frullaniaceae: High Andean species of *Frullania* subgen. *Chonanthelia*: HAARBRINK (1981). *Frullania* in Madagascar: VANDENBERGHEN (1982). Continuation of M of Asian, Pacific and Australasian species of *Frullania*. New Guinea is a speciation center. There occur 77 species, 43 are endemic: (HATTORI 1980a,b, 1981a,b, 1982, 1983). Funariaceae: *Physcomitrium*, *Physcomitrella* and *Goniomitrium* in New Zealand: FIFE (1982). Significant differentiations between populations of *Funaria hygrometrica*, originating in various geographic-climatic regions, but there is no isolation by reproductive barriers: WEITZ and HEYN (1981). *Cygnicollum* nov.gen. with cleistocarpous capsules adapted to arid regions: FIFE and MAGILL (1982). Geocalycaceae: Gondwanalandic taxa: M of *Clasmatocolea* (Lophocoleoideae), originated in the Southern South American sector of Gondwanaland: ENGEL (1980). Brevianthaceae, fam. nov. from Tasmania: ENGEL and SCHUSTER (1981). Leptoscyphoideae of Gondwanaland. *Pedinophyllopsis* and *Pseudolophocolea*: SCHUSTER and ENGEL (1981, 1982). Gigaspermaceae: On the strength of spore morphology *Costesia* and *Neosharpiella* are transferred to the G. The family has a widely scattered predominant Southern Hemispheric distribution: FIFE (1980). *Gigaspermum* in the Mediterranean area: HERRNSTADT et al. (1980). Grimmiaceae: Study of mixed stands and M of the *Rhacomitrium canescens* group: FRISVOLL (1983a). M of *Rhacomitrium* in Svalbard: FRISVOLL (1983b). M of *Schistidium*: BREMER (1980a,b, 1981). Infraspecific taxa of *Grimmia tergestina* in Europe with variance analysis: NOWAK and POELT (1979). Phylogenetic analysis, classification and synopsis of the genera: CHURCHILL (1981). Herbertaceae: Forest species and Paramo species of *Herbertus* in Colombia: REENEN (1982). Jungermanniaceae: Worldwide distribution of *Notoscyphus*, concentrated in South and Southeast Asia: UDAR and KUMAR (1981). Lejeuneaceae: East Asiatic species of *Schiffneriolejeunea* sect. *Saccatae*: GRADSTEIN and TERKEN (1981). *Archilejeunea*, *Drepanolejeunea*, *Schiffneriolejeunea* in India: UDAR and AWASTHI (1981, 1982a,b). *Harpalejeunea*, *Nephelolejeunea* in Australasia: GROLLE (1980a,b). L. in Tasmania: GROLLE (1982). M of *Dicranolejeunea* in Asia: MIZUTANI (1980). Southern Hemisphere Thyamaelloideae: SCHUSTER (1981b). M of *Diplasiolejeunea* in Cuba: REYES (1982). M of Cololejeunoideae in the Indian Ocean incl. Madagascar: TIXIER (1979). Historical development of generic concepts and current state of knowledge: GRADSTEIN (1979). Lepidoziaceae: *Drucella* is placed in a separate tribe Drucellae: BOESEN (1982). *Dendrobazzania* gen. nov., known from Bhutan-Bengal and Queen Charlotte island is a relict and takes an intermediate position between the Lepidozioideae and Bazzanioideae: SCHUSTER and SCHOFIELD (1982). Systematic position of *Kurzia*, *Megalembidium*, *Psiloclada*, *Isolembidium*: SCHUSTER (1980b). Marchantiaceae: The spore morphology and germination indicate the division of *Marchantia* in two subgenera: *Marchantia*, *Chlamidium*: BISCHLER (1982). Metzgeriaceae: The evolutionary progenitor was probably erect and pinnately branched: KUWAHARA (1981). *Metzgeria* species of Colombia: KUWAHARA (1982). Mniaceae: M in different parts of the world: KOPONEN (1980a,b, 1981a,b, 1982, 1983). Mniaceae of China, 56 species are known: LI and CHANG (1979). Orthotrichaceae: M of *Macrocoma*. Analysis of the *Macrocoma tenue* - *M. sullivantii* complex. This species complex consists of an Old World-Southern Hemisphere taxon and a New World and Northern Hemisphere Old World taxon and is probably of Gondwanalandic origin. The species migrated northward and were components of the Arcto-Tertiary flora: VITT (1980a,b, 1982b). M of *Leiomitrium* and *Cardotiella* gen. nov., closely related to *Macrocoma*: VITT (1981c). Generic concept of O.: VITT (1982a). M of New Zealand species of the pantropical genus *Macromitrium*. The genus originated in Gondwanaland: VITT (1983). The four African species of *Macrocoma* belong to the afromontane phytogeographic region: MAGILL and VITT (1981). Pallaviciniaceae: M of *Jensenia* in the Andes of Colombia; dendroid species: GRONDE (1980). Pelliaceae: Evolution and speciation in *Pellia*. *P. endiviifolia* is a relatively primitive taxon. *P. megaspora*, endemic in boreal America, is a distinct species with a unique caryotype. Biochemical data suggest that Japanese populations currently assigned to *P. endiviifolia* may represent an own subspecies: SCHUSTER (1981c), NEWTON (1981), MENDELAK (1981), KRZAKOWA (1981a). Protein markers allow to separate the four *Pellia* species: ZIELINSKI et al. (1981). Plagiochilaceae: *Plagiochila* in the Galapagos islands: INOUE and GRADSTEIN (1980). Taxonomic studies in *Plagiochila*: M of the species of Samoa, Taiwan, Australasia and the Pacific: INOUE (1981, 1982a,b). Generic concept in the Plagiochilaceae: INOUE (1982c). New species of *Plagiochila* in China: CHEN and WU (1979). *Plagiochila* in Madagascar: VANDEN-BERGHEN (1981). Intermediate populations between *P. asplenioides* and *P. porelloides*: KRZAKOWA (1981b). Plagiotheciaceae: New *Plagiothecium* species from Tibet: LI (1981a). Polytrichaceae: Phenotypic variation of *Polytrichum alpinum*. Quantitative analysis of 12 characters: YLI-REKOLA (1980). Porellaceae: M of *Porella* in North America: WON (1983). Pottiaceae: M and keys to *Didymodon*, *Barbula*, *Pseudocrossidium*, *Bryoerythrophyllum* in North America: ZANDER (1978 - 1979, 1979, 1981a,b). *Streptocalypta* in Middle and South America: ZANDER (1982). *Astomum* in Australia, *Viridivellus* is considered to be in the P.: STONE (1980). M of *Tortula* sect. *Rurales* in the eastern Holarctic kingdom. The eastern Mediterranean area and especially South West Asia are speciation centers and centers for adaptations to arid environment: KRAMER (1980). *Pseudaloina*, nov. gen. in Yemen. Xerophytic taxa between *Aloina* and *Aloinella*: DELGADILLO (1982). *Porella* in Chinese boreal orient: GAO and AUR (1978). New taxa from China: LI (1981b). Pterobryaceae: M of *Symphysodontella*; SE Asia, Pacific: MAGILL (1980). M of *Trachyloma*; Australasia. Calyptra is one of the relevant characters: MILLER and MANUEL (1982). Radulaceae: Prior taxonomic subunits of the genus *Radula*: SCHUSTER (1980c). M of *Radula* in Ceylon: ONRAEDT (1983). *Radula* from the forests and Paramos of Colombia, above 2500 m: JANS (1979). Rhizogoniaceae: M in Malaya: MANUEL (1981c). Ricciaceae: M of *Riccia* in Australia: NA-THALANG (1980). *Riccia* species from oriental boreal China: GAO and CHANG (1978). Scapaniaceae: M of *Scapania* in western North America: HONG (1980). *Scapania* subgenus *Plicaticalyx* in eastern Himalayas and south China: AMAKAWA (1981). Seligeriaceae: M of *Seligeria* in New Zealand. Evolution of the species after long-distance dispersal and subsequent evolutionary divergence: VITT and BARTLETT (1983). Sorapillaceae: On the basis of reduced double peristome and pleurocarpous habit a placement in the Isobryales is correct: ALLEN (1981). Sphagnaceae: Northern boreal and arctic species: LANGE (1982). North American Atlantic coastal plain Sphagna, methodological survey: LANE (1981). *Sphagnum capense* group in the afromontane region: EDDY (1979). Splachnaceae: General classification: KOPONEN, A. (1982a). M of Splachnaceae in Australasia and Pacific: KOPONEN, A. (1982b). Splachnobryaceae: Monotypic moss family, related to the families of Funariales: KOPONEN, A. (1981). Thuidiaceae: M in Latin America: GIER (1980). New *Thuidium* species from China: CHANG (1981). M of *Trachyphyllum*, transferred from Entodontaceae to Thuidiaceae: BUCK (1979). Timmiaceae. *Timmia* sect. *Timmiaura*: BRASSARD (1980). Vandiemeniaceae: New family of Metzgeriales. A Metzgerialian taxon with vegetative morphology of *Riccardia* and male branches like *Metzgeria* is described from Tasmania: HEWSON (1982).

d) Cytosystematics

Probably one of the most important discoveries during the period under review was that of mosses with the chromosome number n = 4 (*Hypnodendron dendroides*, *H. comosum*) (RAMSAY 1983). The crucial question is whether this number is primitive (basic) or produced by reduction from a higher number. In these taxa one can still see the transition from an acrocarpous to a pleurocarpous life form and from a parenchymatic to a prosenchymatic leaf-cell net. We must assume that the ancestors of such groups were the point of departure for the pleurocarpous Musci. If this number turns out to be basic, our knowledge of the groups at the point of divergence between the pleurocarpous and modern acrocarpous mosses (Bryales) will have taken an important step forward, revolutionizing the concept of polyploidy in the mosses.

In flowering plants polyploidy is of frequent occurrence and has played an important role in the evolution of species. The situation in the spermatophyta seems to be paralleled by that in Bryophyta. The Bryophyta possess also small chromosomes. The basic numbers, as seen at the moment, seem to be x = 6 or 7 in mosses, or x = 8 or 9 in liverworts. About 90% of mosses studied cytologically are diploid or of higher ploidy, compared with about 10% of liverworts (SMITH and RAMSAY 1982, CROSBY 1980). For general data of polyploidy and the basic numbers compare also ANDERSON 1980, RAMSAY 1983, and SCHUSTER 1981a.

Investigations on the cytotypes of *Atrichum undulatum* show that the triploids are the most frequent (57% of the populations sampled), the haploids the rarest (11%). The cytotypes are – except for the chromosome length – indistinguishable morphologically and cannot be accorded taxonomic recognition (ABDERRAHMAN and SMITH 1982, 1983). The chromosome evolution in Brachytheciaceae (McADAM 1982) shows an increase through polyploidy to 12, then reduction through elimination and fragmentation to 11, 9 and 7. The investigations of BOWERS (1980) on the chromosome numbers and karyotypes of Mniaceae confirm the generic concept of KOPONEN. Only the genus *Pseudobryum* has been reduced to the level of a section in the genus *Plagiomnium*. The survey of De SLOOVER and MESSE (1982) indicates that in Belgium the diploid *Pellia borealis* replaces the haploid *P. epiphylla* at higher elevations.

The chromosome number n = 44 in *Stereophyllum tavoyense* is the highest known for all pleurocarpous taxa (VERMA and KUMAR 1983), n = 27 the highest known for *Lophozia*, namely *Lophozia excisa* in antarctic environment (OCHYRA et al. 1982).

The known data about chromosome counts of Musci and Hepaticae are compiled by FRITSCH (1982) (Musci 1300 taxa, liverworts 700 taxa).

e) Chemosystematics

Although numerous studies have been published (especially in Phytochemistry), the reviewer is not of the opinion that a breakthrough has already been achieved in this still young branch of bryology. Several summary publications have appeared, building on the work described in the article by MARKHAM and PORTER (1978). In this investigation it was shown that the Marchantiales are characterized by flavon-O-glycosides and the Jungermanniales by flavon-C-glycosides.

The conclusions that "the Bryophyta are an unnatural group and a rather tenuous assemblage of the five groups Anthocerotae, Hepaticae, Bryopsida, Andreaeopsida and Sphagnopsida" (SUIRE and ASAKAWA 1979) are not helpful. Various authors (SUIRE and ASAKAWA 1979, 1981, ASAKAWA et al. 1980b, 1981a,b, ASAKAWA 1982, GRADSTEIN et al. 1981, MATSUO 1982) emphasize that the most valuable compounds for chemotaxonomy of bryophytes are the flavonoids and terpenoids, especially the sesquiterpenoids. Mono-, sesqui- and diterpens are produced only by liverworts with oil bodies and probably by some Anthocerotae species, and are significant chemosystematic markers of Hepaticae. Sesquiterpenoids are probably reliable distinctive criteria at the specific and generic level. The change in the composition of lipids and alkanes is partially related to ecological conditions, especially to the moisture of environment. The use of such data as a taxonomic criterion is difficult.

Musci possess in general highly unsaturated fatty acids, triglycerids, stereol mixtures and hopane triterpene hydrocarbons (ASAKAWA et al. 1980a,b, HUNECK 1980, KARUNEN 1982).

A first comprehensive report on biologically active substances from bryophytes, the plant group which was considered almost useless for human life, is given by ASAKAWA (1981). The author reviews the known details on pungency and bitterness, allergenic contact dermatites by *Frullania* species, on anticancer activity (*Polytrichum* spp., *Diplophyllum* spp., *Chiloscyphus* spp.), antimicrobial activity against gram-positive bacteria, and antifungal activity. The isolation and the structure determination of the active substances have not been established.

Flavonoid chemistry. In a survey on distribution of flavonoids in bryophytes, the authors emphasize that hardly any investigation exists on the biogenesis of these compounds: ZINSMEISTER and MUES (1980). Metzgeriaceae, Blasiaceae, Hymenophytaceae are characterized by their flavonoids as main phenolic components, whereas the Pelliaceae and Aneuraceae have characteristic cinnamic acid-type compounds. The families of Jungermanniales can be divided in three groups: (1) with cinnamic acids, (2) with possibly flavanone or flavonol derivatives, (3) flavones as main compounds: MUES (1982a).

Flavone-di-C-glycosides, first detection in a species of liverworts: MUES (1982b). Porellaceae, as the more advanced family, have evolved the ability to synthesize flavones, while this ability is apparently not evolved in the Ptilidiaceae and Mastigiophoraceae: MUES (1982c). *Frullania dilatata* and *F. tamarisci* differ in their flavonoid patterns: MUES et al. (1983). European, North American and Asian samples of *Conocephalum conicum* are clearly distinguished by their flavonoid glycosides. The local Japanese species *C. supradecompositum* differs also on specific level: PORTER (1981). Confirmation of flavonols by *Corsinia*, the only example in the Marchantiales: MARKHAM (1980). Flavonoid patterns indicate the separation of *Apometzgeria* as an individual genus: THEODOR et al. (1981), and Metzgeriaceae and Hymenophytaceae are clearly divergent: THEODOR et al. (1980). *Takakia* flavonoids appear to be unrelated with those of *Haplomitrium gibbsiae*, and an isolated phylogenetic position is suggested for this genus: MARKHAM and PORTER (1979). Among 12 pleurocarpous mosses, only *Pleurozium schreberi* possesses flavonoids: VANDEKERKHOVE (1980).

Enzymatic variations. The enzyme system of populations of *Atrichum angustatum* indicates a significant level of genetic variability: CUMMINS and WYATT (1981). High patterns of variation in isoenzymes and high polymorphism within and between the populations of *Sphagnum pulchrum*: DANIELS (1982). After the serological variability in European *Pellia* species, *P. epiphylla* is the most distinct form and the oldest in Europe: SZWEYKOWSKI et al. (1981a,b). Two forms of *Conocephalum conicum* by enzyme types: ODRZYKOSKI et al. (1981), SZWEYKOWSKI et al. (1980). Isoenzyme polymorphism in populations of *Plagiochila asplenioides*: SZWEYKOWSKI and KRZAKOWA (1979), KRZAKOWA and SZWEYKOWSKI (1979).

3. Morphology and Anatomy

a) Conducting Tissues

In the course of their evolution the Musci have experienced a wide-ranging reduction of the conducting tissues. In some units, however, there has been a differentiation (particularly in the hadrome) into more efficient structures. Whereas in the Polytrichales only the transverse walls of the hydroids are hydrolyzed, this is true of the longitudinal walls in *Timmiella barbuloides*. There is a dissolution of the non-cellulosic polysaccharides. Breaks in the walls occur sporadically, and the middle lamellae are often drawn away (LIGRONE et al. 1980). In *Buxbaumia piperi* most hydroids merge together as a result of the disruption of the walls and form one continuous capillary canaliculus (LIGRONE et al. 1982b). Whereas in the Hepaticae true plasmodesmata-derived pores (perforations) in the hydroid walls are known (e.g., HEBANT 1979, 1980), such structures have not yet been identified in the Musci. FREY and RICHTER (1982) observed structures in the longitudinal walls of *Hypopterygium commutatum* which could correspond to round perforations. The Musci are the only group among the higher plants in which no really unmistakable perforations have yet been found in water-conducting cells. However, the occurrence of perforated hydroids in mosses is very probable because perforated water-conducting cells have evolved independently in all of the other groups (Hepaticae, Pteridophyta, Gymnospermae, Angiospermae) under the same selection pressure. Since it is evident that no perforations are to be found in the Polytrichales and Dawsoniales, which have been the subject of numerous investigations, the search must concentrate on groups in which a further differentiation of the water-conducting system was necessitated by adaptation to certain biotopes or by the evolution of certain life strategies (e.g., dendroids).

It is surprising that the characteristic cells of the leaf rib, the deuters, which are of great systematic importance (FREY 1981), have only very recently been analyzed by electron microscopy (SCHEIRER 1983). These investigations reveal the deuters to be large-diametered parenchymatic elements which exhibit papillate wall ingrowths characteristic of transfer cells. The deuter cytoplasm includes a large number of mitochondria, numerous polysomes and a peripheral network of ER associated with the wall ingrowths, which is typical of transfer cells. The deuters might be involved in intensive short-distance fluxes of solutes.

b) Transfer Cells

The lower part of the embryo differentiates to become a haustorium and penetrates into the tissue of the archegonial pedicel, which often enlarges considerably; sometimes it even reaches deeply into the tissues of the central strand. In this way a connection is established between the sporophyte and the gametophyte. The sporophyte grows largely at the expense of the gametophyte (gonotrophy). The outer walls of the cells that abut on the haustorial cavity in both generations develop labyrinths typical of transfer cells; these wall labyrinths, i.e., knoblike to conelike enlargements of the wall, play an especially important role. This feature is more apparent in the epidermal cells of the sporophyte foot (haustorium). BOPP (1981) and FREY (1981) provide a summary of the data known up to the time of publication; their report is particularly concerned with the investigations of WIENCKE and SCHULZ (1978) and BROWNING and GUNNING (1979a,b,c). The experiments of BROWNING and GUNNING and CAUSSIN et al. (1979, 1983) clearly demonstrate the functional importance of the wall labyrinths for the uptake of water and nutrients in sporophytes. In the haustoria of species of *Polytrichum*, *Dawsonia* and *Dendroligotrichum* there is an additional water-absorbing organ. Other studies in the period under review provide clear evidence of the occurrence of transfer cells in the gametophyte tissue. LIGRONE et al. (1982a) were able to identify differentiated gametophytic transfer cells opposite the middle region of the foot in *Timmiella barbuloides*. In *Buxbaumia piperi* two or three rows of transfer cells are present, which surround the sporophyte foot (LIGRONE et al. 1982b). The abutting gametophytic tissue possesses transfer cells in *Physcomitrium cyathicarpum* as well (LAL and CHAUHAN 1981). In contrast, the haustorium in *Phaeoceros laevis* has a different structure. A placental region surrounds a central bulbous region. The bulbous region consists of large vacuolized cells. In the placental region there is a mixture of transfer cells and branched haustorial cells, belonging to the sporophyte and gametophyte foot respectively (GAMBARDELLA et al. 1981). This organization is a further indication of the independent status of the Anthocerotae.

c) Structural Adaptations

Compared with vascular plants, bryophytes represent an alternative strategy of adaptation to life on land. After the internal water-conducting system had been reduced, numerous external structures arose out of the necessity to "create" an external water-conducting system. PROCTOR (1979) gives a summary of the anatomical structures (leaf-surface papillae, capillary structures) and provides information about water conduction in nonpapillose leaves. BARTHLOTT and SCHULTZE-MOTEL (1981) call attention to a new structure of the external water-conducting system in *Rhacocarpus purpurascens*. All of the cells on each leaf side have perforated wall structures, resulting in a netlike pattern. Up to now, no comparable structures have been found among the bryophytes and the other terrestrial plants. BELL (1982) shows the adaptations in leaf morphology for the species of the Australian arid zone: densely papillose leaf surfaces, revolute or involute leaf margins, costa, putative surface wax deposits, hyaline hair points, and small squarrose cells with a reduced ratio of cell surface area to volume.

d) Oil Bodies

Over 80% of the genera of the Hepaticae possess oil bodies. These oil bodies consist of rather volatile terpenoids, mainly sesquiterpenoids and their derivates, which are not found in this form in any other plants. They are surrounded by a membrane. The genesis of the oil bodies and the nature of the membrane were reported to be unsolved problems (LEHMANN and JASTER 1981, GRADSTEIN 1978). In *Riella helicophylla* and *Ricciocarpus natans* oil-body cells are present as idioblasts and contain a single oil body each. The oil body originates from small vacuoles (LEHMANN and JASTER 1981, KRONESTEDT 1983). In *Riella* the membranes are extremely high in contrast and asymmetric (LEHMANN and JASTER 1981); in *Ricciocarpus natans* the membrane has a tripartite dark-light configuration. Lipophilous globules are embedded in a matrix, which reacts positively for proteins, polysaccharides and lipids. A smooth endoplasmic reticulum is extensively developed and seems to contribute material to the developing oil body (KRONESTEDT 1983). See also the works by APOSTOLAKOS and KATSAROS (1978), GALATIS et al. (1978), KATSAROS and APOSTOLAKOS (1978), KRONESTEDT (1981).

e) Blepharoplast Structure

The blepharoplast ultrastructure of mosses indicates that the male gametes have more characters in common with those of the hepatics than with Anthocerotae; the Sphagnidae should remain as a subclass, *Tetraphis* could well merit subclass rank (DUCKETT and CAROTHERS 1979). Within the Hepaticae there has been a trend from the putatively primitive *Haplomitrium* blepharoblast toward a reduction of structural complexity in the more advanced taxa. DUCKETT et al. (1982) postulate that upon the blepharoplast structure *Haplomitrium* is the closest living survivor to a missing link between bryophytes and tracheophytes. The blepharoplasts of the genera *Sphaerocarpus*, *Riella* and *Geothallus* are very similar and closely related to those reported from the Marchantiales (BROWN et al. 1983, CAROTHERS et al. 1983).

f) Life Forms, Life Strategies

DURING (1979) takes a modern look at the subject of "life strategies". Besides discussing ontogeny and habit, he devotes special attention to the number of spores produced, vegetative propagation and productivity. In *Nanocyperion synusiae*, which according to RAUNKIAER shows only a few differences in its life-form spectrum, it was possible – based on life-strategy spectra for phanerogams and bryophytes – to clarify the patterns and arrive at a differentiated system of the adaptations of habitat (DURING 1980).

In view of their high evolutionary potential the bryophytes have been receiving an increased amount of attention. MÄGDEFRAU (1982) presents a summary of the classical views concerning life forms. DURING (1981), in a study concerned with floristic and sociological aspects, investigates the life strategies of the bryophytes on Lanzarote, where there is a high percentage of colonists. SCHOFIELD (1981) provides a survey and a synthesis of current knowledge on the structure of the moss gametophyte and its adaptation to the environment, and he calls attention to the severe lack of experimental investigations. The great morphological plasticity of the gametophyte, comparable to that of the angiosperms, results in a wide ecological amplitude and thus in the ability to colonize numerous biotopes. VITT (1981a) devotes special attention to the adaptations of the moss sporophyte. The selection pressure which leads to changes in the mechanisms of spore distribution is still difficult to assess at present. VITT perceives a prevailing trend toward reduction of the sporophyte as an adaptation to the environment. Further developments of a conspicuous nature have taken place only in the "entomophilous" species of the Splachnaceae. MOGENSEN (1981) discusses the differences of distribution potential in isospory and anisospory. BISCHLER and JOVET-AST (1981) examine how biochemical and biophysical cytoplasm characteristics in the Marchantiales are correlated with the ecological niches to which they have adapted; here it is a question of traits which are not expressed in morphology. The adaptation of the acrocarpous and pleurocarpous growth forms to various microclimatic conditions was investigated for the first time in a field study by O'KEEFE and FARRAR (1983). LEFORESTIER (1981) describes the adaptive mechanisms in *Fossombronia caespitiformis* by means of which this species survives the dry summer months in the Mediterranean region. FREY (1981), in a discussion of "progressive reduction", describes life strategies which have emerged through selection in the course of adaptation to macroniches (e.g., pleurocarpy, prosenchymatic cell net).

The subject of "life strategies" is touched upon in numerous other investigations. At present, however, there is no comprehensive and convincing way of dealing with this set of topics as a whole. Such a scheme, providing an overall view of the subject, would considerably enhance our knowledge of the evolution, systematics and ecology of the bryophytes.

References

ABDERRAHMAN, S., SMITH, A.J.E.: J. Bryol. 12, 265-271 (1982); - ibid. 12, 479-485 (1983). - ABRAMOVA, A.L., ABRAMOV, I.I.: Conspectus der

Moosflora der Mongolischen Volksrepublik. Leningrad: Nauka 1983. - ALLEN, B.H.: Bryologist 84, 335-338 (1981). - AMAKAWA, T.: Hikobia Suppl. 1, 117-127 (1981). - ANDERSON, L.E.: Cytology and reproductive biology of mosses, pp. 37-76. In: The Mosses of North America, R.J. TAYLOR, A.E. LEVITON (eds.). Lawrence, KS: Allen Press 1980. - APOSTOLAKOS, P., KATSAROS, C.: Protoplasma 97, 13-29 (1978). - ASAKAWA, Y.: J. Hattori Bot. Lab. 48, 277-283 (1980); - ibid. 50, 123-142 (1981); - ibid. 53, 282-293 (1982). - ASAKAWA, Y., SUIRE, C., TOYOTA, M., TOKUNAGA, N., TAKEMOTO, T., HATTORI, S., MIZUTANI, M.: J. Hattori Bot. Lab. 48, 285-303 (1980a). - ASAKAWA, Y., TOKUNAGA, N., TAKEMOTO, T., HATTORI, S., MIZUTANI, M., SUIRE, C.: J. Hattori Bot. Lab. 47, 153-164 (1980b). - ASAKAWA, Y., MATSUDA, R., TAKEMOTO, T., HATTORI, S., MIZUTANI, M., INOUE, H., SUIRE, C., HUNECK, S.: J. Hattori Bot. Lab. 50, 107-122 (1981a). - ASAKAWA, Y., REIKO, M., TOYOTA, M., SUIRE, C., TAKEMOTO, T., INOUE, H., HATTORI, S., MIZUTANI, M.: J. Hattori Bot. Lab. 50, 165-177 (1981b).

BACHURINA, A.F., PARTYKA, L.Y.: Hepaticae and Musci of the Ukraine and of adjacent regions. Kiev: Akademia Nauk USSR 1979. - BARTHLOTT, W., SCHULTZE-MOTEL, W.: Willdenowia 11, 3-11 (1981). - BELL, G.: J. Hattori Bot. Lab. 53, 147-151 (1982). - BISCHLER, H.: Cryptogam. Bryol. Lichênol. 3, 351-364 (1982). - BISCHLER, H., JOVET-AST, S.: Bryologist 84, 208-215 (1981). - BOESEN, D.F.: Lindbergia 8, 77-78 (1982). - BOPP, M.: Advances in Bryology 1, 11-77 (1981). - BOWERS, M.C.: Lindbergia 6, 22-31 (1980). - BRASSARD, G.R.: Lindbergia 6, 129-136 (1980). - BREMER, B.: Lindbergia 6, 1-16 (1980a); - ibid. 6, 89-117 (1980b); - ibid. 7, 73-90 (1981). - BROWN, R.C., CAROTHERS, Z.B., DUCKETT, J.G.: Bryologist 86, 234-243 (1983). - BROWNING, A.J., GUNNING, B.E.S.: J. Exp. Botany 30, 1233-1246 (1979a); - ibid. 30, 1247-1264 (1979b); - ibid. 30, 1265-1273 (1979c). - BRUGGEMAN-NANNENGA, M.A.: Proc. Kon. Ned. Adad. Wet. Ser. C 85, 59-104 (1982). - BUCK, W.R.: Brittonia 31, 379-394 (1979); - J. Hattori Bot. Lab. 48, 71-159 (1980); - Brittonia 35, 248-254 (1983).

CAROTHERS, Z.B., BROWN, R.C., DUCKETT, J.G.: Bryologist 86, 97-105 (1983). - CATCHESIDE, D.G.: Mosses of South Australia. Government Prinks: Woolman 1980. - CAUSSIN, C., DESPEGHEL, J.-P., FAUCHER, M., LEGER, A., BONNEMAIN, J.-L.: C.R. Hebd. Séanc. Acad. Sci. Paris Ser. D, 289, 1329-1334 (1979). - CAUSSIN, C., FLEURAT-LESSARD, P., BONNEMAIN, J.L.: Ann. Bot. 51, 167-173 (1983). - CHANG, S.M.: Flora Tsinlingensis. Tomus III, Bryophyta, Pars 1 Musci. (Flora des Qinling-Gebirges im Süden der Provinz Shensi). Peking: Typis Academiae Scintiarum Sinicae 1978; - Acta Bot. Yunnanica 3, 263-265 (1981). - CHEN, P.-C., WU, P.-C.: Acta Phytotax. Sin. 17(2), 93-95 (1979). - CHOE, D.-M.: Illustrated Flora and Fauna of Korea. Vol. 24: Musci, Hepaticae. Seoul City 1980. - CHOPRA, R.S., KUMAR, S.S.: Mosses of the Western Himalayas and Adjacent Plains. New Delhi: Chronica Botanica 1981. - CHURCHILL, S.P.: A phylogenetic analysis, classification, and synopsis of the genera of the Grimmiaceae, pp. 127-144. In: Advances in Cladistics. Proc. 1st meeting of the Willi Hennig Soc., V.A. FUNK, D.R. BROOKS (eds.). The New York Botanical Garden 1981. - CORLEY, M.F.V.: J. Bryol. 11, 191-208 (1980). - CORLEY, M.F.V., CRUNDWELL, A.C., DÜLL, R., HILL, M.O., SMITH, A.J.E.: J. Bryol. 11, 609-689 (1981). - CROSBY, M.R.: Polyploidy in bryophytes with special emphasis on mosses, pp. 193-197. In: Polyploidy: Biological Relevance, W.H. LEWIS (ed.). New York: Plenum Press 1980. - CRUM, H.A., ANDERSEN, L.E.: Mosses of Eastern North America, 2 vols. New York: Columbia Univ. Press 1981. - CUMMINS, H., WYATT, R.: Bryologist 84, 30-38 (1981).

DANIELS, R.E.: J. Bryol. 12, 65-76 (1982). - DELGADILLO, M.C.: Bryologist 85, 401-404 (1982). - DeSLOOVER, J.L.: Bull. Jard. Bot. Nat. Belg.

53, 97-152 (1983). - DeSLOOVER, J.L., MESSE, V.: J. Hattori Bot. Lab. 53, 153-155 (1982). - DUCKETT, J.G., CAROTHERS, Z.B.: Spermatogenesis in the systematics and phylogeny of the Musci, pp. 385-423. In: Bryophyte Systematics, G.C.S. CLARKE, J.G. DUCKETT (eds.). London, New York: Academic Press 1979. - DUCKETT, J.G., CAROTHERS, Z.B., MILLER, C.C.J.: J. Hattori Bot. Lab. 53, 107-125 (1982). - DURING, H.J.: Lindbergia 5, 2-18 (1979); - Acta Bot. Neerl. 29, 483-496 (1980); - Lindbergia 7, 113-125 (1981).

EDDY, A.: Taxonomy and evolution of *Sphagnum*, pp. 109-121. In: Bryophyte Systematics, G.C.S. CLARKE, J.G. DUCKETT (eds.). London, New York: Academic Press 1979. - EDWARDS, S.R.: J. Bryol. 11, 49-93 (1980). - ENGEL, J.J.: A monograph of *Clasmatocolea* (Hepaticae). Fieldiana Botany, N.S. No. 3. Chicago: Field Museum of Natural History 1980; - Hepaticopsida, pp. 271-304. In: Synopsis and Classification of Living Organisms, S. PARKER (ed.). New York: McGraw Hill 1982. - ENGEL, J.J., SCHUSTER, R.M.: Phytologia 47, 317-318 (1981).

FIFE, A.J.: J. Bryol. 83, 466-476 (1980); - Lindbergia 8, 96-104 (1982). - FIFE, A.J., MAGILL, R.E.: Bryologist 85, 99-103 (1982). - FRAHM, J.-P.: Bryologist 83, 570-588 (1980a); - J. Hattori Bot. Lab. 48, 195-210 (1980b); - Herzogia 5, 499-516 (1981a); - ibid. 5, 517-529 (1981b). - J. Hattori Bot. Lab. 54, 207-225 (1983a); - Lindbergia 9, 99-116 (1983b). - FRAHM, J.-P., FREY, W.: Moosflora. Stuttgart: Ulmer 1983. - FREY, W.: Adv. Bryol. 1, 399-477 (1981). - FREY, W., RICHTER, U.: J. Hattori Bot. Lab. 51, 51-60 (1982). - FRISVOLL, A.A.: A Taxonomic Revision of the *Rhacomitrium canescens* Group (Bryophyta, Grimmiales). Gunneria 41, Bot. Ser. No. 13. Trondheim: Universitetet (1983a); - Lindbergia 9, 41-52 (1983b). - FRITSCH, R.: Index to plant chromosome numbers - Bryophyta. Regnum Vegetabile 108. Amsterdam: Bohn, Scheltema, Holkema 1982.

GALATIS, B., APOSTOLAKOS, P., KATSAROS, C.: Can. J. Bot. 56, 2252-2267 (1978). - GAMBARDELLA, R., LIGRONE, R., CASTALDO, R.: Cryptogam. Bryol. Lichénol. 2, 23-45 (1981). - GANGULEE, H.C.: Mosses of Eastern India and Adjacent Regions. Fascicle 7, Hypnobryales, Leskeineae, pp. 1547-1752. Calcutta: Guha Ray & S. Saraswaty Press 1978; - ibid. Fascicle 8, Hypnobryales (Hypnineae): Floristic Trends: Index: Errata, pp. 1753-2145. Calcutta: Guha Ray & S. Saraswaty Press 1980. - GAO, C., AUR, C.-W.: Acta Phytotax. Sin. 16, 78-90 (1978). - GAO, C., CHANG, K.C.: Acta Phytotax. Sin. 16, 113-118 (1978); - Flora Hepaticarum Chinae Boreali-Orientalis. Peking 1981; - J. Hattori Bot. Lab. 54, 187-205 (1983). - GAO, C., CHANG, K.C., CAO, T.: Acta Bot. Yunnanica 3, 389-399 (1981). - GIER, L.J.: J. Bryol. 11, 253-309 (1980). - GRADSTEIN, S.R.: Acta Bot. Neerl. 27, 425-426 (1978); - The genera of the Lejeuneaceae: Past and present, pp. 83-107. In: Bryophyte Systematics, G.C.S. CLARKE, J.G. DUCKETT (eds.). London, New York: Academic Press 1979. - GRADSTEIN, S.R., TERKEN, L.: Occas. Pap. Farlow Herb. 16, 71-81 (1981). - GRADSTEIN, S.R., MATSUDA, R., ASAKAWA, Y.: J. Hattori Bot. Lab. 50, 231-248 (1981). - GRIFFIN, D. III, GRADSTEIN, S.R., AGUIRRE, C.J.: Acta Bot. Neerl. 3, 175-184 (1982). - GROLLE, R.: J. Hattori Bot. Lab. 47, 237-244 (1980a); - ibid. 48, 161-170 (1980b); - Wiss. Z. Friedrich-Schiller-Univ. Jena, Math. Naturwiss. Reihe 31, 207-277 (1982); - Acta Bot. Fennica 121, 62 pp. (1983a); - J. Bryol. 12, 403-459 (1983b). - GRONDE, K. van der: Proc. Kon. Ned. Akad. Wet. Ser. C 83, 271-278 (1980).

HAARBRINK, J.: Lindbergia 7, 47-57 (1981). - HASEGAWA, I.: Acta Phytotax. Geobot. 30, 15-30 (1979). - HATTORI, S.: J. Hattori Bot. Lab. 47, 85-125 (1980a); - ibid. 47, 165-236 (1980b); - ibid. 49, 147-168 (1981a); - ibid. 49, 359-383 (1981b); - ibid. 51, 203-271 (1982); -

ibid. 54, 133-182 (1983). - HEBANT, C.: Conducting tissues in Bryophyte systematics, pp. 365-383. In: Bryophyte Systematics, G.C.S. CLARKE, J.G. DUCKETT (eds.). London, New York: Academic Press 1979; - J. Hattori Bot. Lab. 47, 63-74 (1980). - HERRNSTADT, I., HEYN, C.C., CROSBY, M.R.: Bryologist 83, 536-541 (1980). - HEWSON, H.J.: J. Hattori Bot. Lab. 52, 163-165 (1982). - HONG, W.S.: Bryologist 83, 40-59 (1980). - HORTON, D.G.: J. Hattori Bot. Lab. 53, 365-418 (1982a); - Nova Hedwigia Beih. 71, 229-237 (1982b); - J. Hattori Bot. Lab. 54, 353-532 (1983). - HUNECK, S.: J. Hattori Bot. Lab. 48, 1-31 (1980).

INOUE, H.: J. Hattori Bot. Lab. 49, 335-357 (1981); - Bull. Natl. Sci. Mus. Tokyo Ser. B 8, 125-144 (1982a); - J. Hattori Bot. Lab. 52, 45-56 (1982b); - Nova Hedwigia Beih. 71, 195-200 (1982c). - INOUE, H., GRADSTEIN, S.R.: Bull. Natl. Sci. Mus. Tokyo Ser. B 6, 7-22 (1980). - IRELAND, R.R.: Moss Flora of the Maritime Provinces of Canada. Bot. Publ. No. 13. Ottawa: National Museum of Natural Sciences 1982. - IWATSUKI, Z.: J. Hattori Bot. Lab. 48, 171-186 (1980); - Advances in Bryology 1, 233-314 (1981).

JAHNS, H.M.: Farne, Moose, Flechten. Mittel-, Nord- und Westeuropa. Bestimmungsbuch. München, Wien, Zürich: BLV 1980. - JANS, E.: Proc. Kon. Ned. Akad. Wet. Ser. C 82, 421-432 (1979). - JANSSENS, J.A.: Bryologist 86, 44-53 (1983). - JOHNSON, A.: Mosses of Singapore and Malaysia. Singapore: University Press 1980.

KARUNEN, P.: J. Hattori Bot. Lab. 53, 255-269 (1982). - KATSAROS, C., APOSTOLAKOS, P.: Can. J. Bot. 56, 2268-2285 (1978). - KOPONEN, A.: Ann. Bot. Fennici 18, 123-132 (1981); - Nova Hedwigia Beih. 71, 239-247 (1982a); - J. Hattori Bot. Lab. 52, 87-91 (1982b). - KOPONEN, T.: Ann. Bot. Fennici 17, 35-55 (1980a); - ibid. 17, 125-162 (1980b); - ibid. 18, 105-111 (1981a); - Acta Bot. Fennica 117, 1-34 (1981b); - J. Hattori Bot. Lab. 52, 75-86 (1982); - Ann. Bot. Fennici 20, 101-104 (1983). - KRAMER, W.: *Tortula* HEDW. sect. *Rurales* DeNOT. (Pottiaceae, Musci) in der östlichen Holarktis. Bryophytorum Bibliotheca, Vol. 21. Vaduz: Cramer 1980. - KRONESTEDT, E.: J. Ultrastruct. Res. 76, 316 (1981); - Nordic J. Bot. 3, 547-558 (1983). - KRZAKOWA, M.: J. Bryol. 11, 447-450 (1981a); - Variation in the *Plagiochila asplenioides* - *Plagiochila porelloides* group, pp. 39-43. In: New Perspectives in Bryotaxonomy and Bryogeography Adam Mickiewicz University, J. SZWEYKOWSKI (ed.). Poznân: Adam Mickiewicz University 1981b. - KRZAKOWA, M., SZWEYKOWSKI, J.: Genetics 93, 711-719 (1979). - KUMAR, S.S.: Hikobia 8, 245-258 (1980). - KUWAHARA, Y.: Bryologist 84, 351-357 (1981); - Proc. Kon. Ned. Akad. Wet. Ser. C 85, 357-380 (1982).

LAL, M., CHAUHAN, E.: Protoplasma 107, 79-83 (1981). - LANDWEHR, J.: Atlas van de Nederlandse Bladmossen. Amsterdam: Koninklijke Nederlandse Naturhistorische Vereniging 1978; - Atlas Nederlandse Levermossen. Amsterdam: Koninklijke Nederlandse Naturhistorische Vereniging 1980. - LANE, D.M.: J. Hattori Bot. Lab. 49, 169-245 (1981). - LANGE, B.: Lindbergia 8, 1-29 (1982). - LEFORESTIER, C.: Cryptogam. Bryol. Lichénol. 2, 223-232 (1981). - LEHMANN, H., JASTER, B.: Protoplasma 106, 109-119 (1981). - LI, X.-J.: Acta Phytotax. Sinica 19, 264-266 (1981a); - Acta Botanica Yunnanica 3, 101-112 (1981b). - LI, X.-J., CHANG, M.: Acta Bot. Yunnanica 1, 32-80 (1979). - LIGRONE, R., CASTALDO, R., GAMBARDELLA, R.: Cryptogam. Bryol. Lichénol. 1, 115-142 (1980). - LIGRONE, R., GAMBARDELLA, R.L., SPOSITO, M.L.: Planta 154, 414-425 (1982a). - LIGRONE, R., GAMBARDELLA, R., CASTALDO, R., GIORDANO, S., LUCIA SPOSITO, M.L.: J. Hattori Bot. Lab. 52, 465-499 (1982b). - LIN, P.-J.: Acta Phytotax. Sin. 17, 92-97 (1979). - LONGTON, R.E.: J. Bryol. 11, 501-520 (1981).

MÄGDEFRAU, K.: Life-forms of Bryophytes, pp. 45-58. In: Bryophyte Ecology, A.J.E. SMITH (ed.). London, New York: Chapman & Hall 1982. - MAGILL, R.E.: J. Hattori Bot. Lab. 48, 33-70 (1980); - Flora of Southern Africa. Bryophyta, Part 1: Mosses, Fasc. 1: Sphagnaceae - Grimmiaceae. Pretoria 1981; - Nova Hedwigia Beih. 71, 273-279 (1982). - MAGILL, R.E., VITT, D.H.: Bothalia 14, 463-466 (1981). - MANUEL, M.G.: Fed. Mus. J. Part II N.S. 26, 163 pp. (1981a); - J. Hattori Bot. Lab. 49, 115-140 (1981b); - Cryptogam. Bryol. Lichénol. 2, 449-455 (1981c); - Nova Hedwigia Beih. 71, 281-289 (1982). - MARGADANT, W.D., DURING, H.: Beknopte flora van Nederlandse Blad en Levermossen. K.N.N.V. Hoogwoud 1982. - MARKHAM, K.R.: Biochem. Syst. Ecol. 8, 11-15 (1980). - MARKHAM, K.R., PORTER, L.J.: Prog. Phytochem. 5, 181-272 (1978); - Phytochemistry 18, 611-615 (1979). - MATSUO, A.: J. Hattori Bot. Lab. 53, 295-304 (1982). - McADAM, S.V.: J. Bryol. 12, 233-258 (1982). - MENDELAK, M.: J. Bryol. 11, 441-446 (1981). - MILLER, H.A.: The phylogeny and distribution of the Musci, pp. 11-39. In: Bryophyte Systematics, G.C.S. CLARKE, J.G. DUCKETT (eds.). London, New York: Academic Press 1979. - MILLER, H.A., WHITTIER, H.O., WHITTIER, B.A.: Prodromus Florae Hepaticarum Polynesiae. Bryophytorum Bibliotheca, Vol. 25. Vaduz: Cramer 1983. - MILLER, N.G., MANUEL, M.G.: J. Hattori Bot. Lab. 51, 273-321 (1982). - MIZUTANI, M.: J. Hattori Bot. Lab. 48, 235-242 (1980). - MOGENSEN, G.S.: Bryologist 84, 187-207 (1981). - MUES, R.: Ber. Dtsch. Bot. Ges. 95, 115-125 (1982a); - J. Hattori Bot. Lab. 51, 61-68 (1982b); - ibid. 53, 271-281 (1982c). - MUES, R., STRASSNER, A., ZINSMEISTER, H.D.: Cryptogam. Bryol. Lichénol. 4, 111-127 (1983).

NA-THALANG, O.: Brunonia 3, 61-140 (1980). - NEWTON, M.E.: J. Bryol. 11, 433-440 (1981). - NOWAK, H.: Revision der Laubmoosgattung *Mitthyridium* (MITTEN) ROBINSON für Ozeanien (Calymperaceae). Bryophytorum Bibliotheca, Vol. 20. Vaduz: Cramer 1980. - NOWAK, H., POELT, J.: Willdenowia 9, 111-129 (1979).

OCHI, H.: J. Fac. Educ. Tottori Univ., Nat. Sci. 29, 49-154 (1980); - ibid. 30, 21-55 (1981); - ibid. 31, 11-47 (1982a); - J. Hattori Bot. Lab. 52, 65-73 (1982b). - OCHYRA, R., PRZYWARA, L., KUTA, E.: J. Bryol. 12, 259-263 (1982). - ODRZYKOSKI, I., BOBOWICZ, M.A., KRZAKOWA, M.: Variation in *Conocephalum conicum* - the existence of two genetically different forms in Europe, pp. 29-32. In: New Perspectives in Bryotaxonomy and Bryogeography, Adam Mickiewicz University, J. SZWEYKOWSKI (ed.), Ser. Biol. 20. Poznán: Adam Mickiewicz University 1981. - O'KEEFE VAN DER LINDEN, J., FARRAR, D.R.: Bryologist 86, 1-13 (1983). - ONRAEDT, M.: Cryptogam. Bryol. Lichénol. 4, 71-85 (1983). - ORBAN, S.: Acta Bot. Acad. Sci. Hung. 27, 169-177 (1981).

PORTER, L.J.: Taxon 30, 739-748 (1981). - PROCTOR, M.C.F.: Structure and eco-physiological adaptation in Bryophytes, pp. 479-509. In: Bryophyte Systematics, G.C.S. CLARKE, J.G. DUCKETT (eds.). London, New York: Academic Press 1979.

RAMSAY, H.P.: Lindbergia 9, 89-92 (1983). - REENEN, G.B.A. van: Lindbergia 8, 110-120 (1982). - REESE, W.D.: Bryologist 86, 23-30 (1983). - RENLIANG, H.: Bryologist 86, 193-233 (1983). - REYES, D.M.: Acta Bot. Acad. Sci. Hung. 28, 145-180 (1982).

SCHEIRER, D.C.: Am. J. Bot. 70, 987-992 (1983). - SCHOFIELD, W.B.: Bryologist 84, 149-165 (1981). - SCHUSTER, R.M.: The Phylogeny of the Hepaticae, pp. 41-82. In: Bryophyte Systematics, G.C.S. CLARKE, J.G. DUCKETT (eds.). London, New York: Academic Press 1979; - The Hepaticae and Anthocerotae of North America, Vol. IV. New York: Columbia Univ. Press 1980a; - J. Hattori Bot. Lab. 48, 337-421 (1980b); - Nova Hedwigia 32, 637-693 (1980c); - Paleoecology, origin, distribution through time, and evolution of Hepaticae and Anthocerotae, pp. 129-191. In: Paleobotany, Paleoecology, and Evolution, Vol. 2, K.J. NIKLAS (ed.). New York: Praeger 1981a; - Phytologia 47, 301-308 (1981b); - J. Bryol. 11, 411-431 (1981c); - Nova Hedwigia 36, 1-16 (1982); - (ed.): New Manual of Bryology, 2 Vols. Nichinan-shi: The Hattori Botanical Laboratory 1983. - SCHUSTER, R.M., ENGEL, J.J.: Phytologia 47, 309-312 (1981); - Lindbergia 8, 65-74 (1982). - SCHUSTER, R.M., SCHOFIELD, W.B.: Bryologist 85, 231-238 (1982). - SEPPELT, R.D.: J. Hattori Bot. Lab. 51, 99-150 (1982a); - ibid. 52, 107-112 (1982b). SHAW, A.J.: J. Hattori Bot. Lab. 50, 1-18 (1981a); - Can. J. Bot. 59, 2359-2378 (1981b). - SHLYAKOV, R.N.: The Hepaticae of the Northern USSR, Vols. 3-5. Leningrad: Nauka 1980-1982. - SJÖGREN, E.: Mem. Soc. Broteriana 26, 273 pp. (1978). - SMITH, A.J.E.: Adv. Bot. Res. 6, 195-276 (1978); - Towards an experimental approach to Bryophyte taxonomy, pp. 195-206. In: Bryophyte Systematics, G.C.S. CLARKE, J.G. DUCKETT (eds.). London, New York: Academic Press 1979. - SMITH, A.J.E., RAMSAY, H.P.: J. Hattori Bot. Lab. 52, 275-281 (1982). - STONE, I.G.: J. Bryol. 11, 231-243 (1980). - SUIRE, C., ASAKAWA, Y.: Chemotaxonomy of bryophytes: a survey, pp. 447-477. In: Bryophyte Systematics, G.C.S. CLARKE, J.G. DUCKETT (eds.). London, New York: Academic Press 1979; - Adv. Bryol. 1, 167-231 (1981). - SZWEYKOWSKI, J., KRZAKOWA, M.: Bull. Acad. Polon. Sci. Sér. Sci. Biol. 27, 37-41 (1979). - SZWEYKOWSKI, J., ODRZYKOSKI, I., ZIELINSKI, R.: Bull. Acad. Polon. Sci. Sér. Sci. Biol. 28, 437-449 (1980). - SZWEYKOWSKI, J., PRUS-GLOWACKI, W., MENDELAK, M.: J. Hattori Bot. Lab. 50, 269-278 (1981a). - SZWEYKOWSKI, J., ZIELINSKI, R., MENDELAK, M.: Bull. Acad. Polon. Sci. Sér. Sci. Biol. 24, 9-19 (1981b).

THEODOR, R., ZINSMEISTER, H.D., MUES, R., MARKHAM, K.R.: Phytochemistry 19, 1695-1700 (1980); - ibid. 20, 1851 1852 (1981). - TIXIER, P.: Bull. Trim. Acad. Malgache 55, 173-247 (1979). - TOWNSEND, C.C.: J. Bryol. 12, 1-6 (1982).

UDAR, R., AWASTHI, U.S.: Geophytology 1, 72-79 (1981); - J. Hattori Bot. Lab. 53, 419-437 (1982a); - Lindbergia 8, 55-59 (1982b). - UDAR, R., KUMAR, A.: J. Hattori Bot. Lab. 49, 247-260 (1981).

VANDEKERKHOVE, O.: Mitt. Z. Pflanzenphysiol. 100, 369-372 (1980). - VANDENBERGHEN, C.: Flore des hépatiques et des anthocérotes de Belgique. Meise 1979; - Bull. Jard. Bot. Nat. Belg. 51, 41-103 (1981); - Cryptogamie, Bryol. Lichénol. 3, 207-224 (1982). - VERMA, S.K., KUMAR, S.S.: Bryologist 86, 121-129 (1983). - VITAL, D.M.: Erpodiaceae (Musci) do Brazil. Diss. Sao Paulo 1980. - VITT, D.H.: Bryologist 83, 405-436 (1980a); - ibid. 83, 437-450 (1980b); - ibid. 84, 166-186 (1981a); - Sphagnopsida, Bryopsida, pp. 305, 307-338, 1078-1079. In: Synopsis and Classification of Living Organisms, S.P. PARKER (ed.). New York: McGraw-Hill 1981b; - J. Hattori Bot. Lab. 49, 93-113 (1981c); - Nova Hedwigia Beih. 71, 261-268 (1982a); - J. Hattori Bot. Lab. 52, 153-159 (1982b); - ibid. 54, 1-94 (1983). - VITT, D.H., BARTLETT, J.K.: Bryologist 86, 106-113 (1983). - VRIES, A. de, ZANTEN, B.O. van, DIJK, H. van: Lindbergia 9, 73-80 (1983).

WALTHER, K.: Bryophytina, Laubmoose. In: A. ENGLER Syllabus der Pflanzenfamilien, 13. Aufl., J. GERLOFF, J. POELT (eds.). Berlin, Stuttgart: Borntraeger 1983. - WATSON, E.V.: British Mosses and Liverworts, 3. ed. Cambridge: University Press 1981. - WEITZ, S., HEYN, C.C.: Bryologist 84, 315-334 (1981). - WIENCKE, C., SCHULZ, D.: Bryophytorum Bibliotheca Vol. 13, pp. 147-167. Vaduz: Cramer 1978. - WON, S.H.: Bryologist 86, 143-155 (1983).

YLI-REKOLA, M.: Ann. Bot. Fennici 17, 277-291 (1980).

ZANDER, R.H.: Phytologia 41, 11-32 (1978-1979); - ibid. 44, 177-214 (1979); - Cryptogam. Bryol. Lichénol. 2, 1-22 (1981a); - ibid. 2, 379-422 (1981b); - Lindbergia 8, 161-165 (1982). - ZANDER, R.H., CLEEF, A.M.: Proc. Kon. Ned. Akad. Wet. Ser. C 85, 627-634 (1982). - ZIELINSKI, R., PRUS-GLOWACKI, W., MENDELAK, M.: Chemical variations in the central European *Pellia* taxa, pp. 25-27. In: New Perspectives in Bryotaxonomy and Bryogeography, Adam Mickiewicz University, J. SZWEYKOWSKI (ed.). Poznán: Adam Mickiewicz University 1981. - ZINSMEISTER, H.D., MUES, R.: Rev. Latinoamer. Quim. 11, 23-29 (1980).

Professor Dr. WOLFGANG FREY
Institut für Systematische Botanik
und Pflanzengeographie der
Freien Universität Berlin
Altensteinstr. 6
D-1000 Berlin 33 (Dahlem)

V. Systematics of the Pteridophytes

By Karl U. Kramer

1. Systematics

By far the most important contribution to the systematics of the ferns and fern allies as a whole is by TRYON and TRYON (1982). It is an attempt at integrating the accumulated knowledge, with the addition of many new data and interpretations, and gives an extensive account of all pteridophyte families and genera as represented in tropical America, with at least brief references to taxa confined to the Old World. Keys to families, genera, and in some cases also species are provided. The taxonomy and affinities of every major group are discussed. Much attention is given to cytotaxonomic data and – this is the major original contribution – to spore characters as observed by SEM technique. The abundant, mostly newly prepared and largely excellent illustrations make the work unique. The family system is modern, and many genera of doubtful validity are sunk in others. Maps show the overall (American) distribution of many genera. A handbook that has no equal.

A general overview of the present status of fern taxonomy, its achievements and its principal challenges and most urgent problems, with extensive bibliography, was given by BIR (1982).

On the basis of the morphology of stems, roots, leaves, and sporangia of all pteridophytes, living and fossil, KATO (1983) arrived at a biphyletic classification, with Lycopodiopsida and Psilotopsida forming one, Filicopsida and Equisetopsida another subdivision of the Tracheophyta. Psilotales are definitely not related to ferns. Ophioglossales stand far apart from the other ferns, as also argued before (KATO 1982).

Hardly any comprehensive monographs of entire fern genera have been published during the last 2 years. The one notable exception is the beautifully executed monograph of *Platycerium* by HENNIPMAN and ROOS (1982). A large number of characters was examined and critically evaluated, and by means of cladistic taxonomy a new scheme of affinities was produced, with new phytogeographical implications. The genus is supposed to have originated in Africa, whence the sole American species is also derived. At present the majority of the 15 recognized species are in Australasia. A sequel is HENNIPMAN and ROOS (1983), with implications for the entire family Polypodiaceae.

A major revisional contribution is HOLTTUM's (1982a) account of the Thelypteridaceae for Flora Malesiana, one of the centers of species concentration. In all, 22 genera with 439 species are dealt with. The amount of information in this volume, dedicated to the memory of CARL F.A. CHRISTENSEN, is enormous. Another center of the family is in Andean South America. The species from part of this area have now been revised in A.R. SMITH's treatment for the Flora of Ecuador (1983). His classification above the species level is the opposite of HOLTTUM's,

Progress in Botany, Vol. 46

as he recognized only a single genus, *Thelypteris* , in which all of the 100 (!) local species are accomodated. This is again a local monograph, rather than a mere contribution to a flora.

A series of contributions to the systematics of *Botrychium* are by WAGNER and WAGNER (1982a, 1983b) and by WAGNER et al. (1981). The genus presents unusual difficulties, as many species are distinguished only by subtle characters, occur in mixed populations, and are very difficult to cultivate. New methods for overcoming these difficulties were outlined by WAGNER and WAGNER (1983a).

Further steps toward a completion of a revision of *Anemia* are by MICKEL (1981, 1982; the latter with data on cytotaxonomy and hybridization), of the neotropical tree-ferns by CONANT (1983) in his revision of *Alsophila* (sensu R.M. TRYON) for the Americas.

Morphological-anatomical work on *Metaxya* and *Lophosoria* (LUCANSKY 1982) showed these two monotypic genera to be quite distinct from one another as well as from Cyatheaceae s.str.

R.M. TRYON's system of tree-fern classification was strongly challenged by HOLTTUM (HOLTTUM and EDWARDS 1983), both as to circumscription and as to tenability of the genera recognized by the former. A revision of the tree-ferns of the Mt. Roraima area is also given. HOLTTUM (1982b) gave a new revision of the western Pacific species group of *Cyathea decurrens*. Surprisingly, it is most closely related to the neotropical *C. multiflora* alliance.

Several contributions to the knowledge of the Old World Hymenophyllaceae were given by IWATSUKI. The species referred to *Macroglena* by COPELAND were shown to belong to four different groups (1981). *Sphaerocionium* was revised for Asia and the Pacific, with a new subdivision of the genus (1982a). The family was reclassified (1982b); now there are eight genera, between the two of the traditional system, the six of MORTON (1968), and the 34 of COPELAND (1947). *Microgonium* and its African relatives were revised by PICHI SERMOLLI (1982b).

PARRIS has continued her work on Grammitidaceae: various notes on *Grammitis* (1980), the *G. poeppigiana-magellanica* complex at the southern end of the world (1981); *Grammitis* in New Guinea, with a multi-access key and other methodological innovations (1983).

Polypodiaceae are also in the focus of attention: *Neolepisorus* emended and revised (CHING and SHING 1983); a new subdivision of *Pyrrosia* (SHING 1983); a reclassification of the Drynarioid ferns in two tribes by CHANDRA (1982a) and a re-assessment of their characters (1982b). On the basis of venation and trichome characters MITSUTA (1982) concluded that they are much closer to the Polypodioid (s. str.) than to the Microsorioid group of genera. PRICE (1983) examined and discussed the characters of the group of *Polypodium pectinatum*, simultaneously emended, and placed it in a distinct, new genus: *Pecluma*.

Stenosemia, now restricted to a single species, was reexamined by CHANDRA (1983) and found to be very close to *Tectaria*. Twenty-three (!) new species of *Coniogramme* from China were described by SHING (1981). *Cornopteris crenulato-serrulata* was placed in the new genus *Neoathyrium* by CHING and WANG (1982a). New species of various genera of Aspidiaceae were described by the same authors (1981).

There are several regional revisions, e.g., *Lindsaea* in India (DIXIT and GHOSH 1983), *Cheilanthes* in France (BADRÉ et al. 1982), Thelypteridaceae in Europe (HOLTTUM 1983), *Onychium* in the Himalaya (KHULLAR and SHARMA 1980). KORNAŚ et al.

(1982) examined the genus *Actiniopteris* in Zambia with methods of numerical taxonomy. The five species recognized by PICHI SERMOLLI proved to be valid, but *A. dimorpha* is perhaps hybridogenous; the new ssp. *diversiformis* may constitute a hybrid swarm which would then be the third known instance in ferns.

B. JOE-HOSHIZAKI continued her series of papers on cultivated ferns: *Davallia* (s. str.) (1981), *Polypodium* (s. lat.) (1982).

2. Bibliography, Collections, Nomenclature

A brief survey of the history of fern taxonomy, up to current projects, with special emphasis on the part taken by CHRISTENSEN, was given by HOLTTUM (dedication in 1982a).

After having relocated part of FÉE's personal fern herbarium in the collection of the Botanical Garden of Rio de Janeiro, WINDISCH (1982) inventoried it and gave a list of the type specimens contained in it and a list of the names based on them. A valuable list of critical identifications of PLUMIER's fern plates (1705) was given by LELLINGER and PROCTOR (1983). PRICE (1982) reexamined the ferns described by HARRINGTON from the collections made by J.B. STEERE, mostly in the Philippines, and gave lectotypifications of all species; the set in the Michigan herbarium is most eligible for the choice of hololectotypes. The currently accepted names for the taxa are also given.

Polypodium virginianum must be typified by a plate illustrating - probably - the tetraploid cytotype. *P. vinlandicum* LÖVE and LÖVE, based on *P. vulgare* L. var. *americanum* HOOKER, applies to the same cytotype and is therefore synonymous (CRANFILL and BRITTON 1983).

PANIGRAHI and BASU (1982) argued that *Cheilanthes pteridioides* (REICH.) C. CHR. must again be called *Ch. fragrans* (L. f.) SWARTZ, and that the species usually called *Ch. mysurensis* WALLICH ex HOOKER is correctly named *Ch. swartzii* WEBB ex BERTH. According to SMITH and FRASER-JENKINS (1982) the correct name for *Dryopteris paleacea* is *D. wallichiana*. For a list of conserved family names of ferns as approved by the Nomenclature Subcommittee, see PICHI SERMOLLI (1981). Further notes on the nomenclature of pteridophyte family names are by the same author (1982a).

Vols. 7 and 8 of the *Bibliography of American Pteridology* were again compiled by HAUFLER (1981, 1982).

3. Floristics

A series of modern fern floras, or modern treatises of ferns in comprehensive floras, can again be announced.

A s i a : The third volume of KURATA and NAKAIKE's monumental series *Illustrations of Pteridophytes of Japan* has appeared (1983); it contains Aspidiaceae (s. l.) and Thelypteridaceae. A comparable treatment, also with Japanese text, but with photographs of herbarium specimens as illustrations, and without distribution data, is NAKAIKE's voluminous *New Flora of Japan: Pteridophyta* (1982a). Almost one half of Vol. 1 of the profusely illustrated *Flora of Tibet* (WU 1983) is devoted to pteridophytes; it contains numerous descriptions of new species, by various authors. A treatment of pteridophytes is also contained in Vol. 1 of the *Flora of Kiangsu*

Province, China (Anon. 1977). A booklet with many colored photographs of pteridophytes of Taiwan was published by KUO (1982). Additions to the series of local fern floras of the Indian subcontinent are a pteridophytic flora of the Garhwal Himalaya (BIR et al. 1982a) and the fern chapter in DEB's *The Flora of Tripura State* (1981); notably the former is modern and critical. Additions to the pteridophytic flora of the Andaman and Nicobar Islands by BALAKRISHNAN (1982), of Kalimantan by IWATSUKI and KATO (1983). A list of the ferns known from India and not included in BEDDOME's handbook was given by NAIR and DIXIT (1981). The first instalment of an enumeration of the ferns of Nepal is a treatment of *Polystichum*, with key (NAKAIKE 1982b). BIR et al. (1982b) assembled all data available on the ecology and distribution of the Polypodiaceae (s. str.) of India, with special emphasis on the Himalaya and the mountains of S. India. The strongest floristic affinity is, not surprisingly, with China.

The discovery of the East African-South Arabic *Cheilanthes coriacea* in Iran was reported by FRASER-JENKINS (1981).

Australia-Pacific: A new treatment of the pteridophytes of the Sydney area is by TINDALE (1982). A popular booklet with colored illustrations and concise descriptions of common pteridophytes of New Zealand was published by CHINNOCK (1981). FOSBERG and CANFIELD (1980) gave notes on the ferns of Palau; FOSBERG et al. (1982) published a checklist of all Micronesian pteridophytes, with synonymy.

Europe: The distribution of the pteridophytes (and all other vascular plants) of Switzerland was mapped in great detail by WELTEN and SUTTER (1982).

PAGE's fern flora of the British Isles (1982a) contains many novel aspects, including blastogeny, phenology, and biosystematic data, with emphasis on ecology; it deals with all pteridophytes. Biosystematic data also form one of the many innovations in the pteridophyte volume of the third edition of HEGI's *Illustrierte Flora von Mitteleuropa* (DOSTAL et al. 1984); it was completely rewritten, and all maps and most illustrations are also new.

The presence of *Asplenium bourgaei* in the Dodekanesos Islands, Greece, was established by GREUTER et al. (1983), of *Asplenium balearicum* in Italy by NARDI (1983). RASBACH et al. (1982) found *Dryopteris villarii* to be still common at its type locality in S.E. France. The population is morphologically similar to *D. submontana* and is diploid, as always tacitly assumed but not proven. The names assigned to the three species of the *D. villarii* complex are correctly applied.

Africa: A major contribution to the fern flora of Africa is JACOBSEN's work *The Ferns and Fern Allies of Southern Africa* (1983) which covers all of Africa south of 16/17° S. Lat. It contains a wealth of ecological data, but the taxonomic treatment, its presentation, and especially the keys and illustrations leave much to be desired.

New records of pteridophytes from the northern parts of Nigeria and the Cameroons, notably *Marsilea* spp., are by KORNAŚ (1983).

America: The pteridophytes are again included in a new edition of *The Flora of New England* (SEYMOUR 1982); but the treatment might be a bit more modern. A special booklet is devoted to the pteridophytic flora of the driftless area in the American Middle West with interesting distribution maps (PECK 1982). The distribution of exclusively gametophytic colonies of *Vittaria* and *Trichomanes* in the north-eastern U.S.

proved to be much wider than anticipated (FARRAR et al. 1983); they occur in moist crevices and grottos roughly up to the limit of the last glaciation.

The third North American subspecies of *Adiantum pedatum* was newly described by CODY (1983) and further discussed by GAWLER (1983). *Dryopteris × australis* is a backcross of *D. celsa (= D. goldiana × ludoviciana)* and *D. ludoviciana*. Its distribution was reported by WAGNER and MUSSELMAN (1982). It occurs, unexpectedly, mostly outside the area of *D. ludoviciana*.

An important modern treatment of the pteridophytes is contained in the new flora of the Bahamas. It is by CORRELL and CORRELL (1982) and richly illustrated. Additions to and notes on the fern flora of Jamaica are by PROCTOR (1982).

The pteridophytic flora of Guatemala and Belize was concluded with the third instalment dealing with "water ferns" and fern allies; it follows the same excellent style as the preceding volumes, with equally fine illustrations, general index (STOLZE 1983). A modernized fern treatment is also present in the new edition of the flora of the Buenos Aires area, Argentina (CABRERA and ZARDINI 1978). See also SMITH (1983) and TRYON and TRYON (1982), discussed in Sect. 1.

4. Geography and Ecology

A new distribution pattern in Japanese plants, depending on a specially calculated temperature mean, is called the "*Gleichenia japonica* type" by NAKAIKE (1983). OHASHI et al. (1982) argued that the Himalaya is an important center of origin for temperate and tropical elements; the genus *Cornopteris* (mapped) is an example.

The distribution of the North American *Woodwardia areolata* is strongly governed by winter temperatures, in the second place by edaphic factors (CRANFILL 1983). CARLSON and WAGNER (1982) reported on detailed studies of the distribution of all North American species of *Dryopteris* and tried to correlate the patterns with the geological and cytotaxonomic (allopolyploids !) history of the taxa. The area of *Pilularia americana* is highly disjunct between the South and Middle West of the U.S. and the Pacific States; it is expanding by conquest of new anthropogenic habitats like impoundments and quarries (DENNIS and WEBB 1981). CUSICK (1983) discovered *Vittaria* gametophyte colonies without sporophytes in north-eastern Ohio, in the previously glaciated coniferous belt; otherwise they were not recorded outside the mixed-mesophytic forest zone.

The distribution patterns of the pteridophytes of Margarita Island off Venezuela were arranged and discussed by ORTEGA (1982).

PAGE and CLIFFORD (1981) discussed the diversity and affinity of the Australian fern flora, poor in relation to the size of the continent. Pantropical genera are often poorly represented. An attempt is made at distinguishing recent arrivals from long-resident taxa; the same genus may contain species of a modern, palaeotropic, and of a more ancient, austro-temperate element. The latter consists of many endemic and/or isolated species. Parallels are to be found in the conifers.

SCHELPE (1983) described some unusual distribution patterns in African ferns. The importance of random long-distance dispersal is considered great.

The ecology of the Mt. Roraima tree-ferns was described by EDWARDS (HOLTTUM and EDWARDS 1983). WINDISCH and PEREIRA-NORONHA (1983) dealt with the ecology and reproduction of an East Brazilian *Plagiogyria* species. ROUX (1982) gave an account of the various types of environment and vegetation of Transkei, South Africa, with their specific pteridophytes (many errors in scientific names). The natural environment of the rare *Phanerosorus* (Matoniaceae) in Borneo is described by WALKER and JERMY (1982). The resemblance of the fern floras of Kalimantan and of Mt. Kinabalu, Sabah, was pointed out by IWATSUKI et al. (1983); they also gave some new data on rheophytic ferns found in the former area.

Detailed ecological and phytogeographical studies of a Scottish and part of an Irish county were made by WILLMOT (1981, 1983).

Autecological and ecophysiological work on pteridophytes is still rather scarce. QUIRK and CHAMBERS (1981) studied the drought tolerance of gametophytes of seven Australian species of *Cheilanthes* and found it to be independent of age and considerably higher than that of gametophytes of non-xeric ferns. PÉREZ-GARCÍA and RIBA (1982) found a significant correlation between the germination of spores of five tree-fern species and the temperature régime within their natural areas.

KLEKOWSKI and KLEKOWSKI (1982) demonstrated a much increased mutation rate for ferns growing in an environment contaminated by polychlorinated biphenyls.

TANNER (1983) made a pilot study of leaf phenology and demography of three species of Jamaican tree-fern. The life-span of the leaves was shorter than that of palms, and varied between the species. There is little correlation between age and height of trunk; the average increase is about 1 m per 15 years.

KOPTUR et al. (1982) analyzed the nectar of five neotropical species of *Polypodium* and of three ferns from other alliances and found it to be poor in sucrose and rather poor in amino acids. It is still not clear whether the nectar serves to attract an ant bodygard or whether it is just an excretion product. Modern work on the nectaries of *Pteridium* in Britain (PAGE 1982b) yielded new data on their structure, correlation with habitat, seasonality, and visitors, mainly ants. These may be attracted to establish colonies in the vicinity of bracken stands, which may be more significant to the plant than more intensive patrolling of expanding leaves. Other possible associations between ferns and ants are discussed by PATERSON (1982).

Gradually some more data accumulate on the interactions between animals (mostly insects) and ferns. OTTOSSON and ANDERSON (1983) investigated the insect fauna of six British fern species – mainly Hemiptera, Hymenoptera, Diptera, and Lepidoptera – and found it to be significantly poorer than that of flowering plants. This is probably due to the widespread presence of chemical defenses. On the other hand, HENDRIX and MARQUIS (1983) and ROWELL et al. (1983), working in Costa Rica, concluded that ferns were quite often attacked by herbivorous animals, e.g., grasshoppers.

5. Morphology and Anatomy

Regularly branched tree-ferns are very rare. OHBA (1982) described a new species from the Bonin Islands whose stems are regularly branched, with a type of branching not recorded before.

Many ferns with thickened vein ends have been credited with hydathodes, but anatomical and – especially – physiological evidence was often lacking. SPERRY (1983) showed the presence of true hydathodes in *Blechnum lehmannii* from Mexico. Secretion functions only in the presence of sufficient root pressure and in the absence of Casparian strips.

NISHIDA and NISHIDA (1982) examined the rhizome histology of two species of *Loxsomopsis*, showing that they are distinct; certain Cretaceous fossils are very similar.

WAGNER and JOHNSON (1983) drew attention to the presence of thickened petiole bases serving as storage organs in some North American ferns, e.g., *Asplenium platyneuron* and species of *Onoclea*, *Matteuccia*, *Athyrium* and *Dryopteris*. This organ is very aptly named "trophopod".

BIR and TRIKHA (1980) studied the rhizome and petiole anatomy of a number of genera of Polypodiaceae. The close coherence of this family in its modern sense was borne out, only *Loxogramme*, often placed in or near Grammitidaceae, stands a bit apart. A subdivision of the family on anatomical grounds was not practicable.

BHAMBIE and MADAN (1982) examined the venation of the three genera of Ophioglossaceae in great detail. They concluded that in this character, *Helminthostachys* is intermediate between *Botrychium* and *Ophioglossum*.

Studies of prothallia and sporelings of *Schizaea dichotoma* by BIERHORST (1983) showed that the young sporelings continue to live subterraneously and saprophytically for quite some time before producing the first leaf, which is not embryonic in origin.

Ontogenetic studies of *Anemia phyllitidis* leaves lead THORSPECKEN and HAGEMANN (1983) to the conclusion that the genus possesses true sori with branched receptacles. Studies of two species of *Lygodium* by MUELLER (1982, 1983) showed that the dichotomy of the primary pinnae is of a hitherto undescribed type, not involving division of the apical cell. The leaf is truly indeterminate and shows striking points of analogy to twining species of angiosperms, i.a., in its circumnutation movements and in the possibility to replace the damaged apex by an otherwise dormant bud.

BIR et al. (1981) examined the stomatal patterns of a large number of Indian *Asplenium* species. The group of *A. nidus* stands apart in this character, among others. Epidermal cell patterns were included in a further study of the same alliance by BIR et al. (1983). YADAV and BHARDWAJA (1982) discovered some new stomatal types in *Ophioglossum* spp. Details of stomatal arrangement in Athyrioid ferns sometimes permit to distinguish closely related species (BIR et al. 1980).

PUROHIT and SHARMA (1982) and PUROHIT et al. (1983) described vessel and tracheary elements of various Indian ferns.

SEM studies of fern spores yield increasingly important data on fern classification; see especially TRYON and TRYON (1982). Work on *Cibotium* spores by GASTONY (1982) has shown how poorly the genus is known taxonomically. Its spores resemble those of *Thyrsopteris* and *Cystodium* most, but also those of *Metaxya* and *Lophosoria*, otherwise mostly placed in or near Cyatheaceae. The spores of *Gymnocarpium* are not species-characteristic but show similar ranges of variation in the few species of this small genus. Spore size is correlated with ploidy level (PRYER and BRITTON 1983).

Electron microscopy often reveals the presence of small globules on the surface of fern spores. As shown by LUGARDON (1981), they have the same fine sculpture as the exospore. Moreover, they must be homologous to the so-called Ubisch bodies of seed plants, underscoring the relationship between the outer exospore and the exine tectum. LUGARDON and HUSSON (1982) studied the spores of heterosporous ferns ("Hydropteridales") and found the outer wall to be so different from the perispore of homosporous ferns that they coined the new term "epispore". The exospore of all species studied resembles the so-called blechnoid type of the homosporous leptosporangiates and stresses the derivation of the "water ferns" from the former.

Studies of prothallial morphology have become fewer, the prothallium being used more for experimental morphogenetical work. NAGMANI and RAGHAVAN (1983) investigated the germination pattern of five species in three genera of Polypodiaceae. It proved to accord with the *Vittaria*, rather than with the *Gleichenia* type, as previously held.

6. Chemotaxonomy

Chemosystematic work on ferns is expanding and yielding increasingly valuable results. A good overview, with many examples and important bibliography, is by COOPER-DRIVER and HAUFLER (1983). Details about how to apply electrophoretic techniques to chemosystematic studies of ferns were given by SOLTIS et al. (1983).

Studies on the flavonoids of Osmundaceae have demonstrated that *Todea* is chemically intermediate between *Osmunda* and *Leptopteris* (SOBEL and WHALEN 1983). Numerical analysis of chromatograms of North American *Gymnocarpium* species by PRYER et al. (1983) demonstrated the usefulness of this technique, bearing out, i.a., the hybrid nature of *G. intermedium* and the heterogeneity of *G. jessoense* ssp. *jessoense*.

The system of Gleicheniaceae as currently envisaged is strongly supported by phytochemical data (WALLACE et al. 1983). *Stromatopteris* forms part of the family. BIERHORST's idea about affinity to Psilotaceae could not at all be confirmed.

A broad search for phloroglucinols in leptosporangiate ferns (WIDÉN et al. 1983) yielded significant results. They are almost confined to ferns having internal glands and are most widespread in *Dryopteris* and *Arachniodes*, much less so in *Polystichum*; *Ctenitis* and *Tectaria* are variable. The distinctness of *Nothoperanema* from *Dryopteris* is confirmed. *Rumohra* is linked by its phloroglucinols to Aspidiaceae, the compounds being quite absent from Davalliaceae, also from the Athyrioid group of ferns. Search for flavonols in eastern North American *Dryopteris* taxa also gave taxonomically meaningful results (PETERSEN and FAIRBROTHERS 1983). On the other hand, work by WALLACE et al. (1982) showed that flavonoid-like compounds are usable for taxonomic interpretation only in some groups like Hymenophyllaceae and *Elaphoglossum*, not or hardly in others like *Marsilea*.

WOLLENWEBER and his co-workers continued their research on "mealy" exudates of Gymnogrammoid ferns, with many taxonomic implications; see WOLLENWEBER (1982), HITZ et al. (1982), JAY et al. (1982), WOLLENWEBER et al. (1982), WOLLENWEBER and YATSKIEVYCH (1982), WOLLENWEBER et al. (1983), and SEIGLER and WOLLENWEBER (1983).

Investigations of isozymes in chloroplasts of *Athyrium filix-femina* showed this fern, despite its high chromosome number of 2n = 80, to be functionally diploid (GASTONY and DARROW 1983).

7. Cytotaxonomy, Biosystematics, Hybridization

Athyrium filix-femina is a very widespread and variable, preferably outbreeding fern. SCHNELLER and SCHMID (1982) showed that a certain range of variability occurs throughout various populations in Switzerland, but probably owing to genecological variation the populations are not homogeneous. Evidence for genetic heterozygosity in *Pellaea andromedifolia* was found through isoenzyme analysis by GASTONY and GOTTLIEB (1982). VASUDEVA and BIR (1982) made chromosome counts of pteridophytes in a hill range in Central India. Few species had not been counted before, but many counts are new for India. The rate of polyploidy was higher than in northern and lower than in southern India (42%). The incidence of apomixis is surprisingly low.

An account of the ferns of Elden Mountain in Arizona is mainly important for its chromosome counts and observations on meiosis (WINDHAM 1983). The first chromosome count for the Marattiaceous genus *Macroglossum* is by WALKER (1981), for *M. alidae* from Sarawak: 2n = 80. A species of *Angiopteris* from the same area had the same number, widespread though not universal in the Marattiaceae (or Marattiales). The count of n = 35 for the Australian *Dictymia brownii* is a somewhat aberrant number in the Polypodiaceae (WALKER and PAGE 1982).

Some interesting instances of hybridization have come to light. HOLTTUM (1984) described a hybrid between a species of *Cyathea* from Mexico and another from Norfolk Island. A natural hybrid between *Pteris bahamensis* and the introduced, closely related *Pt. vittata* was found in Florida and described by WAGNER and NAUMAN (1982). An intermediate hybrid between *Pellaea bridgesii* and *P. mucronata* from California provides good evidence that the former, often regarded as of uncertain affinity, is a member of sect. *Pellaea*, like the latter species (WAGNER et al. 1983). A new species of *Angiopteris* from Yünnan may be a hybrid between *Angiopteris* and *Archangiopteris* (CHING and WANG 1982).

SCHNELLER (1981) showed that there is intergeneric incompatibility between *Dryopteris* and *Athyrium*, the spermatozoids of the one not being able to enter the archegonia of the other.

Data on intrageneric hybridization in *Hypolepis* were compiled and discussed by BROWNSEY (1983). Some species are probably aggregates. The affinities of the genus are also discussed, as are those of the "Dennstaedtiales", but too much weight is attached to chromosome numbers alone.

New hybrids have turned up in *Asplenium* (BADRÉ et al. 1981), *Cheilanthes* (RASBACH and REICHSTEIN 1982, RASBACH et al. 1983a), *Dryopteris* (GIBBY and WIDÉN 1983), and *Woodsia* (BROOKS 1982). The biosystematics of *Cheilanthes* (s. str.) and its hybrids in Macaronesia and the Mediterranean region is extensively treated by VIDA et al. (1983). There are four ancestral diploids, *C. hispanica*, *C. maderensis*, *C. persica*, and *C. pulchella*, and three allotetraploids derived from them. Their probable places of origin are discussed. Nonfertile hybrids are partly of natural occurrence, partly experimentally produced. Five new natural hybrids from this complex were described by RASBACH et al. (1983b). *Notholaena lanuginosa* consists of two cytotypes. The type specimen belongs to the tetraploid form; the diploid is newly described (BADRÉ and REICHSTEIN 1983).

New records of *Asplenium* hybrids from the Iberian Peninsula by RIVAS MARTÍNEZ et al. (1982). Notes on the distribution of three cytotypes of *Asplenium trichomanes* in North America by MORAN (1982). The correct name for *Asplenium platyneuron* × *Camptosorus rhizophyllus* is × *Asplenosorus* (or *Asplenium* ×) *ebenoides*, not *A. hendersonii* (which is *Phyllitis scolopendrium* × *A. ?marinum*) (WALTER et al. 1982). A large colony in Alabama is described, where backcrosses also occur.

In connection with the problem of the origin of heterspory, DUCKETT and PANG (1984) studied the behavior of cultured and subcultured gametophytes of *Platyzoma microphyllum* and *Equisetum giganteum*. The former was known to produce small spores giving rise to male gametophytes, and large spores developing into female but eventually bisexual gametophytes. It was now shown that under experimental conditions male gametophytes may also produce archegonia. *E. giganteum* is isosporous, three-quarters of the spores producing female gametophytes later also bearing antheridia, one-quarter growing into male gametophytes, eventually also bearing archegonia. Thus sex determination takes place during gametophytic development, not during sporogenesis. Larger spores of *P.* yield larger gametophytes which tend to be female; similarly, larger prothallial lobes of *E.* when subcultured tend to develop into archegonium-bearing prothalli. It seems feasible that regular, environment-induced unisexuality of gametophytes was the first step toward heterospory.

8. Folklore, Uses

The first book on the use of pteridophytes in South America, especially their role in folk medicine, was produced by MURILLO (1983). It is richly illustrated and includes a taxonomic section. Most of the data were compiled from the literature; they apply chiefly to Colombia.

9. Fern Allies

The taxonomy of the Chinese Lycopodiaceae was revised by CHING (1981, 1982a,b). The family was split in two, and nine genera in all were recognized; keys; new species. There are now over 60 species in China. *L clavatum* is much more narrowly circumscribed than formerly.

TAKAMIYA and KURITA (1983) investigated chromosome numbers of *Lycopodium* (s.l.) in Japan. The alleged difference in base number between Section *Selago* and Section *Phlegmaria* of Subgen. *Urostachya* is incorrect; all numbers are probably based on x = 17. The base number in Sections *Complanata* and *Lycopodium* of Subgen. *Lycopodium* are 23 and 34, respectively; their connection remains obscure. The hypothetical base numbers of x = 11, 13, and 17, which served as arguments for dividing the genus, do not exist in reality. *L. lucidulum* was newly counted as n = 67 by BEITEL and WAGNER (1982); taxonomic implications are discussed.

The only major contribution to the taxonomy of *Selaginella* is WONG's treatment of the species in the Malay Peninsula (1983). It breaks away from ALSTON's tradition as it provides descriptions and illustrations (some in color) as well as keys. There are interesting field observations and notes on morphology. Twenty-eight species are now recorded, 9 endemic. A key to the Philippine species based on vegetative characters was published by MENDOZA and ZAMORA (1981).

Forked, even branched veins in *Selaginella* was reported by MUKHERJEE and SEN (1981) and WAGNER et al. (1982c).

DENGLER's work on *Selaginella* leaves (1983) demonstrated that the anisophylly is not due to late divergence in the development of ventral vs dorsal leaves but to divergent development at inception. The general pattern of development of the two leaf types is similar, but the smaller, dorsal leaves mature earlier.

A newly described type of microsporangium, with peculiar dehiscence, occurs in many species of *Selaginella* Series *Articulatae* and constitutes another of several distinctive features of this group (SOMERS 1982).

Annual species of *Selaginella* are very rare, and so is apomixis in the genus. KORNAŚ and JANKUN (1983) reported on the triploid African *S. tenerrima* which is both annual and apomictic; this is to be regarded as an adaptation to seasonal drought.

The distribution of the South American species of *Isoëtes* and – especially – the history of their discovery is expounded in great detail by FUCHS (1982). Anatomical studies of the secondary cortex and its development in this genus disclosed traits reminiscent of *Stigmaria* (KARRFALT 1982). The structure of the ligule was described and discussed by SHARMA and SINGH (1984). New chromosome counts for neotropical *Isoëtes* species gave numbers consistently based on x = 11; the highest number as yet recorded is 2n = 132 for *I. novo-granadensis* (dodecaploid). *"Stylites" andicola*, now placed in *Isoëtes*, has 2n = 44 (HICKEY 1984). The same number was found in the North Italian *I. malinverniana*, doubtfully native but, if introduced, of unknown provenance (SCHNELLER 1982). Chromatographic studies of phenolic compounds in species of *Isoëtes* reflect to some extent on their affinity (KOTT and BRITTON 1982).

The two subgenera of *Equisetum* diverge also in certain properties of their DNA, as shown by BENEDICH and ANDERSON (1983). – See also TRYON and TRYON (1982) and DUCKETT and PANG (1984), reviewed elsewhere.

References

Anon.: Flora of Kiangsu Province I, 502 pp (Chinese). Kiangsu: People's Publ. Co. 1977.

BADRÉ, F., REICHSTEIN, T.: Willdenowia 13, 361-367 (1983). - BADRÉ, F., BOURDRIE, M., PRELLI, R., SCHNELLER, J.: Adansonia IV. 3, 473-481 (1981). - BADRÉ, F., TRYON, A.F., DESCHARTRES, R.: Webbia 36, 1-38 (1982). - BALAKRISHNAN, N.P.: Bull. Bot. Surv. India 22, 136-140 ("1980") (1982). - BEITEL, J.M., WAGNER, F.S.: Am. Fern J. 72, 33-35 (1982). - BENEDICH, A.J., ANDERSON, R.S.: Plant Syst. Evol. 143, 47-52 (1983). - BHAMBIE, S., MADAN, P.: Fern Gaz. 12(4), 215-223 (1982). - BIERHORST, D.W.: Am. J. Bot. 70, 1057-1062 (1983). - BIR, S.S.: Aspects Plant Sci. 6, 183-245 (1982). - BIR, S.S., TRIKHA, C.K.: Aspects Plant Sci. 3, 139-157 (1980). - BIR, S.S., SATIJA, C.K., GOYAL, P., JAIN, A.B.: Aspects Plant Sci. 3, 123-137 (1980). - BIR, S.S., SATIJA, C.K., GOYAL, P., KAUR, S.: Biol. Bull. India 3, 24-34 (1981). - BIR, S.S., SATIJA, C.K., VASUDEVA, S.M., GOYAL, P.: Pteridophytic Flora of Garhwal Himalaya, 83 pp. Dehra Dun: Jishore 1982a. - BIR, S.S., SATIJA, C.K., BIR, P.: Aspects Plant Sci. 6, 1-24 (1982b). - BIR, S.S., GOYAL, P., SATIJA, C.K.: Acta Bot. Indica 11, 140-145 (1983). - BROOKS, R.E.: Am. Fern 72, 79-84 (1982). - BROWNSEY, P.J.: Am. Fern J. 73, 97-108 (1983).

CABRERA, A.L., ZARDINI, E.M.: Manual de la flora de los alrededores de Buenos Aires, 2nd ed. Buenos Aires: Ed. Acme 1978. - CARLSON, T.M., WAGNER, W.H: Contrib. Univ. Mich. Herb. 15, 141-162 (1982). - CHANDRA, S.: Fern Gaz. 12(4), 225-229 (1982a); - Gard. Bull. Sing. 34, 229-239 ("1981") (1982b); - Fern Gaz. 12(5), 275-282 (1983). - CHING, R.C.: Acta Bot. Yunnan. 3, 291-305 (1981); - ibid. 4, 119-128 (1982a); - ibid. 4, 213-226 (1982b). - CHING, R.C., SHING, K.H.: Acta Phytotax. Sin. 21, 266-276 (1983) (Chin.; Engl. summ.). - CHING, R.C., WANG, Z.R.: Acta Phytotax. Sin. 19, 118-130 (1981); - ibid. 20, 73-77 (1982a) (Chin. and Engl.); - ibid. 20, 349-350 (1982b). - CHINNOCK, R.J.: Common Ferns and Fern Allies, 79 pp. Mobil New Zealand Nature Series, Reed, Wellington, N.Z. 1981. - CODY, W.J.: Rhodora 85, 93-96 (1983). - CONANT, D.S.: J. Arnold Arbor. Harv. Univ. 64, 333-382 (1983). - COOPER-DRIVER, G.A., HAUFLER, C.: Fern Gaz. 12(5), 283-294 (1983). - COPELAND, E.B.: Genera Filicum. Chronica Botanica, 247 pp. Waltham, MA 1947. - CORRELL, D.S., CORRELL, H.B.: Flora of the Bahama Archipelago. Vaduz: Cramer 1982. - CRANFILL, R.: Am. Fern. J. 73, 46-52 (1983). - CRANFILL, R., BRITTON, D.M.: Taxon 32, 557-560 (1983). - CUSICK, A.W.: Am. Fern J. 73, 33-38 (1983).

DEB, D.B.: The Flora of Tripura State I. New Delhi: Today & Tomorrow's 1981. - DENGLER, N.G.: Am. J. Bot. 70, 181-192, 193-206 (1983). - DENNIS, W.W., WEBB, D.H.: Sida 9, 19-24 (1981). - DIXIT, R.D., GHOSH, B.: Proc. Ind. Acad. Sci. (Plant Sci.) 92, 233-258 (1983). - DOSTÁL, J., REICHSTEIN, T., FRASER-JENKINS, C.R., KRAMER, K.U.: Pteridophyta, Bd. I,1, 309 pp. In: Illustrierte Flora von Mitteleuropa, 3rd ed., G. HEGI (ed.). Berlin: Parey 1984. - DUCKETT, J.G., PANG, W.C.: Bot. J. Linn. Soc. 88, 11-34 (1984).

FARRAR, D.R., PARKS, J.C., McALPIN, B.W.: Rhodora 85, 83-92 (1983). - FOSBERG, F.R., CANFIELD, J.E.: Micronesica 16, 189-200 (1980). - FOSBERG, F.R., SACHET, M.H., OLIVER, R.: Micronesica 18, 23-82 (1982). - FRASER-JENKINS, C.R.: Fern Gaz. 12(3), 155-156 (1981). - FUCHS, H.P.: Proc. Kon. Ned. Akad. Wetensch. C 85, 205-260 (1982).

GASTONY, G.J.: Can. J. Bot. 955-972 (1982). - GASTONY, G.J., DARROW, D.C.: Am. J. Bot. 70, 1409-1415 (1983). - GASTONY, G.J., GOTTLIEB, L.D.: Am. J. Bot. 69, 634-637 (1982). - GAWLER, S.C.: Rhodora 85, 389-390 (1983). - GIBBY, M., WIDÉN, C.J.: Fern Gaz. 12(5), 267-270 (1983). - GREUTER, W., PLEGER, R., RAUS, T., ZIMMER, B., GREUTER, J.J.: Fern Gaz. 12(5), 271-274 (1983).

HAUFLER, C.H.: Bibliography of American Pteridology, Vol. 7, 1981; Vol. 8, 1982. Kansas: Lawrence. - HENDRIX, S.D., MARQUIS, R.J.: Biotropica 13, 108-111 (1983). - HENNIPMAN, E., ROOS, M.C.: Verh. Kon. Ned. Akad. Wetensch. Afd. Natuurk. II. 80, 1-126 (1982); - Verh. Naturwiss. Ver. Hamb. N.F. 26, 321-342 (1982). - HICKEY, R.J.: Am. Fern J. 74, 9-18 (1984). - HITZ, C., MANN, K., WOLLENWEBER, E.: Z. Naturforsch. 37c, 337-339 (1982). - HOLTTUM, R.E.: Thelypteridaceae, pp. 331-509. In: Flora Malesiana II, 1(5) (1982a); - Kew. Bull. 37, 383-388 (1982b); - Acta Bot. Malac. 8, 47-58 (1983); - Am. Fern J. 74, 19-21 (1984). - HOLTTUM, R.E., EDWARDS, P.J.: Kew Bull. 38, 155-158 (1983).

IWATSUKI, K.: Hikobia Suppl. 1, 59-66 (1981); - J. Fac. Sci. Univ. Tokyo III. 13, 203-215 (1982a); - Acta Phytotax. Geobot. 33, 149-159 (1982b) (Jap.; Engl.summ.). - IWATSUKI, K., KATO, M.: Acta Phytotax. Geobot. 34, 55-65 (1983). - IWATSUKI, K., KATO, M., OKAMOTO, M., UEDA, K., DARNAEDI, D.: pp. 31-69 in: Taxonomical and Evolutionary Studies on the Biota in Humid Tropical Malesia, with Reference to Diversity of the Species, T. HIDAKA (ed.), 1983.

JACOBSEN, W.B.G.: The Ferns and Fern Allies of Southern Africa, 542 pp. Durban, Pretoria: Butterworths 1983. - JAY, M., VIRICEL, M.R., FAVRE-BONVIN, J., VOIRIN, B.: Z. Naturforsch. 37c, 721-723 (1982). - JOE-HOSHIZAKI, B.: Baileya 21, 1-76 (1981); - ibid. 22, 1-99 (1982).

KARRFALT, E.E.: Bot. Gaz. 143, 439-445 (1982). - KATO, M.: Acta Phytotax. Geobot. 33, 168-178 (1982) (Jap.); - J. Fac. Sci. Univ. Tokyo III. 13, 263-283 (1983). - KHULLAR, S.P., SHARMA, S.S.: Aspects Plant Sci. 6, 63-87 (1980). - KLEKOWSKI, J., KLEKOWSKI, E.: Am. J. Bot. 69, 721-727 (1982). - KOPTUR, S., SMITH, A.R., BAKER, I.: Biotropica 14, 108-113 (1982). - KORNAŚ, J.: Acta Soc. Bot. Polon. 52, 321-335 (1983). - KORNAŚ, J., JANKUN, A.: Bothalia 14, 647-651 (1983). - KORNAŚ, J., DZWONKO, Z., HARMATA, K., PACYNA, A.: Bull. Jard. Bot. Nat. Belg. 52, 265-309 (1982). - KOTT, L.S., BRITTON, D.M.: Am. Fern J. 72, 15-18 (1982). - KUO, C.M.: Pteridophytes of Taiwan, 138 pp. (Chin.). Taipei: Taiwan Prov. Dept. Educ. 1982. - KURATA, S., NAKAIKE, T.: Illustrations of Pteridophytes of Japan, Vol. 3, 728 pp. Tokyo: Univ. of Tokyo Press 1983.

LELLINGER, D.B., PROCTOR, G.B.: Taxon 32, 565-571 (1983). - LUCANSKY, T.W.: Am. Fern J. 72, 19-29 (1982). - LUGARDON, B.: Pollen et Spores 23, 93-124 (1981). - LUGARDON, B., HUSSON, P.: C.R. Acad. Sci. Paris III, 294, 789-794 (1982).

MENDOZA, R.C., ZAMORA, C.V.: Nat. Appl. Sci. Bull. 33, 175-220 (1981). - MICKEL, J.T.: Brittonia 33, 413-429 (1981); - ibid. 34, 388-413 (1982). - MITSUTA, S.: Acta Phytotax. Geobot. 33, 264-275 (1982) (Jap., Engl. summ.). - MORAN, R.: Am. Fern J. 72, 5-11 (1982). - MORTON, C.V.: Contrib. U.S. Nat. Herb. 38(5), 153-214 (1960). - MUELLER, R.J.: Bot. Gaz. 143, 319-330 (1982); - Am. J. Bot. 70, 682-690 (1983). - MUKHERJEE, R.N., SEN, U.: Fern Gaz. 12(3), 175-177 (1981). - MURILLO, M.T.: Usos de helechos en Suramérica con especial referencia a Colombia, 156 pp. Inst. Cienc. Nat., Mus. Hist. Nat., Bibl. J. TRIANA 5, Bogotá, 1983.

NAGAMI, R., RAGHAVAN, V.: Bot. Gaz. 144, 67-72 (1983). - NAIR, N.C., DIXIT, R.D.: J. Bombay Nat. Hist. Soc. 78, 443-462 (1981). - NAKAIKE, T.: New Flora of Japan. Pteridophyta, 808 pp. Tokyo: Shibundo 1982a; - Rep. Cryptog. Stud. Nepal 135-158 (1982b); - Mem. Nat. Sci. Mus. 16, 105-110 (1983) (Jap.; Engl. summ.). - NARDI, E.: Webbia 36, 217-223 (1983). - NISHIDA, M., NISHIDA, H.: Acta Phytotax. Geobot. 33, 302-307 (1982) (Jap.; Engl. summ.).

OHASHI, H., TATEISHI, Y., MURATA, G., KATO, M.: J. Jap. Bot. 57, 1-9 (1982) (Jap.; Engl. summ.). - OHBA, H.: J. Jap. Bot. 57, 321-327 (1982). - ORTEGA, M.F.J.: Mem. Soc. Cienc. Nat. La Salle 42, 135-150 (1982). - OTTOSSON, J.G., ANDERSON, J.M.: J. Anim. Ecol. 52, 385-406 (1983).

PAGE, C.N.: The Ferns of Britain and Ireland, 447 pp. Cambridge: University Press 1982a; - Fern Gaz. 12(4), 233-240 (1982b). - PAGE, C.N., CLIFFORD, H.T.: Chap. 16 in: Ecological Biogeography of Australia I, A. KEAST (ed.). The Hague: Junk 1981. - PANIGRAHI, G., BASU, S.K.: Taxon 31, 102-104 (1982). - PARRIS, B.S.: N.Z. J. Bot. 18, 145-147 (1980); - Fern Gaz. 12(3), 165-168 (1981); - Blumea 29, 13-222 (1983). - PATERSON, S.: Fern Gaz. 12(4), 243-245 (1982). - PECK, J.H.: Ferns and Fern Allies of the Driftless Area of Illinois, Iowa, Minnesota, and Wisconsin, 140 pp. Milwaukee, WI: Public Museum Contrib. Biol. Geol., Vol. 53, 1982. - PÉREZ-GARCIA, B., RIBA, R.: Biotropica 14, 281-287 (1982). - PETERSEN, R.L., FAIRBROTHERS, D.E.: Bot. Gaz. 144, 104-109 (1983). - PICHI SERMOLLI, R.E.G.: Taxon 30, 163-168 (1981); - Webbia 35, 223-237 (1982a); - ibid. 35, 241-260 (1982b). - PRICE, M.G.: Contrib. Univ. Mich. Herb. 15, 197-204 (1982); - Am. Fern J. 73, 109-116

(1983). - PROCTOR, G.R.: Am. Fern J. 72, 107-114 (1982). - PRYER, K.M., BRITTON, D.M.: Can. J. Bot. 61, 377-388 (1983). - PRYER, K.M., BRITTON, D.M., McNEILL, J.: Can. J. Bot. 61, 2592-2602 (1983). - PUROHIT, S.N., SHARMA, B.D.: Phytomorphology 30(4), 400-402 ("1980")(1982). - PUROHIT, S.N., SINGH, R., SHARMA, B.D.: Phytomorphology 31(1/2), 55-59 ("1981") (1983).

QUIRK, H., CHAMBERS, T.C.: Fern Gaz. 12(3), 121-129 (1981).

RASBACH, H., REICHSTEIN, T.: Webbia 35, 261-273 (1982). - RASBACH, H., REICHSTEIN, T., SCHNELLER, J.: Bot. Helv. 92, 33-40 (1982). - RASBACH, H., RASBACH, K., REICHSTEIN, T., SCHNELLER, J.: Farnblätter 10, 1-13 (1983a); - Webbia 37, 43-62 (1983b). - RIVAS MARTÎNEZ, S., IRANZO, J., SALVO, A.E.: Collect. Bot. 31, 87-95 (1982). - ROUX, J.P.: Fern Gaz. 12(4), 191-196 (1982). - ROWELL, C.H.F., ROWELL-RAHIER, M., BRAKER, H.E., COOPER-DRIVER, G., GÔMEZ, L.D.: Biotropica 15, 207-216 (1983).

SCHELPE, E.A.C.L.E.: Bothalia 14, 417-419 (1983). - SCHNELLER, J.J.: Plant Syst. Evol. 137, 45-56 (1981); - Webbia 35, 307-309 (1982). - SCHNELLER, J.J., SCHMID, B.W.: Adansonia IV, 4, 215-228 (1982). - SEIGLER, D.S., WOLLENWEBER, E.: Am. J. Bot. 70, 790-798 (1983). - SEYMOUR, F.C.: The Flora of New England, 2nd ed. Phytologia Memoirs V, Plainfield, N.J.: 1982. - SHARMA, B.D., SINGH, R.: Am. Fern J. 74, 22-28 (1984). - SHING, K.H.: Acta Bot. Yunnan. 3, 217-238 (1981); - Am. Fern J. 73, 73-78 (1983). - SMITH, A.R.: 18. Polypodiaceae-Thelypteridoideae, 148 pp. In: Flora of Ecuador, G. HARLING, B. SPARRE (eds.). Stockholm: Publ. House, Swed. Res. Councils 1983. - SMITH, A.R., FRASER-JENKINS, C.R.: Taxon 31, 326-329 (1982). - SOBEL, G.L., WHALEN, M.D.: Fern Gaz. 12(5), 295-298 (1983). - SOLTIS, D.E., HAUFLER, C.H., DARROW, D.C., GASTONY, G.J.: Am. Fern J. 73, 9-27 (1983). - SOMERS, S.: Am. Fern J. 72, 88-92 (1982). - SPERRY, J.S.: Am. Fern J. 73, 65-72 (1983). - STOLZE, R.G.: Fieldiana: Botany N.S. 12, 1-91 (1983).

TAKAMIYA, M., KURITA, S.: Acta Phytotax. Geobot. 34, 66-79 (1983). - TANNER, E.V.J.: Bot. J. Linn. Soc. 87, 213-227 (1983). - THORSPECKEN, A., HAGEMANN, W.: Plant Syst. Evol. 143, 133-150 (1983). - TINDALE, M.D.: Pteridophytes. In: Flora of the Sydney region, 3rd ed. N.C.W. BEADLE, O.D. EVANS, R.C. CAORLIN (eds.). Reed, French Forest, N.S.W. 1982. - TRYON, R.M., TRYON, A.F.: Ferns and Allied Plants with Special Reference to Tropical America, 857 pp. New York, Heidelberg, Berlin: Springer 1982.

VASUDEVA, S.M., BIR, S.S.: Aspects Plant Sci. 6, 119-181 (1982). - VIDA, G., MAJOR, A., REICHSTEIN, T.: Acta Bot. Malac. 8, 101-126 (1983).

WAGNER, W.H., BEITEL, J.M., WAGNER, F.S.: Science 218, 793-794 (1982c). - WAGNER, W.H., JOHNSON, D.M.: Taxon 32, 268-269 (1983). - WAGNER, F.S., MUSSELMAN, L.J.: Castanea 47, 182-190 (1982). - WAGNER, W.H., NAUMAN, C.E.: Am. Fern J. 72, 97-102 (1982). - WAGNER, W.H., WAGNER, F.S.: Contrib. Univ. Mich. Herb. 15, 315-324 (1982a); - Michig. Bot. 21, 75-88 (1982b); - Am. Fern J. 73, 53-62 (1983a); - Taxon 32, 51-63 (1983b). - WAGNER, W.H., WAGNER, F.S., LEONARD, S.W., MESLER, N.R.: Castanea 46, 311-322 (1981). - WAGNER, W.H., SMITH, A.R., PRAY, T.R.: Madroño 30, 69-83 (1983). - WALKER, T.G.: Fern Gaz. 12(3), 161-163 (1981). - WALKER, T.G., JERMY, A.C.: Fern Gaz. 12(4), 209-213 (1982). - WALKER, T.G., PAGE, C.N.: Fern Gaz. 12(4), 197-207 (1982). - WALLACE, J.W., MARKHAM, K.R., GIANNASI, D.E., MICKEL, J.T., YOPP, D.L., GÔMEZ, L.D., PITILLO, J.D., SOEDER, R.: Am. J. Bot. 69, 356-362 (1982). - WALLACE, J.W., POZNER, R.S., GÔMEZ, L.D.: Am. J. Bot. 70, 207-211 (1983). - WALTER, K.S., WAGNER, W.H., WAGNER, F.S.: Am. Fern J. 72, 65, -75 (1982). - WELTEN, M., SUTTER, R.: Verbreitungsatlas der Farn- u. Blütenpflanzen der Schweiz. I. Basel: Birkhäuser 1982. - WIDEN, C.J., SARVELA, J., BRITTON, D.M.: Acta Bot. Fenn. 20, 407-417 (1983). - WILLMOT, A.: Fern Gaz. 12(3), 133-154 (1981); - ibid. 12(5), 249-265 (1983). - WINDHAM, M.D.: Am. Fern J. 73, 85-93 (1983). - WINDISCH, P.G.: Am. Fern J. 72, 56-60 (1982). - WINDISCH, P.G., PEREIRA-NORONHA, M.: Am. Fern J. 73, 79-84 (1983). - WOLLENWEBER, E.: pp. 215-224 in: The Plant Cuticle. Linn. Soc. Symp. Ser., Vol. 10, D.F. CUTLER, K.L. ALVIN, C.E. ORICE (eds.). London: Academic Press 1982. - WOLLENWEBER, E., YATSKIEVYCH, G.J.: Nat. Prod. 45, 216-219 (1982). - WOLLENWEBER, E., RÜEDI, P., SEIGLER, D.S.: Z. Naturforsch. 37c, 1283-1285 (1982). - WOLLENWEBER, E., MARX, D., FAVRE-BONVIN, J., BRASSY, C.: Z. Naturforsch. 38c, 146-147 (1983). - WONG, K.M.: Gard. Bull. Sing. 35, 107-135 ("1982") (1983). - WU, C.Y. (ed.): Flora Xizangica, I, 791 pp. Peking: Science Press 1983.

YADAV, B.L., BHARDWAJA, T.N.: Phytomorphology 30(4), 403-406 ("1980") (1982).

Professor Dr. KARL U. KRAMER
Botanischer Garten und Institut für
System. Botanik der Universität
Zollikerstr. 107
CH-8008 Zürich

E. Geobotany

I. The History of Flora and Vegetation During the Quaternary

By Burkhard Frenzel

1. Paleoclimatology

The basis for valid paleoecological investigations is a profound knowledge of the evolution of climate. Moreover, paleoclimatological studies enable us - perhaps - to draw conclusions as to the future development of climate, provided the reasons for and the possible periodicity of changes in climate during the past are understood. So paleoclimatology has become an intensively investigated discipline. Of course it should be known whether the solar irradiance is variable or not (WILLSON et al. 1981, KERR 1982), and tectonic uplifting or downwarping may play an important role in the evaluation of possible climatic influences on the biosphere. So it is worthwhile to note that according to VOZOVIK (1976) the mean elevation of the watersheds in the Central Caucasus Mts. increased from late Pliocene times (Lower Akchagyl: 2000 m), and from the very beginning of the Pleistocene (Upper Akchagyl, 2700 m) to the Middle Pleistocene (end of the Apsheron, 3500 m), to 4400 m in recent times. The same topic has been analyzed by ALESHINSKAYA et al. (1976 a,b) in regard of the Tien Shan Mts., in connection with changes in synchronous paleoecological conditions. Other important boundary conditions are the surface areas of the inland ice-masses (for the last glaciation see DENTON and HUGHES 1981; for the whole of the Pleistocene, the continental shelf areas included: GROSVALD 1983), as well as the changes in the CO_2-content of the global atmosphere in an interglacial-glacial-interglacial cycle. As OESCHGER (1980) has pointed out, the CO_2-content of air bubbles in Greenland and Antarctic inland ice dating from the last pleniglacial seems to have amounted to 200-230 ppm only, as opposed to 300 ppm in modern inland ice. From these figures it is concluded that the rising CO_2-content of the the global atmosphere might have influenced the melting of the glaciers during late glacial times. The same problem is dealt with theoretically and inversely by FLOHN (1980), who deduces for certain periods of the past, the climate of which seems to be relatively well known, the mean CO_2-content of the atmosphere: Upper Tertiary: approximately 755 ppm; last interglacial: between 508 and 555 ppm; Holocene climatic optimum about 6000 before present (b.p.): some 455 ppm. These calculations lead to the well-known and repeatedly discussed problem of changes in the global climate, caused by increasing amounts of various gases (FLOHN 1980, BRYAN et al. 1982, KUKLA and GAVIN 1981, SCHNEIDER et al. 1980) whose real nature is not well understood and their amounts are unknown. From a paleobiological point of view, as well as for climatological considerations, the fact that the amount of dust in the atmosphere was much higher during the Pleistocene/Holocene boundary than at present seems to be of some importance: In Antarctica 3 to 6 times as much, in Greenland up to 12 times as much as in recent times (THOMPSON and MOSLEY-THOMPSON 1981). For the industrial period in Central Europe see SCHÜEPP (1980). Very valuable summaries of various physical and chemical methods involved in paleoclimatological reconstructions are given by DANSGAARD (1980)

Progress in Botany, Vol. 46

and OESCHGER (1980). The significance of pollen-analytical work in reconstructing past climates is stressed by BOYARSKAYA and MURATOVA (1976) for the basin of the river Aldan in Central Yakutia. Yet I doubt that it can really be concluded from observations like this that changes of climate had indeed preceded changes in vegetation.

In paleoclimatological studies of the Quaternary two problems seem to be of utmost interest. They are the general trend of the evolution of the global climate and the occurrence and length of its periodic changes. According to WOODRUFF et al. (1981) the Earth has been nearly unglaciated up to about 16.5 and 14.8 million years b.p. Then the strong accumulation of ice seems to have begun. This view is to some extent corrected by KERR (1981a), emphasizing regional differences in Antarctica. At any rate, this development was complicated by the closing of the Panama Street at about 4 to 3 million years ago, when not only a floral and faunal exchange between North and South America was enabled or enhanced, but when present-day circulation patterns of the oceans came into existence, which are so decisive for ecological and climatological conditions not only in the oceans themselves (KEIGWIN 1982), but on the continents, too. Various observations point to the fact that the global climate seems to have been warmer during the first 1,6 million years of the ice-age than during the last 700,000. This seems to have caused the Arctic Ocean to be free of a perennial ice cover at that time (HERMANN and HOPKINS 1980, PETERSON and LOHMANN 1982, WORSLEY and HERMAN 1980). Yet even the assumption of a vast ice cover of the Arctic Ocean and the North Atlantic during full-glacial times of the last glaciation is questioned by OLAUSSON (1982). As to geochemical evidence of the circulation pattern of the North Atlantic during the last 200,000 years see BOYLE and KEIGWIN (1982). Perhaps the general trends of the global climate discussed here might be correlated with long-lasting changes in the excentricity of the Earth's orbit (400,000 to 413,000 years: KERR 1981b). This leads to the problem of periodicity of changes in climate. As has been pointed out earlier, the wealth of information now available favors the view of Milankovitch climate cycles, i.e., of changes in the Earth's orbit as being of prime importance for the major Pleistocene and Holocene changes of climate (e.g., KERR 1981b, KUTZBACH 1981, OERLEMANS 1981, RUDDIMAN and McINTYRE 1981a, SERGIN 1980). Yet it may be questioned whether these astronomical changes alone suffice for explaining reliably what had happened in the past. So the role of the oceans in amplifying the triggering astronomical effects should be taken into consideration (RUDDIMAN and McINTYRE 1981b) as well as feedback (SERGIN 1980) and stochastic processes (HASSELMANN 1980). By doing so, various programs of climate modeling have been developed, including deep ocean circulation (WATTS and HAYDER 1983), CO_2-contents (THOMPSON and SCHNEIDER 1982), air-sea interactions (NAMIAS and CAYAN 1981, REITER 1980), or interrelations between the Earth's ice caps and climate (STAUFFER 1980). In using these models one realizes how great are the handicaps caused by relying first only on relatively simple boundary conditions, so that the necessity of using much better geographical or geomorphological and glaciological parameters is repeatedly stressed (OERLEMANS 1981, THOMPSON and SCHNEIDER 1982, REITER 1980). Moreover, it is felt that the influence of man, even in preindustrial times, on climate is not well understood so that our suggestions as to spontaneous changes of climate might be to some extent not so well founded as one would have expected (KELLOGG 1980, MITCHELL 1980). Here, also other most interesting influences exist: According to FLOHN (1980), strong volcanic eruptions cause a relatively long-lasting decrease in temperature, and nearly all glacier advances during the Little Ice Age seem to have been correlated with strong volcanic eruptions. On the other hand, SHUKLA and MINTZ (1982) have pointed out the strong importance of the amount of evapotranspiration from the continents, i.e., from vegetation as well, for the amount of precipitation, temperature and wind activity. This stresses the value of paleovegetation maps, as have been prepared by GERASIMOV and VELICHKO (1982) for the last 100, 000 years in Europe, by DELCOURT and DELCOURT (1981) for North America, by TSUKADA (1983) for Japan, and by FRENZEL (1983) for Europe.

The repeatedly held view of surging Antarctic glaciers being of prime importance for the initiation of a glacial period is now questioned by FLOHN (1980), SPJELDNAES (1981), and STAUFFER (1980). Also the nutritional and the physical properties of the oceans are influenced by the influx of terrigenous material. Since the terrestrial vegetation changed so strongly during an interglacial to glacial cycle during the last 125,000 years and since vegetation is of some importance for the magnitude of soil erosion, sediment transport, etc., the observations of KRYLOV et al. (1976) on the sedimentation of terrigenous minerogene material during various phases of the last glaciation in the Mediterranean Sea are of some interest: A very strong influx of terrestrial material seems to have taken place there during the strong melting of inland ice masses, i.e., in late-glacial and middle-Weichselian times. This had influenced not only the Black Sea and the northern parts of the Mediterranean Sea but the coasts of some parts of Northern Africa as well. On the other hand, the remarkable influx of eolian dust from the Sahara in Holocene times can be traced by this means as well.

Most of our suggestions about the surface temperatures of the oceans during Pleistocene times originate from paleoecological investigations of the former behavior of Foraminifera. So critical considerations as to annual changes in the habitats used by the Foraminifera (FAIRBANKS and WIEBE 1980, LOUBERE 1982) are welcome, as well as investigations on the intensity and geographical distribution of carbonate dissolution processes in the deep sea (BALSAM 1982). Here interrelations with deep-sea currents seem to become important, and these were studied to some extent by STUIVER et al. (1983), mostly in respect of water exchange processes. And of course regional peculiarities in the configuration and bottom-surface conditions of shallow parts of the oceans must be duly taken into consideration, as could be shown by BRUNNER (1982) during a reconstruction of the changes in temperature and salinity of the Gulf of Mexico for the last interglacial and glacial maxima, stressing even seasonal differences.

In general, it seems to be easier to understand the reasons for the beginning of a glaciation than for the disappearance of the inland ice masses if no strong amelioration of climate at that time can be observed. Yet that is the very situation. So the interrelations between ice cover of oceans and continents, thickness and surface area of fresh melt- and rainwater on top of the much more haline ocean waters, and freezing processes of the uppermost ocean-water layers during late-glacial times acquire some importance. These problems are dealt with by JONES and RUDDIMAN (1982) and by RUDDIMAN and McINTYRE (1981b), who constructed maps of the position of the polar front from full-glacial to late-glacial times in the Northern Atlantic and of the distribution of melt- and rainwater on the ocean's surfaces from 16,000 to 8,000 b.p.

The history of climate during the last glaciation has been investigated by MESSERLI (1980) for Central and Northern Africa, and by FRENZEL (1980a,b, 1983) for Europe. In these studies the seasonal changes of temperature and to some extent those of rainfall as well have been emphasized. The European data are shown by FRENZEL (1980b) on temperature and rainfall maps, and an attempt is made to evaluate the water-saturation deficit. The problems of the then existing weather conditions are discussed by PIAGET (1980). During the last glaciation the global sea level seems to have lowered to about - 117 m (NARDIN et al. 1981), or about 100-110 m in the Black Sea region (OSTROVSKY et al. 1980). It is held by NARDIN et al. (1981) that the global sea level rose between 18,000 and 13,000 b.p. to about - 20 m, this being followed by a decrease in sea level to - 40 m between 12,000 and 11,000 b.p. The concomitant ecological changes in the shelf areas of the northern part of the Black Sea are discussed at some length by OSTROVSKY et al. (1980), who point out regional and local deviations from the general picture, most of all as a consequence of the configuration of coastlines and the interaction with rivers, and they stress that stratigraphical schemes in marine environments can only be constructed if the then existing paleogeographical and paleoecological conditions are understood. It is a well-known fact that these investigations must be restricted to tectonically stable areas. This is stressed again by POLACH et al. (1981) as regards the Australian Great Barrier Reef. Here it can be shown that the eustatic rise of sea level reached present-day conditions at about 6000 b.p.. At about 3,700 b.p. the postglacial maximum in sea level (somewhat more than 1 m above present-day level) had been reached, whereas at 2,370 b.p. recent conditions prevailed again. It is reasonable to take into consideration that isostatic movements can seriously blur the general picture, as was shown by GRAY et al. (1980) for the west coast of Ungava Bay, Quebec. Yet in general the rising sea level had experienced much more oscillations than were found by POLACH et al. (1981). For the German coasts see, e.g., MENKE (1976), where a certain synchrony between sea level changes and the overall picture of changes in Holocene vegetation is stressed. Yet LINKE (1982) points to the fact that, moreover, even regional peculiarities caused by periods of repeated floods due to phases of increased storm activity should be taken into consideration, as these floods strongly influence the coastal and near perimarine ecological conditions. In all these investigations phases of transgression or regression of the sea are interpreted as evidencing changes of climate (see also KERR 1980). Yet studies on the present-day changes of sea level (ETKINS and EPSTEIN 1982, GORNITZ et al. 1982, ROBOCK et al. 1983, FAURE et al. 1980) point to the even today existing strong difficulties in deciphering the true causes of these changes, i.e., melting of ice versus expansion of the water masses due to increasing temperatures, deformation of the geoid, etc. So, though the paleoecological consequences of rising or decreasing sea levels can be followed quite easily in changing edaphic conditions of the perimarine belt, the question as to their general trends and causes remains rather open to debate.

It seems that the climate history of the Holocene might be evaluated much easier than that of the last glaciation or even of older parts of the Pleistocene. Yet doing so it soon becomes evident that much more accuracy is needed. Some of these problems have already been discussed in connection with the reasons for changing sea level. According to KLIMANOV and ELINA (1980) the mean annual temperature in Karelia seems to have been about 2^{O} to $4^{O}C$ lower in 8,500 b.p. than today, these differences being the smallest in the north. It seems that most of all the summer temperatures were depressed, as compared with present-day conditions. Precipitation is held to have reached approximately the same values as in recent times. About 5,500 b.p. the mean annual temperature is said to have been approximately 2 to $3^{O}C$ higher than it is today, with clear regional differences against the general picture. Precipitation may have equaled or even surpassed recent conditions. And even about 3,500 b.p. the mean annual temperatures seem to have lain about 1^{O} to $2^{O}C$ above the present values, with slightly reduced precipitation as compared with 5,500 b.p. Nearly just on the opposite side of the globe, in New Zealand, there seem to have occurred phases of glacier advances in the years 0 to 430, 1,000 to 2,500, 4,000, and 7,600 to 9,500 b.p. (CHINN 1981). This is in general in phase with changes in climate in the Northern Hemisphere. And these changes can even be felt near the equator. According to TALBOT (1981) a relatively dry and windy climate with dune formation in the Accra plain, Southeast Ghana, seems to have been geomorphologically and paleoecologically active between about 4,500 and 3,800 b.p.. Yet it may be questioned whether these observations on a sandy coast must be interpreted in terms of general changes of climate only.

SCHWEINGRUBER (1980) gives a valuable account of principles and methods of dendroclimatological investigations. In this paper and in the investigations of KOLISHCHUK (1979) on the dendroclimatology of the Ukrainian Carpathians the difficulties in the correct climatological interpretation of changing widths of tree rings caused by years of fruition or diseases are extensively discussed. The dendroclimatological reconstructions can be checked to some extent by observations of the history of glaciers and of periglacial processes, as being evidenced by GORDON (1981) for West Greenland, MATTHEWS (1977) for southern Norway, and by FURRER et al. (1980) for the central Alps. Yet, as can be deduced from observations in the South Orkney Islands, these changes of climate do not seem to have been remarkable ones, since they cannot be detected in reactions of the there existing ice cap (HARKNESS 1981). This might stimulate a critical reappraisal of the values given by KLIMANOV and ELINA (1980) for Karelia, if regional deviations from the general pattern are not taken into consideration. MAYEWSKI et al. (1980) stress these very differences for the Himalayan and Trans-Himalayan glaciers within the last century. These peculiarities point to Holocene circulation patterns of the atmosphere, which were analyzed by LAMB (1980) and by WANNER (1980). Valuable observations are recorded by PFISTER (1980), who used phenologically sensitive plants and clearly climatically induced changes in the hydrosphere for the reconstruction of climate in Switzerland since 1500 A.D. Moreover, it can be shown that phases of increased floods of the river Nile (1,070 to 1,180 and 1,350 to 1,470 A.D.) were connected with an increased precipitation in Equatorial East Africa, those of reduced runoff (930 to 1,070 and 1,180 to 1,350 A.D.), on the other hand, with phases of cold climates in Europe (HASSAN 1981). JANKUHN (1980) stresses the interrelations between changes of climate, ecological conditions, and human activity for the last 4000-5000 years in regions of extreme climatical conditions, as for instance in Scandinavia.

2. History of the Last Glaciation

Intimately connected with paleoclimatological problems is the history of the last glaciation. It seems to be now generally accepted that this glaciation lasted from about 115,000 to 10,000 b.p. Yet it is still a matter of debate whether during this long-lasting period only one advance of the inland ice took place, as is generally held in Europe, or whether several glacier advances occurred. This problem has been discussed already somewhat earlier (e.g., Fortschritte, 40, 429-431). Up till now it could not be solved. So SJØRRING (1982) and RASMUSSEN (1982) report on only one general advance of the Scandinavian inland ice, i.e., after about 24,000 b.p. in Denmark. The same is held by MENKE (1982) for northwestern Germany, MILLER (1982) for southwestern Scania, CHEBOTAREVA and MAKARYCHEVA (1982) for the

Russian Plain, VAN HUSEN (1982) for the Eastern Alps, and by KOZARSKI (1980) for the Great Polish Lowlands. On the other hand ZARRINA (1982) describes for southern Finland traces of an Early Weichselian (Waldaian) glacier advance and reports on a Middle-Weichselian advance of the inland ice-masses even into the surroundings of Yaroslavl and Uglich. This is held to have been the strongest advance of the Scandinavian inland ice. LUNDQVIST (1982) discusses traces of Early and Middle-Weichselian glacier advances in Sweden; SCHLÜCHTER (1982) does so for central Switzerland, yet most of the investigations were done along the lower Weichsel river, between Thorn and Danzig (DROZDOWSKI 1980, GALON 1982, MAKOWSKA 1980, MOJSKI 1980, RZECHOWSKI 1980). The situation is even more complicated by various attempts at dating precisely the interstadial periods of the last glaciation (e.g., VOZNYACHUK and SHULIYA 1976, WYSOCZANSKI-MINKOWICZ 1982), where I feel that sometimes the accuracy of the dating methods used may have been exaggerated, e.g., if the Eemian interglacial is divided into two interglacials, lasting from 135,120 to 128,305 and from 121,490 to 114,680 b.p. respectively (WYSOCZANSKI-MINKOWICZ 1982). Nevertheless, in North America observations again favoring the view of strong early or Middle-Weichselian (Wisconsinan) glaciations were reported (DREIMANIS 1982, DALRYMPLE et al. 1982), and it has been suggested from paleobotanical material that in California the mean annual temperature of the early phases 5 d and 5 b of the last glaciation (according to the deep-sea stratigraphy) must have been 7° to 8°C lower than during the Holocene, thus equaling pleniglacial conditions (ADAM and WEST 1983). Moreover, DANSGAARD et al. (1982), BIRCHFIELD et al. (1981), and NELSON (1980) point to an extremely strong ice accumulation during these earliest parts of the last glaciation, which according to DANSGAARD et al. (1982) must have caused during the first 5 to 10,000 years after the last interglacial a lowering of the global sea level of about 70 m. These calculations and observations are paralleled by estimations made by DODGE et al. (1983) on the situation of the global sea level during Early Weichselian interstadial phases, pointing even then to a lowering of sea level about 108,000 and 81,000 b.p. (interstadial phases) by approximately 10 and 13 m respectively. Yet, in this paper the sea-level changes of the beginning of the last glaciation are explained in some 30% by changes in ice volumes, only the rest being determined by temperature effects. On the other hand, it should be admitted that central south Norway must in general have been free of ice about 46,000 to 19,000 b.p. if several ^{14}C-datings on mammoth bones, which were found there during the last years, give approximately real ages (GARNES and BERGERSEN 1977, see also BERGLUND et al. 1976). The terrestrial ecological conditions during Early and Mid-Weichselian times must have been very detrimental for plant growth. So MENKE (1982) points to the nearly complete lack of a continuous vegetation cover during Middle-Weichselian times in northwestern Germany, and KOZARSKI (1980) reports on strong denudation processes caused by the impact of frost on the soils of the northern Polish lowlands. Comparable observations have been made in the southern part of the Netherlands (VANDENBERGHE 1982) and in southern Germany as well (FRENZEL 1983). This can even be traced by the rodent fauna, the evolution of which in time and space has been reconstructed by CHALINE (1982). The history of the Late-Weichselian glaciation (23,000 to 10,000 b.p.), the southeasternmost end-moraine systems of which are investigated by BOTHNER and SPIKER (1980) in New England, is being thoroughly discussed for the whole of the Northern Hemisphere by KIND (1982).

3. The Paleoecology of the Last Glaciation in the Northern Hemisphere

Dealing with paleoecological problems of the glacial periods the question as to recent equivalents of then existing vegetation types or landscape zones arises. According to VASILYEV (1982a) the periglacial zone of the Northern Hemisphere, dating from pleniglacial times, has found no modern equivalents. Perhaps present-day arctic deserts on several isles in the Arctic Ocean may serve as comparisons, yet here from time to time increased precipitation is noticed, which seems to have been lacking completely in full-glacial conditions. STARKEL (1979) points to the fact that, in contrast to earlier assumptions, during the glacial periods no zonal displacements of complete vegetation zones occurred but that always new plant communities were found by changing sets of competing plant species. So the exact evaluation of the then existing conditions is hampered. This bad situation is even worsened by the fact that the thanatocoenoses strongly deviated from the phytocoenoses, as could be shown by GLASER (1981), who pointed out that from late-glacial North-American sites in general only leaves of *Dryas intergrifolia, Vaccinium uligionosum* var. *alpina, Salix herbacea* with only rare fruits are known. The same can be observed in present-day lake and snowfield sediments in the surroundings of Mt.McKinley, Alaska, this being caused by transport and accumulation mechanisms, but not by the composition of the contributing flora itself. So for an understanding of full-glacial conditions a thorough study of possible present-day somewhat comparable ecological situations might be worthwhile (e.g., hydrology of a small lake in the Canadian Arctic: WOO (1980); temperature regime under various subalpine plant communities of the Transili-Alatau, Central Asia: SEVERSKIY (1978); forage patterns of herbivores in the Arctic: BATZLI and JUNG (1980), BATZLI and SOBASKI (1980), WHITE and TRUDELL (1980), McKENDRICK et al. (1980). Another starting point for answering these difficult questions is the investigation of the techniques and economies of prehistoric man. This seems to have been analyzed for the first time on a continental scale by STEPANOV (1976): During the time of about 20,000 to 16,000 b.p. ephemeral settlements of small hunting hords seem to have occurred in Eastern Europe near to the ice margin in sheltered ravines only, from which the big valleys were easily reached. All the other, bigger settlements seem to have been situated near the southern boundary of the seasonal migrations of the big herbivores. Here, gathering of fruits and hunting seem to have been practiced from perennial settlements. Judging by the implements used, life and economies in Siberia were strongly affected at that time by the severe climatic conditions. Here, in general, reindeer was hunted, whereas in the Ukraine wild horse (river Dniester) or bison (near the Asow Sea) seems to have been of general importance. In the Krim and the Caucasian Mts. no battues seem to have been practiced, this pointing to the much better ecological conditions there. On the other hand, along the river valleys at the foot of the Central Asian Mts. even better conditions must have prevailed, this being deduced from the then practiced division of labor between hunting and implement manufacturing groups. Between 16,000 and 10,000 b.p. mammoth was hunted most of all in the central part of the Russian Plain, bison being of prime importance in the steppe regions of the Ukraine. The large list of animals hunted otherwise points to the improving conditions. In Siberia roving hords seem to have followed the game tracks along the rivers, hunting mammoth, reindeer, horse, bison, muskox, fox, lemming. As to the Pleistocene and Holocene faunas in northeastern Kazachstan: see KOZHAMKULOVA and NURMAMBETOV (1978). MADEYSKA (1979) reported on comparable observations like those made by STEPANOV (1976) for the environments of Cracow, southern Poland. Here the evolution of species-rich faunas can be followed from the presumably tripartite last interglacial (two warm-climate periods, divided from one another by an immigration of a tundra fauna) over the beginning of the last glaciation with its relatively rich forest and steppe faunas to the ensuing tundra fauna of the first part of the pleniglacial. In contrast to the earlier phases, during pleniglacial times human activities were lacking there. In the second part of the pleniglacial (pleniglacial B) climate must have become extremely dry. The pleniglacial ornis includes besides prevailing tundra and steppe birds those of coniferous forest as well (*Lyrurus tetrix, Falco columbaria, Strix uralensis)*. From this it is concluded that in climatically and edaphically suitable places some groves of coniferous trees may have survived the times of utmost severety of climate. On the occurrence of dated remains of large herbivores of the last glaciation: BERGLUND et al. (1976, southern Sweden and adjacent countries); FOLLESTAD and OLSSON (1979, eastern Norway); LONG and MULLER (1981, southern part of North America); CHALINE (1982, rodents of Western and Central Europe). According to WILLIAMS et al. (1981) the insect fauna of the famous Scarborough Formation, Toronto, dating from the beginning of the last glaciation, points to approximately 4,5°C lower July temperatures at that time and to a boreal coniferous forest much poorer in deciduous species than at present. On the other hand, the faunal history of the last 30,000 years in the Grand Cañon, Arizona, allows the reconstruction of an astonishingly slow transition from full- to late-glacial parkland vegetation, in which the amount of desert plant communities was much smaller than at present, and to treeless conditions from about 8,500 b.p. onward at lower elevations (MEAD 1981). On the other hand, in Cañon Creek, near Fairbanks, Alaska, a relatively rich fauna, comprising rodents, mammoth, *Equus lambei*, *Camelops hesternus*, bison, mountain sheep, wolf, *Lepus* cf. *othus* or *arcticus*, *Rangifer* sp., points to an open, treeless vegetation there from about 40,000 to 28,140 b.p. (WEBER et al. 1981). JANZEN and MARTIN (1982) stress that the propagation of *Gomphotheres* species must have been influenced by the late-glacial extinction of 15 big herbivorous species in Central America.

Of great importance for the desert and steppe regions of the world is the question about former changes in the water system, i.e. the problem of the so-called pluvial climates. VASILYEV (1982c) warns against a too simple correlation of pluvials with glacial periods, because a connection between glaciations and phases of an improved water system in present-day arid regions did not exist. This can perhaps be concluded from the history of the Caspian Sea, too, since it is held, on the strength of absolute datings, that transgressions were in general correlated with interglacials, the only exception being the last glaciation (VASILYEV, 1982b). According to HAYNES (1982) the eastern Sahara experienced about 300,000 ± 100,000 years ago a phase of relatively moist climates. It was followed by hyperarid conditions when man had to leave this area. Relatively moist conditions prevailed anew there presumably during the last interglacial and the first part of the last northern glaciation, up to about 35,000 b.p., when climate changed to extremely arid conditions again. They seem to have been followed by a threepartite moister period between 10,000 and 5 000 b.p. by a period which had consisted of two moister phases (precipitation of most 100 to 300 mm per year) and an intercalated drier one. Approximately comparable conclusions were drawn by GEYH and JÄKEL (1974) from the time-distribution pattern of available [^{14}C]-datings in the Sahara. Yet the method of this investigation has been questioned by ROGNON (1975). The Namib Desert seems to have experienced between 36,000 and 28,000 b.p. and about 21,000 b.p. a little bit moister conditions than those at present. The ensuing period must have had a drier climate again, and during the Holocene there seem to have existed regional differences in available moisture (VOGEL and VISSER 1981). Moister conditions in present-day arid regions are repeatedly explained by an increased precipitation. Yet BENSON (1981) discusses the great influence of changes in evapotranspiration on the position of the level of Lake Lahontan in the western United States: The changes between an extremely low lake-level about 15,000 b.p. and a very high position about 14,000 to 11,000 b.p. can explained alone by changes in air temperature and cloudiness of the sky. Lake-level changes of the Issyk Kul, Tien Shan, Central Asia, are extensively described for the Pleistocene by ALESHINSKAYA et al. (1976a,b) and for the Holocene by TROFIMOV (1978). Most of all, the papers published by ALESHINSKAYA and her co-workers contain also valuable short remarks on the changes in vegetation.

4. Vegetation History of the Last Glaciation

In vast regions of the Northern Hemisphere the full-glacial vegetation is in general described as a steppe or, at least, as a tundra-steppe. These, of course, are only names the real meaning of which may be questionable. WALKER et al. (1981) discuss the question whether the present-day dune vegetation, characterized by *Artemisia* and Gramineae, near Prudhoe Bay, Northern Alaska, might be equivalents of this type of full-glacial vegetation. This would mean that this type of an open pleniglacial vegetation has survived till now. The same is held by CWYNAR and RITCHIE (1980), who point to the fact that within the Yukon Territory, Alaska, the treeless vegetation, in which horse, bison, mammoth, camel, saiga and man found refuge between 30, 000 and 14,000 b.p., strongly resembled present-day conditions if one considers only the then existing plant genera and species: *Aconitum, Bupleurum triradiatum*, Caryophyllaceae, *Dryas, Astragalus, Oxytropis, Hedysarum, Androsace, Lesquerella arctica, Pedicularis, Phlox, Polemonium, Polygonum alaskanum, Saussurea, Selaginella sibirca, Lagotis glauca*, and many others. Yet the pollen influx has been much less than at present. From this it is concluded: "We consider that the Beringian herb zone represents a sparse, discontinuous vegetation of herbaceous tundra on upland sites and local sedge-grass meadows on lowlands. "About 14,000 to 12,500 b. p. the pollen influx of the herb vegetation strongly increased, thus bettering browsing conditions for the big herbivores, provided that the beginning of paludification did not exert a negative influence at that time. On the other hand, YURTSEV (1981) emphasizes the close relationship between the glacial treeless vegetation in Yakoutia and Alaska and recent steppe biocoenoses of a relict character there. At any rate, during full-glacial times of the last glaciation the ecological condition of the Alaskan arctic coastal plain must have strongly resembled a desert. This is indicated by large dune systems, which were formed by easterly winds, just as they are today (CARTER 1981). The central part of Alaska seems to have been one of the most interesting refuge areas for more demanding plant and animal species during the last glaciation. This is discussed anew for tree species by HOPKINS and SMITH (1981; containing a map of the immigration of spruce), SCHWEGER and JANSSENS (1980), and THORSON et al. (1981). And even on Queen Charlotte Islands, British Columbia, small refuge areas existed about 15,400 and 16,000 b.p., enabling *Chara, Juncus*, Caryophyllaceae, *Rumex, Potamogeton filiformis, Salix reticulata* and *Picea* to thrive there (WARNER et al. 1982). However, dealing with full- or late-glacial refuge areas for the forest flora a misinterpretation of pollen diagrams, caused by long-distance transport, is always possible. To avoid this, LEOPOLD et al. (1982) use lignin precursors or derivatives within the sediments investigated, since it is held that these substances should indicate the former presence of various systematic groups: vanillyl-phenols suggesting the presence of gymnosperms, syringyl-phenols to angiosperms, while cinnamyl-phenols are typical of nonlignified tissues of various provenance. Bearing this in mind, the full- and late-glacial history of Lake Washington can be very accurately reconstructed. The results, attesting to a treeless vegetation during full-glacial times, are confirmed by BARNOSKY (1981) as to the southern Puget Lowland, Washington, during 25,000 to 17,000 b.p. and during 15,000 to 13,500 b.p., whereas HEUSSER and HEUSSER (1980) can show that between the Columbia River and the Mt. St.Helens a forest, composed of western hemlock, *Abies amabilis* and Cupressaceae, clothed the mountains, this having been later on replaced by a park tundra, whose most important tree species were, e.g., *Picea engelmanni, Tsuga mertensiana, Pinus contorta* and *P. monticola,* together with a large share of grasses, Compositae, and other herbs. This relatively open vegetation, with only some copses as tree vegetation, seems to have been distributed relatively far during full-glacial times. So LEGG and BAKER (1980) report on a forest-steppe vegetation in Boulder County, Colorado, where the percentage of the arboreal pollen was only very low (13% to 30%), whereas *Artemisia,* Gramineae, Chenopodiaceae and Amaranthaceae, Tubuliflorae, and *Juniperus* seem to have been of prime importance, together with *Koenigia islandica!* In the eastern Gran Cañon most of the vertical vegetation belts seem to have been depressed at that time by about 600 to 1,000 m in comparison with present-day conditions. The recent vegetation belts began to be formed there about 15,000 to 11,000 b.p. It is interesting to note that some "glacial relict species" like some firs, could be found there up till 8,500 b.p., this pointing to the fact that the change in vegetation from full-glacial to Holocene conditions lasted for a long time and that it is relatively improbable that complete plant societies were exchanged simultaneously (COLE 1982). In this paper, just as in that of LANNER and van DEVENDER (1981), *Neotoma* middens are used for the reconstruction of the former floras and vegetation types. Between 30,000 and 11,000 b.p. *Pinus remota* and *P. edulis* were common elements in Trans-Pecos Texas and at lower elevations on the northern rim of the Chihuahua-desert, whereas *P. cembroides* could not be found there, though it is at present widely distributed in the Davis and Chisos Mts. It seems that during the Late Pleistocene the recent northern Chihuahua desert was dominated by *Pinus edulis, P. remota, Juniperus* and oaks, the modern desert scrub immigrating only during the Holocene (LANNER and van DEVENDER 1981). These changes in flora and vegetation are investigated intensively for various vertical belts in the Pueblo-Tlaxcala region, Mexico, by OHNGEMACH and STRAKA (1983). Evidently, not always do botanical and geomorphological observations yield the same results, as regards the paleoecological development of this most interesting mountainous area. A comprehensive description of the paleoecological meaning of the huge quantity of various types of palynomorphs met with during this long-lasting work should be offered. The Late Pleistocene and Holocene vegetation history on the west central Mexican Plateau and in the Basin of Mexico is investigated by WATTS and BRADBURY (1982). Here the impact of man and the ecological evolution of Lake Chalco, as brought about by Agricultural activities in its surroundings, are thoroughly discussed.

The CLIMAP-reconstructions of climate during full-glacial times of the last glactiation were based on transfer functions, originating in the present-day and former distribution of certain Foraminifera species in various parts of the oceans. From here deductions to the synchronous terrestrial climates were drawn, indicating only minor changes in climate during full-gacial times in the southeastern part of the United States. Yet WATTS and STUIVER (1980) can convincingly demonstrate that at that time climate in northern Florida must have been appreciably cold, dry, and windy. Moreover, the groundwater table even about 12,000 b.p. lay in southwestern Florida 26 m deeper than at present. These, taken together, caused an open forest-steppe with pine, abundant herbs of the steppe vegetation, oak and hickory widespread there. Cold climate conditions seem to have ended there about 14,600 b.p., when it became possible for the demanding tree species to spread slowly, starting from small refuge areas. Synchronously, the temperature seems to have risen in northeastern Minnesota as well (BIRKS 1981), so that the previously prevailing tundra vegetation, rich in various herb species, was replaced by *Artemisia* and dwarf-shrub communities with immigrating spruce and larch. The differences between a temperature rising since about 14,300 b.p. and the still oscillating glaciers are explained by repeatedly stronger snow accumulation during wintertime. A comparable full-glacial open vegetation could be proven by the occurrence of various animal species and genera, characteristic of steppe-like plant communities for the time of the Clovis-culture in eastern Missouri: *Sciurus*, *Marmota*, *Spermophilus*, *Geomys*, *Microtus pennsylvanicus*, *Synaptomys cooperi*, together with *Mammut americanum* (GRAHAM et al. 1981). On the other hand, in central Indiana relatively open boreal forests, dominated by white spruce, in which mastodon, caribou and giant beavers lived, could be reconstructed by WHITEHEAD et al. (1982). Here coniferous trees were replaced by hardwood taxa about 13,000 b.p., when paludification of vast areas began. From this it is concluded that the spreading of the bog vegetation may have caused mastodon to become susceptible to disease, extreme weather conditions, and the increasing hunting pressure of man, leading ultimately to its extinction. The steppe vegetation in Wyoming's intermontane basins was threatened during full-glacial times by a climate with a mean annual temperature which, according to synchronous periglacial phenomena, must have been some 11° to 16°C lower than at present (MEARS 1981). For very detailed survey of the full-glacial vegetation east of the Rocky Mountains, see WRIGHT 1981.

MIURA and YAMANAKA (1980) report on the vegetation history during Late Pleistocene times on the Hakkoda Mts., northern Honshu, Japan. Here, about 1200 m a.s.l. the dominating *Abies-Picea* forests seem to have been interrupted at that time relatively often by grass and herb communities. *Fagus*, which is of some importance in the present-day montane belt, could not be found in sediments of about 17,000 b.p.. The interpretation of the pollen diagrams is somewhat obscured by the fact that posibble reworking of older pollen floras is not discussed by the authors. As to a small vegetation map of Northern Eurasia about 20,000 b.p. see AVENARIUS et al. (1976). BOLIKHOVSKAYA (1976), investigating palynologically the vegetation history of loesses and fossil soils in the Russian Plain, stresses the big methodological problems, which were encountered during these studies. This holds true even for present-day surface samples, in which the composition of the pollen flora strongly deviates from that of the surrounding vegetation. Nevertheless, it can be shown that during the Moskva-glaciation and during the first part of the last glaciation the Ukraine seems to have been the realm of a very species-rich forest steppe. Only during full-glacial times of the last glaciation seems the share of the steppe-vegetation to have increased seriously. On the other hand, the pollen flora of the soil dating from the last interglacial period favors the view of Ukrainian forest steppes, very rich in demanding tree-species, which were replaced finally by the steppe vegetation. So in this paper the important position of refuge areas of tree species in the southern Ukraine during a glacial period is stressed. The same view is held by STARKEL (1980) as to copses of a coniferous vegetation in the surroundings of Cracow, southern Poland, during full-glacial times of the last glaciation. Moreover, in this paper the changing activity of the southern Polish rivers during the last glaciation is emphasized, together with the history of soil erosion and downwash phenomena during stadial and interstadial times of the last glaciation. On the very interesting problem of loess formation and pedogenesis during interstadial times in the Polish loess areas, together with short remarks on the synchronous vegetation, see MARUSZCZAK (1980), on Russian loesses SEVOSTYANOV (1981) and KRIGER (1981).

The Swiss vegetation history of the last glaciation is intensively investigated by WELTEN (1982), with remarks on the vegetation of the two preceding interglacial periods. The facts presented in this paper will of course stimulate further research, since a tremendous amount of most interesting observations is reported. As to general trends in the development of glacial and interglacial vegetation types in southern Germany and adjacent countries, see FRENZEL (1983). The concomitant problems of Quaternary geology, together with notes on plant and animal remains found in these sediments, are discussed comprehensively by HANTKE (1978, 1980) most of all for Switzerland. It is an admirable compilation of nearly all observations published up to now.

Parallel to the observations on the last-glacial history of climate in southwestern Africa, already cited, SCOTT (1982) reports on the vegetation history of the Transvaal bushveld, South Africa: A relatively mesic bushveld with several montane forests in Central Transvaal about 35,000 b.p. was replaced there till 25,000 b.p. by a drier bushveld, this in turn giving way to an open grassland vegetation with some type of "macchia" between 25,000 and 11,000 b.p., then, with increasing rapidity, the Kalahari thornveld could establish itself in a climate which was becoming warmer.

Synchronously with the northern glaciation South America experienced a strong glaciation too (PORTER 1981), the temperature and precipitation conditions of which were analyzed by HEUSSER and STRECTER (1980). According to HEUSSER (1983) the warm-climate periods in the environs of Laguna de Tagua Tagua, Chile, were always characterized by a steppe-like vegetation, whereas during cold-climate phases *Nothofagus dombeyi* and *Podocarpus andinus* became the most prominent plants. From the fossil vegetation types met with it is concluded that during the cold-climate periods the mean summer temperature was at least about 5°C lower than at present, the amount of precipitation being significantly higher than today. It must be suggested that the tracks of the cyclones moved to the north at that time, the present-day Etesian climate only beginning to form in Holocene times. On the other hand, during full-glacial times climate on Isla Chiloë was colder and drier than at present. The strong increase in the amount of charcoal about 11,380 and 6,130 b.p. points to an increasing activity of man. Mastodons occurred there already about 18,000 b.p., yet they became extinct about 11,380 b.p.. This is explained by HEUSSER (1983) with the increasing influence of man, together with spontaneous changes in browsing conditions. In the Cordillera Pelada, province of Valdivia, Chile, the paludification began about 10,425 b.p. Maximal warmth seems to have been felt there about 9,000 b.p., this causing a decrease in available moisture (HEUSSER 1982). - The paleocological evolution of Lake Valensia, Venezuela, is investigated intensively by BRADBURY et al. (1981), using algal floras, pollen, cladocera, ostracoda, etc. From about 13,000 to 10,500 b.p. the lake seems to have been surrounded by arid savannas, trees nearly completely lacking there. Bogs and shallow lakes were common. Repeatedly, haline conditions prevailed. From about 10,500 to 8,700 b.p. a forest vegetation began to immigrate, consisting of *Bursera* later on of *Spondias*, Moraceae, *Cecropia*, and *Trema*. *Rapanea* and Melastomataceae occurred repeatedly. The grassland was replaced by some type of chaparral and later on by semi-evergreen or even deciduous forests. These early phases of the vegetation history point to the fact that the area cannot have served as refuge for mesic taxa during the last glacial period. At about 8,700 b.p. present-day conditions began. Though there did occur some changes in the vegetation even later on, the influence of man can be relatively reliably only traced back to about 1800 b.p. when the increasing amount of clastic material in the lake points to strong anthropogene disturbances of the spontaneous vegetation.

References

ADAM, D.P., WEST, G.J.: Science 219, 168-170 (1983). - AGADZHANYAN, A. K., DOBRODEEV, O.P. (eds.): Problemy obschej fizicheskoy geografii i paleogeografii, Moskovsk. Universit., 1976. - ALEKSEEV, M.N., TSEJTLIN, S.M. (eds.): Problemy geologii i istorii chetvertichnogo perioda (antropogena). Moskva: Nauka 1982. - ALESHINSKAYA, Z.V., BONDAREV, L. G., CHIGAREV, N.V., SHUMOVA, G.M.: 198-210 in: see ref. AGADZHANYAN and DOBRODEEV 1976a. - ALESHINSKAYA, Z.V., KULIKOV, O.A., MOTUZKO, A.N., FAUSTOV, S.S., SHUMOVA, G.M.: 211-225 in: see ref. AGADZHANYAN and DOBRODEEV 1976b. - AVENARIUS, I.G., MURATOVA, M.V., SPASSKAYA, J.J.: 329-342 in: see ref. AGADZHANYAN and DOBRODEEV 1976.

BALSAM, W.L.: Science 217, 929-931 (1982). - BARNOSKY, C.W.: Quat. Res. 16, 221-239 (1981). - BATZLI, G.O., JUNG, H.-J.G.: Arct. Alp. Res. 12, 483-499 (1980). - BATZLI, G.O., SOBASKI, St.: Arct. Alp. Res. 12, 501-510 (1980). - BENSON, L.V.: Quat. Res. 16, 390-403 (1981). - BERGLUND, B.E., HÅKANSSON, S., LAGERLUND, E.: Boreas 5, 177-191 (1976). - BIRCHFIELD, G.E., WEERTMAN, J., LUNDE, A.: Quat. Res. 15, 126-142 (1981). - BIRKS, H.J.B.: Quat. Res. 16, 322-355 (1981). - BOLIKHOVSKAYA, N.S.: 257-277 in: see ref. AGADZHANYAN and DOBRODEEV 1976. - BOTHNER, M.H., SPIKER, E.C.: Science 210, 423-425 (1980). - BOYARSKA, T.D., MURATOVA, M.V.: 323-328 in: see ref. AGADZHANYAN and DOBRODEEV 1976. - BOYLE, E.A., KEIGWIN, L.D.: Science 218, 784-787 (1982). - BRADBURY, J.P., LEYDEN, B., SALGADO-LABOURIAU, M., LEWIS, W.M., SCHUBERT, C., BINFORD, M.W., FREY, D.G., WHITEHEAD, D.R., WEIBEZAHN, F.H.: Science 214, 1299-1305 (1981). - BRUNNER, CH.A.: Quat. Res. 17, 105-119 (1982). - BRYAN, K., KOMRO, F.G., MANABE, S., SPELMAN, M.J.: Science 215, 56-58 (1982).

CARTER, L.D.: Science 211, 381-383 (1981). - CHALINE, J.: Quat. Stud. Poland 3, 5-13 (1982). - CHEBOTAREVA, N.S., MAKARYCHEVA, I.A.: Quat. Stud. Poland 3, 15-20 (1982). - CHINN, T.J.H.: Arct. Alp. Res. 13, 33-45 (1981). - COLE, K.: Science 217, 1142-1145 (1982). - CWYNAR, L. C., RITCHIE, J.C.: Science 208, 1375-1377 (1980).

DALRYMPLE, G.B., BURKE, R.M., BIRKELAND, P.W.: Quat. Res. 17, 120-122 (1982). - DANSGAARD, W.: 237-245 in: see ref. OESCHGER, MESSERLI and SVILAR 1980. - DANSGAARD, W., CLAUSEN, H.B., GUNDESTRUP, N., HAMMER, C.U., JOHNSEN, S.F., KRISTINSDOTTIR, P.M., RECH, N.: Science 218, 1273-1277 (1982). - DELCOURT, P.A., DELCOURT, H.R.: 123-165 in: Geobotany II, ed. L.R. ROMANS. New York: Plenum 1981. - DENTON, G.H., HUGHES, T.J.: The Last great Ice Sheets, 484 pp. New York: Wiley 1981. - DODGE, R.E., FAIRBANKS, R.G., BENNINGER, L.K., MAURASSE, F.: Science 219, 1423-1425 (1983). - DREIMANIS, A.: Quat. Stud. Poland 3, 21-28 (1982). - DROZDOWSKI, E.: Quat. Stud. Poland 2, 13-20 (1980).

ETKINS, R., EPSTEIN, E.S.: Science 215, 287-289 (1982).

FAIRBANKS, R.G., WIEBE, P.H.: Science 209, 1524-1526 (1980). - FAURE, H., FONTES, J.C., HEBRARD, L., MONTEILLET, J., PIRAZZOLI, P.A.: Science 210, 421- 423 (1980). - FLOHN, H.: 3-17 in: see ref. OESCHGER,

MESSERLI and SVILAR 1980. - FOLLESTAD, B.A., OLSSON, J.U.: Boreas 8, 307-312 (1979). - FRENZEL, B.: 45-63 in: see ref. OESCHGER, MESSERLI and SVILAR 1980, 1980a; - Veröff. Joachim-Jungius Ges. Wiss. Hamburg 44, 9-46 (1980b); - 91-166 in: Urgeschichte in Baden-Württemberg, ed. H.J. MÜLLER-Beck. Stuttgart: Theiss 1983. - FURRER, G., GAMPER-SCHOLLENBERGER, B., SUTER, J.: 91-107 in: see ref. OESCHGER, MESSERLI and SVILAR 1980.

GALON, R.: Quat. Stud. Poland 3, 37-48 (1982). - GARNES, K., BERGERSEN, O.F.: Boreas 6, 135-147 (1977). - GERASIMOV, I.P., VELICHKO, A. A. (eds.): Paleogeografiya Evropy za poslednie sto tysyach let, 155 pp. Moskva: Nauka 1982. - GEYH, M.A., JÄKEL, D.: Paleogeogr. Paleoclimat. Palaeoecol. 15, 205-208 (1974). - GLASER, P.H.: Arct. Alp. Res. 13, 173-182 (1981). - GORDON, J.E.: Arct. Alp. Res. 13, 47-62 (1981). - GORNITZ, V., LEBEDEFF, S., HANSEN, J.: Science 215, 1611-1614 (1982). - GRAHAM, R.W., HAYNES, C.V., JOHNSON, D.L., KAY, M.: Science 213, 1115-1117 (1981). - GRAY, J., BOUTRAY, B. de, HILLAIRE-MARCEL, C., LAURIOL, B.: Arct. Alp. Res. 12, 19-30 (1980). - GROSVALD, M.G.: Pokrovnye ledniki kontinentalnykh shelfov, 216 pp. Moskva: Nauka 1983.

HANTKE, R.: Eiszeitalter. Die jüngste Erdgeschichte der Schweiz und ihrer Nachbargebiete, I, Klima, Flora, Fauna, Mensch, Alt- und Mittel-Pleistozän, Vogesen, Schwarzwald, Schwäbische Alb, Adelegg, 468 pp. Thun: Ott 1978; - Eiszeitalter. Die jüngste Erdgeschichte der Schweiz und ihrer Nachbargebiete, II, Letzte Warmzeiten, Würm-Eiszeit, Eisabbau und Nacheiszeit der Alpen-Nordseite vom Rhein- zum Rhône-System. 703 p. Thun: Ott 1980. - HARKNESS, D.D.: Radiocarbon 23, 252-304 (1981). - HASSAN, F.A.: Science 212, 1142-1145 (1981). - HASSELMANN, K.: 259-260 in: see ref. OESCHGER, MESSERLI and SVILAR 1980. - HAYNES, C.V.: Science 217, 629-633 (1982). - HERMANN, Y., HOPKINS, D.M.: Science 209, 557-562 (1980). - HEUSSER, C.J.: Quat. Res. 17, 71-92 (1982); - Science 219, 1429-1432 (1983). - HEUSSER, C.J., HEUSSER, L.E.: Science 210, 1007-1009 (1980); - HEUSSER, C.J., STRECTER, S.ST.: Science 210, 1345-1347 (1980). - HOPKINS, D.M., SMITH, P.A.: Quat. Res. 15, 217-249 (1981). - HUSEN, D. van: Quat. Stud. Poland 3, 49-55 (1982).

JANKUHN, H.: Veröff. Joachim-Jungius Ges. Wiss. Hamburg 44, 115-139 (1980). - JANZEN, D.H., MARTIN, P.S.: Science 215, 19-27 (1982). - JONES, G.A., RUDDIMAN, W.F.: Quaternary Research 17, 148-172 (1982).

KEIGWIN, LL.: Science 217, 350-353 (1982). - KELLOGG, W.W.: 18-30 in: see ref. OESCHGER, MESSERLI and SVILAR 1980. - KERR, R.A.: Science 209, 483-486 (1980); - Science 213, 427-428 (1981a); - Science 213, 1095-1096 (1981b); - Science 217, 917-919 (1982). - KIND, N.V.: 213-226 in: see ref. ALEKSEEV and TSEJTLIN 1982. - KLIMANOV, V.A., ELINA, G.A.: DAN SSSR 252, 419-423 (1980). - KOLISHCHUK, V.G.: Probl. Bot. 14 (2), 11-17 (1979). - KOZHAMKULOVA, B.S., NURMAMBETOV, E.I.: Bjull. Kom. po izuch. chetvert. per. 48, 31-38 (1978). - KOZARSKI, ST.: Quat. Stud. Poland 2, 21-35 (1980). - KRIGER, N.I.: 5-8 in: Zakonomernosti formirovaniya prosadochnykh svoistv lyossovykh pord Srednej Azii i Yuzhnogo Kazakhstana, eds. N.I. KRIGER, N.E. KOTELNIKOVA, S.I. LAVRUSHEVICH, V.V. SEVOSTYANOV. Moskva: Nauka 1981. - KRYLOV, A.YA., SHIMKUS, K.M., EMELYANOV, E.M., SILIN, YU.P., TSOVBUN, YA.M.: 264-271 in: Opredelenie absolyutnogo vozrasta rudnykh mestorozhdenii i molodykh magmaticheskikh porod, eds. G.D. AFANASYEV, A.P. VINOGRADOV, L.N. OVCHINNIKOV, E.K. GERLING, T.B. PEKARSKAYA. Moskva: Nauka (1976). - KUKLA, G., GAVIN, J.: Science 214, 497-503 (1981). - KUTZBACH, J.E.: Science 214, 59-61 (1981).

LAMB, H.H.: 149-174 in: see ref. OESCHGER, MESSERLI and SVILAR 1980. - LANNER, R.M., DEVENDER, TH.R. van: Quat. Res. 15, 278-290 (1981). -

LEGG, TH.E., BAKER, R.G.: Arct. Alp. Res. 12, 319-333 (1980). - LEOPOLD, E.B., NICKMANN, R., HEDGES, J.I., ERTEL, J.R.: Science 218, 1305-1307 (1982). - LINKE, G.: Probleme der Küstenforschung im südlichen Nordseegebiet 14, 123-157 (1982). - LONG, A., MULLER, A.B.: Radiocarbon 23, 191-217 (1981). - LOUBERE, P.: Quat. Res. 17, 314-324 (1982). - LUNDQVIST, J.: Quat. Stud. Poland 3, 57-59 (1982).

MADEYSKA, T.: Quat. Stud. Poland 1, 15-28 (1979). - MAKOWSKA, A.: Quat. Stud. Poland 2, 37-55 (1980). - MARUSZCZAK, H.: Quat. Stud. Poland 2, 57-76 (1980). - MATTHEWS, J.A.: Boreas 6, 1-24 (1977). - MAYEWSKI, P.A., PREGENT, G.P., JESCHKE, P.A., AHMAD, N.: Arct. Alp. Res. 12, 171-182 (1980). - McKENDRICK, J.D., BATZLI, G.O., EVERETT, K.R., SWANSON, J.C.: Arct. Alp. Res. 12, 565-578 (1980). - MEAD, J. I.: Quat. Res. 15, 311-326 (1981). - MEARS, B.: Quat. Res. 15, 171-198 (1981). - MENKE, B.: Probleme der Küstenforschung im südlichen Nordseegebiet 11, 145-161 (1976); - Quat. Stud. Poland 3, 61-68 (1982). - MESSERLI, B.: 64-90 in: see ref. OESCHGER, MESSERLI and SVILAR 1980. - MILLER, U.: Quat. Stud. Poland 3, 69-77 (1982). - MITCHELL, J.M.: 31-42 in: OESCHGER, MESSERLI and SVILAR 1980. - MIURA, O., YAMANAKA, M.: Ecol. Rev. 19, nr. 3, 167-174 (1980). - MOJSKI, J.E.: Quat. Stud. Poland 2, 77-98 (1980).

NAMIAS, J., CAYAN, D.R.: Science 214, 869-876 (1981). - NARDIN, TH.R., OSBORNE, R.H., BOTTJER, D.J., SCHEIDEMANN, R.C.: Science 213, 331-333 (1981). - NELSON, A.R.: Arct. Alp. Res. 12, 265-286 (1980).

OERLEMANS, J.: Quat. Res. 15, 77-85 (1981). - OESCHGER, H.: 209-236 in: see ref. OESCHGER, MESSERLI and SVILAR 1980. - OESCHGER, H., MESSERLI, B., SVILAR, M. (eds.): Das Klima, Analysen und Modelle, Geschichte und Zukunft. Berlin, Heidelberg, New York: Springer 1980. - OHNGEMACH, D., STRAKA, H.: Beiträge zur Vegetations- und Klimageschichte im Gebiet von Pueblo-Tlaxcala; Pollenanalysen im Mexiko-Projekt, 165 pp. Wiesbaden: Steiner 1983. - OLAUSSON, G.: Quat. Stud. Poland 3, 79-89 (1982). - OSTROVSKIY, A.B., IZMAJLOV, YA.A., SKRYABINA, N.G.: 151-157 in: Ekostratigrafiya i ekologicheskie sistemy geologicheskogo proshlogo. Trudy XXII. sess. vsesoyuzn. paleontol. obshchestva, eds. D.L. STEPANOV, L.J. KHOZATSKIY. Leningrad: Nauka 1980.

PETERSON, L.C., LOHMANN, G.P.: Quat. Res. 17, 26-38 (1982). - PFISTER, CH.: 175-190 in: see ref. OESCHGER, MESSERLI and SVILAR 1980. - PIAGET, A.: 108-133 in: see ref. OESCHGER, MESSERLI and SVILAR 1980. - POLACH, H.A., McLEAN, R.F., THOM, B.G., STODDART, D.R., HOPLEY, D., WILKIE, ST.: Radiocarbon 23, 1-13 (1981). - PORTER, ST.C.: Quat. Res. 16, 263-292 (1981).

RASMUSSEN, L.A.: Quat. Stud. Poland 3, 91-96 (1982). - REITER, F.E.: 275-285 in: see ref. OESCHGER, MESSERLI and SVILAR 1980. - ROBOCK, A., HANSEN, J., GORNITZ, V., LEBEDEFF, S., MOORE, E., ETKINS, R., EPSTEIN, E.: Science 219, 996-998 (1983). - ROGNON, P.: Palaeogeogr., Palaeoclimat., Palaeoecol. 17, 339-342 (1975). - RUDDIMAN, W.F., McINTYRE, A.: Science 212, 617-627 (1981a); - Quat. Res. 16, 125-134 (1981b). - RZECHOWSKI, J.: Quat. Stud. Poland 2, 107-120 (1980).

SCHLÜCHTER, CHR.: Quat. Stud. Poland 3, 97-102 (1982). - SCHNEIDER, S.H., KELLOGG, W.W., RAMANATHAN, V.: Science 210, 6-8 (1980). SCHÜEPP, M.: 191-206 in: see ref. OESCHGER, MESSERLI and SVILAR 1980. - SCHWEGER, CH.E., JANSSENS, J.A.P.: Arct. Alp. Res. 12, 309-317 (1980). - SCHWEINGRUBER, F.H.: 246-256 in: see ref. OESCHGER, MESSERLI and SVILAR 1980. - SCOTT, L.: Quat. Res. 17, 339-370 (1982). - SERGIN, V.YA.: Science 209, 1477-1482 (1980). - SEVERSKIY, E.V.: 73-85 in: Kriogennye yavleniya vysokogorej, ed. A.P. GORBUNOV. Novosibirsk: Nauka 1978. - SEVOSTYANOV, V.V.: 62-68 in: Zakonomernosti formirovaniya prosadochnykh svoystv ljossovykh porod Sredney Azii i Yuzhnogo Kazakhstana, eds. N.I. KRIGER, N.E. KOTELNIKOVA, S.I. LAVRUSHEVICH, V.V. SEVOSTYANOV. Moskva: Nauka 1981. - SHUKLA, J., MINTZ, Y.: Science 215, 1498-1501 (1982). - SJØRRING, ST.: Quat. Stud. Poland 3, 103-109 (1982). - SPJELDNAES, N.: Science 214, 687-688 (1981). - STARKEL, L.: Quat. Stud. Poland 1, 81-94 (1979); - ibid. 2, 121-135 (1980). - STAUFFER, B.: 286-296 in: see ref. OESCHGER, MESSERLI and SVILAR 1980. - STEPANOV, V.P.: 300-322 in: see ref. AGADZHANYAN and DOBRODEEV 1976. - STUIVER, M., QUAY, P.D., OSTLUND, H.G.: Science 219, 849-851 (1983).

TALBOT, M.R.: Quat. Res. 16, 201-220 (1981). - THOMPSON, L.G., MOSLEY-THOMPSON, E.: Science 212, 812-815 (1981). - THOMPSON, ST.L., SCHNEIDER, ST.H.: Science 217, 1031-1033 (1982). - THORSON, R.M., DIXON, E. J., SMITH, G.S., BATTEN, A.R.: Quat. Res. 16, 404-417 (1981). - TROFIMOV, A.K.: Byull. Kom. po izuch. chetvert. per. 48, 79-85 (1978). - TSUKADA, M.: Quat. Res. 19, 212-235 (1983).

VANDENBERGHE, J.: Quat. Stud. Poland 3, 111-118 (1982). - VASILYEV, YU. M.: 172-180 in: see ref. ALEKSEEV and TSEJTLIN 1982a; - 200-204, ibid. 1982b; - 204-213, ibid. 1982c. - VOGEL, J.C., VISSER, E.: Radiocarbon 23, 43-80 (1981). - VOZNYACHUK, L.N., SHULIYA, K.S.: 280-285 in: Opredelenie absolyutnogo vozrasta rudnykh mestorozhdeniy i molodykh magmaticheskikh porod, eds. G.D. AFANASYEV, G.D. VINOGRADOV, L.N. OVCHINNIKOV, E.K. GERLING, T.B. PEKARSKAYA. Moskva: Nauka 1976. - VOZOVIK, YU. I.: 234-244 in: see ref. AGADZHANYAN and DOBRODEEV 1976.

WALKER, D.A., SHORT, S.K., ANDREWS, J.T., WEBBER, P.J.: Arct. Alp. Res. 13, 153-172 (1981). - WANNER, H.: 117-124 in: see ref. OESCHGER, MESSERLI and SVILAR 1980. - WARNER, B., MATHEWES, R.W., CLAGUE, J.J.: Science 218, 675-677 (1982). - WATTS, R.G., HAYDER, M.E.: Science 219, 387-388 (1983). - WATTS, W.A., BRADBURY, J.P.: Quat. Res. 17, 56-70 (1982); - WATTS, W.A., STUIVER, M.: Science 210, 325-327 (1980). - WEBER, F.R., HAMILTON, TH.D., HOPKINS, D.M., REPENNING, CH.A., HAAS, H.: Quat. Res. 16, 167-180 (1981). - WELTEN, M.: Pollenanalytische Untersuchungen im Jüngeren Quartär des nördlichen Alpenvorlandes der Schweiz, Beiträge zur Geologischen Karte der Schweiz, N.F. 156, 174 pp. + 44 pp. Bern: Stömpfli 1982. - WHITE, R.G., TRUDELL, J.: Arct. Alp. Res. 12, 511-529 (1980). - WHITEHEAD, D.R., JACKSON, ST.T., SHEEHAN, M.C., LEYDEN, B.W.: Quat. Res. 17, 241-257 (1982). - WILLIAMS, N.E., WESTGATE, J.A., WILLIAMS, D.D., MORGAN, A., MORGAN, A.V.: Quat. Res. 16, 146-166 (1981). - WILLSON, R.C., GULKIS, S., JANSSEN, M., HUDSON, H.S., CHAPMAN, G.A.: Science 211, 700-702 (1981). - WOO, M.K.: Arct. Alp. Res. 12, 227-235 (1980). - WOODRUFF, F., SAVIN, S.M., DOUGLAS, R.G.: Science 212, 665-668 (1981). - WORSLEY, TH.R., HERMAN, Y.: Science 210, 323-325 (1980). - WRIGHT, H.E.: Quat. Res. 15, 113-125 (1981). - WYSOCZANSKI-MINKOWICZ, T.: Quat. Stud. Poland 3, 129-145 (1982).

YURTSEV, B.A.: Reliktovye stepnye komplesksy severo-vostochnoj Azii, 168 pp. Novosibirsk: Nauka 1981

ZARRINA, E.P.: Quat. Stud. Poland 3, 147-155 (1982).

Professor Dr. Dr. h.c. BURKHARD FRENZEL
Botanisches Institut (210) der Universität,
D-7000 Stuttgart 70 (Hohenheim), FRG

II. Vegetation Science (Sociological Geobotany)

By RÜDIGER KNAPP

1. Fundamental Perspectives, Books, Monographic Volumes and Bibliographies

Multilateral studies and analysis of vegetation and plant communities continue to encompass the majority of actual sociological geobotany. They culminate often in comprehensive vegetation mapping or in inductive investigations on successions and on other syndynamical connections, including experimental approaches. All these studies and their results generate the basis of ever-increasing interconnections of vegetation research and sociological geobotany with other fields of theoretical and applied sciences. Among the structural features and the social interrelations in vegetation, the biocoenotic connections between plant communities and animals recently became more and more emphasized (e.g., FURTADO and MORI 1982, HOFFMANN and THIELE 1982, AHMAD 1983, DENNO and McCLURE 1983, KINZELBACH 1983, STRONG 1983, LAZARE et al. 1984; examples of special contributions in sects. 3a,b). Such connections are also included in more general volumes on community interrelations (MOONEY and GODRON 1983, ANDERSON and KIKKAWA 1984; also GIGON 1983, SCHMIDT 1983a,b; on tropical rain forests: MEDINA et al. 1984, WHITMORE 1984, also COLEY 1983; on mangrove vegetation: KOSTERMANS and SASTROUTOMO 1982, TEAS 1983). Other new volumes consider numerical aspects (GAUCH 1983, GREIG-SMITH 1983), allelopathy (GRODZINSKY 1982, 1983, WALLER and CHOU 1982), long-term vegetation changes in arid western North America (ROGERS 1982) and applications in land use (ELLENBERG 1983). Condensed introductory texts have been presented by WALTER and BRECKLE (1983), KREEB (1983), DEMANGEOT (1984) and GODRON (1984). Additional books will be quoted in context with the related topics in the next sections.

New bibliographies comprehend publications on the vegetation of various countries and states (Spain: IZCO 1982a; Germany, Egypt, Malawi, Botswana: KNAPP 1982; Cameroon, Sudan, Ethiopia, Kenya, Uganda, Tanzania: KNAPP 1983a: India: MEHER-HOMJI and GUPTA 1982, 1983; Pakistan: MEHER-HOMJI 1982), geobotany of *Potamogeton alpinus* (WIEGLEB 1983a), fire influences (ECKELS et al. 1983), allelopathy, competition and related implications, genetics and evolution within vegetation, Rodentia and plant communities, vegetation protection and conservation (KNAPP 1983a).

The rapidly growing number of publications on vegetation science demands a selection of alternating particular branches in our annual reviews in Progress in Botany, but also the avoidance of repeated quotation of certain publications in succeeding volumes. Thus, the reviews in the successive volumes of Progress in Botany should be regarded as an entity; a quotation missing in the present report may be found included in a later or earlier year.

Progress in Botany, Vol. 46

2. General Results and Methods

a) Sampling Methods and Taxon Analysis in Relevé Surveys

Essentially amplified informations in sampling of vegetational stands can be achieved by means of more detailed taxon differentiation and of improved quantification. Inclusion of microspecies, subspecies and varieties, but also of hybrids in relevé surveys ("Vegetationsaufnahmen") can highly improve the possibilities of syntaxonomic differentiation and the characterization of plant communities and of their various indicative values (GÉHU 1982, DEN HARTOG and VAN VIERSSEN 1982, WEBER 1983, KNAPP 1984a, LANDOLT 1984a, R. TÜXEN 1984a). A decisively augmented quantification is attainable by means of a more precise coverage assessment. Determination of cover percentages (KNAPP 1984a, PAKARINEN 1984a, RAMÍREZ 1984a) or application of cover scales with ten or more classes approaching direct percentage measurement (LONDO 1984a, OKSANEN 1984a, RABOTNOV 1984a, also SYKES et al. 1983) have proved to be well applicable within large numbers of relevé surveys. Cover percentage can be exactly measured and checked by means of certain point frequency determinations (FLOYD and ANDERSON 1982, KNAPP 1983b, 1984a). The augmented survey time and the greater experience necessary in applying these methods seem to be completely justified, namely under consideration of the very extended and sumptuous numerical procedures and computer programs which are becoming usual in the treatment of the relevés (e.g., GERDOL and PICCOLI 1982, FEOLI 1982, GOODALL and JOHNSON 1982, TÖRÖK and PODANI 1982, VAN GROENEWOUD 1983, KUHN 1983, MANN and SHUGART 1983, STOEHR and BÖCKER 1983, ZHANG 1983, CHESSEL and GAUTIER 1984a, FEOLI et al. 1984).

Appropriate methods afford extensions of relevé sampling to underground parts of plants (BOOKMAN and MACK, 1982, TĚSAROVÁ et al. 1982, KUTSCHERA-MITTER 1984a), among them living seeds in the soil (NAKAGOSHI et al. 1983, PUTZ 1983, SMITH and KADLEC 1983, STIEPERAERE and TIMMERMAN 1983, NUMATA 1984a), to fossil vegetation deposits (FRITZ and WILMANNS 1982, KOROBKOV and FILIN 1982, SCHWAAR 1982a,b, 1984a, KNÖRZER 1984a), to dispersal units (SCHULTKA and HILGER 1983, LUFTENSTEINER 1984a) and to grid devices (GÉHU 1984, READER and LIETH 1984a). Appropriate survey methods and methodical adaptions have been developed for communities of plants with special life forms, e.g., aquatic angiosperms (SCOPPOLA 1982, KÖCK 1983, LUTHER 1983, WIEGLEB 1983b, SZAFRANSKI and APEMA 1983, KÁRPÁTI and KÁRPÁTI 1984a), fungi (ARNOLDS 1982, WINTERHOFF 1984a) and phytoplankton (e.g., ILMAVIRTA 1984, KENNISH and LUTZ 1984, REYNOLDS 1984).

b) Global Classification of the World's Vegetation

There is a revived interest in the classification of the vegetation of the world in a limited, easily comprehensible number of chorological units, mostly emphasizing dominant physiognomic types of terminal successional stages. New studies contribute to the universal knowledge of such fundamental vegetational units in certain regions: e.g., in parts of Central Asia (LI et al. 1980, CHANG 1981, ZHANG et al. 1981, STANYUKOVICH 1982, SONG 1983, ZHENG DU 1983) and of north-eastern Asia (KHARKEVICH et al. 1983, KHOLOD 1983), in Indonesia and Malaysia (MIYAWAKI 1982, PROCTOR et al. 1983), in north-western Africa (BARBERO et al. 1982, DJEBAILI et al. 1982, DEIL 1984, PELTIER 1984), in eastern Africa (PIGNATTI and WARFA 1983, SCHULTKA and HILGER 1983, WHITE 1983), in Latin America (BROWN et al. 1982, GOLDBERG 1982, VAN DER HAMMEN et al. 1983, JORDAN 1983, KLÖTZLI 1983, LEÓN and ANDERSON

1983, OHNGEMACH and STRAKA 1983, ROIG and BOELCKE 1983; with emphasis on interrelations with agricultural influences and land utilization in the Andes: ELLENBERG 1981, SEIBERT 1982, BECK 1983, HANAGARTH and FISEL 1983, RUTHSATZ 1983b) and in subarctic wetland vegetation of North America (e.g., JASENIUK and JOHNSON 1982, PALA and BOISSONNEAU 1982, SIMS et al. 1982) (on various herbaceous formations: DESCOINGS 1983). New volumes on such units treat desert and semi-desert vegetation (WEST 1983), extra-tropical broad-leaved evergreen forests (OVINGTON 1983, also FUJIWARA 1981, IEHARA et al. 1983) and tropical rain forests (MEDINA et al. 1984, WHITMORE 1984). Reconsiderations on the principal thermal vegetational zones and their nomenclature (SCHROEDER 1983) reveal a prevalent acceptance of the terms "arctic," "boreal," "tropical" and "antarctic." The regions between the freezing limit (Frostgrenze) and the occurrence of minimum temperatures of about -10°C in the northern hemisphere (example: Mediterranean region) are currently named mostly "meridional" or "subtropical." "Meridional" seems to be preferable, since this term is not in use in a deviating sense. More problematic is the nomenclature of the areas between the meridional and the boreal zones (example: most of Central Europe and of eastern North America), named frequently "temperate" or "nemoral." The term nemoral designates originally mainly areas of rich deciduous broad-leaved forests. Thus, it is burdened by the disadvantage that great parts of the zones concerned are not covered naturally by such vegetation. Therefore, the term temperate seems to be more adequate. Also the regions between the tropical and antarctic zones have been named in rather divergent ways, partly due to their vegetational differences (e.g., TOMASELLI 1981, KLING and GLAWTON 1982, KLÖTZLI 1983). The term "austral," already quite frequently used for these regions, has the advantage of avoiding confusion with characters and names of physiognomic vegetational types or with any deviate floristic-chorological definitions. Additionally, the term austral emphasizes adequately the differences of the vegetational zonations on the southern and northern hemispheres respectively, due to special evolutionary history, to divergences in the climatic oceanity and to other reasons.

c) Diversity, Numbers and Dominance Structure of Species in Vegetation

Problems of diversity and related fields in vegetation have been objects of many numerical and biomathematic studies during recent periods (in the last years, e.g. ABUGOV 1982, FISCHER 1982, HAEUPLER 1982, OKSANEN and AHTI 1982, REJMÂNEK 1982, WHEELER and GILLER 1982, BUSING and CLEBSCH 1983, OKSANEN 1983, TER BRAAK 1983), resulting also in several thoughtful theories and considerations. A multitude of models and of formulae have been proposed within the complex of diversity studies. One of the most important groups of parameters concerns the number of taxa (species, etc.) in homogeneous vegetation and in uniform or equivalent sample area sizes. For instance, the alpha diversity is related to this parameter group. Another essential group concerns the deviation of the quantitative representation of the taxa (coverage, abundance, weight, etc.) in the studied vegetation stand from the maximum entropy. This maximum accords with equal percentages of all taxa. This concept has been termed equity, equitability, evenness (sensu stricto), also dominance structure. HAEUPLER (1982) has computed the evenness in ample numbers of relevés and has suggested its application also for table arrangements. Mostly, greatest evenness values (60-80, etc.) exist in plant communities with high species numbers without pronounced dominants (e.g., mixed forests on rich soils, certain herb-rich grasslands). Low evenness values (15-35, etc.) occur in communities of various vegetational formations with few species dominated pronouncedly by a single taxon.

d) Population Death in Vegetation

Simultaneous death of whole populations of certain species within the vegetation of particular areas becomes increasingly actual. In connection with air contamination, it attracts now also large-scale public interest. Such phenomena are not only caused by recent emission disturbances, but are included also in natural successional cycles or series. An example is the death of bamboo populations following their flowering and fruiting. MUELLER-DOMBOIS (1981a,b, 1982, 1983) reviews such phenomena, these being rather frequent in the indigenous vegetation of the Hawaiian Islands. Partly, abnormal high or low amounts of rains cause drought or oxygen deficiencies respectively in the root horizons, lethal for certain species populations. But in several cases, such simultaneous death due to age reasons ("cohort senescence") seems to be the most obvious explanation (examples: *Metrosideros polymorpha*, Myrtaceae; *Dubautia scabra*, Compositae; *Canavalia kauaiensis*). The reason of this uniform age is rapid invasion on sites opened by Hawaiian volcanism. Other population death is implied by regionally new diseases imported from remote continents. The impacts of the *Castanea dentata (=americana)* dieback due to the chestnut blight invasion have been thoroughly studied in the forests of eastern North America. This disease has now invaded also Europe, causing enormous vegetational changes in vast areas dominated by *Castanea sativa*. Another example is the dieback of *Ulmus* species (e.g., LARSON and BARNES 1982, HUENNEKE 1983).

3. Examples of Interrelations Between Vegetation and Animals

a) Interrelations Between Vegetation and Ants

Several ant influences can induce patchiness of vegetation (creation of vegetational mosaics). Ant mounds often imply such developments (e.g., CULVER and BEATTIE 1983). The mainly tropical leaf cutting ants (*Atta*, *Acromyrmex* et al., BOWERS and PORTER 1981) are capable to release such structures by foraging concentrated on particular territories and by limited places with substrates rich in nutrients on abandoned nests and on dump material remaining from former ant-fungus gardens. The species composition of plant communities can be altered by selective leaf cutting of certain taxa. Also the relations of ants and certain epiphytes (ant-epiphytes) can cause changes of vegetation structure (HUXLEY 1982, RIVERO and BARARD 1983). Another most important ant influence is seed harvesting, namely in climatically rather dry regions (A. ANDERSEN 1982, BRIESE 1982, CAMPBELL 1982, DAVISON 1982, MAJER 1982). Since the harvested seed amounts can be enormous, certain species are collected selectively and therophytes often are most concerned due to their life cycle; the composition of plant communities and the successions can be pronouncedly influenced. An example in effects on trees has be found in *Eucalyptus regnans*: 60% of its seeds are consumed in normal years by ants, occasionally even the entire seed crop. Studies on seed dispersal by ants (myrmecochory) become now focused on its vegetation structure effects (BEATTIE and CULVER 1981, DAVIDSON and MORTON 1981, HEITHAUS 1981, MILEWSKI and BOND 1982, WESTOBY et al. 1982).

b) Vegetation and Small Mammals

Small mammals (mainly Rodentia and Lagormorpha) can highly influence vegetation, for instance by selective foraging (BERGMAN 1982, BOOR-

MAN and FULLER 1982, BOWERS 1982, CLARK and INNIS 1982, SENZOTA 1982, LEUTERT 1983; belowground: D.C. ANDERSEN 1982) and by collecting or storing of seeds (seed predation, D.H. BOUCHER 1982, HALVORSON 1982, JANZEN 1982, SULLIVAN and SULLIVAN 1982, McCLOSKEY 1983). Most important impacts of these animals are the induction of vegetational mosaics in connection with their burrowing activities and their extreme influences around the colonies and runway paths (D.C. ANDERSEN 1982, BEST 1982, REICHMAN et al. 1982, SENZOTA 1982, TRUSZKOWSKI 1982, DETLING and PAINTER 1983, LEUTERT 1983). Intricate connections can exist between the elements of vegetational mosaics and the various life expressions of small mammals (differentiated habitat uses: ABRAMSKY and SELLAH 1982, GAINES and JOHNSON 1982, HALLETT 1982, THOMPSON 1982, YAHNER 1982, CROWELL 1983, KINCAID et al. 1983), but also between structures of tropical forests and tree-inhabiting rodents (EMMONS and GENTRY 1983). A parallel analysis of small mammal coenoses (including small Rodentia and Insectivora) and plant communities has been initiated in Brandenburg, using methods of modern syntaxon description and differentiation (PASSARGE 1982). The small mammal coenoses have proved to be also dependent on site conditions, but adequate coincidence with plant community units could be assessed only on the level of higher vegetational syntaxa (orders, classes). This seems to be partly caused by the low species numbers in the small mammal coenoses (between 3.6 and 5.9 only). But certain small mammal species occur simultaneously on sites which have no plant species in common, indicating broader site amplitudes.

4. Vegetation Science as Basis of Nature Protection and Nature Conservation

Intensive connections between vegetation science and nature conservation have existed for long periods. But recently the quantity of vegetation studies on nature protection problems has been much increased (bibliography: KNAPP 1983a). The rapid rarefication or even extinction of several plant communities during the last decades (e.g., RAABE et al. 1982, SUKOPP and ELVERS 1982, WATTEZ and GÉHU 1982, WITTIG and POTT 1982) have induced multiplied endeavors to conserve the last remainders of endangered vegetation units (foundations and methods: DIERSCHKE 1981, ARNOLD et al. 1982, LEWIS 1982, MALATO-BELIZ 1982, MIYAWAKI 1982, MÜHLENBERG 1982, TRAUTMANN 1982, COETZEE 1983, GRIGGS and JAIN 1983, HABER 1983, LIETH and BARTH 1983, SCHIEFER 1983, SIIRA 1983, VORONTZOVA et al. 1983, VOSER-HUBER 1983, WAY 1983, SCHMIDT 1984), combined with studies to locate their sites and status including biotope mapping (e.g., WILMANNS and KRATOCHWIL 1982, AUHAGEN and SUKOPP 1983, MÜLLER and WALDERT 1983, SCHIECHTL and STERN 1983, SUKOPP 1983). Also improved understanding of former utilization practices is often necessary for reasonable conservation measures, e.g., on lopping ("Schneitel-Wirtschaft", BURRICHTER and POTT 1983, POTT 1983). Vegetation protection becomes urgent not only in densely inhabited regions, but also in certain remote areas, e.g., of northern Asia (MALYSHEV and SOBOLEVSKAYA 1983, NAUMENKO and NUKHIMOVSKAYA 1983, STOIKO 1983, ZOLOTUKHIN 1983). Endangered plant communities are, for instance, several wetland vegetation units and certain Atlantic European heaths with *Erica* and *Calluna* (e.g., BARENDREGT 1982, REININGHAUS and SCHMIDT 1982, F. RUNGE 1982b, GIMINGHAM and DE SMIDT 1983, NOIRFALISE 1983, TÜXEN and SOLOMONS 1983), but even particular weed associations of arable fields (RODI 1982, FISCHER 1983). Vegetational changes within protection areas become more and more evident (F. RUNGE 1982b, WESTHUS 1983). Therefore, it requires special efforts and actions to maintain the protected vegetation in their desirable status, namely in conditions of small conservation areas. Thus, several vegetation studies of particular nature protection areas contribute to the detection of the locally necessary measures for this purpose (e.g., HENNEKENS et al. 1982, HOFMANN et al. 1982, F. RUNGE 1982a, SCHLÜTER and BALLER 1982, BENNERT and KAPLAN 1983, GLAVAČ 1983, HEINRICH 1983, PAELINCKX and SOETENS 1983, PROCTOR et al. 1983, RUTHSATZ 1983a, SCHWABE-BRAUN 1983, TÜXEN 1983, WERGER et al. 1983, WILMANNS 1983a, WOLFF 1983, ZHANG 1983).

5. Vegetation of Temperate and Adjacent Regions

a) Steppe Vegetation of Pontic Regions, Central and Northern Asia

The steppe vegetation of these ample areas continues to be intensively studied (LI et al. 1980, BOIKOV 1981, BOIKO 1983, KOVÁŘ and VOLKOVA 1981, ANTONOVA 1983, HILBIG and H.D. KNAPP 1983, MATVEYEVA 1983, SLEMNEV et al. 1983, TOPOLNY et al. 1983, WALTER et al. 1983), partly with emphasis on communities representing relicts from former climatic conditions (BERMAN et al. 1981, LAVRENKO 1981, NAMZALOV and ERSHOVA 1981, YURTSEV 1981, POLOZOVA 1983) or on the dynamics and successional processes (SHALANDINA 1981, CIBANOVA 1982, TITOV and NESKRIABINA 1982). Relations between steppes and forests and their consequences on the potential natural vegetation assumed for certain areas include still open problems; some attributes supposed to be evidential (e.g., certain soil characters) are now questioned as convincing arguments, exemplified by considerations on parts of Romania (NIEDERMAIER 1983). Also conservation and protection problems are amply discussed, since the original steppes are converted more and more into agricultural land, even in extended parts of northern and central Asia (IVAN 1981, MALYSHEV and SOBOLEVSKAYA 1983, NAUMENKO and NUKHIMOVSKAYA 1983, STOIKO 1983, ZOLOTUKHIN 1983).

b) Xeric Grassland of the Alps and of Southern European High Mountains

Certain xeric grasslands in high mountain areas below and around the timberlines, often on steep slopes and very rich in species, become now better understood, partly due to improved studies on the often dominant *Festuca* taxa, but also on other critical genera, such as *Alchemilla, Campanula, Carduus, Centaurea, Hieracium, Koeleria, Phyteuma, Thymus*, etc.. These grasslands have been studied recently in the southern Alps (e.g. *Festucion variae* s.l.: VERGER 1982, KAPLAN 1983) and in the Pyrenees (e.g., *Festucion eskiae, Festucion paniculatae:* MOLERO and VIGO 1981, NÈGRE et al. 1982). Partly comparable vegetation exists also in northern Alps (BESLER and BORNKAMM 1982, SPATZ 1983, ZUMBÜHL 1983). The sites of these plant communities are intermediate between extremely acid and calcareous substrate characters. Due to the climatic situations, this vegetation can include simultaneously xerothermic taxa (of *Festuco-Brometea*, etc.) and species distributed mainly above timerline. These conditions explain the high species diversity (often more than 60 vascular plant taxa on 20 m^2). Syndynamically, these grasslands represent often rather durable intermediate stages, which can be stabilized by means of relatively slight mowing (once per year) or pasturage, thus becoming permanent communities ("Dauergesellschaften").

c) Syntaxonomy of Plant Communities in Central and Western Europe and in Some Adjacent Areas

Syntaxonomic synopses and studies focused on classification following the SIGMA methods reconsider *Fagus sylvatica* communities and related European deciduous forests (DIERSCHKE and SONG 1982, MULLER 1982, THIÊBAUT 1982, JAHN 1983a,b). They emphasize frequently site-indicative species groups (ANTONIETTI 1983, KRIEBITZSCH and HASEMANN 1983, MORAVEC 1983a,b, PASSARGE 1983, ZEIDLER 1983), but also other attributes, e.g., *Acer* participation (BÊGUIN and THEURILLAT 1982, FAJMANOVÂ 1982, HUSOVÂ 1982), influences of aluminium in the soil (M. RUNGE 1983, 1984), symphenology (DIERSCHKE 1983) or symphytosociology (MEDWECKA-KORNÂS 1983). Other European vegetational groups treated syntaxonomically include anthropogenic grassland (DIETL 1982, BALÂTOVÂ-TULAČKOVÂ 1983, HUNDT 1983, HUNDT and HÜBL 1983, ILIJANIĈ and ŠEGULJA 1983, MEISEL 1983, NEUHÄUSL 1983, OBERDORFER and MÜLLER 1983, RAUS 1983), fountain communities (*Montio-Cardaminetea*, GIL and VARO 1982, HADAČ 1983), dune and salt marsh vegetation (GÉHU and DE FOUCAULT 1982, USLU et al. 1982, SIIRA 1984) and ruderal or related communities (BRANDES 1982, 1983, MUCINA 1982a,b, BORNKAMM and HENNIG 1983, BRAUN-BLANQUET and SUTTER 1983, FORSTNER 1983, KRAUSE 1983, OBERDORFER and MÜLLER 1983; in comparison with other regions: OBERDORFER 1983). Also parts of the vegetation of southern Scandinavia and adjacent islands have been viewed under syntaxonomic objectives (F. RUNGE 1983, VEVLE 1983, WESTHOFF et al. 1983). In connection with subdivisions and general characterizations of this life-form group, the importance of lianas is considered within particular European vegetational classes and orders (WILMANNS 1983b).

6. Vegetation of Meridional and Adjacent Regions

a) Vegetation of Mediterranean Islands: Endemism and Syntaxonomy

The indigenous vegetation of the Mediterranean islands includes several endemic taxa (e.g., Corsica: DIERSCHKE 1981, Sicily: POLI-MARCHESE 1982, Cyprus: LEIPPERT 1981). But in many cases, these endemics have been recognized rather late. For instance, most of the plant taxa endemic in the Balearic and Pityusic Islands have been described after about 1924, albeit two ample works had treated before this year especially their flora (HAEUPLER 1983; also COSTA 1982, DE BOLÔS and MOLINIER 1984, KUHBIER et al. 1984). Thus, the present status of endemism of 9.2% (exceeding for instance the 7.6% of Corsica) is mainly a result of rather recent taxonomic studies. Since the endemics are mostly excellent character taxa of associations, alliances or even higher units, the vegetational (syntaxonomic) classification is intricately connected with the status and the progress of the regional taxonomic treatment.

b) Vegetation of Mountains Rising Above Meridional Plains

A situation partly similar to the just discussed conditions in the islands (endemism, relation between syntaxonomy and taxonomy) occurs in certain mountain areas isolated from regions with comparable climates in present times and - even more important - did so continuously during more or less long former periods. Examples in the Mediterranean and neighboring areas can be found in the Iberian Peninsula (C. BOUCHER 1982, IZCO 1982b, MALATO-BELIZ 1982, MAYOR et al. 1982), in central and southern Italy (BIONDI and BLASI 1982, DE FAVERI and NIMIS 1982), in the southern-most Alps (Alpes Maritimes: FOCQUET 1982; Monte Baldo: GERDOL and PICCOLI 1982): on the Crimea (DIDUKH 1983) and in the Alborz, Iran (KLEIN 1982a,b).

7. Tropical Vegetation

a) Savanna Vegetation in Africa

The African savanna vegetation is determined by interconnections of several biotic and abiotic factors, partly highly instable. It has been tried to simulate this interplay by means of an "extended savanna model" (NOY-MEIR 1982, WALKER and NOY-MEIR 1982). The multitude of important factors is somewhat obscured in the regions south of the Sahara, where several conditions are rather uniform and differences in the amounts of available water (e.g., due to various lengths of the rain periods of various quantities of annual precipitations) are predominant determinants of the vegetation structures (e.g., BAKHAREVA 1982, BARRY 1982, WHITE 1983, KNAPP 1984b). Recognition of the interconnections of various factors have been also partly delayed by the fact that many savanna studies have been specially focused only a certain part of the whole biocoenosis (e.g., on wildlife ungulate influences, on utilization of cattle, etc.). Therefore, it is important that the diversified studies in the Lamto savanna (Ivory Coast; LAMOTTE 1982, MENAUT and CÉSAR 1982) are augmented by a similar project in the Nylsvley savanna (Transvaal; HUNTLEY and MORRIS 1982), including, besides detailed syntaxa analysis, studies on the energy budgets (GANDAR 1982, DE JAGER and HARRISON 1982), nitrogen conditions (BATE and GUNTON 1982), litter decomposition (MORRIS et al. 1982) and the various vertebrate and invertebrate animal groups. The influences of large herbivores (ungulates, etc.) on African savanna structures have been again discussed intensively (e.g., BELL 1982, CUMMING 1982, HUNTLEY 1982, McNAUGHTON 1983, PELLEW 1983, HATTON and SMART 1984). Deficient nutritive values of the herbage on poor soils or in humid climates and the relatively high contents of secondary compounds, deterrent to herbivores, found in plants in the just mentioned conditions seem to be of pervasive significance for low numbers of such animals in areas with these sites and for resulting features of vegetation structure (BELL 1982, OWEN-SMITH 1982).

b) Dry Woody Vegetation and Grasslands in Tropical South America

In South American regions with dry periods of three or more months annually (BUCHER 1982, COOPER 1982, COUTINHO 1982, EITEN 1982, MEDINA 1982, ALHO CLEBER 1983, SARMIENTO 1983), dry woodlands are often prevalent. They are partly interrupted by grassland more or less devoid of trees or shrubs (e.g., the Brasilian campo limpo), due to special water conditions or other reasons (e.g., COOPER 1981, EITEN 1982). But also genuine savannas (sensu stricto) with more or less scattered trees and shrubs (covering less than about 10%) in vegetation otherwise dominated by grasses can be highly important in South America (e.g., in Venezuela: BELLO and TORRES 1981, BOISSIÈRE and ARIAS 1981, GONZALEZ JIMÉNES et al. 1981, SAN JOSÉ and FARIÑAS 1983; on certain white sands in Amazonia: ANDERSON 1981). Several authors think that in South America such genuine savanna vegetation is less common and dry woodlands or grasslands are more important than in climatically comparable regions of Africa.

References

ABRAMSKY, Z., SELLAH, O.: Ecology 63, 1242-1247 (1982). - ABUGOV, R.: Ecology 63, 289-293 (1982). - AHMAD, S. (ed.): Herbivorous Insects. 257 pp. New York, London: Academic Press 1983. - ALHO CLEBER, J.R.: Trop. Ecol. 23, 125-133 (1983). - ANDERSEN, A.: Geobotany 4, 31-43 (1982). - ANDERSEN, D.C.: Am. Naturalist 119, 18-28 (1982). - ANDERSON, A.B.: Biotropica 13, 199-210 (1981). - ANDERSON, D.J., KIKKAWA, J. (eds.): Community Ecology. 400 pp. Oxford: Blackwell 1984. - ANTONIETTI, A.: Tuexenia 3, 297-306 (1983). - ANTONOVA, L.A.: Bot. Zh. 68, 916-924 (1983). - ARNOLD, W., BÖSCH, B., SCHMID, H.: Forstwiss. Cbl. 101, 311-343 (1982). - ARNOLDS, E.: Bibl. Mycol. 90, 1-501 1982). - AUHAGEN, A., SUKOPP, H.: Natur Landschaft 58, 9-15 (1983).

BAKHAREVA, S.N.: Bot. Zh. 67, 333-338 (1982). - BALÁTOVÁ-TULÁČKOVÁ, E.: Tuexenia 3, 227-240 (1983). - BARBÉRO, M., BENABID, A., QUÉZEL, P., RIVAS-MARTÍNEZ, S., SANTOS, A.: Doc. Phytosociol. 6, 311-338 (1982). - BARENDREGT, A.: Phytocoenologia 10, 425-462 (1982). - BARRY, J.P.: Écol. Mediterr. 8, 99-124 (1982). - BATE, G.C., GUNTON, C.: Ecol. Stud. 42, 498-513 (1982). - BEATTIE, A.J., CULVER, D.C.: Ecology 62, 107-115 (1981). - BECK, S.: Diss. Bot. 80, 1-186 (1983). - BÉGUIN, C., THEURILLAT, J.-P.: Saussurea 13, 17-33 (1982). - BELL, B.J.: 193-216 in: see ref. HUNTLEY and WALKER 1982. - BELLO, L.F., TORRES, R.: Bol. Soc. Venez. Ci. Nat. 36, 343-344 (1981). - BENNERT, H.W., KAPLAN, K.: Decheniana 136, 5-14 (1983). - BERGMAN, G.: Mem. Soc. Fauna Flora Fennica 58, 49-52 (1982). - BERMAN, D.I., BRUKHANOV, V.A., GELMAM, V.G., KOROTKOVA, E.I.: Bot. Zh. 66, 556-558 (1981). - BESLER, W., BORNKAMM, R.: Tuexenia 2, 135-162 (1982). - BEST, T.L.: J. Mammalogy 63, 532-535 (1982). - BIONDI, E., BLASI, C.: Doc. Phytosociol. 7, 435-442 (1982). - BOIKO, M.F.: Bot. Zh. 68, 1112-1121 (1983). - BOIKOV, T.G.: Bot. Zh. 66, 714-719 (1981). - BOISSIÈRE, J.E., ARIAS, I.A.: Bol. Soc. Venez. Ci. Nat. 36, 341 (1981). - BOLÒS, O. DE, MOLINIER, R.: Monogr. Biol. 52, 185-222 (1984). - BOOKMAN, P.A., MACK, R.N.: Ecology 63, 640-646 (1982). - BOORMAN, L.A., FULLER, R.M.: J. Ecol. 70, 345-356 (1982). - BORNKAMM, R., HENNIG, U.: Flora 172, 267-316 (1983). - BOUCHER, C.: Ecol. Mediterr. 8, 69-84 (1982). - BOUCHER, D.H.: Oecologia 49, 409-414 (1982). - BOWERS, M.A.: J. Mammalogy 63, 361-367 (1982). - BOWERS, M.A., PORTER, S.D.: Ecology 62, 273-275 (1981). - BRANDES, D.: Doc. Phytosociol. 6, 131-154 (1982); - Phytocoenologia 11, 31-115 (1983). - BRAUN-BLANQUET, J., SUTTER, R.: Tuexenia 3, 319-323 (1983). - BRIESE, D.T.: Geobotany 4, 11-24 (1982). - BROWN, D.E., MINCKLEY, W.L., PASE, C.P., TURNER, R.M.: Desert Pl. 4, 27-287 (1982). - BUCHER, E.H.: 48-79 in: see ref. HUNTLEY and WALKER 1982. - BURRICHTER, E., POTT, R.: Tuexenia 3, 443-454 (1983). - BUSING, R.T., CLEBSCH, E.E.G.: Bull. Torrey Bot. Club 110, 304-310 (1983).

CAMPBELL, M.H.: Geobotany 4, 25-30 (1982). - CHANG, D.H.S.: Mountain Res. Dev. 1, 29-48 (1981). - CHESSEL, D., GAUTIER, C.: 61-76 in: see ref. KNAPP (ed.) 1984a. - CIBANOVA, N.A.: Bot. Zh. 67, 229-231 (1982). - CLARK, W.R., INNIS, G.S.: J. Wildlife Manag. 46, 1018-1035 (1982). - COETZEE, B.J.: Diss. Bot. 69, 1-456 (1983). - COLEY, P.D.: Ecol. Monogr. 53, 209-233 (1983). - COOPER, A.: Trop. Ecol. 22, 99-115 (1981); - Trop. Ecol. 23, 193-207 (1982). - COSTA, M.: Colloq. Phytosociol. 6, 355-364 (1982). - COUTINHO, L.M.: 273-291 in: see ref. HUNTLEY and WALKER 1982. - CROWELL, K.L.: Oikos 41, 442-454 (1983). - CULVER, D. C., BEATTIE, A.J.: Ecology 64, 485-492 (1983). - CUMMING, B.J.: 217-245 in: see ref. HUNTLEY and WALKER 1982.

DAVIDSON, D.W., MORTON, S.R.: Oecologia 50, 357-366 (1981). - DAVISON, E.A.: Geobotany 4, 1-6 (1982). - DEIL, U.: Diss. Bot. 74, 1-179 (1984). - DEMANGEOT, J.: Les milieux naturels du globe. 250 pp. Paris: CNRS

1984. - DENNO, R.F., McCLURE, M.S. (eds.): Variable Plants and Herbivores. 712 pp. New York, London: Academic Press 1983. - DESCOINGS, B.: Adansonia B 5, 53-62 (1983). - DETLING, J.K., PAINTER, E.L.: Oecologia 57, 65-71 (1983). - DIDUKH, Y.P.: Bot. Zh. 68, 1556-1566 (1983). - DIERSCHKE, H.: Ber. Symp. Int. Ver. Vegetationskd. 1972, 521-532 (1981); - Verh. Ges. Ökologie 11, 71-84 (1983). - DIERSCHKE, H., SONG, Y.: Ber. Symp. Int. Ver. Vegetationskd. 1981, 513-539 (1982). - DIETL, W.: Schweiz. Landwirtsch. Forschung 21, 85-110 (1982). - DJEBAILI, S., AIDOUD, F., DAHMANI, M.: Dipart. Amb. Terr. Trasp. Emilia Romagna Stud. Doc. 30, 135-157 (1982).

ECKELS, E., RIESS, W., KNAPP, R.: Excerpta Bot. B Sociol. 23, 1-6 1983). - EITEN, G.: 25-47 in: see ref. HUNTLEY and WALKER 1982. - ELLENBERG, H.: Desarrollar sin destruir. 55 pp. La Paz: Instituto Ecológico UMSA 1981; - Ökologische Beiträge zur Umweltgestaltung. (Neudruck.) 111 pp. Stuttgart: Ulmer 1983. - EMMONS, L.H., GENTRY, A.H.: Am. Nat. 121, 513-524 (1983).

FAJMANOVÁ, E.: Preslia 54, 259-269 (1982). - FAVERI, R. DE, NIMIS, P.L.: Ecol. Mediterr. 8, 85-98 (1982). - FEOLI, E.: Dipart. Amb. Terr. Trasp. Emilia Romagna Stud. Doc. 30, 83-108 (1982). - FEOLI, E., LAGONEGRO, M., ORLOCI, L.: Tasks Veget. Sci. 10, 1-143 (1984). - FISCHER, A.: Tuexenia 2, 219-312 (1982); - Phytocoenologia 11, 331-383 (1983). - FLOYD, D.A., ANDERSON, J.E.: Vegetatio 50, 185-186 (1982). - FOCQUET, P.: Doc. Phytosociol. 7, 1-188 (1982). - FORSTNER, W.: Wiss. Mitt. Niederösterreich. Landesmuseum 2, 19-132 (1983). - FRITZ, W., WILMANNS, O.: Ber. Dtsch. Bot. Ges. 95, 1-18 (1982). - FUJIWARA, K.: Bull. Inst. Environ. Sci. Technol. Yokohama Nat. Univers. 7, 67-133 (1981). - FURTADO, J.I., MORI, S. (eds.): Tasek Bera. 413 pp. The Hague: Junk 1982.

GAINES, M.S., JOHNSON, M.L.: Oikos 39, 63-70 (1982). - GANDAR, M.V.: 514-534 in: see ref. HUNTLEY and WALKER 1982. - GAUCH, H.G.: Multivariate Analysis in Community Ecology. 304 pp. Cambridge: Cambridge University Press 1983. - GÉHU, J.-M.: Doc. Phytosociol. 6, 303-309 (1982); - Handbook Veget. Sci. 4, 121-128 (1984). - GÉHU, J.-M., FOUCAULT, B.DE.: Doc. Phytosociol. 7, 387-398 (1982). - GERDOL, R., PICCOLI, F.: Phytocoenologia 10, 487-528 (1982). - GIGON, A.: Ber. Geobot. Inst. ETH Stiftg. Rübel Zürich 50, 149-177 (1983). - GIL, J.-A., VARO, J.: Doc. Phytosociol. 6, 369-386 (1982). - GIMINGHAM, C.H., SMIDT, J.T.DE.: Geobotany 5, 185-199 (1983). - GLAVAČ, V.: Tuexenia 3, 389-406 (1983). - GODRON, M.: Écologie de la végétation terrestre. 197 pp. Paris: CNRS 1984. - GOLDBERG, D.E.: Ecology 63, 942-951 (1982). - GONZALEZ JIMÉNES, E., ESCOBAR, A., PARRA, R.R.: Bol. Soc. Venez. Ci. Nat. 36, 305-324 (1981). - GOODALL, D.W., JOHNSON, R.W.: Vegetatio 49, 197-208 (1982). - GREIG-SMITH, P.: Quantitative Plant Ecology. 3rd ed. 374 pp. Oxford: Blackwell 1983. - GRIGGS, F.T., JAIN, S.K.: Biol. Conserv. 27, 171-193 (1983). - GRODZINSKY, A.M. (ed.): Rol allelopatii v rastenievodstve. 184 pp. Kiev: Naukova Dumka 1982; - Rol toksinov rastitelnogo i mikrosialnogo proishoshchdenija w allelopatii. 152 pp. Kiev: Naukova Dumka 1983. - GROENEWOUD, H. van: Tuexenia 3, 523-528 (1983).

HABER, W.: Z. Kulturtech. Flurbereinig. 1983, 150-158 (1983). - HADAČ, E.: Folia Geobot. Phytotax. 18, 339-361 (1983). - HAEUPLER, H.: Diss. Bot. 65, 1-268 (1982); - Tuexenia 3, 271-288 (1983). - HALLETT, J.G.: Ecology 63, 1400-1410 (1982). - HALVORSON, C.H.: Ecology 63, 423-433 (1982). - HAMMEN, T. van der, PEREZ PRECIADO, A., POLIDORO PINTO, E. (eds.): Studies in Tropical Andes Ecosystems, Vol. 1, 1-345 (1983). - HANAGARTH, W., FISEL, U.: Verh. Ges. Ökologie 11, 185-193 (1983). - HARTOG, C. den, VIERSSEN, W. van: Doc. Phytosociol. 7, 383-386 (1982).

- HATTON, J.C., SMART, N.O.E.: Afr. J. Ecol. 22, 23-30 (1984). - HEINRICH, W.: Landschaftspfl. Naturschutz Thüringen 20, 2-21 (1983). - HEITHAUS, E.R.: Ecology 62, 136-145 (1981). - HENNEKENS, S., HILLEGERS, H., SCHAMINÉE, J.: Levende Nat. 84, 47-54 (1982). - HILBIG, W., KNAPP, H.D.: Flora 174, 1-89 (1983). - HOFFMANN, H.-J., THIELE, H.U.: Decheniana Beih. 27, 1-279 (1982). - HOFMANN, A., MONDINO, G.P., SCOTTA, M.: Dipart. Amb. Terr. Trasp. Emilia Romagna Stud. Doc. 30, 211-223 (1982). - HUENNEKE, L.F.: Bull. Torrey Bot. Club 110, 170-175 (1983). - HUNDT, R.: Verh. Ges. Ökologie 11, 195-206 (1983). - HUNDT, R., HÜBL, E.: Tuexenia 3, 331-342 (1983). - HUNTLEY, B.J.: 101-119 in: see ref. HUNTLEY and WALKER 1982. - HUNTLEY, B.J., MORRIS, J.: 433-455 in: see ref. HUNTLEY and WALKER 1982. - HUNTLEY, B.J., WALKER, B.H. (eds.): Ecology of Tropical Savannas. Ecological Studies, Vol. 42. Berlin, Heidelberg, New York: Springer 1982. - HUSOVÁ, M.: Folia Geobot. Phytotax. 17, 133-135 (1982). - HUXLEY, C.R.: Geobotany 4, 63-73 (1982).

IEHARA, T., MIYATA, I., SUGIMURA, Y.: Jpn. J. Ecol. 33, 109-120 (1983). - ILIJANIĆ, L., ŠEGULJA, N.: Acta Bot. Croatica 42, 63-82 (1983). - ILMAVIRTA, V.: Handbook Veget. Sci. 4, 207-226 (1984). - IVAN, D.: Ber. Symp. Int. Ver. Vegetationskd. 1971, 353-368 (1981). - IZCO, J.: Excerpta Bot. B Sociol. 22, 279-310 (1982a); - Ecol. Mediterr. 8, 289-299 (1982b).

JAGER, J.M. de, HARRISON, T.D.: 456-475 in: see ref. HUNTLEY and WALKER 1982. - JAHN, G.: Forst-Holzw. (Hannover) 38, 142-145 (1983a); - Vortr. Tagungen Arbeitsgem. Forstl. Standorts-Vegetationskd. 9, 1-23 (1983b). - JANZEN, D.H.: Am. J. Bot. 69, 1258-1268 (1982). - JASENIUK, M.A., JOHNSON, E.A.: Can. J. Bot. 60, 2581-2593 (1982). - JORDAN, E.: Tuexenia 3, 101-116 (1983).

KAPLAN, K.: Ber. Geobot. Inst. ETH Stiftg. Rübel Zürich 50, 97-118 (1983). - KÁRPÁTI, I., KÁRPÁTI, K.V.: 195-206 in: see ref. KNAPP (ed.) 1984a. - KENNISH, M.J., LUTZ, R.A. (eds.): Ecology of Barnegat Bay, New Jersey. 300 pp. Berlin, Heidelberg, New York: Springer 1984. - KHARKEVICH, S.S., AZBUKINA, Z.M., KRYLOV, A.G.: Bot. Zh. 68, 1138-1142 (1983). - KHOLOD, S.S.: Bot. Zh. 68, 761-769 (1983). - KINCAID, W.B., CAMERON, G.N., CARNES, B.A.: Ecology 64, 1471-1480 (1983). - KINZELBACH, R. (ed.): Verh. Ges. Ökologie 10, 1-662 (1983). - KLEIN, J.-C.: Phytocoenologia 10, 463-486 (1982a); - Doc. Phytosociol. 6, 191-202 (1982b). - KLING, H.J., GLAWTON, R.: Geogr. Rundschau 34, 461-470 (1982). - KLÖTZLI, F.: Tuexenia 3, 47-65 (1983). - KNAPP, R.: Excerpta Bot. B Sociol. 22, 261-278, 311-319 (1982); - ibid. 23, 39-131, 155-215, 225-300 (1983a); - Tuexenia 3, 477-483 (1983b). - (ed.) Handbook of Vegetation Science, Vol. 4, 370 pp. The Hague, Boston, Lancaster: Junk 1984a; - Afrika-Kartenwerk W 7 (map) Berlin, Stuttgart: Borntraeger 1984b. - KNÖRZER, K.H.: 249-258 in: see ref. KNAPP (ed.) 1984a. - KÖCK, U.-V.: Hercynia 20, 148-177 (1983). - KOROBKOV, A.A., FILIN, V.R.: Bot. Zh. 67, 1351-1361 (1982). - KOSTERMANS, A.Y., SASTROUTOMO, S. (eds.): BIOTROP Special Publ. 17, 1-226 (1982). - KOVÁŘ, P., VOLKOVA, V.G.: Folia Geobot. Phytotax. 16, 125-132 (1981). - KRAUSE, A.: Decheniana 136, 20-29 (1983). - KREEB, K.H.: Vegetationskunde. 331 pp. Stuttgart: Ulmer 1983. - KRIEBITZSCH, W.U., HASEMANN, A.: Verh. Ges. Ökologie 11, 221-237 (1983). - KUHBIER, H., ALCOVER, J.R., GUERA D'ARELLANO TUR, C. (eds.): Biogeography and Ecology of the Pityusic Islands. 704 pp. The Hague, Boston, Lancaster: Junk 1984. - KUHN, N.: Tuexenia 3, 499-522 (1983). - KUTSCHERA-MITTER, L.: 129-160 in: see ref. KNAPP (ed.) 1984a.

LAMOTTE, M.: 415-430 in: see ref. HUNTLEY and WALKER 1982. - LANDOLT, E.: 7-22 in: see ref. KNAPP (ed.) 1984a. - LARSON, J.L., BARNES, W.J.:

Trans. Wis. Acad. Sci. 70, 52-60 (1982). - LAVRENKO, E.M.: Bot. Zh. 66, 313-327 (1981). - LAZARE, J.J., MARTY, R., DAJOZ, R. (eds.): Ecology of Mountain and High Altitude Areas. 420 pp. Bordeaux: GABAS 1984. - LEIPPERT, H.: Bundesanstalt Geowiss. Rohstoffe Hannover Techn. Coop. Project Vol. C 3, 1-91 (1981). - LEÒN, R.J.C., ANDERSON, D.L.: Tuexenia 3: 67-84 (1983). - LEUTERT, A.: Veröff. Geobot. Inst. ETH Stiftg. Rübel Zürich 79, 1-126 (1983). - LEWIS, R.L.: Creation and Restoration of Coastal Plant Communities. 232 pp. Boca Raton: CRC Press 1982. - LI, B., YONG, S., LIU, Z., SUN, H., ZENG, S.: Acta Bot. Sinica 22, 270-279 (1980). - LIETH, H., BARTH, H.: Verh. Ges. Ökologie 11, 265-276 (1983). - LONDO, P.: 45-50 in: see ref. KNAPP (ed.) 1984a. - LUFTENSTEINER, H.W.: 185-194 in: see ref. KNAPP (ed.) 1984a. - LUTHER, H.: Acta Bot. Fennica 123, 1-23 (1983).

MAJER, J.D.: Geobotany 4, 31-43 (1982). - MALATO-BELIZ, J.: Coleçao Parques Natur. 10, 1-92 (1982). - MALYSHEV, L.I., SOBOLEVSKAYA, K.A.: Bot. Zh. 68, 541-545 (1983). - MANN, L.K., SHUGART, H.H.: Vegetatio 52, 77-89 (1983). - MATVEYEVA, E.P.: Bot. Zh. 68, 1692-1694 (1983). - MAYOR, M., NAVA, H.S., ALONSO, J.R.: Bol. Ci. Natur. IDEA (Oviedo) 29, 9-28 (1982). - McCLOSKEY, R.T.: Oikos 41, 233-238 (1983). - McNAUGHTON, S.J.: Ecol. Monogr. 53, 291-320 (1983). - MEDINA, E.: 308 -335 in: see ref. HUNTLEY and WALKER 1982. - MEDINA, E., MOONEY, H. A., VÀSQUES- YÀNES, C. (eds.): Tasks Veget. Sci. 12, 1-254 (1984). - MEDWECKA-KORNÀS, A.: Verh. Ges. Ökologie 11, 217-226 (1983). - MEHER-HOMJI, V.M.: Excerpta Bot. B Sociol 22, 251-254 (1982). - MEHER-HOMJI, V.M., GUPTA, R.K.: Excerpta Bot. B Sociol 22, 255-259 (1982); - ibid. 23, 133-142, 217-224 (1983). - MEISEL, K.: Tuexenia 3, 407-416 (1983). - MENAUT, J.C., CÈSAR, J.: 80-100 in: see ref. HUNTLEY and WALKER 1982. - MILEWSKI, A.V., BOND, W.J.: Geobotany 4, 89-98 (1982). - MIYAWAKI, A. (ed.): Bull. Inst. Environm. Sci. Technol. Yokohama Nat. Univers. 8, 219-378 (1982). - MOLERO, J., VIGO, J.: Trab. Inst. Bot. Barcelona 6, 1-82 (1981). - MOONEY, M., GODRON, M. (eds.): Disturbance and Ecosystems. 310 pp. Berlin, Heidelberg, New York: Springer 1983. - MORAVEC, J.: Preslia 55, 97-122 (1983a); - Verh. Ges. Ökologie 11, 291-304 (1983b). - MORRIS, J.W., BEZUIDENHOUT, J.J., FURNISS, P. R.: 535-553 in: see ref. HUNTLEY and WALKER 1982. - MUCINA, L.: Vegetatio 49, 267-275 (1982a). - Folia Geobot. Phytotax. 17, 21-47 (1982b). - MUELLER-DOMBOIS, D.: Hawaii Bot. Sci. Paper 41, 1-93 (1981a); - Vegetatio 46, 131-140 (1981b); - Hawaiian Bot. Soc. Newsletter 20, 2-6 (1982); - Tuexenia 3, 117-130 (1983). - MÜHLENBERG, M.: Natur Landschaft 57, 295-296 (1982). - MÜLLER, N., WALDERT, R.: Natur Landschaft 58, 419-429 (1983). - MULLER, S.: Doc. Phytosociol. 7, 267-358 (1982).

NAKAGOSHI, N., NEHIRA, K., NAKANE, K.: Mem. Fac. Int. Arts Sci. Hiroshima Univ. IV 8, 87-110 (1983). - NAMZALOV, B.B., ERSHOVA, E.A.: Bot. Zh. 66, 1320-1324 (1981). - NAUMENKO, A.T., NUKHIMOVSKAYA, Y.D.: Bot. Zh. 68, 270-273 (1983). - NÈGRE, R., BAUDIÈRE, A., SERVE, L.: Doc. Phytosociol. 6, 443-475 (1982). - NEUHÄUSL, R.: Verh. Ges. Ökologie 11, 305-310 (1983). - NIEDERMAIER, K.: Tuexenia 3, 241-258 (1983). - NOIRFALISE, A.: Comm. Centre Ecol. Forest. Rur. (Gembloux) 38, 1-15 (1983). - NOY-MEIR, I.: 591-609 in: see ref. HUNTLEY and WALKER 1982. - NUMATA, M.: 161-170 in: see ref. KNAPP (ed.) 1984a.

OBERDORFER, E.: Andrias (Karlsruhe) 2, 53-63 (1983). - OBERDORFER, E., MÜLLER, T.: Süddeutsche Pflanzengesellschaften, Vol. 3. 455 pp. Stuttgart: Fischer 1983. - OHNGEMACH, D., STRAKA, H.: Beiträge zur Vegetations- und Klimageschichte im Gebiet von Puebla-Tlaxcala. 166 pp. Wiesbaden: Steiner 1983. - OKSANEN, J.: Ann. Bot. Fennici 20, 151-155 (1983). - OKSANEN, J., AHTI, T.: Ann. Bot. Fennici 19, 275-301 (1982). - OKSANEN, L.: 51-54 in: see ref. KNAPP (ed.) 1984 a. - OVINGTON, J.E.

(ed.): Ecosystems of the World, Vol. 10, 1-241 Amsterdam, New York: Elsevier 1983. - OWEN-SMITH, N.: 359-404 in: see ref. HUNTLEY and WALKER 1982.

PAELINCKX, D., SOETENS, R.: Bull. Soc. Bot. R. Belg. 116, 74-92 (1983). - PAKARINEN, P.: 35-44 in: see ref. KNAPP (ed.) 1984a. - PALA, S., BOISSONNEAU, A.: Naturaliste Can. 109, 653-659 (1982). - PASSARGE, H.: Tuexenia 2, 257-286 (1982); - ibid. 3, 485-498 (1983). - PELLEW, R.A. P.: Afr. J. Ecol. 21, 41-74 (1983). - PELTIER, J.P.: Feddes Repert. 95, 89-96 (1984). - PIGNATTI, S., WARFA, A.M.: Tuexenia 3, 159-168 (1983). - POLI-MARCHESE, E.: Ecol. Mediterr. 8, 339-354 (1982). - POLOZOVA, T.G.: Bot. Zh. 68, 1503-1511 (1983). - POTT, R.: Oldenburger Jahrb. 83, 357-376 (1983). - PROCTOR, J., ANDERSON, J.M., CHAI, P., VALLACK, H.W.: J. Ecol. 71, 237-260 (1983). - PUTZ, F.E.: Ecology 64, 1069-1074 (1983).

RAABE, E.W., BROCKMANN, C., DIERSSEN, K.: Mitt. Arbeitsgem. Schleswig-Holstein 32, 1-317 (1982). - RABOTNOV, T.A.: 55-60 in: see ref. KNAPP (ed.) 1984a. - RAMÍREZ, C.: 181-184 in: see ref. KNAPP (ed.) 1984a. - RAUS, T.: Tuexenia 3, 259-270 (1983). - READER, R.J., LIETH, H.: 171-180 in: see ref. KNAPP (ed.) 1984a. - REICHMAN, O.J., WHITHAM, T.G., RUFFNER, G.A.: Ecology 63, 687-695 (1982). - REININGHAUS, D., SCHMIDT, M.: Landschaft Stadt 14, 164-185 (1982). - REJMÁNEK, M.: Folia Geobot. Phytotax. 17, 329-333 (1982). - REYNOLDS, C.S.: The Ecology of Freshwater Phytoplankton. 384 pp. Cambridge, London: Cambridge University Press 1984. - RIVERO, J., BARARD, M.A.: J. Bromeliad Soc. 33, 7-29 (1983). - RODI, D., in: Münsingen, p. 659-672. Sigmaringen: Thorbecke 1982. - ROGERS, G.F.: Then and Now. 152 pp. Salt Lake City: University of Utah Press 1982. - ROIG, F.A., BOELCKE, O.: Tuexenia 3, 85-100 (1983). - RUNGE, F.: Die Naturschutzgebiete Westfalens. 4. Aufl. 331 pp. Münster: Aschendorff 1982a. - Tuexenia 2, 69-72 (1982b); - ibid. 3, 199-206 (1983). - RUNGE, M.: Verh. Ges. Ökologie 11, 339-350 (1983); - Düsseldorfer Geobot. Kolloq. 1, 3-10 (1984). - RUTHSATZ, B.: Tuexenia 3, 365-388 (1983a); - Ber. Dtsch. Bot. Ges. 96, 535-576 (1983b).

SAN JOSÉ, J.J., FARIÑAS, M.R.: Ecology 64, 447-453 (1983). - SARMIENTO, G.: Ecology of Neotropical Savannas. 200 pp. Cambridge: Harvard University Press 1983. - SCHIECHTL, M., STERN, R.: Veröff. Österr. MaB Programm 7 (map) (1983). - SCHIEFER, J.: Natur Landschaft 58, 295-300 (1983). - SCHLÜTER, H., BALLER, A.: Landschaftspflege Naturschutz Thüringen 19, 1-11 (1982). - SCHMIDT, W.: Verh. Ges. Ökologie 11, 373-384 (1983a); - Ber. Dtsch. Bot. Ges. 96, 511-533 (1983b); - Natur Landschaft 59, 47-55 (1984). - SCHROEDER, F.-G.: Tuexenia 3, 31-46 (1983). - SCHULTKA, W., HILGER, H.H.: Beitr. Biol. Pflanzen 58, 333-356 (1983). - SCHWAAR, J.: Telma 12, 47-65 (1982a); - Tuexenia 2, 163-172 (1982b); - 259-275 in: see ref. KNAPP (ed.) 1984a. - SCHWABE-BRAUN, A.: Veröff. Naturschutz Landschaftspflege Baden-Württ. 55, 56, 167-237 (1983). - SCOPPOLA, A.: Doc. Phytosociol. 6, 1-130 (1982). - SEIBERT, P.: Naturwiss. Rundschau 35, 147-157 (1982). - SENZOTA, R.B. M.: Afr. J. Ecol. 20, 241-252 (1982). - SHALANDINA, V.T.: Bot. Zh. 66, 52-64 (1981). - SIIRA, J.: Ann. Bot. Fennici 20, 269-280 (1983); - Aquilo Ser. Bot. 20, 1-13 (1984). - SIMS, R.A., COWELL, D.W., WICKWARE, G.M.: Can. J. Bot. 60, 2608-2623 (1982). - SLEMNEV, N.N., BOLD, D., KAZANTSEVA, T.I., FEDOROVA, I.T., JYAKUNIN, G.N.: Bot. Zh. 68, 1533-1537 (1983). - SMITH, L.M., KADLEC, J.A.: J. Appl. Ecol. 20, 673-684 (1983). - SONG, Y.: Tuexenia 3, 131-158 (1983). - SPATZ, G.: Tuexenia 3, 325-330 (1983). - STANYUKOVICH, K.V.: Ekologiya 1982 (4), 85-88 (1982). - STIEPERAERE, H., TIMMERMAN, C.: Bull. Soc. Bot. R. Belg. 116, 62-73 (1983). - STOEHR, M., BÖCKER, R.: Göttinger Flor. Rundbr. 17, 24-39 (1983). - STOIKO, S.M.: Bot. Zh. 68, 1574-1582 (1983). - STRONG, D.R.: Insects on Plants. 320 pp. Oxford: Blackwell 1983. - SUKOPP, H.,

Schriftenr. Dtsch. Rat Landespflege 41, 69-73 (1983). - SUKOPP, H., ELVERS, H. (eds.): Landschaftsentwickl. Umweltforschung Berlin 11, 1-374 (1982). - SULLIVAN, T.P., SULLIVAN, D.S.: J. Appl. Ecol. 19, 33-46 (1982). - SYKES, J.M., HORRILL, A.D., MOUNTFORD, D.M.: J. Ecol. 71, 437-450 (1983). - SZAFRANSKI, F., APEMA, A.K.: Bull. Soc. Bot. R. Belg. 116, 107-117 (1983).

TEAS, H.J. (ed.): Tasks Veget. Sci. 8, 1-200 (1983). - TER BRAAK, C. J.F.: Ecology 64, 454-462 (1983). - TESAŘOVÁ, M., FIALA, K., STUDENÝ, V.: Folia Geobot. Phytotax. 17, 427-430 (1982). - THIÉBAUT, B.: Vegetatio 50, 25-42 (1982). - THOMPSON, S.D.: Ecology 63, 1303-1321 (1982). - TITOV, Y.V., NESKRIABINA, E.S.: Bot. Zh. 67, 1385-1389 (1982). - TOMASELLI, R.: Stud. Trent. Sci. Natur. 58, 471-484 (1981). - TOPOLNY, F.F., IGNATENKO, O.S., SOBAKINSKYKH, V.D.: Bot. Zh. 68, 354-360 (1983). - TÖRÖK, K., PODANI, J.: Doc. Phytosociol. 6, 339-354 (1982). - TRAUTMANN, W.: Natur Landschaft 57, 8-9 (1982). - TRUSZKOWSKI, J.: Acta Theriol. 27, 305-345 (1982). - TÜXEN, J.: Tuexenia 3, 423-436 (1983). - TÜXEN, J., SOLOMONS, J.C.: Jahrb. Naturwiss. Ver. F. Lüneburg 36, 101-137 (1983). - TÜXEN, R.: 3-6 in: see ref. KNAPP (ed.) 1984a.

USLU, T., GÉHU, J.-M., GIMINGHAM, C.H.: Doc. Phytosociol. 6, 477-494 (1982).

VERGER, J.-P.: Doc. Cartograph. Ecol. 25, 51-66 (1982). - VEVLE, O.: Tuexenia 3, 169-178 (1983). - VORONTZOVA, L.I., VASILJEVA, V.D., LOMAKINA, G.A.: Bot. Zh. 68, 1714-1716 (1983). - VOSER-HUBER, M.I.: Diss. Bot. 68, 1-158 (1983).

WALKER, B.H., NOY-MEIR, I.: 556-590 in: see ref. HUNTLEY and WALKER 1982. - WALLER, G.R., CHOU, C.H. (eds.): Proceedings on the Seminar on Allelochemicals and Pheromones. 260 pp. Taipei: Nat. Science Foundation 1982. - WALTER, H., BOX, E.O., HILBIG, W.: Ecosystems of the World, Vol. 5, 193-236. Amsterdam, New York: Elsevier 1983. - WALTER, H., BRECKLE, S.-W.: Ökologie der Erde, Vol. I. 238 pp. Stuttgart: Fischer 1983. - WATTEZ, J.R., GÉHU, J.-M.: Doc. Phytosociol. 6, 263-278 (1982). - WAY, J.M. (ed.): Monographs of the British Crop Protection Council Vol. 26, 1-219, 1983. - WEBER, H.E.: Tuexenia 3, 359-364 (1983). - WERGER, M.J.A., LOUPPEN, J.M.W., EPPINK, J.H.M.: Vegetatio 52, 141-150 (1983). - WEST, N.E. (ed.): Temperate Deserts and Semi-Deserts. 522 pp. Amsterdam, New York: Elsevier 1983. - WESTHOFF, V., SCHAMINÉE, J., SYKORA, K.V.: Tuexenia 3, 179-198 (1983). - WESTHUS, W.: Hercynia 20, 385-391 (1983). - WESTOBY, M., RICE, B., SHELLEY, J.M., HAIG, D., KOHEN, J.L.: Geobotany 4, 75-87 (1982). - WHEELER, B.D., GILLER, K.E.: J. Ecol. 70, 179-200 (1982). - WHITE, F.: Natural Resources Res. Ser. 20, 1-356 (1983). - WHITMORE, T.C.: Tropical Rain Forest of the Far East. 2nd ed. 400 pp. Oxford: Oxford University Press 1984. - WIEGLEB, G.: Excerpta Bot. B. Sociol. 23, 1-37, 145-153, (1983a); - Verh. Ges. Ökologie 11, 479-488 (1983b). - WILMANNS, O.: Ber. Dtsch. Bot. Ges. 96, 281-289 (1983a); - Tuexenia 3, 343-358 (1983b). - WILMANNS, O., KRATOCHWIL, A.: Schriftenr. Dtsch. Rat Landespflege 41, 99-108 (1982). - WINTERHOFF, W.: 227-248 in: see ref. KNAPP (ed.) 1984a. - WITTIG, R., POTT, R.: Decheniana 135, 14-21 (1982). - WOLFF, P.: Abh. Delattina 12, 5-74 (1983).

YAHNER, R.H.: J. Mammalogy 63, 440-445 (1982). - YURTSEV, B.A.: Relict Steppe Complexes of Northeastern Asia. 167 pp. Novosibirsk: Nauka 1981.

ZEIDLER, H.: Tuexenia 3, 455-462 (1982). - ZHANG, J., WANG, J., CHEN, W., LI, B.: Sci. Sinica 24, 1154-1165 (1981). - ZHANG, L.: Acta Phytogeogr. Suecica 73, 1-92 (1983). - ZHENG DU: Erdkunde 37, 34-46 (1983). - ZUMBÜHL, G.: Veröff. Geobot. Inst. ETH Stiftg. Rübel Zürich 81, 1-101 (1983). - ZOLOTUKHIN, N.I.: Bot. Zh. 68, 689-694 (1983).

Professor Dr. RÜDIGER KNAPP
Botanisches Institut I
der Universität
Senckenbergstr. 17-25
D-6300 Giessen, FRG

F. Special Topics

I. Symbiosen: Mykorrhiza

a) Ektomykorrhiza

Von FRANZ H. MEYER

1. Symbiosepartner

Die Endogonaceae waren bisher nur als Partner in der Endomykorrhiza bekannt. Für die Douglasie beschreiben CHU-CHOU und GRACE (1979) eine Ektomykorrhiza, an der *Endogone flammicorona* beteiligt ist.

GIOVANETTI und FONTANA (1981) erzielten mit *Tuber macrosporum* Ektomykorrhizen an einigen Fagales. *Cenococcum graniforme (=C. geophilum)*, ein Mykorrhizapilz mit einem äußerst großen Wirtsspektrum, war in einem *Abies-amabilis*- Bestand der mit Abstand dominierende Mykorrhizapilz (VOGT et al. 1981). Auch in *Eucalyptus*-Anpflanzungen im Südosten der USA trat dieser Pilz häufig auf (ROSE et al. 1981). Auf Halden in Ontario wurde eine neue *Inocybe*- Art *(I. immigrans)* entdeckt, die mit *Betula papyrifera, Populus balsamifera* und *Salix* sp. in Symbiose lebt (MALLOCH 1982). Mit *Hebeloma alpinum* und *H. marginatulum* konnten DEBAUD et al. (1981) an *Dryas octopetala* Ektomykorrhizen erzielen.

Einige Angaben liegen vor über Gastromycetales als Pilzpartner der Ektomykorrhiza. So beschreiben CHU-CHOU und GRACE (1981, 1982) *Hysterangium, Hydnangium* und *Hymenogaster* als Pilzpartner bei *Eucalyptus*-Arten und GAIE und HEINEMANN (1980) *Astraeus hygrometricus* an *Betula pendula*. *Pisolithus arhizus (= P. tinctorius)*, ein weltweit besonders auf armen Böden verbreiteter Mykorrhizapilz, ist in Schweden ausgesprochen selten (ARVEBY 1980).

In den letzten Jahren sind weitere Belege für Pflanzengattungen mit Ektomykorrhiza erbracht worden. Ektomykorrhizen treten nicht nur bei allen Pinaceen, einigen Cupressaceen, den Fagaceen, Betulaceen, Juglandaceen und Salicaceen auf, sondern kommen auch bei folgenden Gattungen vor:

Ulmaceae: *Ulmus*
Dipterocarpaceae: *Monotes* (HÖGBERG 1982)
Cistaceae: *Helianthemum, Cistus* (ROSELL 1981)
Tiliaceae: *Tilia*
Euphorbiaceae: *Poranthera* (WARCUP 1980)
Ericaceae: *Arbutus, Arctostaphylos* (MOLINA und TRAPPE 1982)
Rosaceae: *Crataegus, Dryas* (DEBAUD et al. 1981)
Malus, Pyrus, Sorbus
Myrtaceae: *Eucalyptus, Leptospermum* (WARCUP 1980)
Mimosaceae: *Acacia* (WARCUP 1980)
Caesalpiniaceae: *Afzelia, Anthonotha, Brachystegia, Gilbertiodendron, Julbernardia* (HÖGBERG 1982), *Monopetalanthus, Paramacrolobium*
Sapindaceae: *Allophylus*
Aceraceae: *Acer*
Rhamnaceae: *Pomaderris* (WARCUP 1980)

Progress in Botany, Vol. 46

Goodeniaceae: *Goodenia* (WARCUP 1980)
Stylidiaceae: *Stylidium* (WARCUP 1980)
Asteraceae: Gattungen der Tribus Inuleae (WARCUP und McGEE 1983).

Besonders in Baumschulen und auf für Ektomykorrhizen weniger günstigen Standorten wurden wiederholt Ektendomykorrhizen beobachtet. DANIELSON (1982) zieht aus anatomischen und physiologischen Merkmalen eines isolierten Ektendomykorrhizapilzes den Schluß, daß er zu den Ascomyceten gehören müsse. *Vaccinium*- Arten zeichnen sich in der Regel durch Endomykorrhizen aus. In Florida entdeckten LEE et al. (1982) an *V. ashei* eine Ektendomykorrhiza, deren Pilzpartner *Elaphomyces persoonii* sein dürfte.

2. Anatomie

NYLUND (1980) wies an Ektomykorrhizen von *Picea abies* nach, daß im Bereich des Hartigschen Netzes die Rindenzellen durch den Pilz nicht völlig voneinander isoliert werden, sondern daß Tüpfel mit Plasmodesmen vorhanden sind. Der zwischen den Rindenzellen noch bestehende Stoffaustausch soll die Hypertrophie der Rindenzellen und ihre längere Lebensdauer nach Bildung einer Ektomykorrhiza ermöglichen. Mit Tannin gefüllte Zellen im Bereich des Hartigschen Netzes hatten keine Tüpfel. Dagegen konnten PICHE et al. (1983) an *Pinus-strobus*-Mykorrhizen (Pilzpartner *Pisolithus tinctorius)* im Bereich des Hartigschen Netzes keine Plasmodesmen nachweisen. Nach voller Entwicklung des Hartigschen Netzes wurde das Cytoplasma der Rindenzellen nekrotisch.

Die Feinstruktur der von *Cenococcum geophilum* an *Tilia cordata* gebildeten Mykorrhizen beschreibt PIGOTT (1982) und geht dabei insbesondere auf die Ablagerungen phenolischer Substanzen in den äußeren Rindenzellen ein.

WILSON und HARLEY (1983) verfolgen bei *Fagus sylvatica* die Abwandlungen des Wurzelaufbaus nach Infektion mit Mykorrhizapilzen. Das Volumen der Rindenzellen nimmt nach Infektion im Vergleich zur angrenzenden pilzfreien Region beträchtlich zu. Die Zellvergrößerung ist in erster Linie bedingt durch Ausdehnung der Rindenzellen in radialer Richtung. Die für Kiefernwurzeln charakteristische gabelige Verzweigung halten FAYE und RANCILLAC (1981) dagegen für genetisch bedingt, da sie auch ohne Anwesenheit eines Pilzpartners entstehen kann. Hierfür sprechen auch die Ergebnisse von SARJALA und KUPILA-AHVENNIEMI (1982), wonach die gabeligen Verzweigungen bei einjährigen Kiefern-Sämlingen keine Beziehungen zum Auftreten eines Pilzmantels oder Hartigschen Netzes zeigten.

Über die Feinstruktur der Mykorrhizen von Monotropoideen *(Monotropa, Pterospora* und *Sarcodes)* berichten DUDDRIDGE und READ (1982) sowie ROBERTSON und ROBERTSON (1982). Es wurde kein typisches intrazelluläres Eindringen des Mykorrhizapilzes in die Epidermiszellen der Wirtspflanzen beobachtet, sondern Einstülpungen der Zellwände, in denen sich die Pilzhyphen befinden. Von diesen Wandeinstülpungen gehen labyrinthartige Protuberanzen aus, deren Strukturen an Transfer-Zellen erinnern. Die Protuberanzen treten besonders im Frühsommer zur Zeit der Entwicklung des Blütenschaftes auf und dürfen den Kohlenhydrattransfer vom Pilz zur Wirtspflanze erleichtern. Die Anatomie der Monotropoideen-Mykorrhizen unterscheidet sich damit deutlich von der anderer Ericaceen mit Strukturelementen einer Ektomykorrhiza wie *Arbutus* und *Pyrola*. Den Transferzellen ähnliche Strukturen fanden ASHFORD und ALLAWAY (1982) in der Ektomykorrhiza von *Pisonia grandis* (Nyctaginaceae). GOD-

BOUT und FORTIN (1983) impften *Alnus crispa* und *A. rugosa*, die Rhizothamnien mit *Frankia* enthielten, mit potentiellen Mykorrhizapilzen. Ein Hartigsches Netz entwickelte sich nur teilweise, lediglich nach Impfung mit *Alpova diplophloeus* war es gut ausgeprägt und beschränkte sich auf die äußere Rindenschicht. Diese Autoren diskutieren die anatomischen Charakteristika des Hartigschen Netzes bei Gymno- und Angiospermen in Relation zu den symbiontischen Beziehungen.

3. Physiologie

Wurzeln von Kiefern-Sämlingen fördern auf einem Agar-Medium die Keimung der Basidiosporen des Mykorrhizapilzes *Telephora* beträchtlich, nicht dagegen die Wurzeln von Kräutern und Gräsern (BIRRAUX und FRIES 1981). Nach RUDAWSKA (1982) hemmen Aminosäuren, insbesondere Glutamin- und Asparaginsäure, sowie Harnstoff die Produktion von Auxin und Cytokininen bei Reinkulturen von Mykorrhizapilzen. Ein möglicher Grund für den Rückgang der Ektomykorrhizen nach N-Düngung kann daher der Anstieg des Aminosäuren-Gehaltes der Wurzel sein und die davon ausgehende Beeinträchtigung der Auxinproduktion der Mykorrhizapilze. Bei Sämlingen von *Pinus resinosa* veränderte sich das Verhältnis von Kurz- zu Langwurzeln nach Impfung mit dem Mykorrhizapilz *Pisolithus* zugunsten der Kurzwurzeln (SOHN 1981). Auch aus den Versuchen von FAYE und RANCILLAC (1983) geht hervor, daß die Bildung von Kurzwurzeln durch Mykorrhizapilze stimuliert wird.

PICHE et al. (1981) konnten im Rindengewebe der Kurzwurzeln von *Pinus strobus*, einerlei ob sie mit Mykorrhizapilzen beimpft worden waren oder nicht, keine Stärkekörner nachweisen, reichlich vorhanden waren sie dagegen im Zentralzylinder der Kurzwurzeln und sowohl im Rindengewebe als auch im Zentralzylinder der Langwurzeln. Das Verteilungsmuster der Stärke in den verschiedenen Wurzelarten sollte bedacht werden bei der Diskussion über Beziehungen zwischen Kohlenhydrathaushalt der Wirtspflanze und Mykorrhizabildung. GILTRAP (1982) testete die Fähigkeit von zahlreichen Mykorrhizapilzen zur Produktion von Polyphenoloxidasen. Diese Fähigkeit fehlte den meisten Mykorrhizapilzen oder war nur schwach entwickelt. Einige Pilze jedoch, insbesondere sieben von acht *Lactarius*- Arten, synthetisierten dieses Enzym in gleicher Intensität wie saprophytische Pilze. Bei der Birke zeigten diese *Lactarius*-Arten oft auch eine intrazelluläre Ansiedlung.

Mycel von *Laccaria laccata* und Kulturfiltrate dieses Pilzes induzieren in den Wurzeln von Douglasien-Sämlingen die Anreicherung von phenolischen Substanzen. Diese sind offensichtlich ein Grund für die geringere Infektionsrate der von *Laccaria laccata* beeinflußten Wurzeln mit pathogenen Pilzen wie *Fusarium oxysporum* oder *Pythium irregulare* (SYLVIA und SINCLAIR 1983). Die Rolle des durch Mykorrhizapilze gebildeten Äthylens bei den symbiontischen und pathogenen Wechselbeziehungen zwischen Pilz und Wirt untersuchten GRAHAM und LINDERMAN (1980, 1981).

McCREIGHT und SCHROEDER (1982) testeten das Wachstum von neun Mykorrhizapilzen bei Gegenwart von Cd, Pb und Ni. Alle Pilze wurden in ihrer Entwicklung unterdrückt durch 350 ppm Cadmium, bei 200 ppm Blei wurden fünf Pilzarten unterdrückt und bei 2000 ppm Blei *Cenococcum geophilum*, *Laccaria laccata* und *Suillus luteus*.

REID et al. (1983) wiesen nach, daß bei *Pinus-taeda*-Pflanzen mit Mykorrhizen im Vergleich zu mykorrhizafreien die CO_2-Assimilation beträchtlich erhöht ist, gleichzeitig wird ein höherer Anteil an [^{14}C] der Wurzel zugeleitet und die Wurzelatmung stark angeregt: An 10 Monate alten Pflanzen lag die Menge an Atmungs-$[^{14}C]O_2$ (bezogen auf das Wurzelgewicht) 3,6mal so hoch wie bei Pflanzen ohne Mykorrhiza. Nach FOGEL und HUNT (1983) wurden in einem Douglasien-Bestand in Oregon 73% der Nettoprimärproduktion für Aufbau und Lebensfunktionen der Wurzeln und Mykorrhizen verwandt.

Rhizomorphen des Mykorrhizapilzes *Suillus bovinus* sind befähigt, ihrer Wirtspflanze H_2O zuzuleiten, wie mit Hilfe von $[^3H]_2O$ ermittelt wurde (DUDDRIDGE et al. 1980). An abgeschnittenen Buchenmykorrhizen, die einige Tage im Wasser gehalten wurden, steigt die Atmung nach Phosphatzufuhr (HARLEY und McCREADY 1981). Neun mykorrhizabildende Pilzarten aus fünf Gattungen (*Amanita*, *Boletus*, *Cenococcum*, *Pisolithus* und *Suillus*) synthetisieren Fe-chelatisierende Substanzen, wobei die Menge der gebildeten Siderophoren sich gegenläufig zur Eisenkonzentration im Medium verhält. $CaCO_3$ im Medium steigert die Menge der chelatisierenden Substanzen signifikant (SZANISZLO et al. 1981). Oxalatproduktion durch Mykorrhizapilze halten MALAJCZUK und CROMACK (1982) sowie O'CONNELL et al. (1983) für sehr bedeutungsvoll zur Mobilisierung von Phosphat in *Eucalyptus*- und *Pinus-radiata*-Wäldern, deren Böden Phosphat stark absorbieren bzw. an Phosphat verarmt sind.

VREELAND et al. (1981) injizierten radioaktiven Phosphor in das Phloem von solchen Bäumen (*Abies concolor*, *Pinus jeffreyi*), deren Mykorrhizapilz zugleich eine Symbiose mit der chlorophyllosen *Sarcodes sanguinea* eingeht. Die Injektion erfolgte zur Zeit des stärksten Längenwachstums der *Sarcodes*, und ein Transport markierten Phosphors vom Baum zur *Sarcodes* konnte eindeutig nachgewiesen werden.

4. Ökologie

Wie MASON et al. (1982) sowie LAST et al. (1983) zeigten, liegt beim Auftreten der Fruchtkörper von Mykorrhizapilzen an Birken eine Sukzession vor, derart, daß sich mit zunehmendem Alter der Birken die Pilzpopulation ändert. In Baumschulen tritt häufiger eine Ektendomykorrhiza auf, die nach dem Auspflanzen in den Waldboden verschwindet (THOMAS et al. 1983, THOMAS und JACKSON 1983).

Die bekannte Trockenresistenz der *Cenococcum-geophilum*- Mykorrhiza wurde von PIGOTT (1982) bestätigt. PARKE et al. (1983) verglichen die Nettophotosynthese von Douglasien-Sämlingen mit und ohne Mykorrhiza bei unterschiedlicher Bewässerung. Den Trockenstreß ertrugen Sämlinge mit Mykorrhiza wesentlich besser, die CO_2-Aufnahmerate lag bei ihnen zehnmal höher als bei mykorrhizafreien Sämlingen. *Pinus-ponderosa*-Sämlinge, geimpft mit *Pisolithus tinctorius*, überstanden das Auspflanzen in Prärie-Boden mit deutlich geringeren Ausfällen als solche ohne *Pisolithus* (BAER und OTTA 1981). Auf Sand-Podsol im Zentralamazonas-Gebiet sind die herrschenden Bäume mit Ektomykorrhizen besetzt, gleichzeitig häuft sich Rohhumus an. Im Gegensatz dazu stehen die meisten Standorte der terra firme, wo die Bäume keine Ektomykorrhizen tragen und die Streuzersetzung rasch abläuft. Offensichtlich wird die Streuzersetzung durch die Ektomykorrhiza gebremst, was sich für den Nährstoffkreislauf tropischer Wälder als sehr nützlich erweisen könnte (SINGER und SILVA ARAUJO 1979).

HARVEY et al. (1981) ermittelten, daß die Zahl der Ektomykorrhizen mit dem Humusgehalt von Waldböden ansteigt bis auf etwa 45% organischer Substanz und dann wieder abfällt. Diese Relationen sind auf trockenen

Standorten besser ausgeprägt als auf feuchten. Abbrennen des Humus schmälert den Anteil der Ektomykorrhizen (HARVEY et al. 1980, MALAJCZUK und HINGSTON 1981).

Volldüngung (N,P,K,Ca) verbessert zwar die oberirdische Wuchsleistung der Buche, aber 9 Jahre nach der Düngung war der Anteil der Feinwurzeln nur halb so groß wie auf der Kontrollfläche, und die Mykorrhizen mit viel Außenmycel waren z.T. ersetzt durch glatte, so daß die Aufnahmekapazität herabgesetzt worden ist (BLAISE und GARBAYE 1983). Nach N-Düngung nimmt die Bildung neuer Mykorrhizen ab (BRADY und LOEWENSTEIN 1983, ALEXANDER und FAIRLEY 1983). DIXON et al. (1981) düngten *Quercus*-Sämlinge mit Stickstoff entweder über das Blatt oder über das Wurzelsubstrat. Bei der Blattdüngung waren Fruktose-Gehalt und Mykorrhizabesatz der Wurzeln signifikant höher als bei Düngung des Wurzelsubstrates. Die Autoren empfehlen daher eine N-Düngung über das Blatt, wenn der negative Einfluß des Stickstoffs auf die Mykorrhiza vermieden werden soll.

In erkrankten und absterbenden Tannenbeständen beobachtete BLASCHKE (1981) Verfall von Feinwurzeln und intrazellularen Befall mit parasitischen Phyco- und Zygomyceten. Nach SCHÜTT et al. (1983) ist das Fichtensterben gekoppelt mit einem Rückgang an Mykorrhizen und Absterben von Feinwurzeln. Als eine Ursache für den Feinwurzelverfall diskutiert MEYER (1984) nachlassende Assimilationsleistung der durch Luftverunreinigungen geschädigten Nadeln und den Stickstoffeintrag in den Waldboden, der die Wuchstoffproduktion der Mykorrhizapilze hemmt, worauf der Zufluß löslicher Kohlenhydrate zum Pilz nachläßt. Dieses könnte parasitische Tendenzen beim Pilz auslösen (z.B. Synthese von Zellulase), die zum Verfall von Feinwurzeln führen. MAHONEY et al. (1983) begasten Sämlinge von *Pinus taeda* mit Schwefeldioxid und Ozon, wobei die Hälfte der Versuchspflanzen mit *Pisolithus tinctorius* geimpft worden war. Das Wurzelgewicht der mykorrhizafreien Sämlinge wurde bei allen Begasungen reduziert, nicht jedoch bei den Sämlingen mit Mykorrhiza. Die Mykorrhizabildung wurde durch die Begasung nicht beeinträchtigt.

Bodenentseuchung mit Fungiziden oder Nematiziden schädigt in der Regel auch die Mykorrhizapilzflora (ILOBA 1980). MARX und ROWAN (1981) testeten die Empfindlichkeit der Mykorrhizabildner *Pisolithus* und *Telephora* gegenüber unterschiedlich konzentrierten Bodenentseuchungsmitteln. Nach DIXON et al. (1981a,b) liegt die Trockensubstanzproduktion bei *Quercus-velutina*-Sämlingen deutlich höher, wenn nach Bodenentseuchung mit *Pisolithus* geimpft worden ist und sich Mykorrhizen gebildet haben. Die Widerstandsfähigkeit verschiedener Mykorrhizapilze gegenüber einer Reihe von Herbiziden untersuchten KELLEY und SOUTH (1980) sowie CUDLIN et al. (1980). In der Regel lag die Toleranzgrenze der Mykorrhizapilze höher als die in der Praxis üblichen Anwendungskonzentrationen der Herbizide. Gramoxone wurde besser ertragen als Dithane M-45.

Literatur

ALEXANDER, I.J., FAIRLEY, R.I.: Plant Soil 71, 49-53 (1983). - ARVEBY, A.S.: Sv. Bot. Tidskr. 74, 329-333 (1980). - ASHFORD, A.E., ALLAWAY, W.G.: New Phytol. 90, 511-519 (1982). BAER, N.W., OTTA, J.D.: For. Sci. 27, 277-280 (1981). - BIRRAUX, D., FRIES, N.: Can. J. Bot. 59, 2062-2064 (1981). - BLAISE, T., GARBAYE, J.: Acta Oecol. Plant. 4, 165-170 (1983). - BLASCHKE, H.: Eur. J. For. Path. 11, 375-379 (1981). - BRADY, Th.F., LOEWENSTEIN, H.: Commun. Soil Sci. Plant Anal. 14, 101-114 (1983).

CHU-CHOU, M., GRACE, L.J.: New Zeal. J. For. Sci 9, 344-347 (1979); - ibid. 11, 186-190 (1981); - Soil Biol. Biochem. 14, 133-138 (1982). - CUDLIN, P., MEJSTRIK, V., SASEK, V.: Česka Mykol. 34, 191-198 (1980).

DANIELSON, R.M.: Can. J. Bot. 60, 7-18 (1982). - DEBAUD, J.C., PEPIN, R., BRUCHET, G.: Can. J. Bot. 59, 1014-1020 (1981). - DIXON, R.K., GARRETT, H.E., BIXBY, J.A., COX, G.S., TOMPSON, J.G.: For. Sci. 27, 617-624 (1981a). - DIXON, R.K., WRIGHT,G.M., GARRETT, H.E., COX, G.S., JOHNSON, P.S., SANDER, I.L.: Can. J. For. Res. 11, 487-491 (1981b). - DUDDRIDGE, J.A., MALIBARI, A., READ, D.J.: Nature 287, 834-836 (1980). - DUDDRIDGE, J.A., READ, D.J.: New Phytol. 92, 203-214 (1982).

FAYE, A.D.M., RANCILLAC, M.: New Phytol. 87, 557-566 (1981); - Plant Soil 71, 501-505 (1983). - FOGEL, R., HUNT, G.: Can. J. For. Res. 13, 219-232 (1983).

GAIE, W., HEINEMANN, P.: Bull. Soc. Bot. Belg. 113, 161-165 (1980). - GILTRAP, N.J.: Trans. Br. Mycol. Soc. 78, 75-82 (1982). - GIOVANETTI, G., FONTANA, A.: Allionia 24, 13-18 (1981). - GODBOUT, C., FORTIN, J.A.: New Phytol. 94, 249-262 (1983). - GRAHAM, J.H., LINDERMAN, R.G.: Can. J. Microbiol. 26, 1340-1347 (1980); - Can. J. Bot. 59, 149-155 (1981).

HARLEY, J.L., McCREADY, C.C.: New Phytol. 88, 675-682 (1981). - HARVEY, A.E., LARSEN, M.J., JURGENSEN, M.F.: Can. J. For. Res. 10, 436-440 (1980). - HARVEY, A.E., JURGENSEN, M.F., LARSEN, M.J.: For. Sci. 27, 442-448 (1981). - HÖGBERG, P.: New Phytol. 92, 407-415 (1982).

ILOBA, C.: Flora 170, 127-134 (1980).

KELLEY, W.D., SOUTH, D.B.: Weed Sci. 28, 599-602 (1980).

LAST, F.T., MASON, P.A., WILSON, J.: Plant Soil 71, 9-21 (1983). - LEE, A.J., DAVIES, F.S., KIMBROUGH, J.M.: Hort. Sci. 17, 951-953 (1982).

MAHONEY, M.J., SKELLY, J.M., CHEVONE, B.I.: Phytopathology 73, 370 (1983). - MALAJCZUK, N., CROMACK, K.: New Phytol. 92, 527-531 (1982). - MALAJCZUK, N., HINGSTON. F.J.: Aust. J. Bot. 29, 453-462 (1981). - MALLOCH, D.: Can. J. Bot. 60, 40-45 (1982). - MARX, D.H., ROWAN, S.J.: For. Sci. 27, 167-176 (1981). - MASON, P.A., LAST, F.T., PELHAM, J., INGLEBY, K.: For. Ecol. Manage. 4, 19-40 (1982). - McCREIGHT, J.D., SCHROEDER, D.B.: Environ. Exp. Bot. 22, 1-8 (1982). - MEYER, F.H.: Allg. Forstz. 39, 212-228 (1984). - MOLINA, R., TRAPPE, J.M.: New Phytol. 90, 495-510 (1982).

NYLUND, J.-E.: New Phytol. 86, 373-378 (1980).

O'CONNELL, A.M., MALAJCZUK, N., GAILITIS, V.: Oecologia 56, 239-244 (1983).

PARKE, J.L., LINDERMAN, R.G., BLACK, C.H.: New Phytol. 95, 83-95 (1983). - PICHE, Y., FORTIN, J.A., LAFONTAINE, J.G.: New Phytol. 88, 695-704 (1981). - PICHE, Y., PETERSON, R.L., HOWARTH, M.J., FORTIN, J.A.: Can. J. Bot. 61, 1185-1193 (1983). - PIGOTT, C.D.: New Phytol. 92, 501-512 (1982); - ibid. 92, 513-517 (1982).

REID, C.P.P., KIDD, F.A., EKWEBELAM, S.A.: Plant Soil 71, 415-432 (1983). - ROBERTSON, D.C., ROBERTSON, J.A.: New Phytol. 92, 539-551 (1982). - ROSE, R.W., VAN DYKE, C.G., DAVEY, C.B.: Can. J. Bot. 59, 683-688 (1981). - ROSELL, A.: Collect. Bot. 12, 161-166 (1981). - RUDAWSKA, M.: Acta Physiol. Plant. 4, 11-20 (1982).

SARJALA, T., KUPILA-AHVENNIEMI, S.: Aquilo, Ser. Bot. 18, 16-26 (1982). - SCHÜTT, P., BLASCHKE, H., HOQUE, E., KOCH, W., LANG, K.J., SCHUCK, H.J.: Forstw. Cbl. 102, 158-166 (1983). - SINGER, R., SILVA ARAUJO, I. de J.: Acta Amazonica 9, 25-42 (1979). - SOHN, R.F.: Can. J. Bot. 59, 2129-2134 (1981). - SYLVIA, D.M., SINCLAIR, W.A.: Phytopathology 73, 384-389, 390-397 (1983). - SZANISZLO, P.J., POWELL, P.E., REID, C.P.P., CLINE, G.R.: Mycologia 73, 1158-1174 (1981).

THOMAS, G.W., JACKSON, R.M.: New Phytol. 95, 223-229 (1983). - THOMAS, G.W., ROGERS, D., JACKSON, R.M.: Plant Soil 71, 319-323 (1983).

VOGT, K.A., EDMONDS, R.L., GRIER, C.C.: Holarct. Ecol. 4, 167-173 (1981). - VREELAND, P., KLEINER, E.F., VREELAND, H.: Environ. Exp. Bot. 21, 15-26 (1981).

WARCUP, J.H.: New Phytol. 85, 531-536 (1980). - WARCUP, J.H., McGEE, P.A.: New Phytol. 95, 667-672 (1983). - WILSON, J.W., HARLEY, J.L.: New Phytol. 95, 673-695 (1983).

Professor Dr. FRANZ H. MEYER
Institut für Landschaftspflege
und Naturschutz der Universität
Herrenhäuser Str. 2
D-3000 Hannover, FRG

b) Endomycorrhizae

By F. SCHÖNBECK

1. VA-Mycorrhizae

a) Taxonomy of Endophytes

Synoptic keys for the identification of zygomycetous (vesicular-arbuscular) mycorrhizal fungi have been published by TRAPPE (1982). These include the genera of Endogonaceae, the genera that have affinity with or may be mistaken for Endogonaceae *(Acaulospora, Complexipes, Endogone, Eutrophospora, Gigaspora, Glaziella, Glomus, Modicella, Sclerocystis)*. Moreover keys for the species of the genera known to include mycorrhiza formers are presented *(Acaulospora, Gigaspora, Glomus, Sclerocystis*. The synoptic key offers an alternative to dichotomous keys, because it can be expanded, corrected and new characteristics valuable for species identification can be added without rewriting. This is of advantage because of the proliferation of new species publications (Outlook: 200 species in 1990). Morphological structures are used to identify the VA-fungi, the basis for the taxonomy is formed by the soil-borne resting spores. The spores of *Acaulospora, Eutrophospora* and *Gigaspora* have been termed azygospores, the spores of *Glaziella, Glomus* and *Sclerocystis* are regarded as chlamydospores. A detailed description is given by TRAPPE and SCHENCK (1982).

Six new species in the genus *Glomus, G. intraradices, G. aggregatum, G. fecundisporum, G. tortuosum, G. claroideum* and the new species *Gigaspora albida* have been found in Florida and were described for the first time by SCHENCK and SMITH (1982b).

ROSE and TRAPPE (1980) have reported three endomycorrhizal *Glomus spp.* associated with actinorhizal scrubs.

Two new species, *Acaulospora foveata* and *A. tuberculata*, from tropical America are presented by JANOS and TRAPPE (1982). BERCH and FORTIN (1983) undertake a lectotypification of *Glomus macrocarpus*. As a consequence of the examination of type materials of previously proposed synonyms of this species, three new combinations are made: *Glomus australe* (Berk.) Berch, *Glomus versiforme* (Karst.) Berch, and *Glomus tenebrosum* (Thaxter) Berch.

b) Histology

SWARD (1981a,b) studied the structure of the spores of *Gigaspora margarita*. The dormant spore possesses a complex spore wall composed of four layers, an inner and outer layer that contain polysaccharides, protein and lipid, a laminated chitinuous layer and a so-called cementing layer. In certain areas the cytoplasma forms two dense regions named the synthezing and nuclear centers which contain diffe-

Progress in Botany, Vol. 46

rent proportions of various organelles. Following spore activation there is considerable redistribution of cytoplasma with a subsequent accumulation of cytoplasmic and wall material in the region of the bulbous-shaped attachment, accompanied by changes in the inner wall layer. The ragged appearance of the inner layer adjacent to the developing germ tube suggests that wall degradation is of enzymic rather than of mechanical nature.

During its symbiotic association with the roots of *Vitis vinifera* the VAM fungus *Glomus fasciculatus* exhibits a range of ultrastructural changes in its wall. BONFANTE-FASOLO and GRIPPIOLO (1982) observed a progressive thinning starting from the extraradical phase up to the final intracellular phase, changes in its texture from being lamellar to amorphous and showing different responses to cytochemical reactions, suggesting a gradual simplification of the wall compounds. In the intercellular and arbuscular hyphae chitin and chitosan are not demonstrable. NEMEC (1981) has not found chitin in arbuscular hyphal walls of *Glomus etunicatus* and BONFANTE-FASOLO (1982) has never observed fibrillar structures similar to chitin ribbons in the arbuscular hyphal walls of *G. fasciculatus* associated with *Ornithogalum umbellatum*.

Interface enzymatic activities have been studied in mycorrhizal associations. Alkaline phosphatase is present in the fungus during the growth stimulating phase (GIANINAZZI et al. 1979), peroxidating activity was absent in the matrix and host walls in contact with the fungus (BONFANTE-FASOLO and SCANNERINI 1980). Moreover, ATPases were localized in the host plasmalemma surrounding the finer arbuscule branches in onion and sycamore mycorrhiza (MARX et al. 1982). These observations strengthen the hypothesis that nutrient exchange in VA-mycorrhizas must occur across the living host-fungus interface by an active transport mechanism.

Bacterium-like organelles (BLOs) were found by electron microscopy in the vesicular-arbuscular (VA) mycorrhizal fungus *Glomus caledonius* (MACDONALD and CHANDLER 1981). They occurred free in the cytoplasma of the reproductive spores, in the intercellular hyphae and the thick arbuscule branches in infected roots. They are irregular coccal and embraced by a conspicious cell membrane and a simple cell wall, never enclosed in a host membrane. The fungal cytoplasma did not show any cytopathic effects in the vicinity of the BLOs. The BLOs were found in spores of several isolates of VA-mycorrhizal fungi and appear to be very common (MACDONALD et al. 1982).

TOMMERUP and ABBOTT (1981) found new hyphae within the old hyphae of various VA-mycorrhizal fungi in the cortex of roots which have been dried and stored for 6 months before being returned to normal conditions of growth.

LIM et al. (1983) reported about intrahyphal hyphae in roots of *Trifolium repens* infected with *Glomus fasciculatus*. The intercellular hyphae contain intracellular hyphae that may be either thin- or thick-walled. They possess differentiated protoplasts with nuclei, vacuoles and glycogen deposits. Limited space for growth possibly induce a fungus to proliferate intrahyphally.

Some features of the fine mycorrhizal endophyte *Glomus tenuis* distinguish it from coarse VAM fungi (GIANINAZZI-PEARSON et al. 1981). The unseptate hyphae are smaller and have thinner walls, so that many of them can develop together in the same intercellular space. The walls show a distinct two-layered structure after a test for polysaccharides.

It appears that the vesicles of the Endogonaceae, in addition to their high lipid content, could act as reservoirs for many elements. The principal elements detected by laser-probe mass spectrographs are K, Ca, Mg, and Na (STRULLU et al. 1983).

Development and anatomy of infections by *Glomus mossae*, *G. caledonius* and *G. fasciculatus* on soybean are described by CARLING and BROWN (1982).

c) Physiology and Effects on Plants

Recent evidence indicates that mycorrhizal fungi are capable of limited independent spreading in soil, probably via germination, growth and the formation of secondary spores (WARNER and MOSSE 1980, OCAMPO and HAYMAN 1981). It is suggested that this spreading may be rudimentary saprophytic growth from an organic base. This idea is supported by HEPPER (1983), who reported that for *Glomus caledonius* limited independent growth in vitro is possible.

The endomycorrhizal inoculum usually contains hyphae, vesicles and arbuscules in the root pieces and mostly chlamydospores in the soil. MANJUNATH and BAGYARAJ (1981) reported that chlamydospores are the important component for inoculation of onions with *Glomus fasciculatus*, because plants inoculated with this spores produced more dry matter.

BIERMANN and LINDERMAN (1983b) found that root pieces colonized by species of VAM-fungi forming intraradial vesicles and separated interradial vesicles are highly infective in contrast to extraradial ones.

The inoculum potential and infectivities of six *Glomus* species have been investigated by DANIELS et al. (1981). The differences in inoculum potential are not correlated with differences in spore viability or germinability. Factors influencing development of mycorrhizal infection are the numbers and rate of formation of primary points of infection, the rate of hyphal spread within or on the root and development of secondary infections, and the size of infection unit associated with each entry point (BUWALDA et al. 1982, WILSON and TRINICK 1983a). WILSON and TRINICK (1983b) found that species of *Glomus* differ in forming infection points and secondary infection points.

Studies of the interactions between VAM-fungi showed that irrespective of species the fungus which infected most rapidly in each case had the competitive advantage, probably due to its prior occupancy of the root (ABBOTT and ROBSON 1984).

WARNER and MOSSE (1982) showed the overriding influence of host species and the secondary effects of root density in determing spread of mycorrhiza infection.

JOHNSON et al. (1982b) found that VAM formation is reduced during early stages of flower bud development in *Chrysanthemum*, but not during later bud expansion. These results suggest that reduction of VAM formation during early flower bud develpment resulted from decrease in the availability and net leakage of metabolites necessary to sustain the mycorrhizal fungus during pre- and post-infection.

It is proposed that the mechanism for P inhibition of VAM formation is associated with membrane-mediated decrease in root exudation. GRA-

HAM et al. (1981) confirmed that the rate of exudation is directly related to fundamental changes in root membrane permeability controlled by P and demonstrated that mycorrhizal infection was highly correlated with initial differences in root exudation.

SCHWAB et al. (1983) produced evidence that an amendment in P supply results in changes of content of fructose and arabinose in the exudates of Sudan grass, but has no effect on the sugars of the extract of the grass.

AZCON and OCAMPO (1981) studied the factors affecting the VAM formation of 13 wheat cultivars. They found that absence of mycorrhizal infection in some wheat varieties was associated with lack of sugar exudation from the roots rather than with the sugar content of the roots. VA infection led to a decrease in reducing sugar content of root extracts, and this effect was correlated with the degree of mycorrhizal infection.

Studies by GRAHAM et al. (1982) suggested that isolates of VAM fungi may differ in their capacity to develop an external hyphae system independent of their capacity to colonize the root cortex. In agreement with CLARKE and MOSSE (1981) they found that there is no definite correlation between the amount of fungal infection and its enhancement of plant growth.

The effect of VAM on the phosphate nutrition of plants has been well documented (SMITH 1980, ALLEN et al. 1981a, CLARKE and MOSSE 1981, STRIBLEY et al. 1980). CAPACCIO and CALLOW (1982) investigated the enzymes of polyphosphate synthesis and degradation in mycorrhizal onion roots and in external and internal hyphae of *Glomus mosseae*. PolyP-kinase has been found in extracts of external hyphae and mycorrhizal roots. The activity of both exopolyphosphatase and endopolyphosphatase is greater in mycorrhizal roots, but only detected in internal hyphae.

A recent review by BOWEN and SMITH (1981) examined the uptake of N by mycorrhizal plants. AMES et al. (1983) found that mycorrhizal plants derive more [15N] from organic or inorganic sources when added to the soil. Mycorrhizal plants of *Trifolium subterraneum* have a greater capacity to synthesize nitrate reductase (NADH-dependent) in both roots and shoots than nonmycorrhizal plants when both are growing under phosphate-limiting conditions. OLIVER et al. (1983) concluded that mycorrhizal effects on nitrate reductase activity are largely indirect, mediated via improved phosphate nutrition, because the NADPH-dependent nitrate reductase activity characteristic of fungi is low in mycorrhizal and uninfected plants.

JOHNSON et al. (1982a) suggested that plants grown under long-day photoperiods have greater basipetal transport of photosynthates, which results in improved mycorrhizal infection and plant growth response.

Both growth-stimulating and growth-repressive effects on soybeans occur in dependence on inoculation with different species of *Glomus* and on soil temperature (SCHENCK and SMITH 1982a).

There are some indications that VAM infection can affect water relation in the host plant. A major factor causing increased water transport and reduced resistance to water through-flow may be the increased surface area provided by the hyphae (ALLEN 1982, HARDIE and LEYTON 1981). Lowered stomatal resistance without significant differences in estimated plant resistance led to the conclusion that mycorrhizal infection affected the phytohormon balance which altered stomatal control (LEVY and KRIKUN 1980, ALLEN et al. 1981b).

Onion plants infected with *Glomus etunicatus* are more drought resistant than nonmycorrhizal ones when exposed to serial cycles of soil-water stress (NELSEN and SAFIR 1982). This is attributed to the ability of the VAM fungus to maintain adequate P nutrition in the onions when low soil moisture decreased the mobility of phosphate ions.

Mycorrhizal plants of *Eupatorium odoratum*, *Sorghum bicolor* and *Guizotia abyssinica* show much better growth and mineral uptake with increasing soil O_2 than do the nonmycorrhizal plants (SAIF 1983).

Recently ALLEN et al. (1980, 1982) produced experimental evidence that phytohormon changes appear in *Bouteloua gracilis* resulting from mycorrhizal infection with *Glomus fasciculatus*. Increasing cytokinin activity in both leaves and roots and quantitative and qualitative changes in gibberellin-like substances are reported, moreover abscisic acid concentration decreased in leaves. The altered ratios demonstrated major changes in the physiology of the plant with mycorrhiza formation.

Mycorrhizal finger millet plants produce leaves with stronger veins and more bundle sheat cells containing more chloroplasts than nonmycorrhizal plants, an effect which can perhaps refer likewise to a changed hormonal metabolism (KRISHNA et al. 1981).

TIMMER and LEYDEN (1980) examined the growth of citrus in presence of different rates of copper and phosphate additions and concluded that formation of VAM increased copper uptake by hosts. GILDON and TINKER (1983a,b) studied the effects of heavy metals on mycorrhizal infection and the uptake of copper of infected plants. Commonly, heavy metals reduce the degree of infection of onions with *Glomus mosseae*, but isolates of VAM-fungi from areas heavily contamined seem to be tolerant and could protect plants against the effects of heavy-metal toxicity.

d) Ecology

In a review FOGEL (1980) examined the importance of mycorrhizae on nutrient cycling in forests.

KOSKE and HALVORSON (1981) showed in an ecological study of VAM in a barrier sand dune that areas with greatest vegetation cover by *Ammophila* are associated with greatest spore density and fungal diversity in the rhizosphere.

Experimental evidence that Cruciferae and Chenopodiaceae are not truly VAM families comes from studies with several species, grown together with mycorrhizal host plants. Small amounts of infection, cortical mycelium and vesicles, but no arbuscules, develop in the nonhost when a host nurse plant is present, but far less or even none develop in nonhost grown alone (OCAMPO et al. 1980).

The presence of arbuscules in the Chenopodiaceae has only occasionally been reported in the literature. ALLEN (1983) found that arbuscules in *Atriplex gardneri* (Moq.) Standl. are primarily associated with the fine roots, suggesting that the presence of VAM may be seasonal. SCHWAB et al. (1982) were able to induce arbuscules in the normally nonmycorrhizal plant *Chenopodium quinoa* with the herbicide simazine. The work of CHILVERS and DAFT (1981) confirmed that there is a close relationship between the incidence of root hairs and VAM production in *narcissus*. In agreement with ST JOHN (1980) who worked with tropical trees, they showed that the abundance of mycorrhizal endophytes is directly related to the degree of hairiness of the roots.

Some studies have demonstrated conclusively that VAM fungi can connect the roots of plants of the same species and of different species (HEAP and NEWMAN 1980a). Experiments with [^{32}P] have shown that phosphorus transport is more extensive between mycorrhizal donor and receiver plants than between nonmycorrhizal plants (HEAP and NEWMAN 1980b). WHITTINGHAM and READ (1982) concluded that direct transfer of nutrients between mature and young plants occurs via mycorrhizal hyphae, and their results indicated that this transfer is sufficient to provide measurable growth in sink plants.

Rhizobium spp. and VAM fungi in legume roots form a kind of double symbiosis. Usually mycorrhizal effects on legume nodulation, nitrogen fixation and plant growth are similar to the effect of adding phosphate (MUNNS and MOSSE 1980). However, efficient endophytes enhance growth much more than added superphosphate in pots, at a rate of about 40 kg/ha (HAYMAN and PAGE 1981).

Rhizobium bacteria may enhance the initial establishment of VAM by polysaccharide production, this leading to increased synthesis of polygalacturonase at the infection site, and indicating that root penetration by VAM fungi is related to increased enzyme production near the point of infection (AZÓN-G de AGUILAR et al. 1980).

MANJUNATH et al. (1981) suggested that beneficial effects on onions due to triple inoculation with *Beijerinckia mobilis*, *Aspergillus niger* and *Glomus fasciculatus* could be attributed to the separate activities of these organisms, such as nitrogen fixation, production of growth promoting substances and stronger uptake of phosphorus.

KRISHNA et al. (1982) found that *Glomus fasciculatus* and *Streptomyces cinnamomeous* interact antagonistically when added simultaneously, because *Streptomyces* reduces spore production and development of infection by *Glomus*, while *Glomus* reduces the multiplication of *Streptomyces*.

Several studies indicate that VAM fungi influence plant disease. DAVIS and MENGE (1981) found that *Glomus mosseae* and *Gigaspora margarita* confer tolerance or resistance against *Phytophthora parasitica* on citrus although neither isolated from nor as nutritionally efficient as others on citrus. It is postulated that they are not well adapted to symbiotic association with citrus and elicit a resistance mechanism by the host. BAATH and HAYMAN (1983) reported that mycorrhizal infection of tomato does not affect *Verticillium* wilt. Furthermore MCGRAW and SCHENK (1981) reported increased symptoms of *Fusarium* wilt of tomatoes with VAM present.

Mycorrhiza is affected by pesticides, especially fungicides (SPOKES et al. 1981, NEMEC 1980, OCAMPO and HAYMAN 1980). The endophytes *Glomus fasciculatus*, *G. mossae* and *G. microcarpus* show differential susceptibilities to fungicides. *G. microcarpus* seems to be stimulated by chloroneb (SPOKES et al. 1981).

MENGE (1982) summarized the effects of fumigants and fungicides on VAM fungi. The nematicidial fumigants DBCP, 1,3-D and ethylene dibromide apparently do not reduce mycorrhizal population in soil. DBCP and 1,3-D are reported to increase mycorrhizal infection and spore production both in the field and in the greenhouse.

SCHWAB and MENGE (1982) showed that herbicides that stimulate root exudation also increase mycorrhizal infection. This may explain why DBCP, demosan, sodium azide, ridomil and terrazole have been found to stimulate the infection of treated plants with mycorrhiza.

Growth regulators clearly influence mycorrhiza development without direct action on the fungus. Especially ethylene generators enhanced mycorrhiza infestation (DEHNE and SCHÖNBECK 1981). The effects of mycorrhizae on plant disease were reviewed by DEHNE (1982).

e) Use in Plant Production

MENGE (1983) described the utilization of VAM fungi in agriculture. Commercial use of VA mycorrhizal fungi is currently restricted to situations where the natural populations of the VAM fungi have been destroyed, as in fumigated, chemically treated areas, greenhouses, or disturbed areas such as coal spoils, strip mines or waste areas (KHAN 1981, REEVES et al. 1979, HOEPFNER et al. 1983).

Commercial production of mycorrhizal inoculum is presently being attempted at several locations in the U.S.A. Production of large quantities of spores is difficult (MACDONALD 1981, MERTZ et al. 1979). JOHNSON and MENGE (1981) and HAYMAN (1980) described methods for producing mycorrhizal inoculum for commercial use.

DANIELS and MENGE (1981) presented results suggesting that the VAM fungus *Glomus epigaeus* shows great potential for commercial application. It is capable of producing extremely large numbers of spores in pot cultures, stimulating the dry-weight production of a wide variety of crops. The spores can be stored for at least 3 months with little reduction in germinating capacity.

Generally mycorrhizae are established in seedlings before their transplant to the field (BIERMANN and LINDERMAN 1983a).

2. Ericales - Mycorrhizae

Mycorrhiza formation appears to be the rule in the Ericales (READ 1983). The apparently ubiquitous causal fungi have hyphae with simple septation and Woronin bodies (SCANNERINI and BONFANTE-FASOLO 1983), indicators of ascomycetous lineage.

Pezizella ericae Read is the only endophyte that has been both identified and successfully used to synthesize the ericoid mycorrhiza. Ultrastructural studies of mycorrhizal roots of *Calluna* and *Vaccinium* have confirmed that it is an ascomycete (DUDDRIDGE and READ 1982b). Other fungi have been observed to be closely associated with ericaceous plants (BONFANTE-FASOLO 1980, PETERSON et al. 1980, ENGLANDER and HULL 1980). Among these the most frequently reported is the basidiomycete *Clavaria spp.*, but because the attempts to cultivate the fungus failed, the true nature of this relationship is uncertain (ENGLANDER and HULL 1980). COUTURE et al. (1983) identified *Oidiodendron apiseum* Robak as an endophyte of ericoid mycorrhiza in *Vaccinium spp.*, forming conidiophores under aseptic conditions too.

While most ericaceous species produce either ericoid or arbutiod mycorrhizas, some, for example *Leucothae daviseae*, apparently have the capacity to yield both types (LARGENT et al. 1980).

The fungi of arbutoid mycorrhizae are basidiomycetes, as it has been confirmed in a ultrastructural study (DUDDRIDGE 1980), which showed that the internal hyphae frequently possess dolipore septa. This confirms the observation that the fungi involved are probably similar to those forming ectomycorrhizae in forest trees (MOLINA and TRAPPE 1982).

Recent studies (DUDDRIDGE and READ 1982a) have shown that the achlorophyllous plant *Monotropa* has a distinctive mycorrhizal structure, for which the term monotropoid is proposed. The absence of fungal penetration combined with the presence of transfer-cell characteristics distinguishes this type of mycorrhiza from the arbutoid and ericoid types.

In a detailed chronological analysis of ericoid mycorrhiza it was confirmed that breakdown begins with deterioration of the host rather than of the fungal tissue (DUDDRIDGE and READ 1982b). After penetration of the cortical cells the fungal hyphae proliferate extensively. The host plasmalemma invaginates to envelope each branch of the invading fungus but is separated from the fungal cell wall by the interfacial matrix, made up of pectin material (DUDDRIDGE 1980). Finally, the host cell is almost completely occupied by the fungal hyphae. Numbers of cell organelles proliferate extensively in the ericoid infection units, this suggesting intensive physiological activity. The active lifespan of the individual infection unit is not more than 5 or 6 weeks (READ 1983). The first sign of breakdown is a loss of structural integrity of the host organelles, particularly the mitochondria. Host cytoplasm degenerates and the interfacial matrix becomes progressively wider. Most host organelles have degenerated before fungal deterioration can be seen (READ 1983).

BONFANTE-FASOLO and GIANINAZZI-PEARSON (1982) reported that the external hyphae of *Pezizella ericae* around roots of *Calluna vulgaris* develop a fibrillar sheath, which has a polysaccharid-protein composition, only in the presence of the host root, and they suggest that this may play a role in the recognition and adhesion phenomena in ericoid mycorrhiza.

Mycorrhizal infection enhances the N and P supply to the host plant, but a major effect of the ericoid infection is enhancement of host nitrogen content. This arises from an improved exploitation of available ammonium ions and from the capacity of the mycorrhizal fungi of utilizing simple organic nitrogen compounds, like amino acids (STRIBLEY and READ 1980).

Organic matter seems to play a major role in establishing and maintaining of mycorrhizal colonization. DUCLOS and FORTIN (1983) enhanced mycorrhiza formation by the addition of a small quantity of glucose. Moreover, in the absence of organic matter fungi during the first month after inoculation were unable to colonize the newly forming roots.

Recent studies (BRADLEY et al. 1982) suggest that ericoid mycorrhizal infection may play a role in providing resistance to heavy metals in *Calluna*, *Vaccinium* and *Rhododendron*. It is concluded that exclusion of metals in mycorrhizal plants is attributable to absorption of metallic ions by the actively growing fungi in the host cortical cells.

MOORE-PARKHURST and ENGLANDER (1982) postulated that variability in nursery environments and plant production practices influences the mycorrhizal status of *Rhododendron spp.* in comercial nurseries.

POWELL (1982) inoculated blueberry plants with a *Pezizella ericae* isolate successfully and showed that this stimulated growth and nutrient uptake.

References

ABBOTT, L.K., ROBSON, A.D.: New Phytol. 96, 275-281 (1984). - ALLEN, M.F.: New Phytol. 91, 191-196 (1982); - Mycologia 75, 773-776 (1983). - ALLEN, M.F., MOORE, T.S., CHRISTENSEN, Jr. and M.: Can. J. Bot. 58, 371-374 (1980). - ALLEN, M.F., SEXTON, J.C., MOORE, T.S., CHRISTENSEN, Jr. and M.: New Phytol. 87, 687-694 (1981a). - ALLEN, M.F., SMITH, W. K., MOORE, T.S., CHRISTENSEN, Jr. and M.: New Phytol. 88, 683-693 (1981b). - ALLEN, M.F., MOORE, T.S., CHRISTENSEN, Jr. and M.: Can. J. Bot. 60, 468-471 (1982). - AMES, R.N., REID, C.P.P., PORTER, L.K., CAMBARDELLA, C.: New Phytol. 95, 381-396 (1983). - AZCON, R., OCAMPO, J.A.: New Phytol. 87, 677-685 (1981). - AZON-G DE AGUILAR, C., BAREA, J., OLIVARES, J.: 2nd International Symposium of Microbial Ecology, University of Warwick, Coventry, U.K. Abstr. No. 18.7. (1980).

BAATH, E., HAYMAN, D.S.: 1983. New Phytol. 95, 419-426 (1983). - BERCH, S.M., FORTIN, J.A.: Can. J. Bot. 61, 2608-2617 (1983). - BIERMANN, B.J., LINDERMAN, R.G.: J. Am. Soc. Hort. Sci. 108, 972-976 (1983a); - New Phytol. 95, 97-105 (1983b). - BONFANTE-FASOLO, P.: Trans. Br. Mycol. Soc. 75, 320-325 (1980); - Protoplasma 111, 113-120 (1982). - BONFANTE-FASOLO, P., GIANINAZZI-PEARSON, V.: New Phytol. 91: 691-704 (1982). - BONFANTE-FASOLO, P., GRIPPIOLO, R.: Can. J. Bot. 60, 2303-2312 (1982). - BONFANTE-FASOLO, P., SCANNERINI, S.: Caryologia 33, 126-127 (1980). - BOWEN, G.D., SMITH, S.E.: Ecol. Bull. 33, 232-247 (1981). - BRADLEY, R., BURT, A.J., READ, D.J.: New Phytol. 91, 197-209 (1982). - BUWALDA, J.G., ROSS, G.J.S., STRIBLEY, D.P., TINKER, P.B.: New Phytol. 91, 669-682 (1982).

CAPACCIO, L.C.M., CALLOW, J.A.: New Phytol. 91, 81-91 (1982). - CARLING, D.E., BROWN, M.F.: Phytopathology 72, 1108-1114 (1982). - CHILVERS, M.T., DAFT, M.F.J.: New Phytol. 89, 247-261 (1981). - CLARKE, C., MOSSE, B.: New Phytol. 87, 695-703 (1981). - COUTURE, M., FORTIN, J.A., DALPE, Y.: New Phytol. 95, 375-380 (1983).

DANIELS, B.A., MENGE, J.A.: New Phytol. 87, 345-354 (1981). - DANIELS, B.A., MCCOOL, P.M., MENGE, J.A.: New Phytol. 89, 385-391 (1981). - DAVIS, R.M., MENGE, J.A.: New Phytol. 87, 705-715 (1981). - DEHNE, H. W.: Phytopathology 72, 1115-1119 (1982). - DEHNE, H.W., SCHÖNBECK, F.: Fifth North American Conference on Mycorrhiza, Quebec, Canada, Abstr. No. 18 (1981). - DUCLOS, J.L., FORTIN, J.A.: New Phytol. 94, 95-102 (1983). - DUDDRIDGE, J.A.: Ph. D. thesis. University of Sheffield, Sheffield, U.K. (1980). - DUDDRIDGE, J.A., READ, D.J.: New Phytol. 92, 203-214 (1982a); - Can. J. Bot. 60, 2345-2356 (1982b).

ENGLANDER, H., HULL, R.J.: New Phytol. 84, 661-667 (1980).

FOGEL, R.: New Phytol. 86, 199-212 (1980).

GIANINAZZI, S., GIANINAZZI-PEARSON, V., DEXHEIMER, J.: New Phytol. 82, 127-132 (1979). - GIANINAZZI-PEARSON, V., MORANDI, D., DEXHEIMER, J., GIANINAZZI, S.: New Phytol. 88, 633-639 (1981). - GILDON, A., TINKER, P.B.: New Phytol. 95, 247-261 (1983a); - ibid. 95, 263-268 (1983b). -

GRAHAM, J.H., LEONARD, R.T., MENGE, J.A.: Plant Physiol. 68, 548-552 (1981). - GRAHAM, J.H., LINDERMAN, R.G., MENGE, J.A.: New Phytol. 91, 183-189 (1982).

HARDIE, K., LEYTON, L.: New Phytol. 89, 599-608 (1981). - HAYMAN, D.S.: Nature (London) 287, 487-488 (1980). - HAYMAN, D.S., PAGE, R.J.: Rothamsted Report for 1980, Part 1, p. 203 (1981). - HEAP, A.J., NEWMAN, E.I.: New Phytol. 85, 169-171 (1980a); - ibid. 85, 173-179 (1980b). - HEPPER, C.: New Phytol. 93, 537-542 (1983). - HOEPFNER, E.F., KOCH, B.L., COVEY, R.P.: J. Am. Soc. Hort. Sci. 108, 207-210 (1983).

JANOS, D.P., TRAPPE, J.M.: Mycotaxon 15, 509-514 (1982). - JOHNSON, C.R., MENGE, J.A.: Am. Nurseryman. 155, 79-86 (1981). - JOHNSON, C.R., MENGE, J.A., SCHWAB, S., TING, I.P.: New Phytol. 90, 665-669 (1982a). - JOHNSON, C.R., GRAHAM, J.H., LEONARD, R.T., MENGE, J.A.: New Phytol. 90, 671-675 (1982b).

KHAN, A.G.: New Phytol. 87, 363-370 (1981). - KOSKE, R.E., HALVORSON, W.L.: Can. J. Bot. 59, 1413-1422 (1981). - KRISHNA, K.R., SURESH, H. M., SYAMSUNDER, J., BAGYARAJ, D.J.: New Phytol. 87, 717-722 (1981). - KRISHNA, K.R., BALAKRISHNA, A.N., BAGYARAJ, D.J.: New Phytol. 92, 401-405 (1982).

LARGENT, D.L., SUGIHARA, N., WISHNER, C.: Can. J. Bot. 58, 2274-2279 (1980). - LEVY, J., KRIKUN, J.: New Phytol. 85, 25-31 (1980). - LIM, L.L., FINERAN, B.A., COLE, A.L.J.: New Phytol. 95, 231-239 (1983).

MACDONALD, R.M.: New Phytol. 89, 87-93 (1981). - MACDONALD, R.M., CHANDLER, M.R.: New Phytol. 89, 241-246 (1981). - MACDONALD, R.M., CHANDLER, M.R., MOSSE, B.: New Phytol. 90, 659-663 (1982). - MANJUNATH, A., BAGYARAJ, D.J.: New Phytol. 87, 355-361 (1981). - MANJUNATH, A., MOHAN, R., BAGYARAJ, D.J.: New Phytol. 87, 723-727 (1981). - MARX, C., DEXHEIMER, J., GIANINAZZI-PEARSON, V., GIANINAZZI, S.: New Phytol. 90, 37-43 (1982). - MCGRAW, A.-C., SCHENK, N.C.: Phytopathology 71, 894 (1981). - MENGE, J.A.: Phytopathology 72, 1125-1132 (1982); - Can. J. Bot. 61, 1015-1024 (1983). - MERTZ, S.M., HEITHAUS, J.J., BUSH, R.L.: Trans. Br. Mycol. Soc. 72, 167-169 (1979). - MOLINA, R., TRAPPE, J.M.: New Phytol. 90, 495-509 (1982). - MOORE-PARKHURST, S., ENGLANDER, L.: Can. J. Bot. 60, 2342-2344 (1982). - MUNNS, D.N., MOSSE, B.: pp. 115-125 in: Advances in Legume Science, eds. R.J. SUMMERFIELD, A.H. BUNTING, London: HMSO 1980.

NELSEN, C.E., SAFIR, G.R.: Planta 154, 407-413 (1982). - NEMEC, S.: Can. J. Bot. 58, 522-526 (1980); - ibid. 59, 609-617 (1981).

OCAMPO, J.A., HAYMAN, D.S.: Trans. Br. Mycol. Soc. 74, 413-416 (1980); - New Phytol. 87, 333-343 (1981). - OCAMPO, J.A., MARTIN, J., HAYMAN, D.S.: New Phytol. 84, 27-35 (1980). - OLIVER, A.J., SMITH, S.E., NICHOLAS, D.J.D., WALLACE, W.: New Phytol. 94, 63-79 (1983).

PETERSON, T.A., MUELLER, W.C., ENGLANDER, L.: Can. J. Bot. 58, 2421-2433 (1980). - POWELL, C.L.: J. Am. Hort. Sci. 107, 1012-1015 (1982).

READ, D.J.: Can. J. Bot. 61, 985-1004 (1983). - REEVES, F.B., WAGNER, D., MOORMAN, T.: Am. J. Bot. 66, 6-13 (1979). - ROSE, S., TRAPPE, J.M.: Mycotaxon 8, 297-301 (1980).

SAIF, S.R.: New Phytol. 95, 405-417 (1983). - SCANNERINI, S., BONFANTE-FASOLO, P.: Can. J. Bot. 61, 917-943 (1983). - SCHENCK, N.C., SMITH, G.S.: New Phytol. 92, 193-201 (1982a); - Mycologia 74, 77-92 (1982b). - SCHWAB, S.M., MENGE, J.A., in: MENGE, J.A.: Phytopathology 72, 1125-

1132 (1982). - SCHWAB, S.M., JOHNSON, E.L.V., MENGE, J.A.: Plant Soil 64, 283-287 (1982). - SCHWAB, S.M., MENGE, J.A., LEONARD, R.T.: Plant Physiol. 73, 761-765 (1983). - SMITH, S.E.: Biol. Rev. 55, 475-510 (1980). - SPOKES, J.R., MACDONALD, R.M., HAYMAN, D.S.: Pestic. Sci. 12, 346-350 (1981). - ST JOHN, T.V.: New Phytol. 84, 483-487 (1980). - STRIBLEY, D.P., READ, D.J.: New Phytol. 86, 365-371 (1980). - STRIBLEY, D.P., TINKER, P.B., RANER, J.H.: New Phytol. 86, 261-266 (1980). - STRULLU, D.G., CHAMEL, A., ELOY, J.F., GOURRET, J.:.: New Phytol. 94, 81-88 (1983). - SWARD, R.J. : New Phytol. 87, 761-768 (1981a); - ibid. 88: 661-666 (1981b).

TIMMER, L.W., LEYDEN, R.F.: New Phytol. 85, 15-23 (1980). - TOMMERUP, I.C., ABBOTT, L.K.: Soil Biol. Biochem. 13, 431-433 (1981). - TRAPPE, J.M.: Phythopathology 72, 1102-1108 (1982). - TRAPPE, J.M., SCHENCK, N.C., in: Methods and Principles of Mycorrhizal Research, ed. N.C. SCHENCK. St. Paul, Minnesota: American Phytopathological Society 1982.

WARNER, A., MOSSE, B.: Trans. Br. Mycol. Soc. 74, 407-410 (1980); - New Phytol. 90, 529-536 (1982). - WHITTINGHAM, J., READ, D.J.: New Phytol. 90, 277-284 (1982). - WILSON, J.M., TRINICK, M.J.: New Phytol. 93, 543-553 (1983a); - Aust. Soil Res. 21, 73-81 (1983b).

Professor Dr. F. SCHÖNBECK
Institut für Pflanzenkrankheiten und
Pflanzenschutz der Universität
Herrenhäuser Str. 2
D-3000 Hannover 21, FRG

Subject Index